Soil Quality and Management

Soil Quality and Management

Edited by Henry Wang

SYRAWOOD
PUBLISHING HOUSE
New York

Published by Syrawood Publishing House,
750 Third Avenue, 9th Floor,
New York, NY 10017, USA
www.syrawoodpublishinghouse.com

Soil Quality and Management
Edited by Henry Wang

International Standard Book Number: 978-1-68286-383-1 (Hardback)

Cataloging-in-publication Data

Soil quality and management / edited by Henry Wang.
p. cm.
Includes bibliographical references and index.
ISBN 978-1-68286-383-1
1. Soils--Quality. 2. Soil management. 3. Soil biodiversity. 4. Soil science. I. Wang, Henry.
S591 .S68 2017
631.4--dc23

Printed in China.

TABLE OF CONTENTS

PREFACE

Soil quality is defined as the condition of soil necessary for the survival of any living species. Soil quality incorporates agricultural soil science as well as soil management and soil ecology. Soil quality and soil management, together aim to facilitate sustainable ecological practices that will support plant and animal life as well as the management of related fields of irrigation and sowing practices. This book covers in detail some existent theories and innovative concepts revolving around soil ecology and management. Different approaches, evaluations, methodologies and advanced studies in this field have been included in this text. It is appropriate for students seeking detailed information in this area as well as for experts. It includes contributions of experts and scientists which will provide innovative insights into this field.

The world is advancing at a fast pace like never before. Therefore, the need is to keep up with the latest developments. This book was an idea that came to fruition when the specialists in the area realized the need to coordinate together and document essential themes in the subject. That's when I was requested to be the editor. Editing this book has been an honour as it brings together diverse authors researching on different streams of the field. The book collates essential materials contributed by veterans in the area which can be utilized by students and researchers alike.

Each chapter is a sole-standing publication that reflects each author's interpretation. Thus, the book displays a multi-facetted picture of our current understanding of applications and diverse aspects of the field. I would like to thank the contributors of this book and my family for their endless support.

Editor

1

Structure, Composition and Metagenomic Profile of Soil Microbiomes Associated to Agricultural Land Use and Tillage Systems in Argentine Pampas

Belén Carbonetto[1]*, Nicolás Rascovan[1], Roberto Álvarez[2], Alejandro Mentaberry[3], Martin P. Vázquez[1]*

1 Instituto de Agrobiotecnología de Rosario (INDEAR), Predio CCT Rosario, Santa Fe, Argentina, **2** Facultad de Agronomía, Universidad de Buenos Aires, Buenos Aires, Argentina, **3** Departamento de Fisiología y Biología Molecular y Celular, Facultad de Ciencias Exactas y Naturales, Universidad de Buenos Aires, Buenos Aires, Argentina

Abstract

Agriculture is facing a major challenge nowadays: to increase crop production for food and energy while preserving ecosystem functioning and soil quality. Argentine Pampas is one of the main world producers of crops and one of the main adopters of conservation agriculture. Changes in soil chemical and physical properties of Pampas soils due to different tillage systems have been deeply studied. Still, not much evidence has been reported on the effects of agricultural practices on Pampas soil microbiomes. The aim of our study was to investigate the effects of agricultural land use on community structure, composition and metabolic profiles on soil microbiomes of Argentine Pampas. We also compared the effects associated to conventional practices with the effects of no-tillage systems. Our results confirmed the impact on microbiome structure and composition due to agricultural practices. The phyla *Verrucomicrobia, Plactomycetes, Actinobacteria*, and *Chloroflexi* were more abundant in non cultivated soils while *Gemmatimonadetes, Nitrospirae* and WS3 were more abundant in cultivated soils. Effects on metabolic metagenomic profiles were also observed. The relative abundance of genes assigned to transcription, protein modification, nucleotide transport and metabolism, wall and membrane biogenesis and intracellular trafficking and secretion were higher in cultivated fertilized soils than in non cultivated soils. We also observed significant differences in microbiome structure and taxonomic composition between soils under conventional and no-tillage systems. Overall, our results suggest that agronomical land use and the type of tillage system have induced microbiomes to shift their life-history strategies. Microbiomes of cultivated fertilized soils (i.e. higher nutrient amendment) presented tendencies to copiotrophy while microbiomes of non cultivated homogenous soils appeared to have a more oligotrophic life-style. Additionally, we propose that conventional tillage systems may promote copiotrophy more than no-tillage systems by decreasing soil organic matter stability and therefore increasing nutrient availability.

Editor: Kathleen Treseder, UC Irvine, United States of America

Funding: Funding for this work was provided by Agencia Nacional de Promoción Científica y Tecnológica, Argentina-PAE 37164. The funders had no role in study design, data collection and analysis, decision to publish, or preparation of the manuscript.

Competing Interests: The authors have declared that no competing interests exist.

* E-mail: martin.vazquez@indear.com (MPV); belen.carbonetto@indear.com (BC)

Introduction

Agriculture is facing major challenges nowadays. Production will have to double in the next 50 years in order to face growing food demand and bioenergy needs [1,2]. This must be done without increasing environmental threats such as climate change, biodiversity loss and degradation of land and freshwater. Achieving such a goal represents one of the greatest scientific challenges ever. This is in part because of the trade-offs among economic and environmental goals and because of the insufficient knowledge about the biological, biogeochemical and ecological processes that are relevant for sustainable ecosystem functioning [3,4]. Much has been done during the last decade to gain sufficient information on agricultural ecosystem biology, still, more work needs to be done to gain deeper comprehension and to be able to reduce the negative environmental impacts of agriculture [2,5,6]. The main focus should be oriented to soil degradation. Soil fertility, as the capacity to sustain abundant crop production, needs to be preserved. Nowadays soil fertility is maintained by dependence on external inputs; with increasing water contamination [7]. In this context, the key to understand the behavior of life-supporting elements in soil, such as carbon, nitrogen, and phosphorus lies in the fluxes between their various forms in the environment, which are modulated by biology [8]. Comprehension of soil microorganism dynamics is then essential to understand soil processes that affect fertility. Ecological approaches are being taken into account in soil microbial studies trying to address these questions. These approaches involve diversity and functional analyses of soil communities [9,10]. Scholes & Scholes point out that this complex view is necessary for the comprehension of soil systems and that soil restoration of biological processes is the key to achieving lasting food and environmental security [8].

Argentine Pampas is an important player in this scenario. With a plain area of 50 million ha., nearly 50% of the whole Pampas area is devoted to crop production [11]. Cultivation began in the 19th Century in the central humid portion of the region, in soils of high fertility, and spread in last decades to the south and the semiarid west [12]. Soil degradation (i.e. intense erosion and net loss of nutrients and organic carbon) caused by the use of conventional tillage systems were reported in the Pampas [13–17].

Nowadays between 60 and 80% of production is conducted under conservational no-till practices [18]. Extensive research was done to evaluate the effects of reduced tillage and no-tillage systems on soil physical properties, water content, fertility and crop yields[19]. The main outcome of these analyses points that the adoption of limited tillage systems led to soil improvement, by augmenting organic matter content and soil structure. Still, external fertilization is needed in order to restore nutrient levels and fertility regardless the tillage system employed.

Even though the effects of different tillage practices on soil physical and chemical characteristics have been deeply studied, changes in microbial biodiversity and functioning have been poorly reported in Argentinean Pampas. Most works have studied tillage effects on microbial biomass or specific microbial activities (i.e. utilization of specific substrates, extracellular enzyme production, mineralization, etc.) rather than on full microbiome [13,20,21]. Other studies have focused on the behavior of specific bacterial taxa [22,23]. Reports with an ecological approach (i.e. microbial community analysis) have usually focused on individual effects of land use such as the application of herbicides [24,25]. In these cases, biodiversity variability has been assessed using classical fingerprinting techniques (such as RFLP and DGGE) that lack information about microbial taxonomic identity and only capture the most dominant species in the environment [26,27]. In the last few years, 16S amplicon pyrosequencing has been largely implemented to determine microbial diversity and structure of many different ecosystems worldwide [28,29]. This strategy allows a more exhaustive characterization of community patterns and composition. Moreover, some works have incorporated the use of shotgun metagenomics to study the metabolic potential of soil microbiomes [10,30]. The shotgun approach generates a massive amount of data using random high-trhoughput sequencing of soil isolated DNA. This allows the identification of functional capabilities by gene annotation and the comparison of metabolic profiles between samples. To our knowledge, Figuerola et al. [31] were the only authors studying microbial communities in agronomical soils of Argentine Pampas using high throughput sequencing approaches. They observed differences in microbial community composition of soils under no-tillage systems using 16S pyrosequencing. As a novelty, our efforts focused on assessing the impact of long-term agriculture on Pampas soil microbiomes using both shotgun metagenomics and deep 16S amplicon sequencing approaches. We evaluated the effect of more than a hundred years of agronomical land use on both community features and metabolic profiles of soil microbiomes in comparison with nearby control soils with no agricultural records. We also addressed the differences between the effects of two tillage systems: conventional tillage vs. no-tillage on microbial communities.

Several previous studies of soil microbiomes from different parts of the world showed the effects of agronomical land use on soil microbial communities [10,32–35]. Some of these studies showed differences in trophic strategies between microbial communities related to tillage; and most of them were done in experimental plots. As a novelty, we tested the impact of long term agriculture in soils sampled in production fields in the Argentine Pampas, allowing a deeper insight to the effects of intense land use on soil ecosystems functioning. We confirmed the hypothesis that agronomical practices affected Pampas soil microbiomes by promoting a shift of life-history and trophic strategies. We also showed differences in the effects of contrasting tillage systems (i.e. conventional vs. no- tillage) on community taxonomic and metabolic composition on a long term experiment.

Materials and Methods

Sites description and sampling

Soil samples were taken in production and experimental fields between June and August 2010. To address the effects of agricultural land use on soil microbial communities, three different production farms were sampled in the Rolling Pampas area: "La Estrella", "La Negrita" and "Criadero Klein" (See Rascovan et al and Table S1 for details). Rolling Pampas soils are classified as Typic Argiudolls [36] and mean annual rainfall and temperature were1002 mm and 16.8°C respectively. Two treatments were defined: *cultivated* for production plots, and *no cultivated* for farm-houses parks. Production plots were under cultivation for at least one century under conventional tillage systems, with a mixed rotation of pastures and annual grain crops. During the last 15 years before sampling plots were subjected to continuous crop cultivation under no-tillage systems (i.e. minimal soil disturbance, permanent soil cover, rotations and fertilization). The last crop rotation before sampling was wheat-soybean. Nitrogen and phosphorus fertilizers were applied. Samples were collected one month after soybean harvest. Soil samples were also collected nearby the farmers' houses where no agricultural land use (no tillage nor cultivation) was recorded for the last 30 years except from grass mowing. Parks around farmers' houses are usually considered as undisturbed environments in Argentine Pampas [37]. Soils under no land use were covered with grass and other herbaceous (non-woody) plants common in the region such as *Cirsium sp, Trifolium sp, Micropsis sp, Festuca sp, Dichondra sp, Cyperus sp* and *Taraxacum officinale*. For numerical analyses purposes the three farms are treated as experimental replicates. Four soil samples were taken with an auger from the upper 20 cm soil layer in each farm and treatment. A total of 24 samples were collected in Rolling Pampa soils.

In order to compare effects of contrasting tillage systems, samples were also collected in a 34-year-old experiment located in Balcarce in the Southern Pampas (See Rascovan et al and Table S1 for more details). Samples were taken in experimental plots because no production fields are using conventional tillage for crop production nowadays in the Pampas. Soils in Balcarce are a complex of Typic Argiudolls and Petrocalcic Paleudols and mean annual rainfall and temperatures were 875 mm and 13.8°C respectively. The experiment was carried out in three (175 m^2) experimental plots ($n = 3$). Treatments were defined as: *no tillage* (NT) and *conventional tillage* (CT). NT plots had minimal soil disturbance and permanent soil cover combined with rotations; which have included pastures and grain crops (soybean, corn, wheat) during the last 16 years. CT plots were managed with moldboard plough. Nitrogen fertilization was performed in NT and CT plots (60 kg N ha-1). Last rotation before sampling was corn-soybean. Two sub-samples were collected from all treatment and replicate plots a month after soybean harvest. A total of 12 samples were collected.

Samples were immediately sent to the lab after collection. Samples used for DNA purification were air dried and sieved through 1 mm mesh to thoroughly homogenize, break aggregates and remove roots and plant detritus, then stored at −80°C. DNA purification and library preparation was previously described in Rascovan. et al.[38].

None of the sampling sites is located in protected areas. Permissions were obtained directly from each farm owner or manager: Alejandro Cattaneo at La Negrita and La Estrella, Roberto Klein at Criadero Klein and Guillermo Studdert at Balcarce experiment.

Soil chemical and physical measurements

Soil organic carbon was determined by wet digestion and organic matter was estimated [39]. Nitrate-nitrogen was analyzed by 2 M KCL extraction and the phenoldisulfonic acid method [40]. Extractable phosphorus was determined by the Bray method [41]. The pH was measured in a soil:water ratio 1:2.5. Salinity was estimated by the determination of electrical conductivity [42]. Texture analysis was performed by the hydrometer method [43] and nitrogen was determined by Kjeldahl method [40].

16S amplicon sequencing and shotgun metagenomic datasets

To analyze the effect of agronomical practices on soil microbiomes, sequence data from the previously reported Pampa dataset [38] was used. In order to evaluate microbial community structure and taxonomic composition, a total of 112,800 high-quality filtered 16S rRNA gene amplicon sequences, obtained from the 42 soil DNA samples (replicates and subsamples were included). DNA shotgun metagenomic data was used to analyze metabolic profiles. Shotgun metagenomic data completed a total of 10,445,170 sequences. In this case sequences were obtained from one subsample per sampling replicated plot.

In brief, libraries were prepared as follows: DNA was isolated from 10 g of soil of each of the 42 soil samples using the Power MaxSoil DNA Isolation Kit following the manufacturer's instructions (MO BIO Laboratories, Inc.). For amplicon libraries the V4 hyper variable region of the 16s rRNA gene was amplified. Duplicated reactions were performed using barcoded bacterial universal primers containing Roche- 454 sequencing A and B adaptors and a nucleotide multiple identifier (MID) to sort samples: 563F: 5′-CGTATCGCCTCCCTCGCGCCATCA-GACGAGTGCGTAYTGGGYDTAAAGNG -3′ (where AC-GAGTGCGT is an example, different MIDs for each sample were used) and 802R (5′-CTATGCGCCTTGCCAGCCCGCT-CAGTACCRGGGTHTCTAATCC, 5′-CTATGCGCCTTGC-CAGCCCGCTCAGTACCAGAGTATCTAATTC, 5′-CTAT-GCGCCTTGCCAGCCCGCTCAGCTACDSRGGTMTCTA-ATC, 5'-CTATGCGCCTTGCCAGCCCGCTCAGTACNVG-GGTATCTAATCC) [44]. All amplicons were cleaned using Ampure DNA capture beads (Agencourt- Beckman Coulter, Inc.) and pooled in equimolar concentrations before sequencing on a Genome Sequencer FLX (454-Roche Applied Sciences) using Titanium Chemistry according to the manufacturer's instructions.

Shotgun metagenomic libraries were prepared by nebulization, followed by tagging with GS-FLX-Titanium Rapid Library MID Adapters Kit (454-Roche Applied Sciences) and sequenced with a Genome Sequencer FLX (454-Roche Applied Sciences) using Titanium Chemistry according to the manufacturer's instructions. Sequencing runs were performed in INDEAR sequencing facility.

All the sequences used in the present study are available in The Sequence Read Archive (SRA) under accession number SRA058523 and SRA056866. See Rascovan et al. [38] for more information.

Amplicon sequence processing, OTU classification and taxonomic assignment

Sequence data were quality controlled and denoised with the ampliconnoise.py script of QIIME [45].This script also eliminated chimeras. Sequences obtained from Rolling Pampa soil libraries and Balcarce soil libraries were processed separately. Sequences were clustered into Operational Taxonomic Units (OTUs) using the pick_otus.py script with the Uclust method [46] at 97% sequence similarity. Rolling Pampas samples yielded 2,591sequences on average (ranging from 1,455 to 3,991sequences). Balcarce samples yielded 2,329 reads on average (ranging from 1,211 to 4,755 reads).OTU representative sequences were aligned using PyNast algorithm [47] with QIIME default parameters. Phylogenetic trees containing the aligned sequences were then produced using FastTree [48]. All downstream analyses were determined after each sample was randomly rarefied to 70% the number of reads of the smallest sample (i.e. 1,080 reads for Rolling Pampa libraries and 850 reads for Balcarce libraries). Phylogenetic distances between OTUs were calculated using unweighted and weighted Unifrac [49]. Taxonomic classification of sequences was done with Ribosomal Database Project (RDP) Classifier using Greengenes database using a 50% confidence threshold [50,51].

Microbial community analyses

Unifrac phylogenetic pairwise distances among samples were visualized with principal coordinates analysis (PCoA). Analysis of similarity statistics (ANOSIM) was calculated to test a-priori sampling groups. BIOENV analysis was performed to elucidate which soil properties correlated with community patterns. All calculations were carried out with R packages 'BiodiversityR' and 'Vegan' [52,53]. T- tests were performed with QIIME script otu_category_significance.py, and R scripts in order to elucidate differences in read abundances.

Shotgun metagenomic sequence processing and analysis

SSF files obtained from shotgun sequencing runs were uploaded to the MG-RAST webserver [54] for sequence filter and analyses. Reads more than two standard deviations away from the mean read length were discarded. For dereplication removal MG-RAST used a simple k-mer approach to rapidly identify all 20 character prefix identical sequences. This step is required in order to remove artificial duplicate reads. We obtained an average of 1.28×10^6 filtered reads per sample for Rolling Pampa shotgun libraries, and an average of 304,258 filtered reads per sample for Balcarce libraries.

Filtered high quality sequences were assigned to Cluster of orthologous groups (COG) by the MG-RAST sever pipeline using a similarity-based approach. COGs were assigned with a maximum E value of 10^{-20}, an average alignment of 80 amino acids length and 70% average identity.

Relative abundances were calculated by dividing the number of hits for each COG or COG-category by the total number of filtered reads in each sample. Euclidean distances based on relative abundances were calculated between sample pairs. PCoA visualizations and ANOSIM calculations were performed. All calculations were carried out with R packages 'BiodiversityR' and 'Vegan'. T- tests were performed with QIIME script otu_category_significance.py, and R scripts in order to elucidate differences in COG relative abundances between samples.

Results

Microbiome community changes related to agricultural land use

The PCoA visualization revealed clear differences between cultivated and noncultivated soils (ANOSIM $R = 0.8406$, $p \leq 0.001$; Figure 1A).Similar results were obtained when using Bray Curtis distance matrices (Figure S1).

The soil properties (Table S1) that best explained the phylogenetic variation observed in microbial communities were determined using Clarke and Ainsworth's BIOENV analysis. Our results showed that variables that best correlated with community

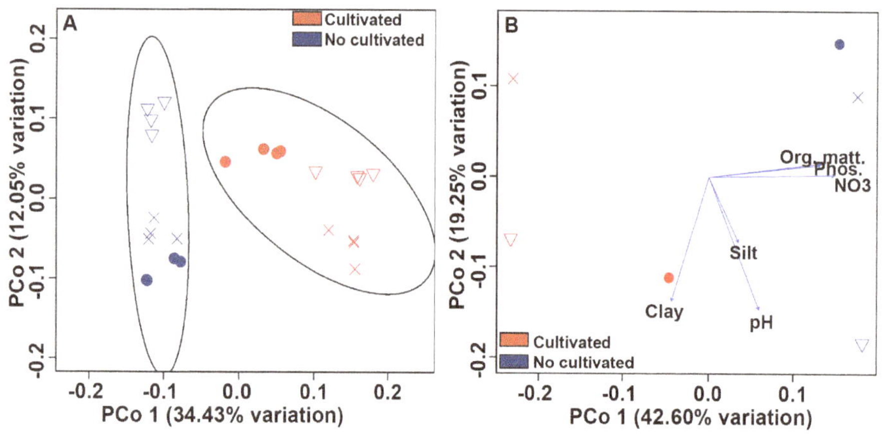

Figure 1. PCoA plots of Pampa production field soil microbiomes based on 97% similarity Weighted Unifrac distance matrices. A) PCoA of cultivated and non-cultivated soil microbiomes. All sub-samples are plotted. Standard error ellipses show 95% confidence areas. B) PCoA biplot of soil properties that best explained variation in community structure. Correlations were calculated using BIOENV on average data of each sampled site (Mantel r = 0.6107, p≤0.05). Circles represent samples from "La Estrella", crosses represent "Criadero Klein" samples and triangles represent "La Negrita" samples.

differences were organic matter, clay and silt content, nitrates, phosphorus and pH (Mantel r = 0.6107, p≤0.05).The PCoA biplot (Figure 1B) showed that organic matter, phosphorus and nitrate levels correlated with the first ordination axis that discriminates between cultivated and non cultivated soils. These three properties were higher in soils under no land use.

Regarding the taxonomic analyses, we observed that members of phyla *Verrucomicrobia, Planctomycetes, Actinobacteria* and *Chloroflexi* were more abundant in non cultivated soils (p≤0.05) (Figure 2). On the other hand, we found that sequences related to *Gemmatimonadetes*, candidate division WS3 and *Nitrospirae* were enriched in cultivated soils (p≤0.05) (Figure 2). No significant differences were found for *Proteobacteria* (for non of the Clases), *Acidobacteria* and *Bacteroidetes* phyla. The ten mentioned taxa represent on average 95% of total sequences of each sample.

Microbiome metagenomic profile changes related to agricultural land use

We found that cultivated and non cultivated soils also clustered apart when metagenomic functional categories were used for the analysis. The first two components of the PCoA explained over 60% of the variability between samples (Figure 3). Standard deviation ellipses overlapped in the ordination plot, indicating that some features are shared between metagenomes. Still, a positive correlation was observed between metabolic and weighted-Unifrac distance matrices (Mantel r: 0.5036 p≤0.05). The analyses of individual COG categories revealed that the relative abundances of COG categories associated with transcription, protein modification, nucleotide transport and metabolism, wall and membrane biogenesis and intracellular trafficking and secretion were higher in cultivated soils (Figure 4, p≤0.05). A deeper analysis inside COG categories revealed that COGs related to Coenzyme A and acetyl-Coa metabolism, energy storage and starvation or quiescence such as, pantothenate kinase, phospho-transacetylase, and trehalose utilization protein were more abundant in non cultivated than in cultivated soils (Figure S2, p≤0.05). On the other hand, COGs related to rapid regulation systems, tricarboxylic acid cycle and nitrogen assimilation such as urease, citrate synthase, glutamate synthase, fumarate hydratase, S-adenosyl- homocysteine hydrolase, S-adenosyl-methionine synthetase, cobalamin biosynthesis protein and ABC transporters, were more abundant in cultivated soils (Figure S2, p≤0.05).

Microbiome community structure and composition related to conventional tillage and no-tillage systems.

To compare the structure of microbiomes under different tillage systems, we collected samples from an experimental field located in Balcarce in the Southern Pampas. The 34-year-old experiment compared two tillage systems: no-tillage (NT) and conventional tillage (CT). Weighted Unifrac analysis showed differences in community structure associated to the tillage system employed

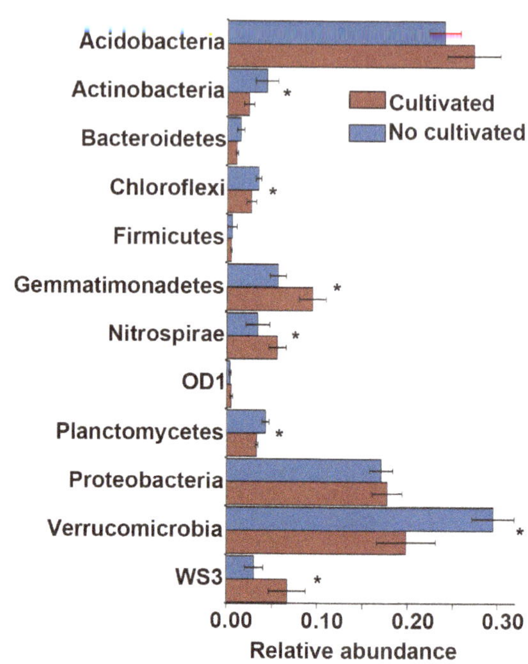

Figure 2. Relative abundances of taxonomic groups in Pampa production field soil microbiomes. Bars represent ± 1 standard error. (*) indicate significant differences (p≤0.05).

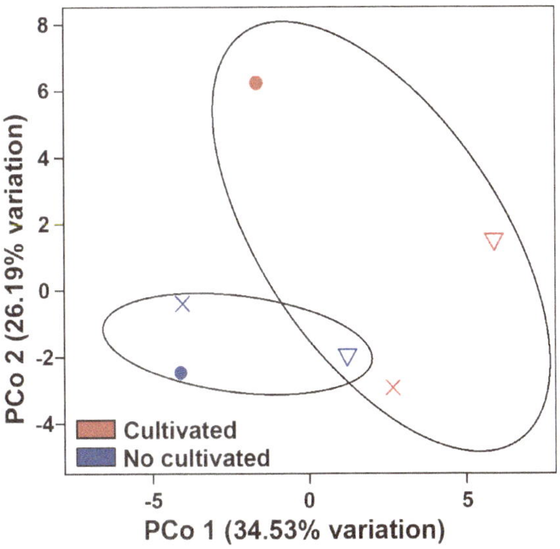

Figure 3. PCoA plot of metagenomic data based on Euclidean distance matrices of COG categories of Pampa production field soil microbiomes. Standard error ellipses show 95% confidence areas. Circles represent samples from "La Estrella", crosses represent "Criadero Klein" samples and triangles represent "La Negrita" samples.

(ANOSIM R = 0.9009, p<0.05, Figure 5A). The first axis explained 31.56% of total variation and separated NT from CT. Additionally, we showed that nitrates were the only soil variable (Table S1) that significantly correlated with community structure (BIOENV analysis, Mantel r = 0.7721, p≤0.01, Figure 5B).

Moreover, microbiomes of CT and NT soils also differed in taxonomic composition. Members of *Acidobacteria*, *Gemmatimonadetes*, candidate division TM7 and class *Gammaproteobacteria* were more abundant in CT soils, while *Nitrospirae*, candidate divisionWS3 and *Deltaproteobacteria* were more represented in NT soils (Figure 6, p≤0.05).

Microbiome metabolic profiles related to conventional tillage and no- tillage systems

Variation in metagenomic profiles between CT and NT microbiomes was analyzed with PCoA based on Euclidean distance matrices of COG abundances. We could not find significant differences in overall profile metabolic structure between tillage systems (Figure S3). Additionally, we did not find significant correlation between Euclidean metabolic matrices and phylogenetic matrices. However, categories related to intracellular trafficking and secretion, amino acid transport and metabolism, and energy production and conversion were shown to be more abundant in soil under CT (Figure 7).

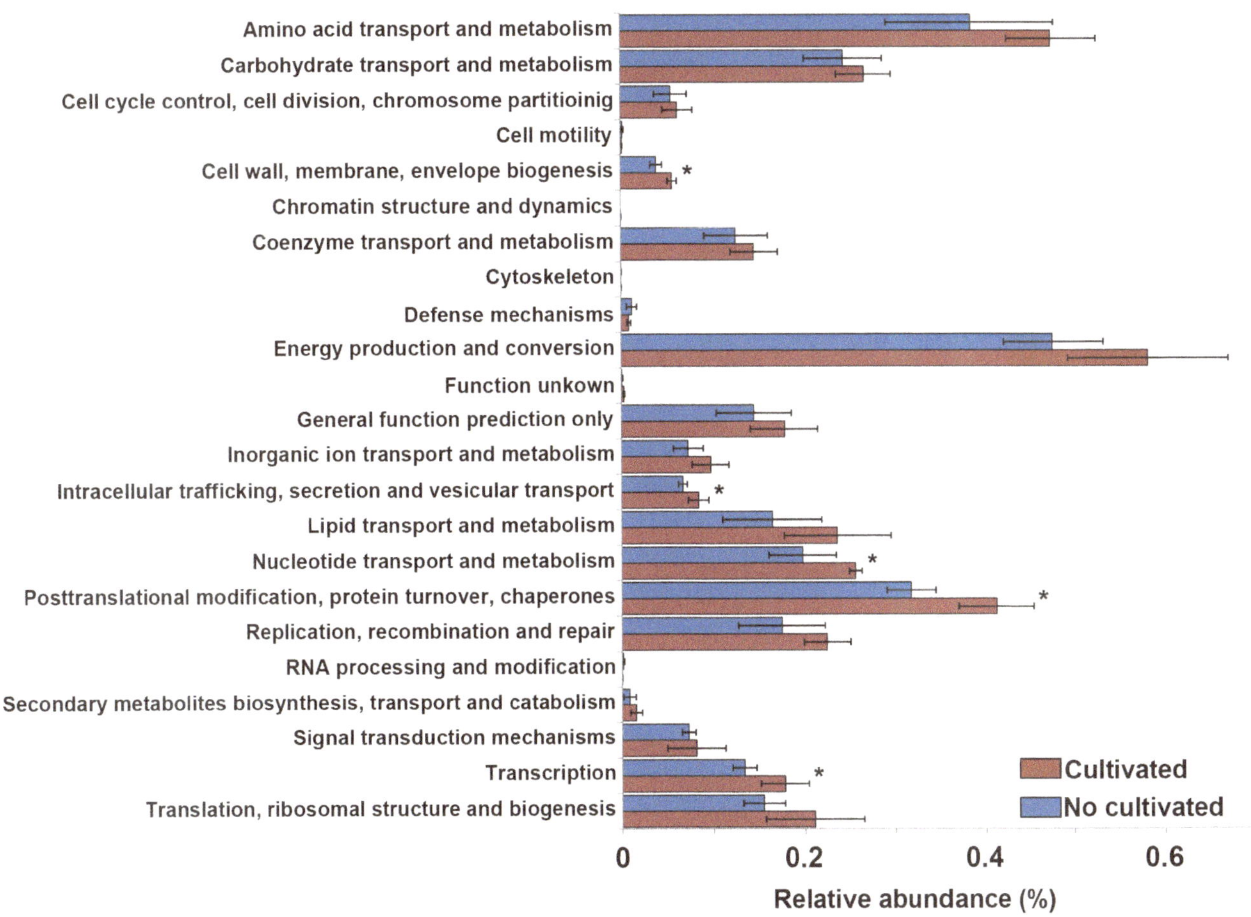

Figure 4. Relative abundances of COG categories in Pampas production field soil microbiomes. Bars represent ±1 standard error. (*) indicate significant differences (p≤0.05).

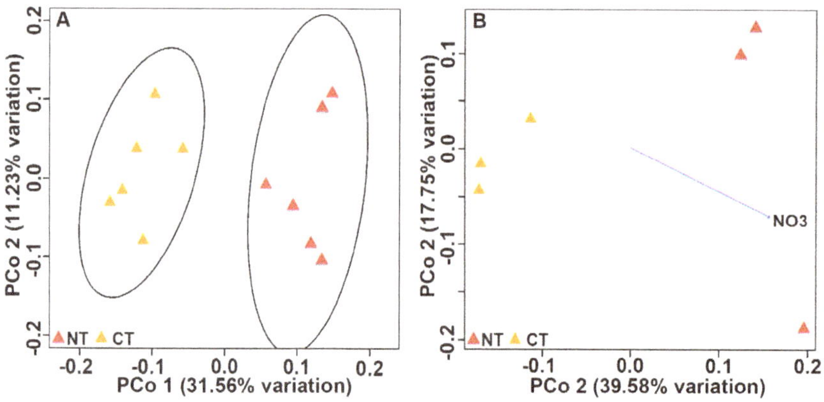

Figure 5. PCoA plots of Balcarce experimental field soil microbiomes based on 97% similarity Weighted Unifrac distance matrices. A) CA: conventional tillage; NT: no-tillage. All sub-samples are plotted. Standard error ellipses show 95% confidence areas. B) PCoA biplot, nitrate was the variable that best explained variation in community structure. Correlations were calculated using BIONEV on average data of each experimental plot (Mantel r = 0.7721, p≤0.01).

Discussion

Much work still needs to be done to get a comprehensive view of the soil microbiomes. Our work is one of the firsts done in the Argentine Pampas at this resolution level, with a combination of metagenomic and phylogentic approaches; and it is aimed to contribute to the comprehension of soil microbiomes function and dynamics. In that context, our results are in agreement with previous works that showed differences in soil microbial community structure and taxonomic composition due to the presence of agricultural land use [9,10,32,33].

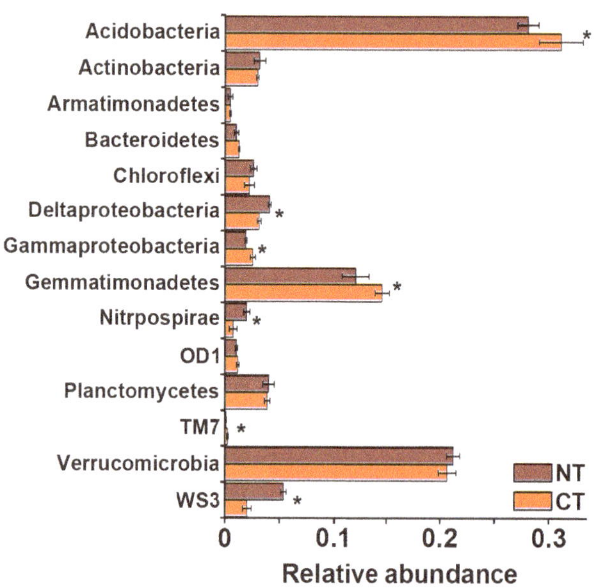

Figure 6. Comparison of relative abundances of taxonomic groups in NT and CT soils. CA: conventional tillage; NT: no-tillage. Bars represent ±1 standard error. (*) indicate significant differences (p≤ 0.05).

Effects of agricultural land use on community composition of soil microbiomes in Argentine Pampas

We observed effects on taxonomic composition at the phylum level. *Verrucomicrobia, Plactomycetes, Actinobacteria* and *Chloroflexi* were more abundant in soils that were never cultivated; while *Gemmatimonadetes, Nitrospirae* and WS3 were more abundant in crop cultivated soils. These results are in agreement with the copiotroph/oligotroph hypothesis [55], that propose that high number of oligotrophic prokaryotes may be found in bogs and soils with high amounts of recalcitrant organic matter [56]. On the other hand, copiotrophic organisms are able to use labile nutrient fractions and to grow at higher rates as a consequnece. In our study, non-cultivated soils are considered to be oligotrophic since they present high levels of organic matter highly rich in humic acids [12] and cultivated soils as copiotrophic environments due to fertilization and the seasonal presence of crop residues, which increases organic matter and nitrogen accessibility [57]. Under this assumption, bacteria in non-cultivated soils are expected to be K selected and to present low growth rates and very efficient nutrient uptake systems with higher substrate affinities. In contrast, bacteria in cultivated soils are expected to be r-selected and to have higher rates of activity per biomass unit, higher turnover rates and faster growth rates. Our results showed a trend toward these statements since a reduction in the abundance of taxa with oligotrophic characteristics, such as *Verrucomicrobia* [34], and *Planctomycetes* [58] were detected in cultivated fertilized soils. Moreover, this is in agreement with recent findings that confirm a correlation between *Verrucomicrobia* abundance patterns and conditions of limited nutrient availability in Prairie Soils in the United States [9]. On the other hand, the relative abundance of phylum *Gemmatimonadetes* was increased in fertilized cultivated soils as previously described for nitrogen-fertilized forest soils [59]. Consistently with our results, these authors observed that nitrogen fertilization was related to a higher abundance of *Gemmatimonadetes* and detected no presence of *Verrucomicrobia*. Little is known about *Gemmatimonadetes* ecology and metabolism since only one representative from this phylum has been isolated and characterized [60]. Even though, their presence in environments with a wide range of nutrient concentrations and redox states suggests versatile metabolisms [61].

We can also say that cultivated soils are more heterogeneous environments than non cultivated soils. Crop rotation, periodic fertilization and pesticide application generate temporal and

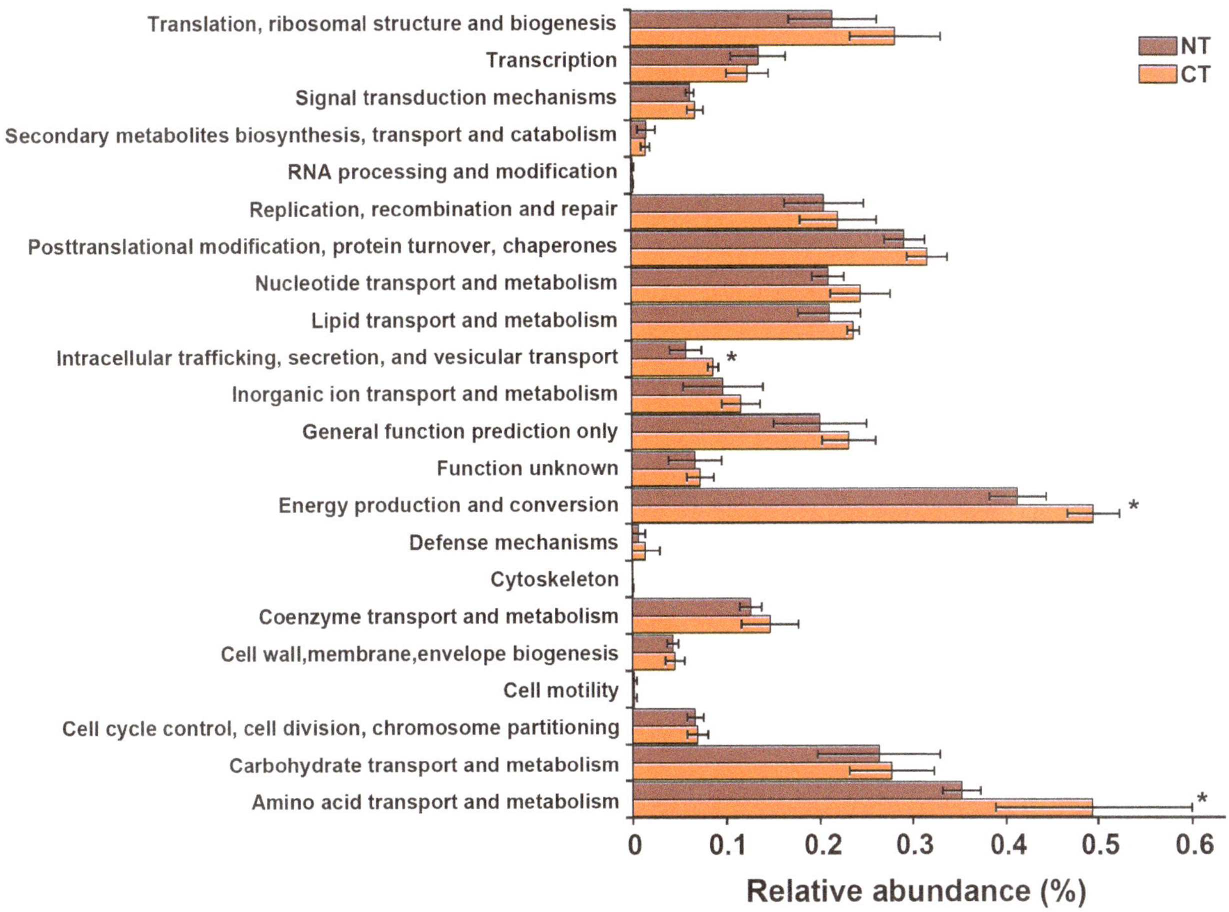

Figure 7. Relative abundances of COG categories in soil microbiomes of tillage systems comparison experiment in Balcarce. CA: conventional tillage; NT: no-tillage. Bars represent ±1 standard erro. (*)indicate significant differences ($p \leq 0.05$).

spatial changes in soil chemical properties, and therefore in nutrient availability and approachability for microorganisms. Bacteria dominating in this type of soils should be adapted to heterogeneity. These microorganisms should be able to fine-tune carbon and nitrogen intakes according to their metabolic needs under frequent external changes. This seems to be the case for *Gemmatimonadetes* suggesting a generalist ecological strategy. The high abundance of *Nitrospirae* may also respond to heterogeneous conditions. It has been proposed that *Nitrospirae* lineages occupy different positions on an imaginary scale reaching from K- to r-strategies [62]. We hypothesize that in cultivated soils, *Nitrospirae* K-strategists would be exploiting nitrite in high N microenvironments, while r-strategist would be mining low concentration areas in a nitrogen gradient enhanced by fertilization [63]. This competition would be less fierce in non cultivated soils due to the highest homogeneity and a less marked nitrogen gradient due to the lack of fertilization and more recalcitrant organic matter forms.

Effects of agricultural land use on metabolic profiles of soil microbiomes in Argentine Pampas

Our results from shotgun metagenomic data also indicated a tendency for the microbiomes of cultivated soils towards adaptation to nutrient heterogeneity. The highest relative abundances of sequences assigned to COGs related to transcription, protein modifications, nucleotide transport and metabolism, wall and membrane biogenesis and intracellular trafficking and secretion in cultivated fertilized soils are consistent with a copiotrophic strategy (i.e. rapid tight metabolism regulation and fast grow rates).Moreover, some of these COG categories were previously shown to be up-represented in copiotrophic marine microorganism genomes [58]. In a deeper look we detected that the relative abundance of sequences assigned to riboswitch regulated genes was higher in cultivated soils (i.e. cobalamin biosynthesis protein, S-adenosyl-methionine synthetase, S-adenosyl- homocysteine hydrolase) [64–66]. These ancient regulators may be playing a central role in a life-style strategy adapted to nutrient heterogeneity since they were described to be the most 'economical' and fast-reacting regulatory systems (no intermediate factors involved) [67]. Moreover, the abundant riboswitch- COGs in cultivated soils were related to synthesis of B vitamins. The levels of these vitamins have already been linked to differences in community composition in marine ecosystems [68]. More studies are needed in order to address this relationship in soil environments; still, our results suggest an important role of B-vitamins in fertilized heterogeneous soils.

Another interesting observation was the highest abundance of glutamate synthase (GOGAT) related COGs in cultivated soils metagenomic profiles. The combined role of GOGAT with glutamine synthetase and glutamate dehydrogenase allows the cells to sense ammonia external levels [69]. The high abundance of this regulation system detected in cultivated soil metagenomes suggests its importance in the detection of N fluxes related with fertilization. Moreover COGs related to the tricarboxylic acid cycle (TCA), such as citrate synthase and succinate dehydrogenase, were more abundant in cultivated soil microbiomes. GOGAT nitrogen assimilation pathway and TCA are related. It has been shown that the concentration of compounds of both pathways changes considerably and rapidly upon nitrogen up shift; in contrast, the concentrations of glycolytic intermediates remains homeostatic [70]. In addition, abundance of urease related COGs were also higher in cultivated soils. The soils sampled in this study have a long history of agricultural land use and have been long fertilized with both type of nitrogen (i.e. ammonia-based and urea-based fertilizers). This kind of environmental pressure finally selected a microbiome adapted to changing N sources and availability.

It is important to mention that these results do not infer that expression or activity of these metabolisms will be necessarily increased in cultivated soil microbiomes, still, the highest abundance of these COGs could be reflected in a highest diversification and specialization. The presence of a highest copy number of strategic genes have already been linked to copiotrophic or oligotrophic life-styles [58].

Microbiomes of non cultivated soils showed a higher abundance of sequences related to Coenzyme A and acetyl-Coa metabolism than microbiomes of cultivated soils. It is known that acetyl-CoA is a fundamental building block and energy source [71]. High acetyl-CoA levels would indicate a "proliferative" or "fed" state, while low acetyl-CoA levels (and high CoA levels) would be indicative of a "quiescent" or "starved" state. Pantothenate kinase (PanK), the key enzyme in CoA syntehsis, was also highest in non cultivated soil microbiomes [72]. In addition, it was stated that some oligotrophs preferentially use lipids as immediate and stored sources of carbon and energy in marine environments [58]. The observed abundance of Pank genes may ensure a correct CoA intracellular level in fasted moments, allowing proper lipid utilization. In addition, non cultivated metagenomic profiles showed higher abundance of sequences related to trehalose utilization. Trehalose is known to serve as energy source in many microorganisms [73]. Members of *Actinobacteria* genera *Mycobacterium* and family *Frankiaceae* are known to produce and/or utilize trehalose [74–76]. As mentioned above, our results showed higher abundance of *Actinobacteria*-related sequences in non cultivated soils. Moreover, sequences classified within family *Frankiaceae* were only present in these soils and the abundance of sequences assigned as *Mycobacterium* was higher than in cultivated soils (not shown). These observations are congruent with an oligotrophic strategy based on the use of storage components for carbon and energy sources.

The responses of the phylogenetic structure and metagenomic profile to agronomic land use were significantly correlated, suggesting some degree of correspondence between these different microbiome features. These results agree with previous observations done in soils from different biomes [9,77]. Moreover, Fierer et al. found similar a similar correlation between metagenomic and phylogenetic data for microbial communities of agricultural soils under different nitrogen gradients in experimental plots [10].

Correlation with soil properties

In addition, soil properties such as organic matter, phosphorus and nitrate levels explained most of the variability observed between cultivated and non cultivated soils. As mentioned, the highest amount of organic matter is found mostly in a recalcitrant form in these soils. Moreover, it was already established that nitrate accumulation exhibits a negative correlation with organic carbon availability [78]. In addition it has been proved that organic matter source and quality played an important role in regulating the magnitude of carbon metabolism and could be as important as nutrient abundance in water environments [79]. Our results are congruent with these observations since non cultivated soils, with highest levels of organic carbon, presented metagenomic profiles with tendencies to oligotrophy and the best explanation for this scenario is the low lability of the recalcitrant forms of carbon and nitrogen.

Assessing the effects of different tillage systems on Pampas soil microbiomes

Our results also showed differences in the structure and composition of soil microbiomes between no-till and conventional tillage soils. Sequences related to phyla *Gemmatimonadetes*, candidate division TM7 and *Acidobacteria*, were highest in CT soils, while the abundances of *Nitrospirae* and candidate division WS3 were highest in NT soils. Tillage is the principal agent producing soil disturbance and subsequent soil structure modification [80,81]. The negative effects of CT in soil stabilization and macroaggregate losses were previously registered in Pampas soils [82]. It has also been shown that NT increases macroaggregate abundance and organic matter content [83].These increments are related with a more recalcitrant organic matter, with increased humic acid contents and nutrient retention [84–86]. The higher abundance of *Nitrospirae* in NT soils reinforces the idea of a community adapted to a better N mining in these environments. Moreover, we observed very low abundance of *Nitrobacter* related sequences in Balcarce soils and no significant difference was observed between CT and NT soils (not shown). Changes in ammonia oxidation due to a decrease in ammonia availability by humic substances have already been proposed in microcosm experiments [87]. If this is the case of Balcarce soils, higher humic acids in NT soils would be decreasing ammonia availability for oxidation into nitrite and therefore decreasing nitrate availability, compared to CT soils. The predicted highest stability of organic matter and abundance of macroaggregates in NT soils are probably generating more marked nitrite gradients than in CT soils. The highest abundance of *Nitrospirae* in NT may be reflecting a highest lineage diversity that would be better adapted to these gradients.

In addition, communities under NT showed higher abundance of sequences related to the order *Syntrophobacterales* (*Deltaproteobacteria*, Figure S4). These are known anaerobic and syntrophic organisms [88,89]. These characteristics may be an advantage in stable soils with higher number of highly humic macro aggregates since syntrophy is known to be important for community functioning in micro-environments with low nutrient levels [90]. On the other hand, CT soils presented higher relative abundance of *Gammaproteobacteria* related sequences. The order *Xanthomonadales* was the main responsible for these differences (Figure S4). This taxon has already been associated to CT practices [32].

Even though we could not find significant differences associated to tillage systems at the community structure level for metabolic profiles, COG categories related to intracellular trafficking and secretion, amino acid transport and metabolism and energy production and conversion were more abundant in CT miro-biomes. These results suggest a tendency of CT microbiomes to a more copiotrophic life-style strategy than NT microbiomes.

Conclusion

Our results are consistent with the hypothesis that microbiomes exhibit different life- history and trophic strategies in Pampean soils under different land uses and tillage systems. Our data suggest that microbiomes of fertilized cultivated soils have more flexible metabolisms adapted to nutrient fluxes with tendencies to copiothropy while microorganisms in non cultivated soils are better adapted to lowest external nutrient availability and homogenous environment. The lowest nutrient accessibility in non cultivated soils may be explained by the higher amount of humic substances, recalcitrant organic matter and the lack of fertilizer amendments. Moreover, NT soils, with most stable structure and highest macroaggregate abundance, presented microbiomes better adapted to recalcitrant environments; while CT microbiomes presented a higher tendency to copiotrophy.

This work is of major contribution to understand how historical changes in soil properties due to agronomical land use have altered the diversity and function of below-ground communities. The importance of high-throughput characterization for the reconstruction of pre- agricultural microbiomes is being reinforced nowadays [9]. Following this direction, our findings will be very useful in future restoration and monitoring programs of Argentine Pampas ecosystems.

Supporting Information

Figure S1 PCoA plots of Pampa production field soil microbiomes based on average Bray Curtis distance matrices. A) PCoA of cultivated and non cultivated soil microbiomes. Standard error ellipses show 95% confidence areas. B) PCoA biplot of soil properties that best explained variation in community structure. Correlations were calculated using BIOENV on average data of each sampled site (Mantel $r = 0.6214$, $p \leq 0.05$). Circles represent samples from "La Estrella", crosses represent "Criadero Klein" samples and triangles represent "La Negrita" samples.

Figure S2 Relative abundances of Cluster of Orthologous groups (COGs) in Pampa production field soil microbiomes. Bars represent ± 1 standard error. Only significant COGs are showed.

Figure S3 Comparison of tillage systems effects on the structure of metabolic profiles. PCoA plot based on Euclidean distance matrices. CA: conventional tillage; NT: no-tillage. Standard error ellipses show 95% confidence areas.

Figure S4 Relative abundances of reads assigned to orders within classes *Gammaproteobacteria* and *Delta-porteobacteria* in NT and CT soils.

Table S1 Soil chemical and physical properties.

Acknowledgments

We acknowledge Germán F. Domínguez and Guillermo A. Studdert from the "Manejo Sustentable del suelo" research group (from Facultad de Ciencias Agrarias, Universidad Nacional de Mar del Plata, Unidad Integrada Balcarce) for providing samples of the tillage systems experiment. We thank Gonzalo Berhongaray, Josefina de Paepe, and Maria Rosa Mendoza for his help during sampling in the Rolling Pampas. We also acknowledge Roxana Colombo and Marcelo Soria for helpful discussion and advice.

Author Contributions

Conceived and designed the experiments: BC NR RÁ AM MV. Performed the experiments: BC NR. Analyzed the data: BC. Contributed reagents/materials/analysis tools: BC NR RÁ AM MV. Wrote the paper: BC NR MV. Obtained permission for sampling: RÁ.

References

1. Foley JA, Ramankutty N, Brauman KA, Cassidy ES, Gerber JS, et al. (2011) Solutions for a cultivated planet. Nature 478: 337–342.
2. Tilman D, Balzer C, Hill J, Befort BL (2011) Global food demand and the sustainable intensification of agriculture. Proc Natl Acad Sci U S A 108: 20260–20264.
3. Tilman D, Cassman KG, Matson PA, Naylor R, Polasky S (2002) Agricultural sustainability and intensive production practices. Nature 418: 671–677.
4. Balmford A, Green RE, Scharlemann JPW (2005) Sparing land for nature: exploring the potential impact of changes in agricultural yield on the area needed for crop production. Global Change Biology 11: 1594–1605.
5. Power AG (2010) Ecosystem services and agriculture: tradeoffs and synergies. Philos Trans R Soc Lond B Biol Sci 365: 2959–2971.
6. Foley JA, Defries R, Asner GP, Barford C, Bonan G, et al. (2005) Global consequences of land use. Science 309: 570–574.
7. Bennett EM, Carpenter SR, Caraco NF (2001) Human Impact on Erodable Phosphorus and Eutrophication: A Global Perspective: Increasing accumulation of phosphorus in soil threatens rivers, lakes, and coastal oceans with eutrophication. BioScience 51: 227–234.
8. Scholes MC, Scholes RJ (2013) Dust Unto Dust. Science 342: 565–566.
9. Fierer N, Ladau J, Clemente JC, Leff JW, Owens SM, et al. (2013) Reconstructing the Microbial Diversity and Function of Pre-Agricultural Tallgrass Prairie Soils in the United States. Science 342: 621–624.
10. Fierer N, Lauber CL, Ramirez KS, Zaneveld J, Bradford MA, et al. (2012) Comparative metagenomic, phylogenetic and physiological analyses of soil microbial communities across nitrogen gradients. ISME J 6: 1007–1017.
11. Satorre EH, Slafer GA (1999) Wheat: Ecology and Physiology of Yield Determination: Taylor & Francis.
12. Hall AJ, Rebella CM, Ghersa CM, Culot JP (1992) Field crop systems of the Pampas. In: Pearson CJ, editor. Field Crop Ecosystems of the World. Amsterdam: Elsevier Science & Technology Books. pp. 413–450.
13. Alvarez R, Díaz RA, Barbero N, Santanatoglia OJ, Blotta L (1995) Soil organic carbon, microbial biomass and CO2-C production from three tillage systems. Soil and Tillage Research 33: 17–28.
14. Bernardos JN, Viglizzo EF, Jouvet V, Lértora FA, Pordomingo AJ, et al. (2001) The use of EPIC model to study the agroecological change during 93 years of farming transformation in the Argentine pampas. Agricultural Systems 69: 215–234.
15. Alvarez R (2001) Estimation of carbon losses by cultivation from soils of the Argentine Pampa using the Century Model. Soil Use and Management 17: 62–66.
16. Hevia GG, Buschiazzo DE, Hepper EN, Urioste AM, Antón EL (2003) Organic matter in size fractions of soils of the semiarid Argentina. Effects of climate, soil texture and management. Geoderma 116: 265–277.
17. Quiroga AR, Buschiazzo DE, Peinemann N (1996) Soil Organic Matter Particle Size Fractions in Soils of the Semiarid Argentinian Pampas. Soil Science 161.
18. Kassam A, Friedrich T, Shaxson F, Pretty J (2009) The spread of Conservation Agriculture: justification, sustainability and uptake. International Journal of Agricultural Sustainability 7: 292–320.
19. Alvarez R, Steinbach HS (2009) A review of the effects of tillage systems on some soil physical properties, water content, nitrate availability and crops yield in the Argentine Pampas. Soil and Tillage Research 104: 1–15.
20. Gomez E, Bisaro V, Conti M (2000) Potential C-source utilization patterns of bacterial communities as influenced by clearing and land use in a vertic soil of Argentina. Applied Soil Ecology 15: 273–281.
21. Aon MA, Cabello MN, Sarena DE, Colaneri AC, Franco MG, et al. (2001) I. Spatio-temporal patterns of soil microbial and enzymatic activities in an agricultural soil. Applied Soil Ecology 18: 239–254.
22. Agaras B, Wall LG, Valverde C (2012) Specific enumeration and analysis of the community structure of culturable pseudomonads in agricultural soils under no-till management in Argentina. Applied Soil Ecology 61: 305–319.
23. Nievas F, Bogino P, Nocelli N, Giordano W (2012) Genotypic analysis of isolated peanut-nodulating rhizobial strains reveals differences among populations obtained from soils with different cropping histories. Applied Soil Ecology 53: 74–82.
24. Zabaloy MC, Garland JL, Gómez MA (2008) An integrated approach to evaluate the impacts of the herbicides glyphosate, 2,4-D and metsulfuron-methyl on soil microbial communities in the Pampas region, Argentina. Applied Soil Ecology 40: 1–12.
25. Zabaloy MC, Gómez E, Garland JL, Gómez MA (2012) Assessment of microbial community function and structure in soil microcosms exposed to glyphosate. Applied Soil Ecology 61: 333–339.

26. Deng W, Xi D, Mao H, Wanapat M (2008) The use of molecular techniques based on ribosomal RNA and DNA for rumen microbial ecosystem studies: a review. Molecular Biology Reports 35: 265–274.
27. Pontes D, Lima-Bittencourt C, Chartone-Souza E, Amaral Nascimento A (2007) Molecular approaches: advantages and artifacts in assessing bacterial diversity. Journal of Industrial Microbiology & Biotechnology 34: 463–473.
28. Sogin ML, Morrison HG, Huber JA, Welch DM, Huse SM, et al. (2006) Microbial diversity in the deep sea and the underexplored "rare biosphere". Proceedings of the National Academy of Sciences 103: 12115–12120.
29. Fortunato CS, Herfort L, Zuber P, Baptista AM, Crump BC (2012) Spatial variability overwhelms seasonal patterns in bacterioplankton communities across a river to ocean gradient. ISME J 6: 554–563.
30. Delmont TO, Prestat E, Keegan KP, Faubladier M, Robe P, et al. (2012) Structure, fluctuation and magnitude of a natural grassland soil metagenome. ISME J 6: 1677–1687.
31. Figuerola ELM, Guerrero LD, Rosa SM, Simonetti L, Duval ME, et al. (2012) Bacterial Indicator of Agricultural Management for Soil under No-Till Crop Production. PLoS ONE 7: e51075.
32. Souza RC, Cantão ME, Vasconcelos ATR, Nogueira MA, Hungria M (2013) Soil metagenomics reveals differences under conventional and no-tillage with crop rotation or succession. Applied Soil Ecology 72: 49–61.
33. Lauber CL, Ramirez KS, Aanderud Z, Lennon J, Fierer N (2013) Temporal variability in soil microbial communities across land-use types. ISME J 7: 1641–1650.
34. Ramirez KS, Craine JM, Fierer N (2012) Consistent effects of nitrogen amendments on soil microbial communities and processes across biomes. Global Change Biology 18: 1918–1927.
35. Ramirez K, Lauber CL, Knight R, Bradford MF (2010) Consistent effects of nitrogen fertilization on soil bacterial communities in contrasting systems. Ecology: 3463–3470
36. Lavado RS (2008) La Región Pampeana: historia, características y uso de sus suelos. In: Alvarez R, editor. Materia Orgánica Valor agronómico y dinámica en suelos pampeanos. Buenos Aires: Editorial Facultad de Ingenieria. pp. 1–11.
37. Berhongaray G, Alvarez R, De Paepe J, Caride C, Cantet R (2013) Land use effects on soil carbon in the Argentine Pampas. Geoderma 192: 97–110.
38. Rascovan N, Carbonetto B, Revale S, Reinert M, Alvarez R, et al. (2013) The PAMPA datasets: a metagenomic survey of microbial communities in Argentinean pampean soils. Microbiome 1: 21.
39. Nelson DW, Sommers LE, Sparks DLE, Page ALE, Helmke PAE, et al. (1996) Total Carbon, Organic Carbon, and Organic Matter. Methods of Soil Analysis Part 3-Chemical Methods: Soil Science Society of America, American Society of Agronomy. pp. 961–1010.
40. Bremner JME, Sparks DLE, Page ALE, Helmke PAE, Loeppert RH (1996) Nitrogen-Total. Methods of Soil Analysis Part 3-Chemical Methods: Soil Science Society of America, American Society of Agronomy. pp. 1085–1121.
41. Kuo SE, Sparks DLE, Page ALE, Helmke PAE, H LR (1996) Phosphorus. Methods of Soil Analysis Part 3-Chemical Methods: Soil Science Society of America, American Society of Agronomy. pp. 869–919.
42. Rhoades JDE, Sparks DLE, Page ALE, Helmke PAE, H. LR (1996) Salinity: Electrical Conductivity and Total Dissolved Solids. Methods of Soil Analysis Part 3-Chemical Methods: Soil Science Society of America, American Society of Agronomy. pp. 417–435.
43. Gee GW, Bauder JWEKA (1986) Particle-size Analysis. Methods of Soil Analysis: Part 1—Physical and Mineralogical Methods: Soil Science Society of America, American Society of Agronomy. pp. 383–411.
44. Cole J, Wang Q, Cardenas E, Fish J, Chai B, et al. (2009) The Ribosomal Database Project: improved alignments and new tools for rRNA analysis. Nucleic acids research: 141–145.
45. Caporaso JG, Kuczynski J, Stombaugh J, Bittinger K, Bushman FD, et al. (2010) QIIME allows analysis of high-throughput community sequencing data. Nat Meth 7: 335–336.
46. Edgar RC (2010) Search and clustering orders of magnitude faster than BLAST. Bioinformatics 26: 2460–2461.
47. Caporaso JG, Bittinger K, Bushman FD, DeSantis TZ, Andersen GL, et al. (2010) PyNAST: a flexible tool for aligning sequences to a template alignment. Bioinformatics 26: 266–267.
48. Price MN, Dehal PS, Arkin AP (2009) FastTree: Computing Large Minimum Evolution Trees with Profiles instead of a Distance Matrix. Molecular Biology and Evolution 26: 1641–1650.
49. Lozupone C, Knight R (2005) UniFrac: a New Phylogenetic Method for Comparing Microbial Communities. Applied and Environmental Microbiology 71: 8228–8235.
50. DeSantis TZ, Hugenholtz P, Larsen N, Rojas M, Brodie EL, et al. (2006) Greengenes, a Chimera-Checked 16S rRNA Gene Database and Workbench Compatible with ARB. Applied and Environmental Microbiology 72: 5069–5072.
51. Wang Q, Garrity GM, Tiedje JM, Cole JR (2007) Naïve Bayesian Classifier for Rapid Assignment of rRNA Sequences into the New Bacterial Taxonomy. Applied and Environmental Microbiology 73: 5261–5267.
52. Kindt R, Coe R (2005) Tree Diversity Analysis: A Manual and Software for Common Statistical Methods for Ecological and Biodiversity Studies: World Agrofirestry Centre.
53. Dixon P (2003) VEGAN, a package of R functions for community ecology. Journal of Vegetation Science 14: 927–930.
54. Meyer F, Paarmann D, D'Souza M, Olson R, Glass EM, et al. (2008) The metagenomics RAST server - a public resource for the automatic phylogenetic and functional analysis of metagenomes. BMC Bioinformatics 9: 386.
55. Fierer N, Bradford MA, Jackson RB (2007) Toward an ecological classification of soil bacteria. Ecology 88: 1354–1364.
56. Dion P, Nautiyal CS (2008) Microbiology of Extreme Soils: Springer.
57. Galantini J, Rosell R (2006) Long-term fertilization effects on soil organic matter quality and dynamics under different production systems in semiarid Pampean soils. Soil and Tillage Research 87: 72–79.
58. Lauro FM, McDougald D, Thomas T, Williams TJ, Egan S, et al. (2009) The genomic basis of trophic strategy in marine bacteria. Proceedings of the National Academy of Sciences 106: 15527–15533.
59. Nemergut DR, Townsend AR, Sattin SR, Freeman KR, Fierer N, et al. (2008) The effects of chronic nitrogen fertilization on alpine tundra soil microbial communities: implications for carbon and nitrogen cycling. Environmental Microbiology 10: 3093–3105.
60. Zhang H, Sekiguchi Y, Hanada S, Hugenholtz P, Kim H, et al. (2003) Gemmatimonas aurantiaca gen. nov., sp. nov., a Gram-negative, aerobic, polyphosphate-accumulating micro-organism, the first cultured representative of the new bacterial phylum Gemmatimonadetes phyl. nov. International Journal of Systematic and Evolutionary Microbiology 53: 1155–1163.
61. DeBruyn JM, Nixon LT, Fawaz MN, Johnson AM, Radosevich M (2011) Global Biogeography and Quantitative Seasonal Dynamics of Gemmatimonadetes in Soil. Applied and Environmental Microbiology 77: 6295–6300.
62. Maixner F, Noguera DR, Anneser B, Stoecker K, Wegl G, et al. (2006) Nitrite concentration influences the population structure of Nitrospira-like bacteria. Environmental Microbiology 8: 1487–1495.
63. Attard E, Poly F, Commeaux C, Laurent F, Terada A, et al. (2010) Shifts between Nitrospira- and Nitrobacter-like nitrite oxidizers underlie the response of soil potential nitrite oxidation to changes in tillage practices. Environmental Microbiology 12: 315–326.
64. Edwards AL, Reyes FE, Héroux A, Batey RT (2010) Structural basis for recognition of S-adenosylhomocysteine by riboswitches. RNA 16: 2144–2155.
65. Loenen WA (2006) S-adenosylmethionine: jack of all trades and master of everything? Biochem Soc Trans 34: 330–333.
66. Winkler WC, Breaker RR (2005) Regulation of bacterial gene expression by riboswitches. Annual Review of Microbiology 59: 487–517.
67. Nudler E, Mironov AS (2004) The riboswitch control of bacterial metabolism. Trends in biochemical sciences 29: 11–17.
68. Sañudo-Wilhelmy SA, Cutter LS, Durazo R, Smail EA, Gómez-Consarnau L, et al. (2012) Multiple B-vitamin depletion in large areas of the coastal ocean. Proceedings of the National Academy of Sciences.
69. Yan D (2007) Protection of the glutamate pool concentration in enteric bacteria. Proceedings of the National Academy of Sciences 104: 9475–9480.
70. Doucette CD, Schwab DJ, Wingreen NS, Rabinowitz JD (2011) α-ketoglutarate coordinates carbon and nitrogen utilization via enzyme I inhibition. Nat Chem Biol 7: 894–901.
71. Cai L, Tu BP (2011) On Acetyl-CoA as a Gauge of Cellular Metabolic State. Cold Spring Harbor Symposia on Quantitative Biology 76: 195–202.
72. Leonardi R, Rehg JE, Rock CO, Jackowski S (2010) Pantothenate Kinase 1 Is Required to Support the Metabolic Transition from the Fed to the Fasted State. PLoS ONE 5: e11107.
73. Elbein AD, Pan YT, Pastuszak I, Carroll D (2003) New insights on trehalose: a multifunctional molecule. Glycobiology 13: 17R–27R.
74. Barabote RD, Xie G, Leu DH, Normand P, Necsulea A, et al. (2009) Complete genome of the cellulolytic thermophile Acidothermus cellulolyticus 11B provides insights into its ecophysiological and evolutionary adaptations. Genome Research 19: 1033–1043.
75. Lopez MF, Fontaine MS, Torrey JG (1984) Levels of trehalose and glycogen in Frankia sp. HFPArI3 (Actinomycetales). Canadian Journal of Microbiology 30: 746–752.
76. Tropis M, Meniche X, Wolf A, Gebhardt H, Strelkov S, et al. (2005) The Crucial Role of Trehalose and Structurally Related Oligosaccharides in the Biosynthesis and Transfer of Mycolic Acids in Corynebacterineae. Journal of Biological Chemistry 280: 26573–26585.
77. Fierer N, Leff JW, Adams BJ, Nielsen UN, Bates ST, et al. (2012) Cross-biome metagenomic analyses of soil microbial communities and their functional attributes. Proceedings of the National Academy of Sciences 109: 21390–21395.
78. Taylor PG, Townsend AR (2010) Stoichiometric control of organic carbon-nitrate relationships from soils to the sea. Nature 464: 1178–1181.
79. Apple JK, del Giorgio PA (2007) Organic substrate quality as the link between bacterioplankton carbon demand and growth efficiency in a temperate salt-marsh estuary. ISME J 1: 729–742.
80. Bayer C, Mielniczuk J, Amado TJC, Martin-Neto L, Fernandes SV (2000) Organic matter storage in a sandy clay loam Acrisol affected by tillage and cropping systems in southern Brazil. Soil and Tillage Research 54: 101–109.
81. Langdale GW, West LT, Bruce RR, Miller WP, Thomas AW (1992) Restoration of eroded soil with conservation tillage. Soil Technology 5: 81–90.
82. Bongiovanni MD, Lobartini JC (2006) Particulate organic matter, carbohydrate, humic acid contents in soil macro- and microaggregates as affected by cultivation. Geoderma 136: 660–665.
83. Plaza-Bonilla D, Cantero-Martínez C, Viñas P, Álvaro-Fuentes J (2013) Soil aggregation and organic carbon protection in a no-tillage chronosequence under Mediterranean conditions. Geoderma 193–194: 76–82.

84. Jiao Y, Whalen JK, Hendershot WH (2006) No-tillage and manure applications increase aggregation and improve nutrient retention in a sandy-loam soil. Geoderma 134: 24–33.
85. Tivet F, de Moraes Sá JC, Lal R, Borszowskei PR, Briedis C, et al. (2013) Soil organic carbon fraction losses upon continuous plow-based tillage and its restoration by diverse biomass-C inputs under no-till in sub-tropical and tropical regions of Brazil. Geoderma 209–210: 214–225.
86. Slepetiene A, Slepetys J (2005) Status of humus in soil under various long-term tillage systems. Geoderma 127: 207–215.
87. Dong L, Córdova-Kreylos AL, Yang J, Yuan H, Scow KM (2009) Humic acids buffer the effects of urea on soil ammonia oxidizers and potential nitrification. Soil Biology and Biochemistry 41: 1612–1621.
88. Sieber JR, McInerney MJ, Gunsalus RP (2012) Genomic Insights into Syntrophy: The Paradigm for Anaerobic Metabolic Cooperation. Annual Review of Microbiology 66: 429–452.
89. McInerney MJ, Sieber JR, Gunsalus RP (2009) Syntrophy in anaerobic global carbon cycles. Current Opinion in Biotechnology 20: 623–632.
90. Kim HJ, Boedicker JQ, Choi JW, Ismagilov RF (2008) Defined spatial structure stabilizes a synthetic multispecies bacterial community. Proceedings of the National Academy of Sciences 105: 18188–18193.

2

Changes in the Bacterial Community of Soybean Rhizospheres during Growth in the Field

Akifumi Sugiyama[1]*, Yoshikatsu Ueda[1], Takahiro Zushi[1], Hisabumi Takase[2], Kazufumi Yazaki[1]

1 Research Institute for Sustainable Humanosphere, Kyoto University, Uji, Kyoto, Japan, **2** Faculty of Bioenvironmental Science, Kyoto Gakuen University, Kameoka, Kyoto, Japan

Abstract

Highly diverse communities of bacteria inhabiting soybean rhizospheres play pivotal roles in plant growth and crop production; however, little is known about the changes that occur in these communities during growth. We used both culture-dependent physiological profiling and culture independent DNA-based approaches to characterize the bacterial communities of the soybean rhizosphere during growth in the field. The physiological properties of the bacterial communities were analyzed by a community-level substrate utilization assay with BioLog Eco plates, and the composition of the communities was assessed by gene pyrosequencing. Higher metabolic capabilities were found in rhizosphere soil than in bulk soil during all stages of the BioLog assay. Pyrosequencing analysis revealed that differences between the bacterial communities of rhizosphere and bulk soils at the phylum level; i.e., Proteobacteria were increased, while Acidobacteria and Firmicutes were decreased in rhizosphere soil during growth. Analysis of operational taxonomic units showed that the bacterial communities of the rhizosphere changed significantly during growth, with a higher abundance of potential plant growth promoting rhizobacteria, including *Bacillus, Bradyrhizobium,* and *Rhizobium*, in a stage-specific manner. These findings demonstrated that rhizosphere bacterial communities were changed during soybean growth in the field.

Editor: Matthew W. Fields, Montana State Univeristy, United States of America

Funding: This study was partly supported by a Kyoto University Step-Up Grant for Young Scientists (AS) and a grant from the Ministry of Agriculture, Forestry, and Fisheries of Japan (Genomics-based Technology for Agricultural Improvement, SFC2001) (AS). The funders had no role in study design, data collection and analysis, decision to publish, or preparation of the manuscript.

Competing Interests: The authors have declared that no competing interests exist.

* Email: akifumi_sugiyama@rish.kyoto-u.ac.jp

Introduction

The rhizosphere is the small region around the roots, defined as "the zone of soil surrounding the root which is affected by it" [1,2], where plants and millions of microbes interact with each other [3]. Rhizosphere microbes were shown to have intense activity and to be important for plant health and growth [4]. For example, mycorrhiza and rhizobia provide phosphorous and nitrogen, respectively, and microbes called plant-growth-promoting rhizobacteria (PGPR) exert both direct and indirect effects on plant growth, such as the prevention of colonization by pathogens and modulation of plant immunity [5–7]. These rhizosphere microbes are regarded as prominent components of sustainable agriculture that reduce the use of fertilizers and pesticides [8]. Plants have been shown to accommodate rhizosphere microbes by providing a platform and nutrients mainly at the root exudates, which account for up to 40% of photosynthates [9–12]. In addition to the climate and chemical properties of soils, resident plants exert influence on rhizosphere microbial communities. Microbial communities have been found to depend on the plant species grown in the same type of soil [13–16], demonstrating a tight interaction between plants and rhizosphere microbial communities [17].

Because rhizosphere microbial communities are important in plant growth and performance, these communities have been extensively studied using both culture-dependent and culture-independent methods [4,17,18]. Recent advances in next generation sequencing methods have enabled in-depth analyses of rhizosphere microbial communities. *Arabidopsis* root bacterial communities have been investigated comprehensively [19–21], and the analysis of soils collected under *Arabidopsis* plants in their natural habitats indicated possible interactions between *Arabidopsis* growth and the microbial communities in these soils [22]. However, despite increases in community-based analyses of rhizosphere bacterial communities [4,17], it remains unclear how plant and bacteria communicate to form rhizosphere bacterial communities from 'reservoir' bulk soil. In particular, few studies have comprehensively analyzed rhizosphere microbial communities during growth using next generation sequencing [16,23,24] although growth-dependent analyses of rhizosphere microbial communities have been performed using methods such as automated ribosomal interspacer analysis (ARISA) and denaturing gradient gel electrophoresis (DGGE) [25–28].

Legume plants include important crop species such as soybeans (*Glycine max*), which supply nutrients rich in protein and oil for human consumption. Legume plants have been used to investigate plant-microbe interactions in the rhizosphere, due to their agricultural importance and ability to form symbiotic relationships with rhizobia and arbuscular mychorrhizal fungi (AMF) [29–31]. Genetic analysis using model legume plants such as *Lotus japonicus* and *Medicago truncatula* revealed pathways leading to symbiosis. Despite the large number of reports analyzing the components of legume-rhizobia and legume-AMF symbiosis, the broad range of rhizosphere microbial species in soil, which could affect legume interactions with rhizobia and AMF, have not been characterized in depth. The mechanisms underlying legume plant interactions with various soil microbes during growth in the field remain especially elusive. Understanding the composition of rhizosphere microbial communities during growth in the field could provide a basis for optimizing agricultural utilization of rhizosphere microbes. For example, DGGE showed that the composition of the soybean rhizosphere changed during growth with alterations in the relative contributions of various phyla, including Proteobacteria, Acidobacteria, Bacteroidetes, Nitrospirae, Firmicutes, Verrucomicrobia and Acidobacteria [27]. To enhance understanding of legume-microbe interactions in the field, and to obtain basic information on soybean rhizosphere bacterial communities for further research, bacterial communities of rhizospheres were analyzed together with bulk soil during soybean growth in a field in Kyoto Prefecture, Japan. The physiological properties of the bacterial communities were analyzed by community-level BioLog substrate utilization assays. In addition, the 16S ribosomal RNA (rRNA) gene, which is the most important target in the study of bacteria, was analyzed by PCR amplicon pyrosequencing. The results of this study suggest that plant growth could affect the composition of rhizosphere bacterial communities in addition to the seasonal effects.

Materials and Methods

Study site and sampling

Field experiments were conducted in an experimental field of Kyoto Gakuen University, Kameoka, Kyoto, Japan (N 34° 99′ 38″, E 135° 55′ 14″). This field has been maintained organically for more than 7 years, and sugar cane was grown there the previous year. Soybean seeds (*Glycine max* cv. Fukujishi) were purchased from Takii Shubyo (Kyoto, Japan) and sown on May 25, 2012. Plants were irrigated as needed, and emerging weeds were manually removed throughout the crop season on a weekly basis. Bulk soil samples were collected before sowing the seeds. Rhizosphere and bulk soil samples were collected on July 9, 2012 (at the beginning of flowering, R2), on August 25, 2012 (after the seeds matured, but pods were still green, R6), and on September 24, 2012 (at full maturity, when the pods were dry, R8). Bulk soil, defined a soil that does not adhere to plant roots, was obtained at least 20 cm from the plants. Bulk soils from five different spots were combined into one sample. Rhizosphere soil, defined as soil that adheres to the plant root after gentle shaking [32], was obtained from five plants, using sterile brushes, and combined into one sample. Both rhizosphere and bulk soil samples were immediately transferred to the laboratory in a cool container (0–10°C) within 2 hours. Each sample was homogenized and passed through a 1 mm sieve, and 1.5 g of each sample was immediately used for the BioLog substrate utilization assay. The remaining soil was kept at −30°C until subsequent DNA extraction. The chemical properties of the soil at initial stage were analyzed using standard methods in the soil analysis laboratory of Otsuka Agritechno Co. Ltd. (Tokyo, Japan). The soil was found to have an electrical conductivity of 0.28 dS/m, a pH of 6.7, and to contain 0.5 ppm NH_4^+, 30 ppm, NO_3^-, 5.7 ppm K, 3.2 ppm S, 3.72 ppm Ca, 7.6 ppm Mg, 0.02 ppm B, 0.58 ppm Fe, 0.01 ppm Cu, 0.03 ppm Zn, and 1.6 ppm Cl. Neither P nor Mn was detected. Detection limit for P and Mn was 0.02 and 0.0003 ppm, respectively. No fertilizers and pesticides were applied during the growth season.

DNA extraction and real time PCR to quantify total bacteria

DNA was extracted from 0.5 g soil using the Soil DNA Isolation Kit (Mo Bio, Carlsbad, CA) according to the manufacturer's protocol. Extracted DNA was quantified using the Qubit Quantification Platform dsDNA HS Assay Kit (Invitrogen, Carlsbad, CA) and stored at −30°C until use.

For the real time PCR to quantify the total bacterial communitieswe used 16S rRNA gene-specific primers (968F and 1401R) essentially as described previously [33]. Amplification reaction was carried out using Thunderbird SYBR qPCR mix (Toyobo, Osaka, Japan) according to the manufacture's protocol. Amplification reactions were as follows: 2 min at 95°C, 40 cycles of 15 sec at 95°C, 30 sec at 50°C, 30 sec at 72°C.

Barcoded pyrosequencing and sequence analysis

For the amplicon pyrosequencing the 16S rRNA gene was amplified using the barcorded primers; 27F, 5′-gcctccctcgcgccatcagnnnnnnnnnnnnAGAGTTTGATCMTGGCTCAG-3′ and 388R, 5′-gccttgccagcccgctcagTCTGCTGCCTCCCGTAGGAGT-3′ [34–36]. The lowercase sequences of each primer are necessary adapters for binding and amplification during pyrosequencing. The uppercase sequences targeted the conserved regions of the 16S rRNA gene [37,38], and the "n" repeats represent unique barcodes, which allow the sorting and assignment of each individual sequence read. The barcode sequences were provided by Operon Co Ltd., and are listed in Table S1. Each 25 µL PCR mixture contained 20 ng soil DNA and 1 U Tks Gflex DNA polymerase (Takara, Japan). PCR amplification was performed in a thermal cycler (iCycler, BioRad, Hercules, CA), with the protocol consisting of an initial denaturation at 94°C for 1 min, followed by 30 cycles of denaturation at 98°C for 10 sec, annealing at 55°C for 15 sec, and extension at 68°C for 30 sec. PCR amplicons were verified by electrophoresis on 1.2% agarose gels. A total of nine PCR amplicons (three PCR replicates x three different DNA samples) were combined for each barcode, and bands were extracted using the Wizard SV Gel and PCR Clean-Up System (Promega, San Luis Obispo, CA) according to the manufacturer's protocol. The DNA concentration of each purified PCR product was measured using the Qubit Fluorometer with Quant-iT dsDNA BR Assay Kit (Invitrogen) according to the manufacturer's protocol. Equal amounts of PCR amplicons harboring 21 different barcode sequences (7 samples × 3 replicates) were combined into a single tube, and ethanol-precipitated. Dried DNA samples were sent to the next-generation sequencing laboratory of Operon Co., Ltd., and analyzed using the GS Junior Titanium System (Roche, Basel, Switzerland).

Sequence reads were imported into Mothur v. 1.29.0 (http://www.mothur.org) for operational taxonomic unit (OTU) generation and classification [39]. All sequence data was de-noised through the default parameters of Mothur (shhh.flows), and de-noised sequence reads were trimmed using the following criteria (trim.seqs): no more than two mismatches for the forward primer sequence, no more than one mismatch in the barcode sequence, no homopolymeric sequences of more than eight nucleotides, a read length of more than 200 bases, and trimming to a common length of 200 base pairs. A bacterial 16S rRNA reference alignment was exported from SILVA [40]. Chimera sequences were detected using the chimera.uchime command in Mothur [41] and were removed (remove.seqs) from further analysis. Sequences were aligned using the Needleman-Wunsch algorithm, pairwise genetic distances were calculated (calc = onegap, countends = T), and sequences were clustered into OTUs using the furthest neighbor algorithm (cutoff = 0.03). All 454 pyrosequenc-

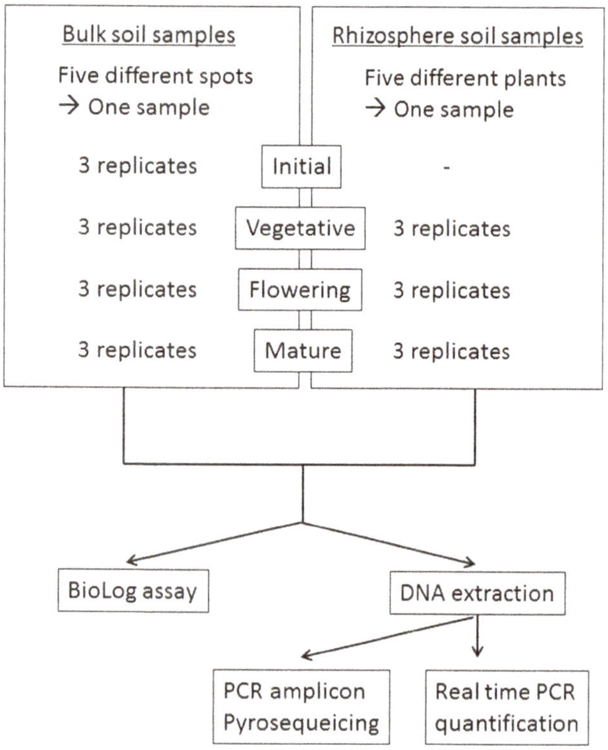

Figure 1. Bulk and rhizosphere sample preparation.

ing reads have been deposited into the DDBJ Sequence Read Archive under accession number DRA001076 [42]. Statistical analyses were performed using PC-ORD Software v. 6.0, as described [22,35,43] and the R statistical package (R Development Core Team).

Microbial community analysis BioLog EcoPlates

BioLog analysis was performed as described [22]. Briefly, 1.5 g aliquots of three soil samples were each added to 15 ml sterile saline solution (0.87% w/v NaCl), and diluted 1:10,000 with the same solution. The diluted solution (150 μL) was added to each well of the BioLog EcoPlate (Biolog, CA), followed by incubation at 25°C for 96 h. Absorbance at 595 nm was recorded with Spectraflour plus (TECAN), and data were analyzed using the average well color development (AWCD) method [44].

Results

Soil sampling and community-level physiological profiling

Soybeans were grown from May to September 2012 in a field in Kyoto Prefecture, Japan. Soil samples were obtained during the initial stage (before the seeds were sown), the vegetative stage (beginning of flowering, R2), the flowering stage (after seeds became mature, but the pods were still green, R6), and in the mature stage (pods were dry, R8), as described in the Materials and Methods section. A flow chart is also shown in Figure 1. The BioLog substrate utilization assay was performed on the day of sampling to avoid changes in bacterial communities during storage of the soil. The BioLog assay was originally developed to identify microbial isolates based on their substrate utilization profiles, and is now widely used to obtain community-level substrate utilization profiles [22,44–47]. Color intensity was determined by calculating the average well color development (AWCD) on each plate. The AWCD of rhizosphere soil was 1.5 to 3-fold higher than that of bulk soil throughout soybean growth when same amount of soil was used (Figure 2), indicating that tested metabolic capabilities of rhizosphere soil were higher than bulk soil and that the physiological profile of the two sets of soil samples differed significantly. Overall, most of the substrates were highly metabolized by rhizosphere soils (Table S2). Total bacterial communities were also quantified using real time PCR, demonstrating that 2.0×10^8 to 5.2×10^8 gene copy number of 16S rRNA were observed in soils (Table S3). When BioLog data were normalized with 16S rRNA gene, higher activities were observed in rhizosphere soil than bulk soils at vegetative and flowering stages, but not in mature stage. Significantly higher activities were observed in mature soils of both bulk and rhizosphere, compared with vegetative and flowering stages (Figure S1).

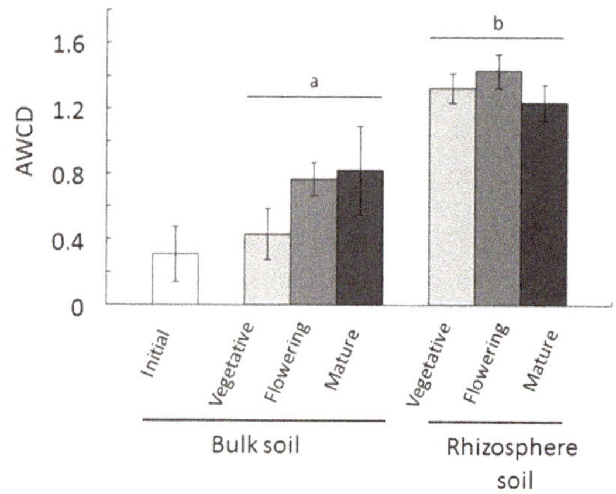

Figure 2. Community level physiological profiling. The BioLog substrate utilization assay was performed to generate community-level substrate utilization profiles. Average well color development (AWCD) after a 96-hour incubation was recorded. Values are mean ± SD (n = 3). The effect of growth stage and interaction between both factors (growth stage and compartment) were not significant in two-way ANOVA, but significant difference was observed between bulk and rhizosphere soil, as shown by the different letters ($p<0.05$).

Phylum-level analysis of bacterial communities during soybean growth

DNA was extracted from both bulk and rhizosphere soil and stored at −30°C. PCR was subsequently performed, and the amplicons were sequenced using the GS Junior Titanium System (Roche). A total of 62,828 sequences from 21 samples were obtained from the pyrosequencing reactions (7 different soil types in triplicate, Table S4). Sequences were analyzed using Mothur v.1.29.0 [39] and The Ribosomal Database Project Pyrosequencing Pipeline.

All sequences could be classified into 17 phyla. Proteobacteria was the dominant phyla in all soils, followed by Actinobacteria and Chloroflexi (Figure 3 and Figure S2). Rhizosphere soil showed changes distinct from bulk soil during soybean growth, with Proteobacteria increasing from 42.8±2.2% in the vegetative stage to 51.1±3.9% in the mature stage, while Acidobacteria decreased from 11.8±2.9% to 7.8±1.6% and Firmicutes decreased from 4.8%±2.2 to 2.1±0.6%. In bulk soil, however, Acidobacteria was increased from 7.1±0.8% to 13.5±1.2%, while Actinobacteria was decreased, from 29.4±0.8% to 16.0±1.2%, during soybean growth.

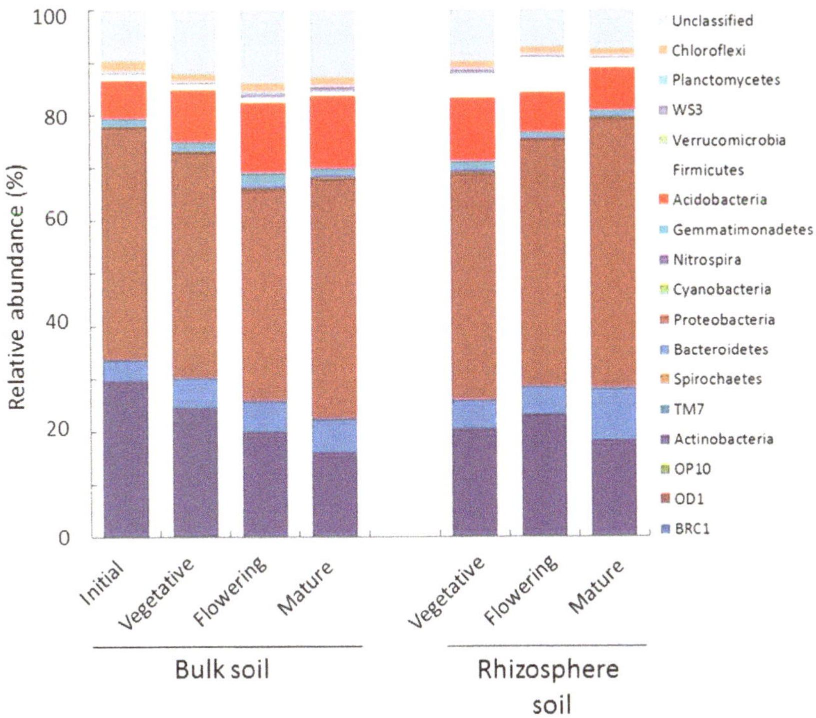

Figure 3. Relative abundance of major bacterial phyla present in bulk and rhizosphere soil during soybean growth, as revealed by pyrosequencing. Each bar represents the average relative abundance of triplicates.

OTU-based analysis of bacterial communities during soybean growth

Principal Component Analysis (PCA) was performed with the relative abundance values of OTUs used to compare bacterial communities at each stage (Figure 4). The bacterial community structures of bulk soil in the vegetative, flowering, and mature stages differed from those of initial bulk soil, but the former three communities clustered together, suggesting that relatively small changes occurred in the bacterial communities of bulk soil during soybean growth. In contrast, the communities of rhizosphere soil differed during all four stages (Figure 4), suggesting that rhizosphere bacterial communities that formed during the vegetative stage were further modified during growth. To characterize the changes in the bacterial communities of bulk and rhizosphere soil during soybean growth, the relative abundances of OTUs were compared during growth, with a list of the 1,000 most abundant OTUs shown in Table S5. The most abundant OTU was found to be *Bradyrhizobium* (Rank 1), followed by Steroidobacter (Rank 2), and unclassified bacteria (Rank 3).

PCA analysis showed that the first principal component (PC1) (28.3%) appeared to represent the difference in bacterial communities between bulk and rhizosphere soil as well as the changes of rhizosphere bacterial communities during growth, whereas PC2 (19.1%) represented the difference in bacterial communities between initial soil and soil after soybean growth. To characterize the changes in bacterial communities during growth in detail, 20 OTUs that showed high absolute loading on PC1 were selected (12 positive and 8 negative), and their relative abundance was compared (Figure 5). Positive (+) and negative (−) loadings correlate positively and negatively with axis, respectively. OTUs with high positive loading on PC1 were more abundant in rhizosphere than in bulk soil. *Bradyrhizobium* (rank 4), *Bacillus* (rank 21 and 170), and *Stenotrophomonas* (rank 158 and 241) were highly increased during the flowering stage. In contrast, OTUs with high negative loading were more abundant in initial bulk soil and less abundant in both bulk and rhizosphere soil after soybean growth. Most of these OTUs were found to be *Massilia* (Figure 5B), a group of Gram-negative Betaproteobacteria widely found in soils, while their effects on resident plant species have not yet been reported[21,48].

The pyrosequencing of 16S rDNA reveals bacterial communities at the genus level; therefore, all OTUs found to belong to each genus (Figure 5) were combined, and the relative abundance of these genera was compared during growth (Figure 6). *Bacillus* was one of the most abundant genera in the soybean field, with a relative abundance in rhizosphere soil higher during the vegetative and flowering stages than during the mature stage. The relative abundances of *Stenotrophomonas* and *Strepromyces* in rhizosphere soil were significantly higher during the flowering stage than during either the vegetative or mature stage. In contrast, the relative abundance of *Massilia* was highest in the initial bulk soil, accounting for up to 6% of the bacterial community in these samples (Figure 6). The relative abundance of these genera showed similar patterns as the representative OTUs in Figure 4, suggesting that a large proportion of community changes occurred at the genus level.

Discussion

Plants have been found to influence rhizosphere bacterial communities, which are important for crop growth and yield [4,49]. In this study, the rhizosphere bacterial communities of soybeans grown in the field were analyzed using both culture-dependent physiological profiling and culture independent 16S rRNA metagenomics. High numbers of bacteria have been reported to flourish in the rhizosphere, due mostly to the supply

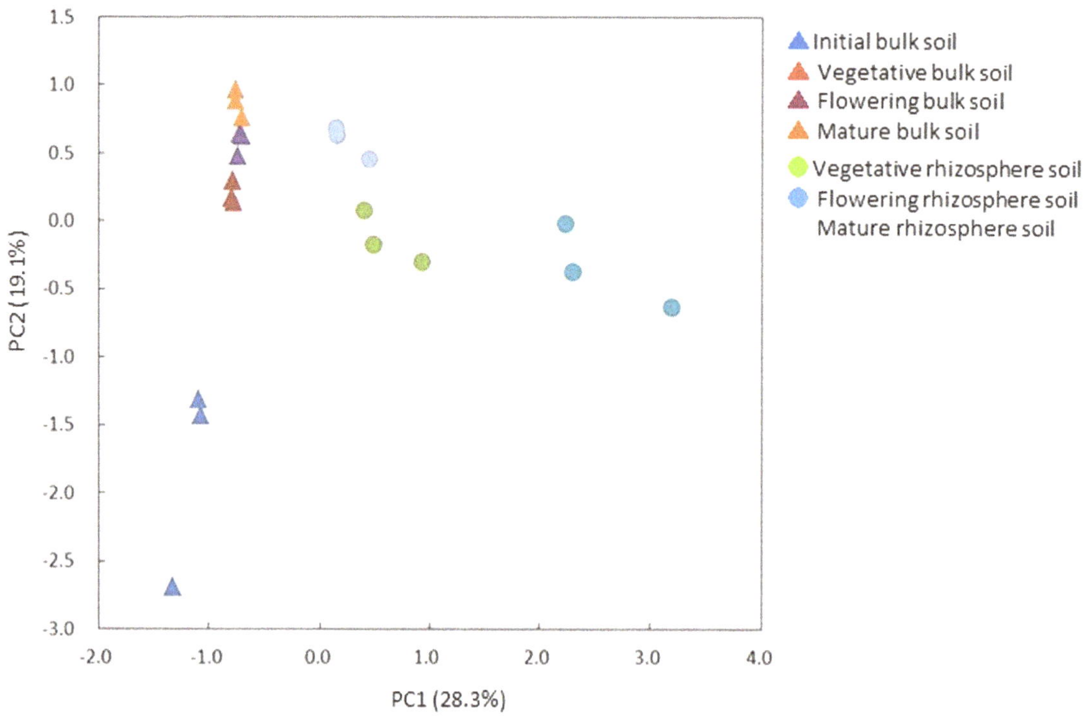

Figure 4. Principal component analysis of the relative abundance of OTUs.

of nutrients and the platform supplied by resident plants [3,17,50]. Using BioLog substrate utilization assays, we found that tested metabolic capabilities of rhizosphere soil were higher than bulk soil during soybean growth, and rhizosphere soil had higher microbiological activity in the vegetative stage (Figure 2). These results suggest that soybeans affected rhizosphere bacterial communities within 6 weeks after sowing (vegetative stage), with further alterations possibly occurring during later growth stages to maintain physiologically active rhizosphere bacterial communities.

A metagenomic 16S rRNA amplicon approach was employed to characterize the bacterial communities of both bulk and rhizosphere soil during soybean growth. Pyrosequencing of PCR amplicons and sequence analyses using Mothur v. 1.29.0 [39] revealed that Proteobacteria was the most abundant phylum in all soil types, with changes in bacterial communities at the phylum level during soybean growth occurring in both rhizosphere and bulk soil (Figure 3). Of the Proteobacteria, the Alphaproteobacteria were the most abundant, followed by Beta-, Gamma-, and Deltaproteobacteria in all soil types (Figure S3). The relative abundance of the Alpha- and Gammaproteobacteria, as well as their ratio, was found to be altered drastically by nitrogen fertilization and soybean nodulation [51]; however, we found only minor changes in these two classes in both bulk and rhizosphere soil during soybean growth. OTU-based analyses revealed that rhizosphere bacterial communities changed as soybean growth stage changed, suggesting the stage-specific formation of unique rhizosphere bacterial communities (Figure 4). In contrast, bulk soil bacterial communities showed small changes during plant development, but differed markedly from initial stage bulk soil bacterial communities. The latter was likely due to seasonal changes, as well as distant effects from resident plant species.

It has been hypothesized that resident plant species accommodate a specific set of rhizosphere bacterial communities, both to optimize growth and to prevent infection by pathogens [4]. In this study, *Bradyrhizobium* was shown to be abundant in rhizosphere soil, presumably due to the chemotaxis of *Bradyrhizobium* to the root exudates of soybean and proliferation around the roots [52]. Other genera that are highly abundant in rhizosphere soil (Figures 5 and 6) also contain potential PGPRs. For example, various species of *Bacillus* have been associated with systemic resistance to pathogens [53], phosphorus solubilization, and the production of antibiotics [54,55]. Furthermore, *Stenotrophomonas maltophilia* was found to reduce nematode densities in soils [56] and *Enterobacter cloacae* to produce indole-3-acetic acid and solubilize phosphate [57,58]. In addition, *Sphingomonas* sp. was found to produce indole-3-acetic acid [59] and *Streptomyces* sp. to reduce the infection of pine roots by *Fusarium* and *Armillaria* [60], although both these genera also contain plant pathogens, including *Sphingomonas melonis* [61] and *Streptomyces scabies* [62,63]. We found that several species of these potential PGPRs were elevated in the rhizosphere of soybean at growth stage-specific manner. This may be due to changes in soybean root exudates during growth, inasmuch as the profiles of root exudates and rhizosphere microbes have been reported to correlate during the growth of *Arabidopsis* [23,24].

In conclusion, this study revealed the changes that occur in the rhizosphere bacterial communities of soybeans grown in the field. Physiological activities measured by BioLog assay suggested that metabolic capabilities of rhizosphere soil were higher than bulk soil during the growth season. Rhizosphere bacterial communities were changed during soybean growth, in a manner distinct from bacterial communities in bulk soil, and rhizosphere bacterial communities contained potential PGPR genera that are both highly abundant and growth stage-specific. Changes observed in rhizosphere bacterial communities are in part due to the seasonal and surface effects as for bulk soil communities, but substantial changes observed for rhizosphere communities suggest the

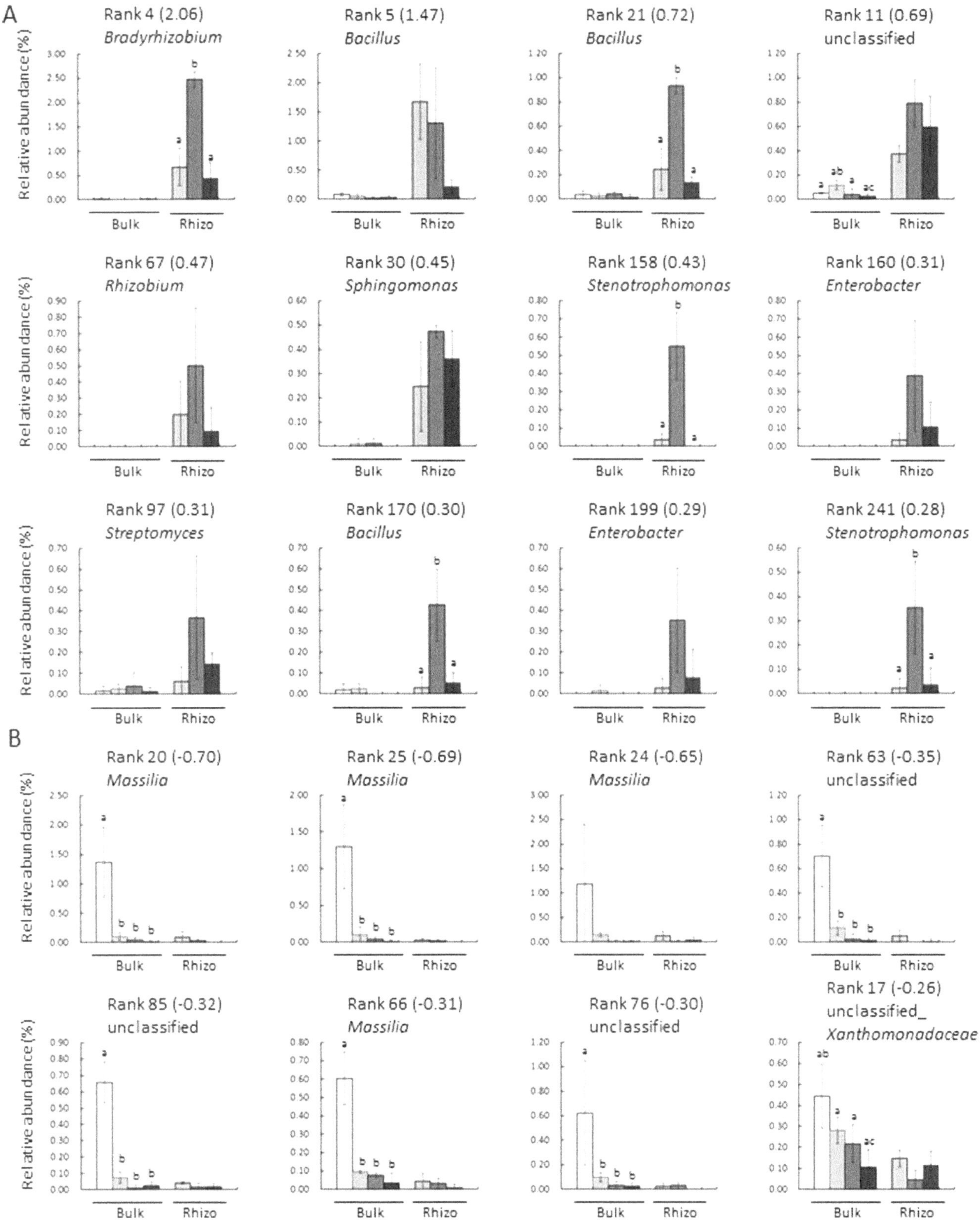

Figure 5. Relative abundance of OTUs with high absolute loading on Principal Component 1 (PC1). (A) OTUs with high positive loading on PC1; (B) OTUs with high negative loading on PC1. White bars represent bulk soil and black bars represent rhizosphere soil. The relative abundance was calculated using the data from pyrosequencing analysis. The numbers in parentheses are loading values. □; initial soil, ■; vegetative stage, ■; flowering stage, ■; mature stage. Values are mean ± SD (n = 3), with significant differences by Tukey's HDS test ($P<0.05$) indicated.

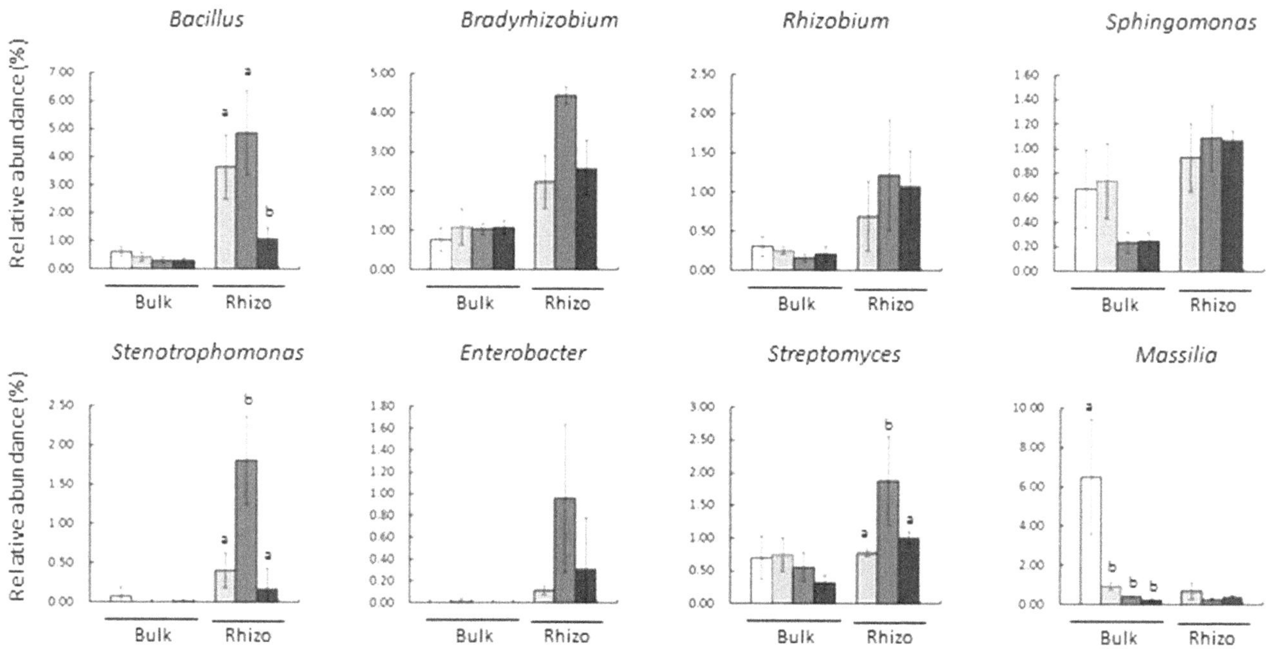

Figure 6. Relative abundance of eight different genera during soybean growth. The relative abundance of the eight genera shown in Figure 4 was compared. □; initial soil, ■; vegetative stage, ■; flowering stage, ■; mature stage. Values are mean ± SD (n = 3), with significant differences by Tukey's HDS test (P<0.05) indicated.

influence from soybean growth. Further studies on both metabolic activities of soybean such as root exudation and the physiological functions of these rhizobacteria on plant growth are warranted to elucidate the reciprocal interactions between plants and rhizosphere microbes in the fields for better utilization of rhizosphere bacteria for sustainable agriculture.

Supporting Information

Figure S1 Community level physiological profiling normalized in 16S rRNA gene. The BioLog substrate utilization assay was performed to generate community-level substrate utilization profiles. Average well color development (AWCD) after a 96-hour incubation was recorded, and normalized with 16S rRNA gene. Values are mean ± SD (n = 3). Different letters indicate significant differences (P<0.05) by one-way ANOVA with Tukey's HSD test.

Figure S2 Relative abundance of each bacterial phylum. The relative abundance of each phyla was compared. □; initial soil, ■; vegetative stage, ■; flowering stage, ■; mature stage. Values are mean ± SD (n = 3)

Figure S3 Relative abundance of each class of Proteobacteria.

Table S1 Barcode sequences used for pyrosequencing primers.

Table S2 Substrate utilization indicated by color development in Biolog Eco Plates.

Table S3 Gene copy numbers of 16S rRNA in soil.

Table S4 Number of sequences reads used for analysis.

Table S5 The 1,000 most abundant OTUs found in the experimental soybean field.

Acknowledgments

We would like to thank Ms. Yuko Kobayashi for her technical assistance, Dr. Jiro Sekiya of Kyotogakuen University for valuable advice, and Otsuka Agri Techno Co. Ltd., for the analysis of soil chemicals. We would also like to thank Dr. Kiwamu Minamisawa of Tohoku University for the critical reading of this manuscript.

Author Contributions

Conceived and designed the experiments: AS HT KY. Performed the experiments: AS. Analyzed the data: AS YU TZ. Wrote the paper: AS KY.

References

1. Darrah PR (1993) The rhizosphere and plant nutrition: a quantitative approach. Plant Soil 155–156: 1–20.
2. Hartmann A, Rothballer M, Schmid M (2008) Lorenz Hiltner, a pioneer in rhizosphere microbial ecology and soil bacteriology research. Plant Soil 312: 7–14.
3. Mendes R, Garbeva P, Raaijmakers JM (2013) The rhizosphere microbiome: significance of plant beneficial, plant pathogenic and human pathogenic microorganisms. FEMS Microbiol Rev 37: 634–663.
4. Berendsen RL, Pieterse CMJ, Bakker PA (2012) The rhizosphere microbiome and plant health. Trends Plant Sci 17: 478–486.
5. Bloemberg GV, Lugtenberg BJJ (2001) Molecular basis of plant growth promotion and biocontrol by rhizobacteria. Curr Opin Plant Biol 4: 343–350.
6. Van der Ent S, Van Hulten M, Pozo MJ, Czechowski T, Udvardi MK, et al. (2009) Priming of plant innate immunity by rhizobacteria and beta-aminobutyric acid: differences and similarities in regulation. New Phytol 183: 419–431.

7. Lugtenberg B, Kamilova F (2009) Plant-growth-promoting rhizobacteria. Annu Rev Microbiol 63: 541–556.
8. Beneduzi A, Ambrosini A, Passaglia LMP (2012) Plant growth-promoting rhizobacteria (PGPR): Their potential as antagonists and biocontrol agents. Genet Mol Biol 35: 1044–1051.
9. Oger PM, Mansouri H, Nesme X, Dessaux Y (2004) Engineering root exudation of lotus toward the production of two novel carbon compounds leads to the selection of distinct microbial populations in the rhizosphere. Microb Ecol 47: 96–103.
10. Bais HP, Weir TL, Perry LG, Gilroy S, Vivanco JM (2006) The role of root exudates in rhizosphere interactions with plants and other organisms. Annu Rev Plant Biol 57: 233–266.
11. Haichar FZ, Marol C, Berge O, Rangel-Castro JI, Prosser JI, et al. (2008) Plant host habitat and root exudates shape soil bacterial community structure. ISME J 2: 1221–1230.
12. Badri DV, Vivanco JM (2009) Regulation and function of root exudates. Plant Cell Environ 32: 666–681.
13. Innes L, Hobbs PJ, Bardgett RD (2004) The impacts of individual plant species on rhizosphere microbial communities in soils of different fertility. Biol Fertil Soils 40: 7–13.
14. Batten KM, Scow KM, Davies KF, Harrison SP (2006) Two invasive plants alter soil microbial community composition in serpentine grasslands. Biol Invasions 8: 217–230.
15. Garbeva P, van Elsas JD, van Veen JA (2008) Rhizosphere microbial community and its response to plant species and soil history. Plant Soil 302: 19–32.
16. Inceoglu O, Al-Soud Wa, Salles JF, Semenov AV, van Elsas JD (2011) Comparative analysis of bacterial communities in a potato field as determined by pyrosequencing. PLoS One 6:e23321.
17. Bakker MG, Manter DK, Sheflin AM, Weir TL, Vivanco JM (2012) Harnessing the rhizosphere microbiome through plant breeding and agricultural management. Plant Soil 360: 1–13.
18. DeAngelis KM, Brodie EL, DeSantis TZ, Andersen GL, Lindow SE, et al. (2009) Selective progressive response of soil microbial community to wild oat roots. ISME J 3: 168–178.
19. Bulgarelli D, Rott M, Schlaeppi K, Ver Loren van Themaat E, Ahmadinejad N, et al. (2012) Revealing structure and assembly cues for Arabidopsis root-inhabiting bacterial microbiota. Nature 488: 91–95.
20. Lundberg DS, Lebeis SL, Paredes SH, Yourstone S, Gehring J, et al. (2012) Defining the core Arabidopsis thaliana root microbiome. Nature 488: 86–90.
21. Bodenhausen N, Horton MW, Bergelson J (2013) Bacterial communities associated with the leaves and the roots of Arabidopsis thaliana. PLoS One 8: e56329.
22. Sugiyama A, Bakker MG, Badri DV, Manter DK, Vivanco JM (2013) Relationships between Arabidopsis genotype-specific biomass accumulation and associated soil microbial communities. Botany 91: 123–126.
23. Chaparro JM, Badri DV, Vivanco JM (2013) Rhizosphere microbiome assemblage is affected by plant development. ISME J, in press.
24. Chaparro JM, Badri DV, Bakker MG, Sugiyama A, Manter DK, et al. (2013) Root exudation of phytochemicals in Arabidopsis follows specific patterns that are developmentally programmed and correlate with soil microbial functions. PLoS One 8: e55731.
25. Mougel C, Offre P, Ranjard L, Corberand T, Gamalero E, et al. (2006) Dynamic of the genetic structure of bacterial and fungal communities at different developmental stages of Medicago truncatula Gaertn. cv. Jemalong line J5. New Phytol 170: 165–175.
26. Houlden A, Timms-Wilson TM, Day MJ, Bailey MJ (2008) Influence of plant developmental stage on microbial community structure and activity in the rhizosphere of three field crops. FEMS Microbiol Ecol 65: 193–201.
27. Xu YX, Wang GH, Jin J, Liu JJ, Zhang QY, et al. (2009) Bacterial communities in soybean rhizosphere in response to soil type, soybean genotype, and their growth stage. Soil Biol Biochem 41: 919–925.
28. Micallef SA, Shiaris MP, Colon-Carmona A (2009) Influence of Arabidopsis thaliana accessions on rhizobacterial communities and natural variation in root exudates. J Exp Bot 60: 1729–1742.
29. Denison RF, Kiers ET (2011) Life histories of symbiotic rhizobia and mycorrhizal fungi. Curr Biol 21: R775–R785.
30. Yokota K, Hayashi M (2011) Function and evolution of nodulation genes in legumes. Cell Mol Life Sci 68: 1341–1351.
31. Jie W, Liu X, Cai B (2013) Diversity of rhizosphere soil arbuscular mycorrhizal fungi in various soybean cultivars under different continuous cropping regimes. PLoS One 8: e72898.
32. Inceoglu O, Salles JF, van Overbeek L, van Elsas JD (2010) Effects of plant genotype and growth stage on the betaproteobacterial communities associated with different potato cultivars in two fields. Appl Environ Microbiol 76: 3675–3684.
33. Uroz S, Ioannidis P, Lengelle J, Cebron A, Morin E, et al. (2013) Functional assays and metagenomic analyses reveals differences between the microbial communities inhabiting the soil horizons of a Norway spruce plantation. PLoS One 8: e55929
34. Badri DV, Quintana N, El Kassis EG, Kim HK, Choi YH, et al. (2009) An ABC transporter mutation alters root exudation of phytochemicals that provoke an overhaul of natural soil microbiota. Plant Physiol 151: 2006–2017.
35. Manter DK, Delgado JA, Holm DG, Stong RA (2010) Pyrosequencing reveals a highly diverse and cultivar-specific bacterial endophyte community in potato roots. Microb Ecol 60: 157–166.
36. McKenzie VJ, Bowers RM, Fierer N, Knight R, Lauber CL (2012) Co-habiting amphibian species harbor unique skin bacterial communities in wild populations. ISME J 6: 588–596.
37. Lane DJ, Pace B, Olsen GJ, Stahl DA, Sogin ML, et al. (1985) Rapid determination of 16S ribosomal RNA sequences for phylogenetic analyses. Proc Natl Acad Sci U S A 82: 6955–6959.
38. Marchesi JR, Sato T, Weightman AJ, Martin TA, Fry JC, et al. (1998) Design and evaluation of useful bacterium-specific PCR primers that amplify genes coding for bacterial 16S rRNA. Appl Environ Microbiol 64: 795–799.
39. Schloss PD, Westcott SL, Ryabin T, Hall JR, Hartmann M, et al. (2009) Introducing Mothur: open-source, platform-independent, community-supported software for describing and comparing microbial communities. Appl Environ Microbiol 75: 7537–7541.
40. Pruesse E, Quast C, Knittel K, Fuchs BM, Ludwig W, et al. (2007) SILVA: a comprehensive online resource for quality checked and aligned ribosomal RNA sequence data compatible with ARB. Nucleic Acids Res 35: 7188–7196.
41. Edgar RC, Haas BJ, Clemente JC, Quince C, Knight R (2011) UCHIME improves sensitivity and speed of chimera detection. Bioinformatics 27: 2194–2200.
42. Kodama Y, Shumway M, Leinonen R; International Nucleotide Sequence Database Collaboration (2012) The Sequence Read Archive: explosive growth of sequencing data. Nucleic Acids Res 40: D54–D56.
43. Sugiyama A, Vivanco JM, Jayanty SS, Manter DK (2010) Pyrosequencing assessment of soil microbial communities in organic and conventional potato farms. Plant Dis 94: 1329–1335.
44. Garland JL, Mills AL (1991) Classification and characterization of heterotrophic microbial communities on the basis of patterns of community-level sole-carbon-source utilization. Appl Environ Microbiol 57: 2351–2359.
45. Girvan MS, Bullimore J, Pretty JN, Osborn AM, Ball AS (2003) Soil type is the primary determinant of the composition of the total and active bacterial communities in arable soils. Appl Environ Microbiol 69: 1800–1809.
46. Shentu JL, He ZL, Yang XE, Li TQ (2008) Microbial activity and community diversity in a variable charge soil as affected by cadmium exposure levels and time. J Zhejiang Univ Sci B 9: 250–260.
47. Frąc M, Oszust K, Lipiec J (2012) Community level physiological profiles (CLPP), characterization and microbial activity of soil amended with dairy sewage sludge. Sensors (Basel) 12: 3253–3268.
48. Ofek M, Hadar Y, Minz D (2012) Ecology of root colonizing Massilia (Oxalobacteraceae). PLoS One 7: e40117.
49. Uroz S, Buee M, Murat C, Frey-Klett P, Martin F (2010) Pyrosequencing reveals a contrasted bacterial diversity between oak rhizosphere and surrounding soil. Environ Microbiol Rep 2: 281–288.
50. Sugiyama A, Yazaki K (2012) Root exudates of legume plants and their involvement in interactions with soil microbes. In: Vivanco JM, Baluska F, editors. Secretions and exudates in biological systems. Berlin Heidelberg: Springer. pp. 27–48.
51. Ikeda S, Okubo T, Kaneko T, Inaba S, Maekawa T, et al. (2010) Community shifts of soybean stem-associated bacteria responding to different nodulation phenotypes and N levels. ISME J 4: 315–326.
52. Barbour WM, Hattermann DR, Stacey G (1991) Chemotaxis of Bradyrhizobium japonicum to soybean exudates. Appl Environ Microbiol 57: 2635–2639.
53. Choudhary DK, Johri BN (2009) Interactions of Bacillus spp. and plants - with special reference to induced systemic resistance (ISR). Microbiol Res 164: 493–513.
54. Govindasamy V, Senthilkumar M, Magheshwaran V, Kumar U, Bose P, et al. (2011) Bacillus and Paenibacillus spp.: potential PGPR for sustainable agriculture. In: Maheshwari DK, editor. Plant growth and health promoting bacteria. Berlin Heidelberg: Springer. pp. 333–364.
55. Kumar A, Prakash A, Johri BN (2011) *Bacillus* as PGPR in crop ecosystems. In: Maheshwari DK, editor. Bacteria in agrobiology: crop ecosystems. Berlin Heidelberg: Springer. pp. 37–59.
56. Insunza V, Alstrom S, Eriksson KB (2002) Root bacteria from nematicidal plants and their biocontrol potential against trichodorid nematodes in potato. Plant Soil 241: 271–278.
57. Richardson AE, Barea JM, McNeill AM, Prigent-Combaret C (2009) Acquisition of phosphorus and nitrogen in the rhizosphere and plant growth promotion by microorganisms. Plant Soil 321: 305–339.
58. de Souza R, Beneduzi A, Ambrosini A, da Costa PB, Meyer J, et al. (2013) The effect of plant growth-promoting rhizobacteria on the growth of rice (Oryza sativa L.) cropped in southern Brazilian fields. Plant Soil 366: 585–603.
59. Tsavkelova EA, Cherdyntseva TA, Klimova SY, Shestakov AI, Botina SG, et al. (2007) Orchid-associated bacteria produce indole-3-acetic acid, promote seed germination, and increase their microbial yield in response to exogenous auxin. Arch Microbiol 188: 655–664.
60. de Vasconcellos RLF, Cardoso EJBN (2009) Rhizospheric streptomycetes as potential biocontrol agents of Fusarium and Armillaria pine rot and as PGPR for Pinus taeda. Biocontrol 54: 807–816.
61. Buonaurio R, Stravato VM, Kosako Y, Fujiwara N, Naka T, et al. (2002) Sphingomonas melonis sp. nov., a novel pathogen that causes brown spots on yellow Spanish melon fruits. Int J Syst Evol Microbiol 52: 2081–2087.

62. Johnson EG, Sparks JP, Dzikovski B, Crane BR, Gibson DM, et al. (2008) Plant-pathogenic Streptomyces species produce nitric oxide synthase-derived nitric oxide in response to host signals. Chem Biol 15: 43–50.

63. Seipke RF, Song L, Bicz J, Laskaris P, Yaxley AM, et al. (2011) The plant pathogen Streptomyces scabies 87-22 has a functional pyochelin biosynthetic pathway that is regulated by TetR- and AfsR-family proteins. Microbiology 157: 2681–2693.

3

Ecomorphological Variation of the Wireworm Cephalic Capsule: Studying the Interaction of Environment and Geometric Shape

Hugo A. Benítez[1,2]*, Thomas Püschel[1], Darija Lemic[3], Maja Čačija[3], Antonela Kozina[3], Renata Bažok[3]

1 Faculty of Life Sciences, University of Manchester, Michael Smith Building, Oxford Road, Manchester, United Kingdom, **2** Instituto de Alta Investigación, Universidad de Tarapacá, Casilla, Arica, Chile, **3** University of Zagreb, Faculty of Agriculture, Department for Agricultural Zoology, Svetošimunska, Zagreb, Croatia

Abstract

Studying the association between organismal morphology and environmental conditions has been very useful to test hypothesis regarding the influence of climate on shape. It has been long recognized that different environments produce dissimilar stress levels in insects, which can be reflected on the ability of an individual to overcome these pressures and spread further. *Agriotes* (Coleoptera: Elateridae) species infest agricultural fields in different parts of Croatia, inhabiting different climatic conditions. Previous biological studies have indicated that there is a relationship between some *Agriotes* biological parameters such as density and climatic conditions such as soil moisture and temperature. However, it is still unknown how these environmental properties influence the wireworm morphological structure. This is highly relevant because the head of this species is directly involved in the mobility in the soil, thus affecting the invasive capacity of this insect. Therefore the aim of this study was to assess the association between different climatic conditions and the morphological variation of *Agriotes* cephalic capsule. Advanced multivariate analysis and geometric morphometric tool were applied to study the covariation between shape and environmental variables. Partial Least Squares methods were used in order to analyse the association between the wireworm head shape and three different climatic conditions: soil type, temperature and rainfall. Our results showed that there is a high covariation between the wireworm head shape and the climatic conditions. It was suggested that the observed shape–environment association could be result of the high plasticity of this species in relation to its invasive capacity.

Editor: Hector Escriva, Laboratoire Arago, France

Funding: This research was supported by the Ministry of Science, Education and Sports of the Republic of Croatia by project: "The spatial distribution of economically important pests with the use of GIS" (178-1782066-2065). The funders had no role in study design, data collection and analysis, decision to publish, or preparation of the manuscript.

Competing Interests: The authors have declared that no competing interests exist.

* Email: hugo.benitez@postgrad.manchester.ac.uk

Introduction

Taxonomic classification and biological diversity analyses have been traditionally based on morphological descriptions [1]. Driven by a mathematical quantitative revolution, morphological studies have experienced a significant revitalization due to the development of shape analysis based on statistical multivariate techniques and novel visualisation methods. Broadly, morphometrics refer to the quantitative analysis of form (i.e. shape and size) and how it covaries with respect to other factors (e.g. ecology, development, biomechanics, and genetics among others) [2–5]. Geometric morphometrics (GMM) is a coordinate-based method, which means that their primary data are 2D or 3D Cartesian coordinates of anatomically distinguishable landmarks (i.e. discrete anatomical loci that are arguably homologous among all the individuals under analysis). Coordinates are better when compared to linear data, because they preserve the spatial information of a structure, providing a relatively complete description of an organism's shape [2,3,6]. One of the great characteristics of GMM is that it allows studying the association between shape and other kinds of data, such as ecological, genetic, biomechanical, or other relevant factors. This is really useful because one of the traditional interests of ecologists is to associate character states or different phenotypic values with environmental data. There are several matters for which morphological analyses play an important role in ecology. For instance, ecomorphological studies have revealed constraints and selective factors affecting the phenotypic response to certain environments [7–9], how morphology influences the ecological distribution of a particular phenotype [10–13] and evolutionary trends such as phylogenetically conserved morphologies [11,14–16]. In all these cases, morphology represents certain organismal aspects that relate and individual to its environment, hence its importance. Indeed, the association between morphology and ecology could provide useful insights about the expression of the phenotype-environment interaction and the related evolutionary history. There is a plethora of available methods to study association between morphological and ecological variables (e.g. regression analysis; canonical correlation analysis; Mantel test; principal coordinate analysis; etc.). However, Partial Least Squares (PLS) method is probably one of the mostly applied when assessing the covariation between shape and other factors [17]. For example, it has been used to relate morphometric and ecological variables [5,13,18–20], morphometric and allele frequency data [21], shape and behavior [18], different parts of the same configuration of landmarks [21–23], and even data from different parts or different views of the same specimens [24,25].

Wireworms are click beetle larvae from the genus *Agriotes* (Coleoptera: Elateridae) that considerably damage field crops, especially potatoes [26–28]. They are long-lived soil insects; most species spend 3–5 years in the larval stage [26,28]. Their body are elongated and hard, characterized by a reddish-brown coloration [29]. They are polyphagous and usually inhabit most kinds of soil. Some wireworm species are serious agricultural pests [28,30,31]. Larvae feed on germinating seeds and plant roots, affecting negatively seedlings and young plants, often causing their death [32,33]. Among over 10,000 species known worldwide [29], there are just around 150 considered dangerous enough to cause significant damages to agricultural cultigens. The five most important *Agriotes* species in Croatia include *A. lineatus* L., *A. sputator* L., *A. obscurus* L., *A. brevis* Cand. and *A. ustulatus* Schall. They can produce significant economic losses in agriculture, due to their feeding behaviour [34]. *Agriotes* "species infest agricultural fields and cause serious economic damages in the continental part of Croatia. Therefore, studying these species is relevant due to the impact that they have on agricultural production.

East continental Croatia is characterized by chernozemic soils, while the central region is defined by ground water gley and alluvial muds soils, and the west part is characterized by luvic and pseudoglay soils [35]. As a result of the dissimilar soils found in each one of these regions, the prevailing microhabitat parameters are different. The most important parameters are soil moisture and temperature, both of which are known to impact upon oviposition, egg survival and larval development [27,30,34,36]. Unlike the other four species, *Agriotes ustulatus* overwinters only as larvae and develops for two (three calendar) years [27], and the higher temperatures positively affect its abundance [27,37]. According to previous literature, *A. ustulatus* was the most frequent species only in the eastern regions of Croatia [32,38,39] but recent studies have shown that this species has also spread to other regions of continental Croatia [37,40].

It is well known that the adaptation over time to a specific environment is the result of both environmental pressures and geographic distance [41–43]. Moreover, it is well documented that adverse temperatures, nutritional stresses, presence of chemicals, population density and many other factors that affect development can lead to an increase in morphological asymmetry as a result of high intraspecific variation (e.g. [42,44,45]). Therefore, it is expected that when environmental conditions change, organisms and populations should adapt to the new conditions [45]. In this context, adaptive variation plays a key role since it reflects historical evolution and determines the population's phenotypic response [46–48]. Moreover, Bouyer et al. [49] have suggested that the influence of environment on an organism's genotype takes more time to manifest than on the phenotype, and as such the study of environmental influence on populations and individuals should be made on phenotypic, rather than genotypic, characters. Hence, we preferred the application of geometric morphometric methods instead of molecular analyses, because our goal was to analyse the association between environmental factors and morphological structure, rather than determining the genetic response to different climatic conditions. Previous studies have analyzed the effect of different ecological factors on insect morphology from diverse perspectives, for instance anti-predator defences [50–52], behaviour and sexual dimorphism [53–57], physiology [58–60] and environmental adaptations [61–64]. However our approach is innovative, because it is a novel attempt to characterize the relationship between a larval morphology with closely related environmental conditions. The aim of this study was to assess the association between different climatic conditions (i.e. temperature, rainfall and soil type) on the morphological differentiation of *Agriotes ustulatus* cephalic capsule. Furthermore, this study applied for the first time geometric morphometric techniques to analyze beetle larvae shape variation.

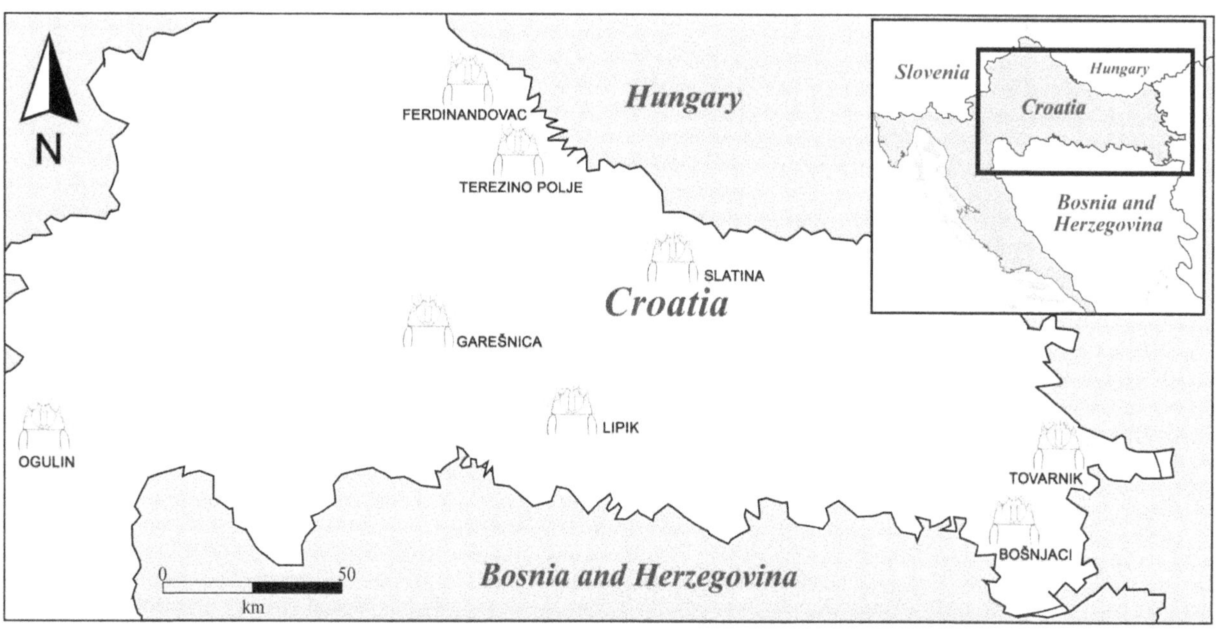

Figure 1. Geographic distribution of the eight Croatian localities of *A. ustulatus* examined in this study.

Table 1. Localities with their corresponding geographic coordinates and environmental variables: n, number of specimens; T, mean annual between 2009–2013 temperature (°C); Rain, Rainfall (mm); Type of soil (group 1: chernozemic soil, group 2: alluvial soil, group 3: ground water gley soil and group 4: luvic and pseudoglay soil).

Location	Samples (n)	Coordinates	Temperature (Average 2009–2013)	Rainfall (Average 2009–2013)	Type of Soil
a. GAREŠNICA	40	45°34′15.39″N; 16°55′56.02″E	11.87	58.83	ground water gley soil
b. LIPIK	35	45°24′59.61″N; 17°9′24.10″E	11.90	70.6	ground water gley soil
c. TOVARNIK	37	45°9′24.43″N; 19°9′14.51″E	12.35	40.99	chernozemic soil
d. BOŠNJACI	39	45°2′54.64″N; 18°44′59.59″E	12.49	60.97	chernozemic soil
e. SLATINA	40	45°42′51.70″N; 17°41′59.18″E	11.26	61.57	alluvial soil
f. TEREZINO POLJE	2	45°56′5.47″N; 17°27′27.82″E	11.513	71.69	alluvial soil
g. FERDINANDOVAC	13	46°3′42.85″N; 17°10′59.92″E	11.078	66.71	alluvial soil
h. OGULIN	46	45°15′27.46″N; 15°12′38.78″E	10.96	129.171	luvic and pseudoglay soil

Materials and Methods

Ethics Statement

"N/A".

Our samples not needed any specific permission to collect (locations/activities) and this article is not involved in any endangered or protected species.

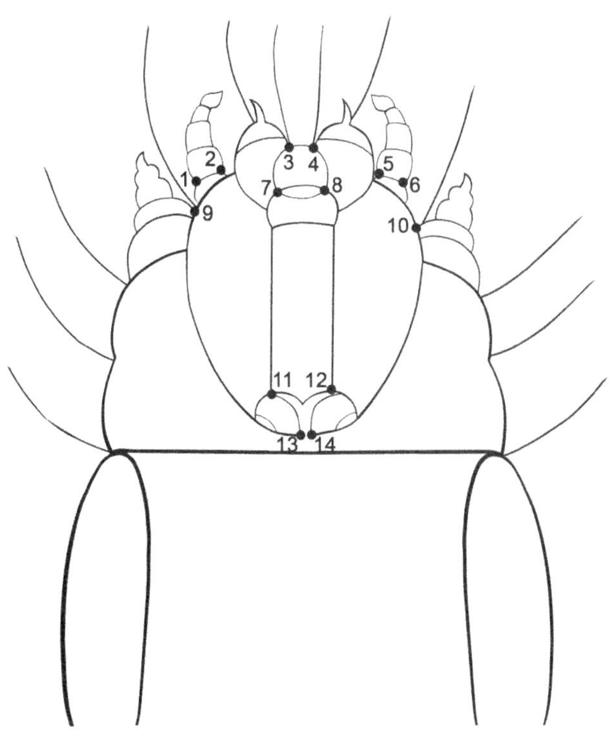

Figure 2. Representation of the 14 morphological landmarks identified on the cephalic capsule of *A. ustulatus*.

Data collection

In this study we analysed morphological differences in 258 wireworms of *Agriotes ustulatus*. The wireworm larvae were sampled during spring (May) and autumn (September) from 2009 until 2013. Although the taxonomic identification of the *Agriotes* species at the larval stage is extremely difficult, the analyzed larvae were positively classified as member of *A. ustulatus* based on the distinctive presence of specific marks on the last abdominal segment. This particular character is an exclusive trait of this species and none of the other members of this genus shows it [65]. They came from eight different locations of continental Croatia. Locations were grouped according to climatic conditions and soil type. Group 1 corresponds to the eastern locations of Tovarnik and Bošnjaci, which had drier weather conditions and chernozemic soils. Group 2 includes Slatina, Terezino Polje and Ferdinandovac, which had lower temperatures as compared to the eastern location but higher moisture and alluvial soils. Group 3 are Garešnica and Lipik locations, which had climate conditions in between the two previous groups and ground water gley soils. Group 4 consisted of one location on the west, Ogulin, which is on higher altitude and had cooler climate (perhumid climate), with luvic and pseudoglay soils (Fig. 1, Table 1). All sampled areas were crop fields, cultivated with oilseed rape, maize, barley, soybean, sugar beet, tobacco, alfalfa and potato. The sampling method consisted in digging a series of ca. five holes (dimensions 25×25×25 cm) on each field [38]. All the excavated soil was manually crumbled to small pieces on a black foil and examined thoroughly to detect any wireworm presence. The collected specimens were preserved in 70% ethanol. Species determination was done by examining the ninth abdominal segment [65]. Only larvae of the approximately same developmental stage (7th–13th instar, larger than 10 mm) of *Agriotes ustulatus* species were selected

Table 2. Measurement error Procrustes ANOVA for both centroid size and shape of *A. ustulatus*, characterised by matching symmetry.

Centroid size					
Effect	**SS**	**MS**	**df**	**F**	**p-value**
Individual	4.682	0.142	33	12.1	<0.0001
Error 1	**0.374**	**0.012**	**32**		
Shape					
Effect					
Individual	0.156	3.95E-04	396	3.09	<0.0001
Side	0.0049	0.0004	12	3.23	0.0002
Ind×Side	0.051	0.0001	396	2.47	<0.0001
Error 1	**0.039**	**5.17E-05**	**768**		

Sums of squares (SS) and mean squares (MS) are in units of Procrustes distances (ie. dimensionless).

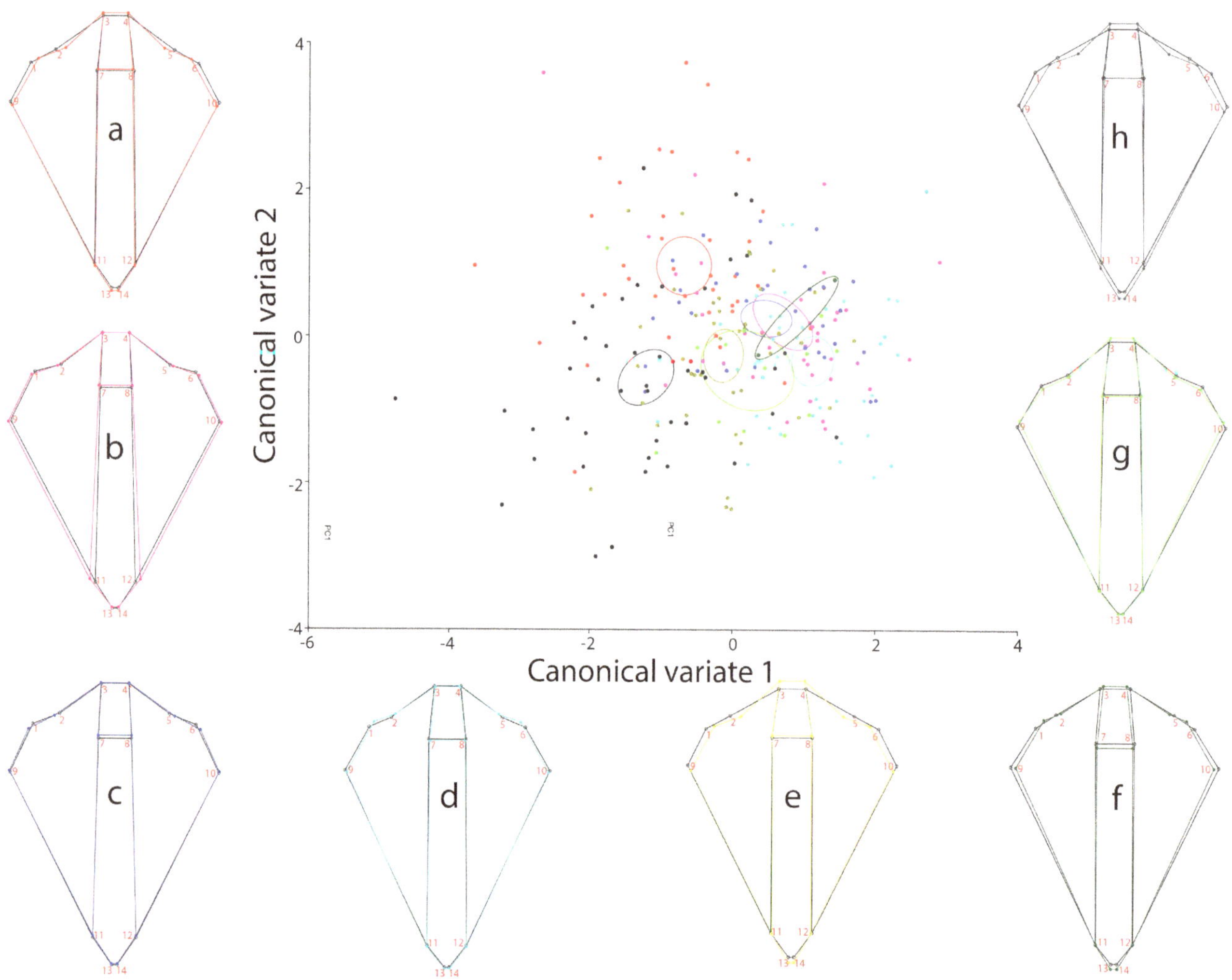

Figure 3. Canonical Variate analysis of the eight *A. ustulatus* populations. The figure shows the firsts two CV axes and the wireframe visualization of the average shape for all populations. a: GAREŠNICA (red) b: LIPIK (pink) c: TOVARNIK (blue) d: BOŠNJACI (light blue) e: SLATINA (dark yellow) f: TEREZINO POLJE (green) g: FERDINANDOVAC (light green) h: OGULIN (black).

Table 3. Procrustes ANOVA for both centroid size and shape of *A. ustulatus*, characterized by object symmetry.

Centroid size					
Effect	**SS**	**MS**	**df**	**F**	**p-value**
Location	7.09	1.013	7	17.31	<0.0001
Individual	14.28	0.058	244		
Shape					
Effect					
Location	0.233	0.0028	84	7.17	<0.0001
Individual	1.136	0.0004	2928	3.78	<0.0001
Side	0.009	0.0008	12	8.1	<0.0001
Ind×Side	0.309	0.0001	3012		

Sums of squares (SS) and mean squares (MS) are in units of Procrustes distances (dimensionless) SH: Shape, CS: centroid size.

for further analysis [27]. All the specimens are housed in the Department for Agricultural Zoology, University of Zagreb.

Landmark Acquisition

Cephalic capsules were photographed using a Leica DFC295 digital camera (3M Pixel) on a trinocular mount of a Leica MZ16a stereo-microscope and saved in JPEG format using the Leica Application Suite v3.8.0 (Leica Microsystems Limited, Switzerland). Fourteen homologue type 1 landmarks (Fig. 2), (Table S1, Table S2) were marked on the wireworm mandible, so that they could be accurately and repeatedly identified. Each landmark was digitised using tpsDIG V2.17 [66] and imported into MorphoJ v1.04d for further statistical analyses [67]. The dataset containing the raw coordinates used in this study is available as supplementary material. Measurement error (ME) has a critical importance when analyzing shape. In order to assess the ME level, the cephalic capsules of 35 individual beetles were digitized twice [68].

Shape Analysis

Once obtained the Cartesian coordinates for all landmarks, the shape information was extracted by performing a full Procrustes fit [69,70] taking into account the object symmetry of this anatomical structure. Procrustes superimposition is a mathematical procedure that removes the variation due to rotation, position, orientation and size [69]. Due to the existence of object symmetry in the analysed structure, reflection was removed by including the original and mirror image of all configurations in the analysis and simultaneously superimposing all of them [71].

Allometry is a key factor influencing shape variation [72,73], therefore allometric effects were assessed by performing a multivariate regression of shape on centroid size, pooling the dataset by location. Then the covariance matrix of the residuals pooled by location was computed to perform the following analyses.

Multivariate Analysis

The main patterns of variation in the shape space were visualized by carrying out a principal component analysis (PCA), calculated from the covariance matrix of the averaged population symmetric component of shape [71]. In order to statistically test if there were differences between the locations, a canonical variate analysis (CVA) of the symmetric component of shape variation was applied [74]. CVA maximizes the differences between groups

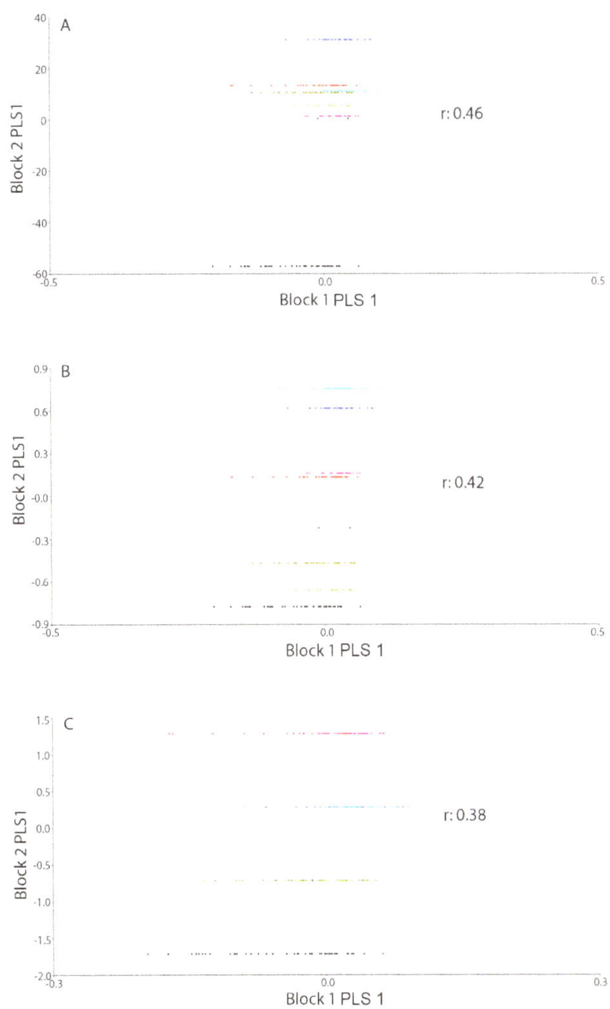

Figure 4. Partial least squares analysis of head shape and environmental condition in *A. ustulatus*. The horizontal axis of the scatter plot is the first PLS axis for head shape and the vertical axis is the first PLS axis for environmental condition for the eight localities. A: temperature, B: Rainfall and C: Soil Type.

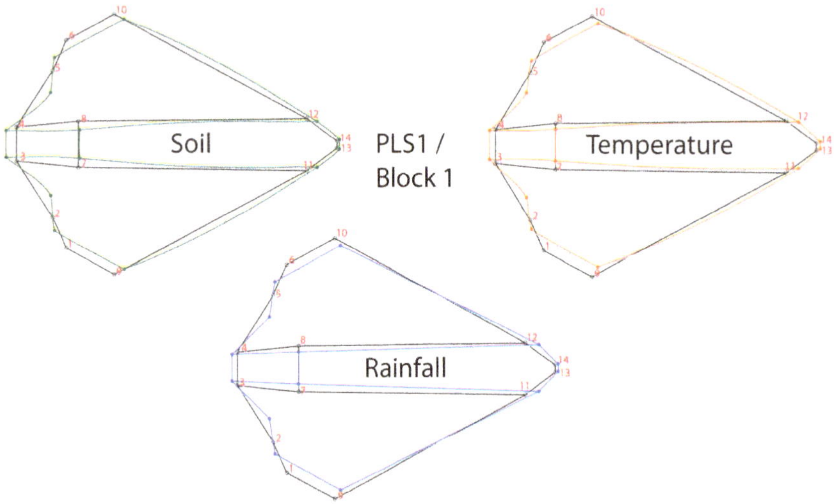

Figure 5. Wireframe visualization of the average shape extracted from the PLS1 associated with the different environmental conditions. The figure colors shows the particular displacements of landmarks relative to each one of the environmental conditions, green: soil, orange: temperature and blue: rainfall.

relative to the variation within groups and it is therefore one of the most used methods to distinguish among groups [74]. The statistical significance of the pairwise differences in mean shapes was assessed with performing a permutation test of the Procustes distance (10,000 permutations per test).

The amount of symmetric variation was assessed by computing a Procrustes ANOVA as applied in other studies to analyse object symmetry [71,75,76]. This test was used to compare the individual-reflection interactions as compared to measurement error, being this latter estimated from the total variation of the entire landmark configuration.

Climatic Correlation

In order to assess the degree of association between shape and climatic variables, a two block partial least square (PLS) was performed. The geographic coordinates and climatic condition (temperature, rainfall and soil type) (Table 1) were extracted from Croatian Meteorological and Hydrological Service (for complete values since 2009 to 2013 see Table S3), for each one of the analysed sites. PLS was preferred due to the fact that climatic variables are strongly collinear [17]. The two-block PLS analysis is a widely used morphometric method, and it is based on the singular value decomposition of the common covariance matrix of the two variables sets, which in our case consist of the shape and climatic conditions matrices [19,20]. PLS analysis estimates the correlated pairs of lineal combinations (singular vectors) between these two sets. These singular vectors are generated as pairs of latent variables (one per block), explaining most of the covariance between the original variable blocks, thus maximizing the covariation between the two sets of variables [17,19,20]. In order to test the significance of the correlation between the pairs of singular vectors (i.e. climatic variable singular vectors and shape singular vectors), a permutation test with 10,000 rounds randomizations was performed.

Mantel Test

In order to statistically test whether there was relationship between shape differences and geographic location, a Mantel test was performed. The correlation level was assessed comparing Procrustes (i.e. the morphometric distances between all the wireworms at each location), and geographic (i.e. the geodesic distances between all the locations where the wireworms were sampled) distance matrices by computing their product moment correlation and the Mantel test statistic (observed Z values compared to their permutation distribution after 10,000 rounds).

Results

The Procrustes ANOVA applied to assess the measurement error showed that the mean square for individual variation exceeded the measurement error; therefore it was negligible (Table 2). The same analysis applied to assess the population differences showed significant differences among localities for both size and shape ($P<0.0001$) (Table 3). The PCA of the residuals of the regression (used to excluded the allometric effects) of the head shape variation showed that the first three PCs accounted for 79.2% (PC1 = 52.8%; PC2 = 17.04%; PC3 = 9.3%) of the total shape variation, hence providing a reasonable approximation of the total amount of variation. The CVA showed a significant differentiation among localities based on Procrustes distances (Fig. 3). The Mantel test indicated that in spite of a relatively high (0.727) correlation value between the morphometric (i.e. Procrustes) and geographic distances (i.e. km), it was not significant (p-value = 0.076; 10,000 permutation rounds). The two-block, partial least-squares analyses of the head shape indicated a moderate association between the shape variables and the environmental ones (Rainfall: $r = 0.422$, p-value<0.001; Temperature: $r = 0.46$, p-value<0.001; Type of Soil: r: 0.38, p-value<0.001) (Fig. 4). A strong and significant amount of covariation between the head shape and the environmental variables was graphically visualised by the first block and PLS1 (Fig. 5).

Discussion

The eight populations were distinguished according to the relative displacement of the symmetric landmarks 1, 2–5, 6 and 9–10. *A. ustulatus* presents a relatively long larval stage as compared to most insects (ca. 3–5 years) [27]. This interesting life history trait has led to several studies focusing on the effect of soil type on population density or invasive capacity [26,36,37,77]. Nonetheless, there have been no studies trying to assess the influence of soil type (and other climatic conditions) on the cephalic capsule shape. It was expected to contribute to the understanding of the influence of environmental factors on the morphological development of wireworms.

The first dimensions of the PCA accounted for most of the morphological variation, showing that the wireworm head shape has a relative regional ordination (wireframe Fig. 3). This means that each population has relatively different head shape with respect to the others, in such a manner that the majority of the observed shape differences are due to regional differences. The CVA results confirm this interpretation, although there was some overlap between the confidence interval ellipses (Fig. 3). The Mantel test was performed in order to assess whether this regional ordination was expressed in a morphoclinal way (i.e. in accordance with a gradual differentiation relative to distance). Our results did not support the latter perspective, but on the contrary showed that there was no lineal relationship between the shape of the wireworms and the distance between the collection sites. Hence, it was possible to discard the existence of a morphoclinal variation in the larvae head shape. Although, this conclusions must be cautiously considered, since it has been shown that the Mantel test is not always the best method to test for linear relationships [78].

The PLS results showed that the shape variation of the head observed in *A. ustulatus* covaries with the environmental factors. Our study has confirmed a high covariation between the wireworm head shape and the analyzed climatic conditions, although the correlation values between PLS axes were moderate. It is important to keep in mind that these correlations are not the decisive factor that has been maximized by the PLS analysis. In the present study, the shape changes associated to the climatic variables were clearly visible in PLS1, showing that the head shape had an expansion of landmarks 3 and 4 (base of the hair on the mentum) due to soil type differences resulting in a morphotype defined by an elongated head (Fig. 5). The average effect of temperature showed a head shape expansion of landmarks on the maxillary palpi (1–6 and 2–5) and a contraction of the mentum, showing a narrowed head that increasingly expands towards the cardo and alocardo (Fig. 5). Finally the rainfall effect was characterized by a contraction of the landmarks of the maxillary palpi and a moderate elongation of the juncture between the submentum and the cardo (Fig. 5). Nevertheless the shape changes associated with the PLS1 axes showed fairly complex local shifts of landmarks (Fig. 5). Both temperature and rainfall showed similar covariation patterns with respect to shape, exhibiting relative displacements of individual head traits rather than large-scale deformations. However, because the landmark configuration does not include all the traits of the head, this finding should be interpreted cautiously.

It is important to always take into account the possible functional consequences of the observed shape differences when an association between morphology and environment is found [20]. Therefore, the observed shape–environment association noticeable in the first pair of PLS vectors could be interpreted in functional terms. The observed covariation between climatic conditions and shape changes could be a result of the high plasticity of this species. Actually, those populations inhabiting ground water gley soil (i.e. soils were drainage is poor due to a high phreatic surface), have more narrow cephalic capsules (Fig. 3b), suggesting a better penetrating capacity. On the other hand, those individuals occupying wetter muds have more expanded heads (Fig. 3h), allowing them to possibly use it as a shovel in order to dig these kinds of soils in an enhanced way. This plasticity may also explain the capacity of this species as invasive pest, occupying a broad range of host plants both as larva and adult [26,36]. Previous studies in a diverse range of organisms have shown that phenotypic plasticity can explain the observed morphological differences, especially when considering the ecological distribution of the analysed species [79,80]. A similar study carried out on the Chrysomelid pest beetle (*Diabrotica virgifera virgifera*), have also established the existence of a relationship between an anatomical structure and soil type [81]. This species exhibit an elongated hind wing morphotype in environments dominated by water gley soil type. By contrast, where a chernozemic or alluvial soil type dominated, a narrow hind wing morphotype was found. Even though the precise mechanism underlying the relationship between *A. ustulatus* and *D. virgifera* phenotypes and soil type is probably different, it is highly suggestive that both are pest species with wide ecological ranges. Perhaps this observed phenotypic plasticity is related to the invasive capacity of this kind of species, although more analyses are needed.

Further geostatistical analyses of spatial patterning and clustering could adjust sampling methods, in order to properly understand the key factors determining population size and pest dynamics [82]. Toepfer et al. [83] have studied the clustering of this group, concluding that this species are environmentally mediated by mortality factors that affect pest populations, which might vary spatially as well as seasonally and due to biological factors such as limited movement of larvae [83]. Hence, it is highly relevant to continue studying the relationship between shape and climatic conditions, but including as well some specific kinematic measurements of the different larvae population displacing through dissimilar soil types. Additionally, future studies should assess the relationship between vegetational structure and climatic conditions (i.e. host plant differences due to climate) and their effect on the morphology of *A. ustulatus* and how this influences its pest dynamics.

Supporting Information

Table S1 Anatomical Description of the 14th Landmarks in *Agriotes* Schall.

Table S2 Raw Landmark dataset.

Table S3 Average monthly temperatures and Rainfall.

Acknowledgments

We thank the two anonymous reviewers that provided us with valuable comments that clearly improved our manuscript. We give great gratitude to Ivana Marušić and Marina Buketa for the technical assistance. We give great gratitude to the Ministry of Science, Education and Sports of the Republic of Croatia by project: "The spatial distribution of economically important pests with the use of GIS" (178-1782066-2065). HB and TP are grateful with the Becas Chile scholarship program, CONICYT Chile.

Author Contributions

Conceived and designed the experiments: HAB TP DL MC AK. Performed the experiments: DL MC AK. Analyzed the data: HAB TP. Contributed reagents/materials/analysis tools: HAB TP DL MC AK RB. Contributed to the writing of the manuscript: HAB TP DL.

References

1. Daly HV (1985) Insect Morphometrics. Annual Review of Entomology 30: 415–438.
2. O'Higgins P (2000) The study of morphological variation in the hominid fossil record: biology, landmarks and geometry. Journal of Anatomy 197: 103–120.
3. Rohlf FJ, Marcus LF (1993) A revolution in morphometrics. Trends in Ecology & Evolution 8: 129–132.
4. Henderson A (2006) Traditional morphometrics in plant systematics and its role in palm systematics. Botanical Journal of the Linnean Society 151: 103–111.
5. Adams DC, Rohlf FJ (2000) Ecological character displacement in *Plethodon*: Biomechanical differences found from a geometric morphometric study. Proceedings of the National Academy of Sciences of the United States of America 97: 4106–4111.
6. Adams DC, Rohlf FJ, Slice DE (2013) A field comes of age: geometric morphometrics in the 21st century. Hystrix-Italian Journal of Mammalogy 24: 7–14.
7. Lauder GV (1982) Historical biology and the problem of design. Journal of Theoretical Biology 97: 57–67.
8. Lande R, Arnold SJ (1985) Evolution of mating preference and sexual dimorphism. Journal of Theoretical Biology 117: 651–664.
9. De Esteban-Trivigno S (2011) Ecomorphology of extinct Xenarthrans: analysis of the mandible using geometric morphometrics methods. Ameghiniana 48: 381–398.
10. De Esteban-Trivigno S (2011) Searching ecomorphological patterns shared by extant ungulates and extinct Xenarthrans. Ameghiniana 48: 189–209.
11. Schluter D (1988) Character Displacement and the Adaptive Divergence of Finches on Islands and Continents. The American Naturalist 131: 799–824.
12. Ricklefs Re (1987) Community Diversity: Relative Roles of Local and Regional Processes. Science 235: 167–171.
13. Klingenberg CP, Ekau W (1996) A combined morphometric and phylogenetic analysis of an ecomorphological trend: Pelagization in Antarctic fishes (Perciformes: Nototheniidae). Biological Journal of the Linnean Society 59: 143–177.
14. Paton DC, Collins BG (1989) Bills and tongues of nectar-feeding birds: A review of morphology, function and performance, with intercontinental comparisons. Australian Journal of Ecology 14: 473–506.
15. Monteiro LR (2013) Morphometrics and the comparative method: studying the evolution of biological shape. Hystrix-Italian Journal of Mammalogy 24: 25–32.
16. Monteiro LR, Nogueira MR (2010) Adaptive radiations, ecological specialization, and the evolutionary integration of complex morphological structures. Evolution 64: 724–743.
17. Rohlf FJ, Corti M (2000) Use of two-block partial least-squares to study covariation in shape. Systematic Biology 49: 740–753.
18. Corti M, Rohlf FJ (2001) Chromosomal speciation and phenotypic evolution in the house mouse. Biological Journal of the Linnean Society 73: 99–112.
19. Monteiro LR, Duarte LC, dos Reis SF (2003) Environmental correlates of geographical variation in skull and mandible shape of the punaré rat *Thrichomys apereoides* (Rodentia: Echimyidae). Journal of Zoology 261: 47–57.
20. Piras P, Marcolini F, Raia P, Curcio M, Kotsakis T (2010) Ecophenotypic variation and phylogenetic inheritance in first lower molar shape of extant Italian populations of *Microtus* (*Terricola*) *savii* (Rodentia). Biological Journal of the Linnean Society 99: 632–647.
21. Klingenberg CP, Zaklan SD (2000) Morphological integration between developmental compartments in the Drosophila wing. Evolution 54: 1273–1285.
22. Bookstein FL, Gunz P, Mitteroecker P, Prossinger H, Schaefer K, et al. (2003) Cranial integration in *Homo*: singular warps analysis of the midsagittal plane in ontogeny and evolution. Journal of Human Evolution 44: 167–187.
23. Drake AG, Klingenberg CP (2008) The pace of morphological change: historical transformation of skull shape in St Bernard dogs. Proceedings of the Royal Society B-Biological Sciences 275: 71–76.
24. Klingenberg CP, Badyaev AV, Sowry SM, Beckwith NJ (2001) Inferring developmental modularity from morphological integration: Analysis of individual variation and asymmetry in bumblebee wings. American Naturalist 157: 11–23.
25. Tabachnick RE, Bookstein FL (1990) The Structure of Individual Variation in *Miocene globorotalia*. Evolution 44: 416–434.
26. Traugott M, Schallhart N, Kaufmann R, Juen A (2008) The feeding ecology of elaterid larvae in central European arable land: New perspectives based on naturally occurring stable isotopes. Soil Biology and Biochemistry 40: 342–349.
27. Furlan L (1998) The biology of *Agriotes ustulatus* Schaller (Col., Elateridae). II. Larval development, pupation, whole cycle description and practical implications. Journal of Applied Entomology-Zeitschrift Fur Angewandte Entomologie 122: 71–78.
28. Tóth Z (1984) Click beetles (Elateridae) in the soils of central Europe - their distribution and description. Part I. (Gen.: Agriotes). Acta Phytopathologica Academiae Scientiarum Hungaricae 19: 13–29.
29. Laibner S (2000) Elateridae of the Czech and Slovak Republics. Zlin: Kabourek.
30. Parker WE, Howard JJ (2001) The biology and management of wireworms (*Agriotes spp.*) on potato with particular reference to the U.K. Agricultural and Forest Entomology 3: 85–98.
31. Tóth M (2013) Pheromones and attractants of click beetles: an overview. Journal of Pest Science 86: 3–17.
32. Kovačević Ž (1960) Problematika zemljišnih štetnika u istočnoj Slavoniji. Sav Polj: 567–580.
33. Brunner N, Kromp B, Meindl P, Pázmándi C, Traugott M (2005) Evaluation of different sampling techniques for wireworms (Coleoptera, Elateridae) in arable land. IOBC/wprs Bulletin 28: 117–122.
34. Benefer C, Herk WG, Ellis JS, Blackshaw RP, Vernon RS, et al. (2013) The molecular identification and genetic diversity of economically important wireworm species (Coleoptera: Elateridae) in Canada. Journal of Pest Science 86: 19–27.
35. Bogunović M, Vidaček Ž, Racz Z, Husnjak S, Sraka M (1996) Namjenska pedološka karta 1:300000 Faculty of Agriculture: University of Zagreb.
36. Benefer CM, Knight ME, Ellis JS, Hicks H, Blackshaw RP (2012) Understanding the relationship between adult and larval *Agriotes* distributions: The effect of sampling method, species identification and abiotic variables. Applied Soil Ecology 53: 39–48.
37. Kozina A (2012) The factors for the temporal and spatial distribution of the economically important maize pests: University of Zagreb.
38. Maceljski M (2002) Poljoprivrednaentomologija. Čakovec: Zrinski.
39. Bažok R (2007) Žičnjaci. Glasilo biljne zaštite 5: 339–344.
40. Bažok R, Igrc-Barčić J (2010) Pheromone applications in maize pest control. New York: Nova Science.
41. Bravi R, Benitez HA (2013) Left-right asymmetries and shape analysis on *Ceroglossus chilensis* (Coleoptera: Carabidae). Acta Oecologica-International Journal of Ecology 52: 57–62.
42. Benitez HA, Sanzana M-J, Jerez V, Parra LE, Hernandez CE, et al. (2013) Sexual Shape and Size Dimorphism in Carabid Beetles of the Genus *Ceroglossus*: Is Geometric Body Size Similar Between Sexes Due to Sex Ratio? Zoological Science 30: 289–295.
43. Alibert P, Moureau B, Dommergues JL, David B (2001) Differentiation at a microgeographical scale within two species of ground beetle, *Carabus auronitens* and *C-nemoralis* (Coleoptera, Carabidae): a geometrical morphometric approach. Zoologica Scripta 30: 299–311.
44. Benitez H, Briones R, Jerez V (2008) Fluctuating asymmetry in two populations of *Ceroglossus chilensis* (Eschscholtz, 1829) (Coleoptera: Carabidae) in agroecosystem of *Pinus radiata* d. Don, Bio-Bio region, Chile Gayana 72: 131–139.
45. Clarke G (1998) Developmental Stability and Fitness: The Evidence Is Not Quite So Clear. The American Naturalist 152: 762–766.
46. Ghalambor CK, McKay JK, Carroll SP, Reznick DN (2007) Adaptive versus non-adaptive phenotypic plasticity and the potential for contemporary adaptation in new environments. Functional Ecology 21: 394–407.
47. Pfennig DW, Wund MA, Snell-Rood EC, Cruickshank T, Schlichting CD, et al. (2010) Phenotypic plasticity's impacts on diversification and speciation. Trends in Ecology & Evolution 25: 459–467.
48. Parsons SA, Joern A (2014) Life history traits associated with body size covary along a latitudinal gradient in a generalist grasshopper. Oecologia 174: 379–391.
49. Bouyer J, Ravel S, Dujardin J-P, De Meeüs T, Vial L, et al. (2007) Population Structuring of *Glossina palpalis gambiensis* (Diptera: Glossinidae) According to Landscape Fragmentation in the Mouhoun River, Burkina Faso. Journal of Medical Entomology 44: 788–795.
50. Stone GN, Schönrogge K (2003) The adaptive significance of insect gall morphology. Trends in Ecology & Evolution 18: 512–522.
51. Haine ER, Moret Y, Siva-Jothy MT, Rolff J (2008) Antimicrobial Defense and Persistent Infection in Insects. Science 322: 1257–1259.
52. Vogelweith F, Thiéry D, Moret Y, Colin E, Motreuil S, et al. (2014) Defense strategies used by two sympatric vineyard moth pests. Journal of Insect Physiology 64: 54–61.
53. Gross P (1993) Insect Behavioral and Morphological Defenses Against Parasitoids. Annual Review of Entomology 38: 251–273.
54. Moczek AP, Emlen DJ (2000) Male horn dimorphism in the scarab beetle, *Onthophagus taurus*: do alternative reproductive tactics favour alternative phenotypes? Animal Behaviour 59: 459–466.
55. Roy D, Seehausen O, Nosil P (2013) Sexual dimorphism dominates divergent host plant use in stick insect trophic morphology. BMC Evolutionary Biology 13: 135.
56. McCullough EL, Emlen DJ (2013) Evaluating the costs of a sexually selected weapon: big horns at a small price. Animal Behaviour 86: 977–985.
57. Benitez HA, Avaria-Llautureo J, Canales-Aguirre CB, Jerez V, Parra LE, et al. (2013) Evolution of sexual size dimorphism and its relationship with sex ratio in carabid beetles of Genus *Ceroglossus* Solier. Current Zoology 59: 769–777.
58. Cook DF (1990) Differences in courtship, mating and postcopulatory behaviour between male morphs of the dung beetle *Onthophagus binodis* Thunberg (Coleoptera: Scarabaeidae). Animal Behaviour 40: 428–436.
59. Nijhout HF, Emlen DJ (1998) Competition among body parts in the development and evolution of insect morphology. Proceedings of the National Academy of Sciences 95: 3685–3689.
60. Castañeda LE, Nespolo RF (2013) Phenotypic and Genetic Effects of Contrasting Ethanol Environments on Physiological and Developmental Traits in *Drosophila melanogaster*. PLoS ONE 8: e58920.
61. Demirci B, Lee Y, Lanzaro GC, Alten B (2012) Altitudinal genetic and morphometric variation among populations of *Culex theileri* Theobald (Diptera: Culicidae) from northeastern Turkey. Journal of Vector Ecology 37: 197–209.

62. Benitez HA, Pizarro-Araya J, Bravi R, Sanzana M-J, M AF (2014) Morphological variation on isolated populations of *Praocis* (*Praocis*) *spinolai*. Journal of Insect Science 14: 1–12.
63. Gurgel-GonÇAlves R, Ferreira JBC, Rosa AF, Bar ME, GalvÃO C (2011) Geometric morphometrics and ecological niche modelling for delimitation of near-sibling triatomine species. Medical and Veterinary Entomology 25: 84–93.
64. Sanzana M-J, Parra LE, Sepulveda-Zuniga E, Benitez HA (2013) Latitudinal gradient effect on the wing geometry of *Auca coctei* (Guerin) (Lepidoptera, Nymphalidae). Revista Brasileira De Entomologia 57: 411–416.
65. Kaupp A, Wurst C (1997) Familie Elateridae – Nachträge und Berichtigungen. In: Klausnitzer B, editor. Die Larven der Käfer Mitteleuropas. Jena: Gustav Fischer Verlag. 330–345.
66. Rohlf FJ (2013) TPSdig, v. 2.17. NY State University at Stony Brook.
67. Klingenberg CP (2011) MorphoJ: an integrated software package for geometric morphometrics. Molecular Ecology Resources 11: 353–357.
68. Arnqvist G, Martensson T (1998) Measurement error in geometric morphometrics: empirical strategies to assess and reduce its impact on measures of shape. Acta Zoologica Academiae Scientiarum Hungaricae 44: 73–96.
69. Dryden I, Mardia K (1998) Statistical Shape Analysis. Chichester: John Wiley and Son.
70. Rohlf FJ, Slice D (1990) Extensions of the Procustes methods for the optimal superimposition of landmarks. Systematic Zoology 39: 40–59.
71. Klingenberg CP, Barluenga M, Meyer A (2002) Shape analysis of symmetric structures: Quantifying variation among individuals and asymmetry. Evolution 56: 1909–1920.
72. Gould SJ (1966) Allometry and size in ontogeny and phylogeny. Biological Reviews of the Cambridge Philosophical Society 41: 587-&.
73. Klingenberg CP (1996) Multivariate allometry. In: Marcus LF, Corti M, Loy A, Naylor GJP, Slice DE, editors. Advances in Morphometrics. 23–49.
74. Campbell NA, Atchley WR (1981) The Geometry of Canonical Variate Analysis. Systematic Zoology 30: 268–280.
75. Klingenberg CP, Monteiro LR (2005) Distances and directions in multidimensional shape spaces: Implications for morphometric applications. Systematic Biology 54: 678–688.
76. Klingenberg CP, McIntyre GS (1998) Geometric morphometrics of developmental instability: Analyzing patterns of fluctuating asymmetry with procrustes methods. Evolution 52: 1363–1375.
77. Kozina A, Čačija M, Igrc Barčić J, Bažok R (2013) Influence of climatic conditions on the distribution, abundance and activity of *Agriotes lineatus* L. adults in sex pheromone traps in Croatia. International Journal of Biometeorology 57: 509–519.
78. Legendre P, Fortin M-J (2010) Comparison of the Mantel test and alternative approaches for detecting complex multivariate relationships in the spatial analysis of genetic data. Molecular Ecology Resources 10: 831–844.
79. Villani MG, Allee LL, Díaz A, Robbins PS (1999) Adaptive strategies of edaphic arthropods. Annual Review of Entomology 44: 233–256.
80. Téllez TR, Møller AP (2006) Fluctuating asymmetry of leaves in *Digitalis thapsi* under field and common garden conditions. International Journal of Plant Sciences 167: 321–329.
81. Benitez HA, Lemic D, Bazok R, Gallardo-Araya CM, Mikac KM (2014) Evolutionary directional asymmetry and shape variation in *Diabrotica virgifera virgifera* (Coleoptera: Chrysomelidae): an example using hind wings. Biological Journal of the Linnean Society 111: 110–118.
82. Liebhold AM, Rossi RE, Kemp WP (1993) Geostatistics and Geographic Information Systems in Applied Insect Ecology. Annual Review of Entomology 38: 303–327.
83. Toepfer S, Ellsbury MM, Eschen R, Kuhlmann U (2007) Spatial clustering of *Diabrotica virgifera virgifera* and *Agriotes ustulatus* in small-scale maize fields without topographic relief drift. Entomologia Experimentalis et Applicata 124: 61–75.

4

Impact of Fertilizing Pattern on the Biodiversity of a Weed Community and Wheat Growth

Leilei Tang[1,2], Chuanpeng Cheng[1], Kaiyuan Wan[1,3]*, Ruhai Li[4], Daozhong Wang[5], Yong Tao[1], Junfeng Pan[1], Juan Xie[1], Fang Chen[1,6]*

1 Key Laboratory of Aquatic Botany and Watershed Ecology, Wuhan Botanical Garden, Chinese Academy of Sciences, Wuhan, China, **2** Hainan Modern Agriculture Inspection and Testing Precaution & Control Center, Agricultural Department of Hainan Province, Haikou, China, **3** Ecological Restoration (ECORES) Lab, Chengdu Institute of Biology, Chinese Academy of Sciences, Chengdu, China, **4** Institute of Plant Protection and Soil Science, Hubei Academy of Agricultural Sciences, Wuhan, China, **5** Institute of Soil and Fertilizer Science, Anhui Academy of Agricultural Sciences, Hefei, China, **6** China Program of International Plant Nutrition Institute (IPNI), Wuhan, China

Abstract

Weeding and fertilization are important farming practices. Integrated weed management should protect or improve the biodiversity of farmland weed communities for a better ecological environment with not only increased crop yield, but also reduced use of herbicides. This study hypothesized that appropriate fertilization would benefit both crop growth and the biodiversity of farmland weed communities. To study the effects of different fertilizing patterns on the biodiversity of a farmland weed community and their adaptive mechanisms, indices of species diversity and responses of weed species and wheat were investigated in a 17-year field trial with a winter wheat-soybean rotation. This long term field trial includes six fertilizing treatments with different N, P and K application rates. The results indicated that wheat and the four prevalent weed species (*Galium aparine, Vicia sativa, Veronica persica* and *Geranium carolinianum*) showed different responses to fertilizer treatment in terms of density, plant height, shoot biomass, and nutrient accumulations. Each individual weed population exhibited its own adaptive mechanisms, such as increased internode length for growth advantages and increased light interception. The PK treatment had higher density, shoot biomass, Shannon-Wiener and Pielou Indices of weed community than N plus P fertilizer treatments. The N1/2PK treatment showed the same weed species number as the PK treatment. It also showed higher Shannon-Wiener and Pielou Indices of the weed community, although it had a lower wheat yield than the NPK treatment. The negative effects of the N1/2PK treatment on wheat yield could be balanced by the simultaneous positive effects on weed communities, which are intermediate in terms of the effects on wheat and weeds.

Editor: Shuijin Hu, North Carolina State University, United States of America

Funding: This work was supported by the China Program of International Plant Nutrition Institute (IPNI-HB-34) and the Opening Project of Hubei Key Laboratory of Wetland Evolution & Ecological Restoration(2011-02). The funders had no role in study design, data collection and analysis, decision to publish, or preparation of the manuscript.

Competing Interests: The authors have declared that no competing interests exist.

* E-mail: kaiyuanwan@126.com (KW); fchen@ipni.ac.cn (FC)

Introduction

Weeds are one of the major constraints to crop yields and quality [1,2]. However, as one of the primary producers within farming systems, weeds are of central importance to the arable system's food web. The weed community provides a range of resources for higher trophic groups, supports a high diversity of insect species and birds [3,4], and therefore plays an important role in the biological diversity of agroecosystems. Weeds also have other ecosystem functions, including nutrient cycling and soil preservation [5]. In addition, there are correlations between an impoverished landscape and the appearance of pests and diseases [6]. Moreover, an overreliance on herbicides has imposed a cost upon society and the environment [7]. The biodiversity of weed communities in a cropland can therefore be an important element for the reliable and sustainable provision of agroecosystem services. However, encouraging in-field biodiversity is unpopular among farmers because of the risk of decreased crop production as a result of weed competition. It is important to match crop production with conservation of biological resources to develop more sustainable systems [8].

Crop nutrient management practices may serve as an important component of more robust weed management programs [9–11]. Fertilization alters soil fertility, thus affecting weed density, nutrient uptake, and biomass yield, which in turn affects species composition and biodiversity [9,12–14]. For example, Mahn [15] observed a general decline in the number of weed individuals and an increase in weed biomass with increasing rates of nitrogen (N) fertilizer. In another study, *Digitaria ischaemum* Shreb (smooth crabgrass) was found to be the dominant species under N + potassium (K) and non-fertilized treatments, *Cyperus rotundus* L (purple nutsedge) dominated under phosphorus (P)+K treatment, and more weed species and higher Shannon' s diversity (H') values were detected in the balanced fertilization treatment [14]. These effects, however, indirectly depend on the light penetration caused by crop competition variation [16]. Balanced fertilization promotes the growth of crops, resulting in closed crop stands and light limitation for the weed communities growing underneath, thereby affecting weed species diversity [14,17].

Varying physiological responses of weed species to soil amendments are one of the explanations for weed community biodiversity [18]. Murphy and Lemerle [18] reported that the type and rate of fertilizers applied and the physiology of the species involved play an important role in weed population shifts. Haas and Streibig [19] also showed that high N levels will favor weed species that possess either physiological shade tolerance (esp. *Stellaria media*) or are able to climb into more favorable light conditions (esp. *Galium aparine*). Pyšek and Lepš [20] and Bengtsson et al. [21] also found that N fertilization favors shade-tolerant, climbing, and competitive weed species, but suppresses other kinds of species and results in a decrease of weed species diversity. With cessation of mineral fertilizer application during the period of conversion in a cropland, van Elsen [6] observed a gradual decline of weeds like *G. aparine*, which need a high nitrate level, and an increase in leguminous weeds (esp. *Vicia spp.*).

Documentation of the effects of a farming system on vegetation diversity is an important step toward understanding ecosystem function in agricultural landscapes [22], but only a handful of studies have examined the effect of fertilization on weed community biodiversity in agroecosystems. Biodiversity promotes ecosystem productivity, sustainability and stability in grassland [23–24]. Based on these reports, it is hypothesized that appropriate fertilization of agroecosystems would not only provide desirable crop productivity but also maintain the biodiversity of weed communities.

Long-term field experiments would ensure that proper data are accumulated and that confounding experimental effects are reduced, thereby providing better insights into the effects of prolonged fertilization over time [25]. It was also hypothesized that weed species could develop physiological adaptive mechanisms under long-term fertilization conditions. This study was conducted in a winter wheat field under continuous fertilization since 1994 with the objective of examining the effects of different N, P and K fertilization patterns on crop growth and yield. Effects on weeds were not examined at initiation, so baseline data are unavailable. Therefore, the objective of this study was to evaluate the cumulative effects of different fertilizing patterns on weed community biodiversity and their adaptive mechanisms with data from one-year sampling in this long-term experiment.

Materials and Methods

Site and Experimental Design

The long-term fertilization field experiment used in this study has been run by the Anhui Academy of Agricultural Sciences since 1994 and is located in Mengcheng county in the Anhui province of China (33°13′38 N, 116°36′58 E). The field studies did not involve endangered or protected species and no specific permits were required. This region has a warm temperate to sub-humid monsoon climate. The mean temperature, precipitation, evaporation and selected soil properties of the experimental site are given in Table 1. Since 1999, there has been a crop rotation of winter wheat (*Triticum aestivum* L.) and soybean (*Glycine max* [L.] Merr.).

Fertilizer treatments consisted of six combinations with different rates of N, P and K (Table 1) applied as urea, calcium superphosphate, and potassium chloride, respectively. All fertilizers were applied by soil surface broadcasting before the sowing of wheat. No fertilizer was applied during the soybean production seasons. The plot size was 20 m^2 (4×5 m) with three replications in a randomized complete block design. Soybean ('Zhonghuang 13') and winter wheat ('Yannong 19') were planted at a density of 249,800 seeds and 4,685,100 plants per hectare on 15 June and 20 October 2010, respectively.

Table 1. Experimental conditions at the field site.

Item	Value
Weather	
Mean annual temperature (1994–2010)	15°C
Rainfall (annual mean, 1994–2009)	872 mm
Rainfall (2010)	821 mm
Evaporation (annual mean, 1994–2010)†	1026 mm
Soil conditions	
Soil type	Lime concretion black soil
Soil pH	8.0
Soil organic matter	9.9 g kg^{-1}
Total soil N	0.79 g kg^{-1}
Soil available P	7.8 mg kg^{-1}
Soil available K	111 mg kg^{-1}
Crop rotation	
Winter wheat-maize cropping system	1994–1998
Winter wheat- Soybean cropping system	1999–2010
Fertilizer treatments	N, P_2O_5, K_2O kg ha^{-1}
Control	0, 0, 0
PK	0, 90, 135
NP	188, 90, 0
NK	188, 0, 135
NPK	188, 90, 135
N1/2PK*	188, 45, 135

†Evaporation was measured using the Penman formula.
*The N1/2PK treatment was designed to further evaluate the effect of P on winter wheat as it was the most critical soil limiting nutrient at the beginning of the experiment.

Herbicides were used according to the weed spectrum and expert recommendations. All treatments had a broad application of herbicides, and manual weeding was implemented before 2008. To better examine the effects of fertilization on weed community biodiversity, only a broadcast application of herbicides was continued in each growing season, and without manual weeding after 2008. Tribenuron methyl (Anhui Research Institute of Chemical Industry, China) at 12 g ai.ha^{-1} for wheat and Acetochlor (Anhui Huilong Group Rmf Agrochemical Co., Ltd, China) at 1500 g ai.ha^{-1} for soybean were used once every growing season in all the experimental plots.

Sampling Procedures

Weed evaluations were conducted three times during wheat growing season on April 10 (177 days after sowing [d]), May 8 (205 d), and May 23 (221 d), 2010. At each sampling date (growing period), three 0.5×0.5 m quadrats were systematically positioned to avoid edge effects and re-sampling of the previously sampled areas in each plot. Five wheat plants and all weeds present in the quadrat were clipped, collected, sorted by species, counted, and oven dried at 70°C for 48 h before weighing. Wheat tillers were also counted, and plant heights of wheat and weed species were measured.

At 205 d and 221 d, the tallest plant of *Galium aparine* (a prevalent broadleaved weed species) in each quadrat was selected for the measurement of internode length. Light measurements were also made in the quadrats (using Minolta Illuminance Meter T-1H) above the wheat and on the soil surface. At 221 d, each individual weed species and wheat plant was sampled for total N, P and K analysis. The N was measured using the Kjeldahl method, and P and K were measured using an inductively coupled plasma optical emission spectrometer (ICP-OES) after the plant samples were digested [26].

Statistical Procedures

Weed density was calculated as the total plant number of a particular weed species per square meter, total weed density was calculated as the total of all weed species per square meter, and winter wheat density was calculated as the total number of tillers per square meter. The species diversity of weed communities was assessed by calculating different indices. For example, Species richness (S) was measured by the mean number of species in each treatment [27]. Shannon's diversity index (H'), Simpson index (D) and Pielou index (E) were calculated for the weed communities using the following equations:

$$H' = \frac{N \ln N - \sum n \ln n}{N}$$

$$D = \sum (n/N)(n/N)$$

$$E = \frac{H'}{\ln S}$$

where N is the total number of individuals per plot and n is the number of individuals per species per plot. Shannon's diversity index (H') takes not only the number of species (community richness) into account but also the relative distribution of each species in the community (community evenness). The Simpson index was applied to measure the degree of dominance of individual weed species.

Rank-abundance plots were used to display species relative abundance data. Abundance distributions provide a complete description of the community diversity and simultaneously show both components of species diversity, species number, and evenness. The relative abundance of a species indicates its degree of dominance or subordination in the weed community (i.e., the greater the relative abundance of a species in the weed community, the higher its dominance, [28]). Relative growth rate (RGR, g m^{-2} d^{-1}), the increase of plant biomass per unit time, was used to compare crop growth with that of weed species.

Statistical analysis was done using SPSS version 16.0. Weed density and biomass data were $\log_e$ (x+1) transformed before analysis to meet homogeneity of variance assumptions. For normally distributed parameters, the General Linear Model Univariate was used, and means were compared based on Tukey's multiple comparison tests $P<0.05$. Otherwise, nonparametric tests were used, and median values were presented.

Results

Structural Changes and Biological Diversity at the Community Level

The results indicated that total weed density was influenced by fertilization ($P<0.0001$), but was not affected by growing periods ($P=0.59$) (Fig. 1). A general decline in the number of weed individuals was observed in the different fertilization treatments except in the NK treatment where the number increased slightly. The PK treatment showed the highest total mean weed density (711 plants m^{-2}) followed by the control and NK treatments. Weed densities in these three treatments were significantly higher than the 201 plants m^{-2} in the treatments with N and P (72% less than that in the PK treatment).

Fertilization increased weed biomass in the winter wheat field (Fig. 2). The PK treatment obtained the highest ($P<0.05$) total weed biomass of 108.03 and 117.94 g m^{-2} at 205 and 221 d, respectively. Treatments with N and P showed a lower weed shoot biomass.

The species diversity of the weed community was modified by fertilizing patterns. Species number and evenness were affected differently by the fertilization treatments, which were shown in (a) the rank-abundance plots and (b) the vertical bar chart (Fig. 3). Among all the recorded species (13 spp.), the species numbers in the PK (11 spp.) and N1/2PK (11 spp.) treatments reached the highest, followed by the control (9 spp.) and NK (9 spp.) treatments, while the lowest value (7 spp.) was observed in the NPK treatment.

The evenness of the weed community also varied with the fertilization treatments (Fig. 3a, b). Fertilizing patterns greatly shaped the equitability in the partitioning of total biomass among species in the community as shown by the slope of the rank-abundance plots. A higher Shannon-Wiener Index (Fig. 3b) and weed community evenness (Pielou Index, Fig. 3a, b) was observed in the PK treatment, followed by the NK and control treatments. However, in these treatments, the Simpson Index was lower compared with that in the NP and NPK treatments (Fig. 3b). Compared with the NPK treatment, the half P rate treatment (N1/2PK) showed a higher equitability of total biomass partitioning (Fig. 3a) with higher Pielou Index and Shannon-Wiener Index values (Fig. 3b).

Changes at the Population Level

Densities of Weed Population. *Galium aparine* L. (tender catchweed bedstraw), *Vicia sativa* L. (common vetch), *Veronica persica* Poir (iran speedwell), and *Geranium carolinianum* L. (carolina geranium) were the most dominant species in the weed community. These four dominant species comprised >90% of the total weed density. Average densities of the four prevalent broadleaved weeds during the three growing periods in the different fertilization treatments are shown in Fig. 4. Densities of these four weed species varied significantly among the treatments. *Galium aparine* densities were greatly affected by fertilization ($P<0.0001$) and growing period ($P=0.02$). No significant difference was detected in the interactions of treatments with growing period ($P=0.43$). The densities of *G. aparine* (138 plants m^{-2}) were higher in the treatments with N and P fertilizers, while no *G. aparine* was found in the control treatment.

Vicia sativa density was significantly affected by fertilization ($P<0.0001$) and growing period ($P=0.008$). However, in contrast to *G. aparine*, *Vicia sativa* grew better in treatments without N and/or P with an average density of ~200 plants m^{-2}, while it had only 10–30 plants m^{-2} in the treatments with N and P. *Geranium carolinianum* density was also affected by fertilization ($P<0.0001$) and growing period ($P<0.001$), and the interactions of fertilization with growing period ($P<0.001$). Only three plants of *G. carolinianum* were found in the NP treatment at 205 d and no plants were found at 221 d. During the early growing period, *G. carolinianum* densities showed no significant differences among the treatments, with an average value of 72 plants m^{-2}. However, its densities responded differently to fertilization treatments in the

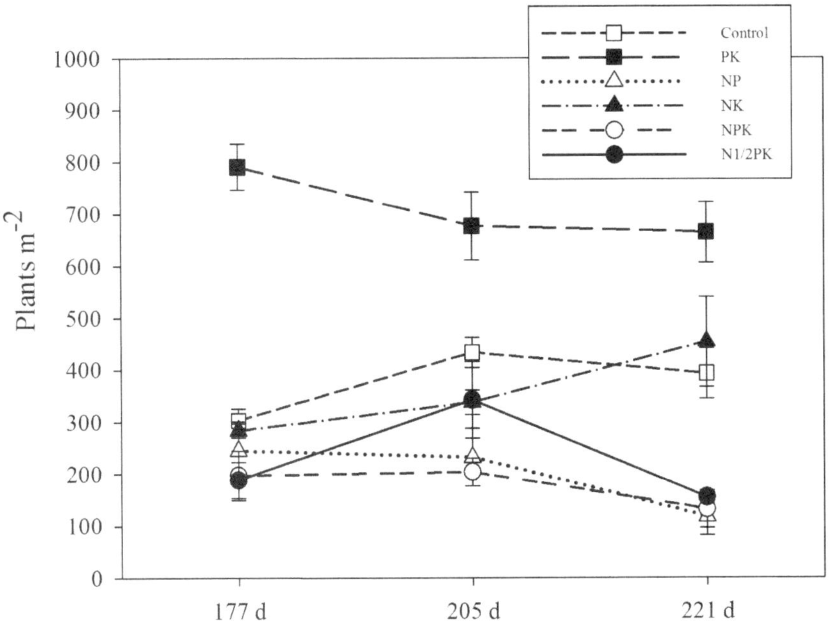

Figure 1. The influence of fertilization on the density of the weed community.

latter growing period, with the same tendency as *V. sativa* in each treatment. *Veronica persica* grew well only in the PK treatment, and few or no plants were detected in the other treatments.

Plant Height of Weed Population. Among the four prevalent broadleaved weed species, *V. persica* grew underneath the crop canopy, and its maximum plant height was <50 cm (Fig. 5D). On the contrary, other species were erect for capturing adequate light. Fertilization significantly affected plant height of *G. carolinianum* (Fig. 5A), *V. sativa* (Fig. 5B) and *G. aparine* (Fig. 5C). *Geranium carolinianum* was taller in the PK and NK treatments, even taller than in the NPK treatment at 205 d. Plant height of *V. sativa* in the PK treatment was at a maximum, reaching 73.8 cm at maturity stage, followed by the NK treatment where it reached 64.1 cm, while lower values were observed in treatments with N and P. Conversely, *G. aparine* showed higher plant heights of 83.8 cm, 85.3 cm and 73.3 cm in the NP, NPK and N1/2PK treatments, respectively. The study confirmed that fertilization had little effect on the number of *G. aparine* shoot internodes. However, fertilization significantly modified the internodal length of *G. aparine*, especially from the 4th to the 12th internode (Table 2).

Shoot Biomass of the Weed Population. Shoot biomass of the four weed species varied with different fertilization treatments (Table 3). Like the weed density, *G. aparine* shoot biomass in the NP treatment was highest at each growing period, followed by the

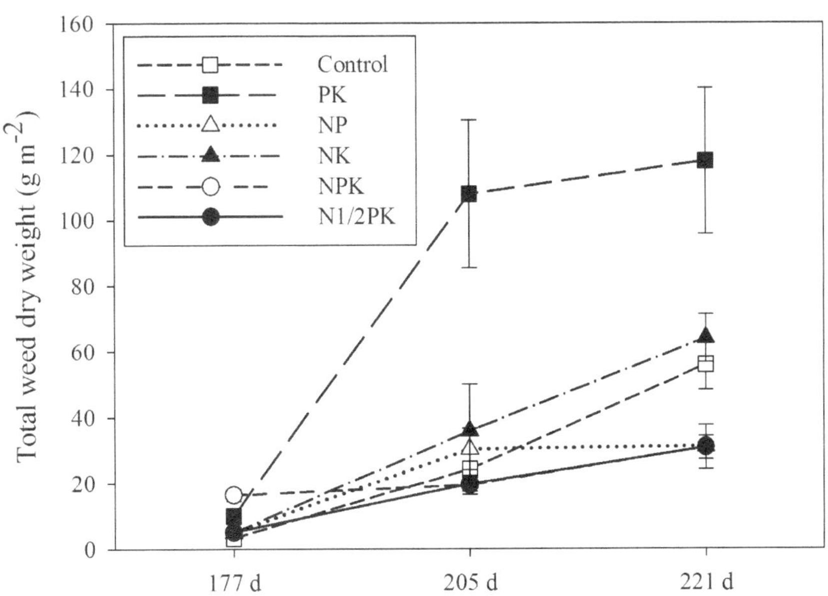

Figure 2. The influence of fertilization on the biomass of the weed community.

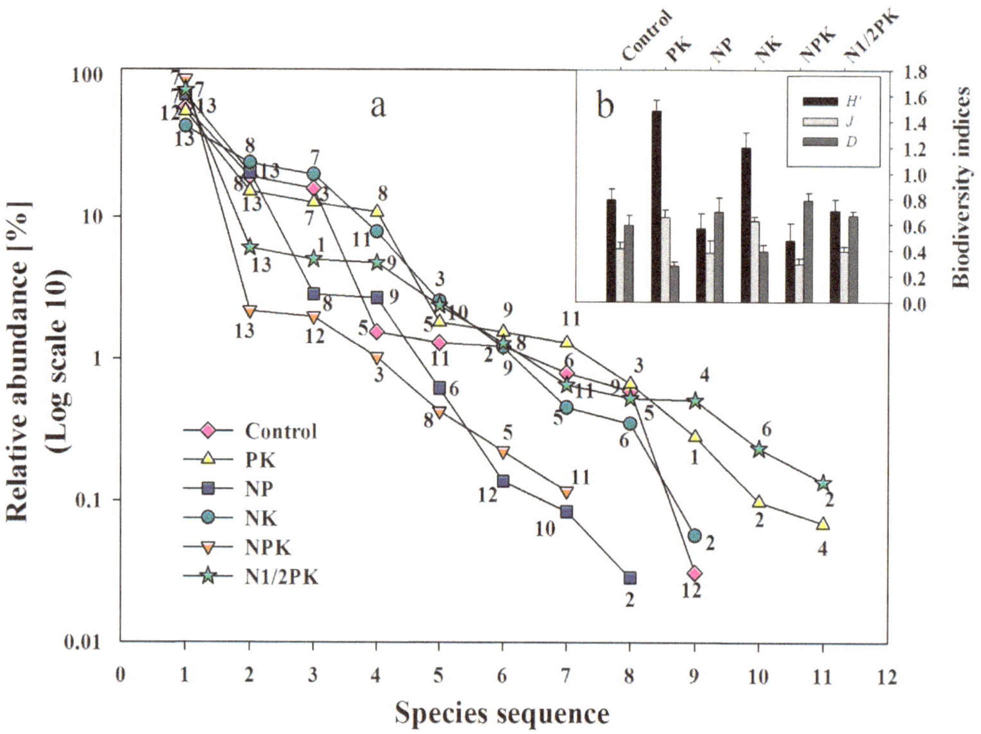

Figure 3. The influence of fertilization on the species diversity of the weed community. (a) Rank-abundance plots corresponding to different fertilization treatments in a winter wheat field. (b) The influence of fertilization on the biodiversity indices of weed communities (*H'*, Shannon-Winner Index; *J*, Pielou Index; and *D*, Simpson index). Species reference: (1) *Avena fatua* L., (2) *Calystegia hederacea* Wall., (3) *Cirsium segetum* Bge., (4) *Cyperus rotundus* L., (5) *Erigeron annuus* (L.) Pers., (6) *Euphorbia helioscopia* L., (7) *Galium aparine* L. var. *tenerum* (Gren.et Godr.) Rcbb., (8) *Geranium carolinianum* L., (9) *Lithospermum arvense* L., (10) *Mazus pumilus* (Burm. f) V. Steenis., (11) *Plantago virginica* L., (12) *Veronica persica* Poir., (13) *Vicia sativa* L.

NPK and N1/2PK treatments. The relative growth rates of *G. aparine* in the treatments without N and/or P were lower, especially in the NK treatment (0.11 g m^{-2} d^{-1}), than that in the NP treatment (0.83 g m^{-2} d^{-1}). The highest proportion of *G. aparine* shoot biomass to total weed biomass was obtained in the NPK treatment (91%), followed by the NP and N1/2PK treatments at

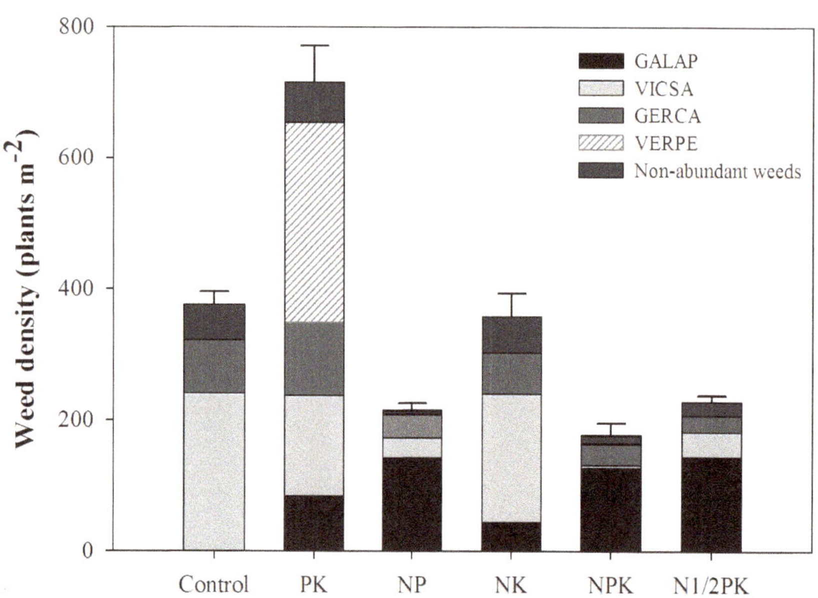

Figure 4. The effects of fertilization on average weed densities. Analyses of variance were performed on $\log_e$-transformed data, but means of untransformed data are shown. Error bars are the SE of the mean.

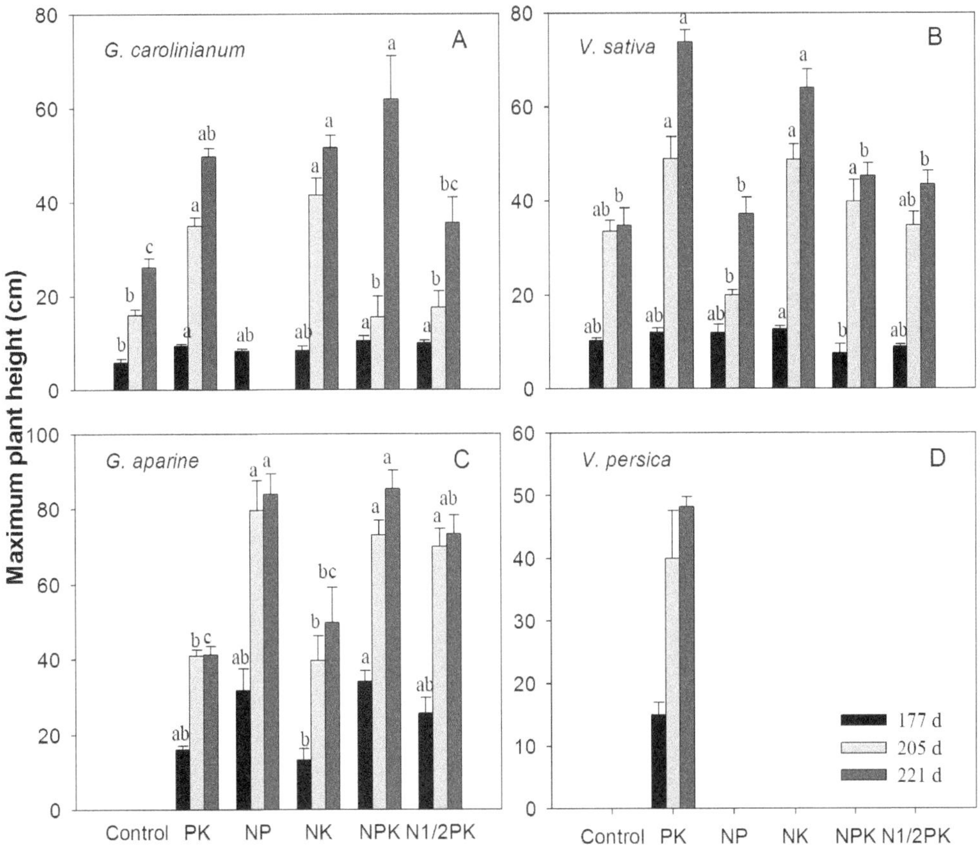

Figure 5. The influence of fertilization on plant height of *G. carolinianum* (A), *V. sativa* (B), *G. aparine* (C) and *V. persica* (D).

89% and 77%, respectively, while treatments without N and/or P had the lowest value (8%). These results indicated that *G. aparine* had an absolute growth advantage under higher soil fertility conditions, and its growth could be significantly promoted by the application of N and P fertilizers.

Biomass accumulation of *V. sativa* followed an opposite trend to that of *G. aparine* for the different fertilization treatments. The average shoot biomass of *V. sativa* in the control, NK and PK treatments were 3.32, 14.81, and 25.17 g m^{-2}, respectively in contrast to values under the NP, NPK and N1/2PK treatments where the values were 0.63, 0.69, and 1.33 g m^{-2} at 177, 205 and 221 d, respectively. The relative growth rate showed that *V. sativa* also grew rapidly in the control, PK and NK treatments (0.50 g m^{-2} d^{-1}). The highest proportion of *V. sativa* shoot biomass to the total weed biomass was observed in the control treatment with 68% and 41% at 205 and 221 d, respectively, followed by the NK and PK treatments, while treatments incorporating N and/or P fertilizers showed the lowest values (4%).

The average shoot biomass of *G. carolinianum* at 177 d reached 5.05 g m^{-2}, which was much higher than the corresponding values for *G. aparine*, *V. sativa* and *V. persica* (1.72, 1.97 and 0.55 g m^{-2}, respectively). The *G. carolinianum* shoot biomass under the NPK treatment (9.11 g m^{-2}) was higher than that under control. However, in the middle and latter growing periods, *G. carolinianum* shoot biomass performed similarly to *V. sativa* with the different fertilization treatments. Specifically, biomass accumulation of *V. persica* was only higher in the PK treatment. Its biomass at 205 d accounted for 81% of the total biomass at 221 d, and it also accounted for >50% of the total shoot biomass of weeds present in the PK treatment.

Nutrient Uptake by Weeds. Nutrient uptake by the four weed species differed among treatments (Table 3). In general, nutrient uptake was higher in *G. aparine* and *V. sativa* than in *G. carolinianum*, suggesting that the two weed species were stronger nutrient competitors to wheat. Nutrient uptake of weed species in different treatments followed a pattern similar to shoot biomass in the corresponding treatments. Average N, P, and K uptakes by *G. aparine* in the NP, NPK, and N1/2PK treatments were 0.353, 0.090, and 0.447 g m^{-2}, respectively, which were 435, 375, and 425% higher than those under the PK treatment, and 45, 158, and 62% higher than those under the NK treatment. The N, P, and K uptakes under the control, PK, and NK treatments were 0.426, 0.076, and 0.361 g m^{-2}, respectively for *V. sativa* and 0.163, 0.052, and 0.187 g m^{-2}, respectively for *G. carolinianum*. For *V. persica*, the highest nutrient uptake was found under the PK treatment, and its P and K uptakes were 127 and 241% higher, respectively than those of *V. sativa*.

Winter Wheat Responses to Fertilization

Winter Wheat Density. Significant differences in wheat density were found among the fertilization treatments ($P<0.001$). Wheat density increased in the treatments that included N and P. Wheat density in the NP treatment was significantly ($P<0.01$) higher than that in the control, PK and NK treatments. However, there were no significant differences in winter wheat density within the NP, NPK and N1/2PK treatments. The average density was

Table 2. Effects of different fertilization treatments on internode length (cm) and number of *G. aparine* plant.

Treatment	Internode number	The 4th Internode	The 5th Internode	The 6th Internode	The 7th Internode	The 8th Internode	The 9th Internode	The 10th Internode	The 11th Internode	The 12th Internode
177 d										
CK	–	–	–	–	–	–	–	–	–	–
PK	9 a	1.5 b	2.0 ab	2.1 bc	2.1 b	2.0 bc	1.3 b	–	–	–
NP	11 a	1.8 b	2.5 ab	2.8 ab	4.3 a	3.9 ab	3.2 ab	–	–	–
NK	9 a	1.4 b	1.6 b	1.4 c	1.6 b	1.5 c	1.0 b	–	–	–
NPK	11a	2.8 a	3.1 a	2.8 ab	2.5 ab	4.0 ab	5.5 a	–	–	–
N1/2PK	10 a	2.4 a	2.9 a	3.2 a	3.3 ab	3.3 a	3.4 ab	–	–	–
205 d										
CK	–	–	–	–	–	–	–			
PK	16 a	1.4 b	1.9 b	2.5 c	3.2 c	3.8 b	4.7 bc	4.7 ab	4.3 bc	4.4 b
NP	18 a	3.8 ab	4.4 ab	5.8 ab	6.1 ab	6.3 a	6.2 abc	5.9 ab	6.1 ab	5.1 ab
NK	16 a	2.6 ab	3.0 ab	3.6 bc	4.1 bc	4.2 b	4.2 c	3.8 b	3.7 c	3.4 b
NPK	18 a	3.7 a	4.9 a	6.2 a	7.0 a	7.2 a	6.9 a	6.1 a	6.7 a	6.8 a
N1/2PK	18 a	3.3 ab	4.3 ab	5.2 ab	6.1 ab	6.4 a	6.3 ab	6.0 a	5.8 ab	5.3 ab

Different letters within each day after sowing (d) indicate significant differences among treatments (Tukey test; $P<0.05$).
The internode lengths (heights) are taken between the previous node and the listed node.

Table 3. Shoot biomass of the four prevalent weed species at 177, 205 and 221 d and nutrient uptake at 221 d in the different fertilization treatments.

Treatments	Weed biomass (g m $^{-2}$)			Biomass proportion (%)[e]		Nutrient uptake (g m $^{-2}$)		
	177 d	205 d	221 d	205 d	221 d	N	P	K
G. aparine								
Control	0.00	0.00	0.00	0.00	0.00	—	—	—
PK	1.39 ab	10.38 bc	10.38 bc	12.72 b	8.80 b	0.066 b	0.019 c	0.085 b
NP	2.85 a	27.05 a	27.63 a	89.06 a	89.13 a	0.331 a	0.093 ab	0.337 b
NK	0.85 b	2.68 c	5.21 c	7.48 b	8.14 b	0.243 ab	0.035 c	0.276 b
NPK	2.83 a	18.24 ab	27.33 a	94.03 a	88.41 a	0.455 a	0.114 a	0.611 a
N1/2PK	2.43 a	15.43 ab	23.42 ab	78.41 a	76.02 a	0.273 ab	0.064 ab	0.392 ab
V. sativa								
Control	2.68 ab	16.42 a	22.75 a	67.71 a	40.72 a	0.365 a	0.042 b	0.352 a
PK	3.73 a	12.41 a	28.58 a	15.20 b	24.24 a	0.538 a	0.133 a	0.400 a
NP	0.48 bc	0.47 b	1.53 b	1.56 c	5.26 b	0.028 b	0.006 c	0.013 b
NK	3.54 a	15.60 a	24.17 a	43.55 a	37.76 a	0.442 a	0.061 b	0.412 a
NPK	0.11 c	0.42 b	0.64 b	2.19 c	2.07 b	0.009 b	0.002 c	0.006 b
N1/2PK	1.31 bc	1.19 b	1.83 b	6.06 bc	5.94 b	0.035 b	0.006 c	0.032 b
G. carolinianum								
Control	0.65 b	4.68 abc	8.68 abc	19.30 a	15.53 ab	0.124	0.016	0.093
PK	5.59 ab	8.83 a	16.42 a	10.82 ab	13.93 abc	0.172	0.107	0.253
NP	7.50 ab	1.06 bc	0.00 d	3.50 ab	0.00 c	—	—	—
NK	2.94 ab	8.62 ab	12.94 ab	24.07 a	20.22 a	0.200	0.034	0.214
NPK	9.11 a	0.08 c	1.75 bcd	0.43 b	5.65 bc	0.024	0.008	0.027
N1/2PK	4.53 ab	0.25 d	0.28 cd	1.29 ab	0.26 c	0.001	0.000	0.002
V. persica								
Control	0.00	0.00	0.00	0.00	0.00	—	—	—
PK	4.68	65.94	81.19	59.69	58.52	0.762	0.415	1.903
NP	0.00	0.00	0.68	0.00	2.25	—	—	—
NK	0.00	0.00	0.33	0.00	0.51	—	—	—
NPK	0.01	0.13	0.12	0.81	0.30	—	—	—
N1/2PK	0.00	0.00	0.00	0.00	0.00	—	—	—

[e]indicates the proportion of shoot biomass within a species to the total weeds biomass per treatment.
Different letters within a species indicate significant differences among treatments (Tukey test; $P<0.05$).

524 tillers per m^2 in the latter three treatments, which was 82% higher than that in the control treatment (data not shown).

Plant Height of Winter Wheat. Fertilization significantly increased wheat height at all the growing periods and the final heights ranged from 60.0 to 79.4 cm at maturity stage (Fig. 6). Wheat heights of the NPK (78.2 cm) and N1/2PK treatments (79.4 cm) were 31 and 33% higher, respectively than the control treatment.

Winter Wheat Biomass, Nutrient Uptakes and Grain Yield. Fertilization significantly increased the shoot biomass accumulation of winter wheat (Table 4). Higher wheat shoot biomasses were observed in the treatments with N and P. The highest wheat shoot biomass was found in the N1/2PK treatment, and significant differences were also found in two of the three growing periods within the treatments containing N and P.

Wheat nutrient uptake is usually positively correlated with its shoot biomass accumulation. However, the wheat P uptake had no significant difference between the N and P treatments in this study (Table 4). The highest N and K uptakes were in the N1/2PK treatment. They were 18 and 78% higher than those in the NP treatment and 35 and 31%, higher respectively than those in the NPK treatment. These results might be attributed to the lower shoot biomass and nutrient accumulations of the weed community in the N1/2PK treatment (Fig. 2, Table 4). It also suggested that reducing P application in the NPK treatment decreased the productivity of weeds, such as *G. aparine* (Table 3), and improved nutrient uptake and biomass accumulation of wheat.

The NPK treatment resulted in the highest wheat yield, followed by the N1/2PK treatment (Fig. 7). In contrast, the control treatment obtained the highest light penetration (55%). The average light transmittance rate in the treatments with N and P was 5% at 205 d, which was 92% lower than the light transmittance value of the control treatment (Fig. 7).

Discussion

Weed community structure changed sharply among the different fertilization treatments and the results concurred with previous research conducted under field conditions by Mahn [15].

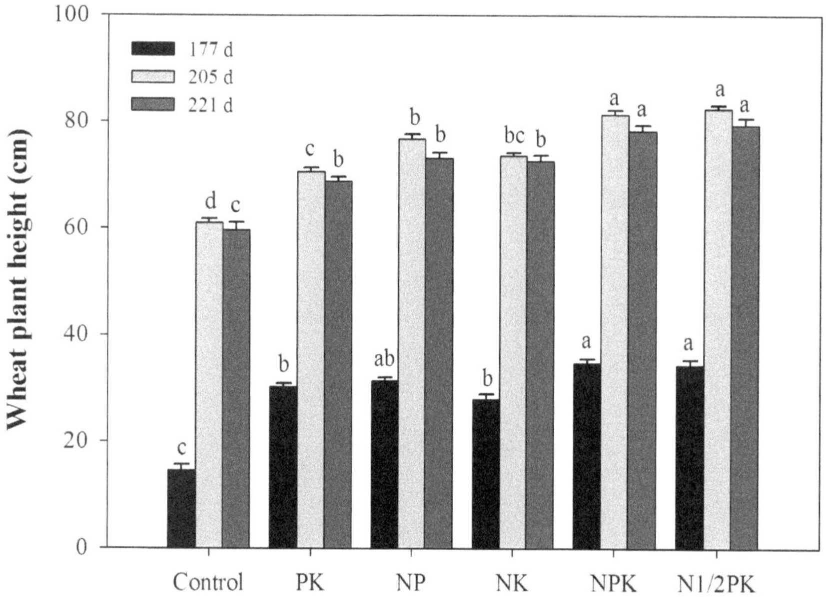

Figure 6. The influence of fertilization on wheat plant height at three growing periods.

The PK treatment showed the highest weed density and shoot biomass, followed by the NK and control treatments, while the treatments with N and P resulted in lower values. Yin et al. [14] and Nie et al. [17] also reported similar results. The structural variability of the weed community across treatments could be directly explained by the different soil fertility conditions. For example, soil available N concentration plays an important role in terminating dormancy by removing germination constraints and promoting the emergence of seedlings [29]. Soil available P was the primary nutrient regulating the species composition and floristic construction of the weed communities, followed by N and K [30]. Moreover, solar radiation reaching weeds was indirectly modulated by different fertilization treatments (Fig. 7). Wheat growth was promoted in the treatments with N and P, especially in the balanced fertilization treatments (Table 4, Fig. 7), increasing its ability to intercept solar radiation (Fig. 7), and thereby affecting the growth of weed species. The structural variability of the weed community may also be significantly affected by rainfall and other environmental factors, such as mean annual temperature. Ochoa-Hueso and Manrique [31] showed that apart from nutrients, plant germination and growth are also influenced by water availability. Pal et al. [32] and Šilc et al. [33] also reported that environmental factors, such as mean annual precipitation and temperature showed significantly positive and negative correlation coefficients with weed species composition, respectively.

The four prevalent weed populations and wheat differed substantially in their responses to different fertilization treatments. *Galium aparine* grew well in high fertility soil conditions, and was taller than wheat. *Vicia sativa* and *G. carolinianum* grew well under the control, NK, and PK treatments, and they were also taller than wheat in these treatments. *Veronica persica* grew well only in the PK treatment. These results show that the different weed species had formed their own adaptive mechanisms. Morphological plasticity enables plants to cope with a wide variety of ecological conditions, and nutrient availability has been widely shown to induce important anatomical responses [34]. For instance, *G. aparine* grew well under higher soil fertility conditions through climbing into more favorable light conditions by internode elongation (Table 2),

Table 4. The effects of different fertilization treatments on wheat shoot biomass at 177, 205 and 221 d, and nutrient uptake and harvest indices at 221 d.

Treatments	Biomass (g m^{-2})			Nutrient uptake (g m^{-2})			Weed parameters percentages (%)[f]				
	177 d	205 d	221 d	N	P	K	weed biomass	N	P	K	Weed number
Control	149 d	405 c	457 d	4.11 c	0.62 d	3.05 d	47.37	52.59	11.59	22.47	81.8
PK	411 c	1003 b	1032 c	5.12 c	2.37 bc	5.56 cd	100.00	100.00	100.00	100.00	100.0
NP	512 bc	1302 a	1503 b	14.52 a	4.10 a	8.12 bc	26.39	25.19	14.49	12.73	72.7
NK	395 c	843 b	1069 c	6.91 bc	1.08 cd	6.02 cd	54.27	65.93	18.84	32.96	81.8
NPK	601 ab	1337 a	1556 b	12.68 ab	3.75 ab	11.05 ab	26.21	36.30	18.84	23.97	63.6
N1/2PK	700 a	1476 a	2234 a	17.17 a	3.26 ab	14.46 a	26.12	25.19	11.59	17.60	100.0

[f]indicates the percentage of the value of a treatment to that in the PK treatment.
Different letters indicate significant differences among treatments (Tukey test; P<0.05).

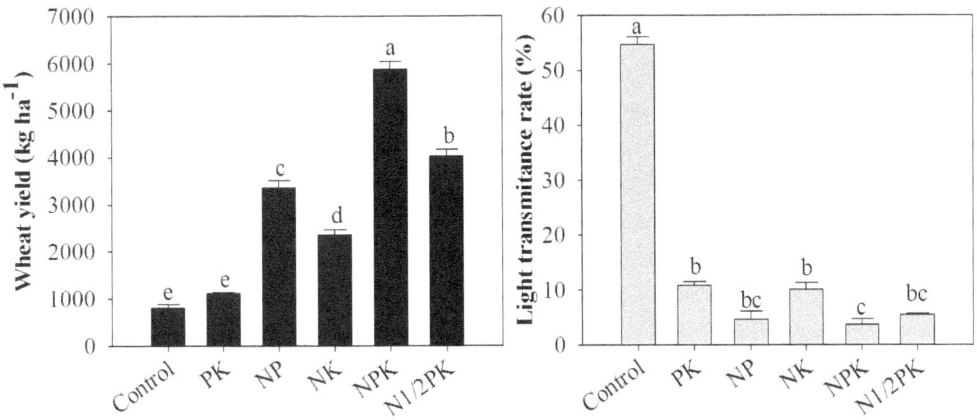

Figure 7. The influence of fertilization on wheat yield and light transmittance at the ground surface.

and responded to soil nutrient supply in the order of P>N>K. In contrast, *V. sativa* and *G. carolinianum* increased their heights for receiving more light under lower soil fertility and limited nutrient conditions. Li et al. [35] reported similar results for *G. carolinianum*, but Yin et al. [13] reported different results for *G. aparine* versus *V. sativa* and *V. persica*. These differences in the effects of fertilization on cropland weeds might be due to herbicide usage. A broadcast application of herbicides was used by Li et al. [35], while no herbicide was used by Yin et al. [13]. Furthermore, environmental conditions, such as plot location, temperature, crop type, precipitation, soil texture, neighboring habitat, and soil pH, which vary in time and space, may also cause variation in weed species responses [36–38].

The PK treatment showed the highest Shannon-Wiener and Pielou Indices, followed by the control and NK treatments. In contrast, these three treatments had lower Simpson Indices than the treatments with N and P. Solar penetration on the ground surface was also one of the reasons for the variation in weed community biodiversity, which is similar to the findings of other studies [16,39,28]. Biomass is often used as an indicator of the amount of resources captured by a crop. Higher biomass and grain yield of wheat in the NP, NPK and N1/2PK treatments indicated that wheat intercepted a larger proportion of light radiation (Fig. 7), thereby reducing the amount reaching the weeds. Wheat yield showed a significant negative correlation with light penetration (Pearson correlations >0.880, $P<0.0001$). The species richness, Shannon-Wiener and Pielou Indices and the Simpson Index of weed communities were proved to have significantly negative and positive linear function relationships with wheat yield [30]. Shading leads to thinning mortality in individuals of subordinate weed species, and consequently reduces evenness and biodiversity among weed species [39,40]. Only those weed species which possess either physiological shade tolerance or are able to climb into more favorable light conditions could grow well [18]. In contrast to other species, *G. aparine* had higher plant density, shoot biomass and nutrient accumulations in the NP, NPK and N1/2PK treatments with lower light penetration. Mahn [15] also reported that the mean biomass values for *G. aparine* in higher N fertilization treatments within a spring barley field were almost two times more than the values under the no N treatment. Under high soil fertility conditions in this study, *G. aparine* could enhance its plant height through internode elongation, especially from the 4th to the 12th internodes for capturing more sun light. Its plants were 14% and 9% taller than wheat plants in the NP and NPK treatments, respectively (Fig. 5). Therefore, weed vegetation under higher soil fertility conditions was predominantly composed of a few species, such as *G. aparine*, resulting in a higher Simpson Index and lower Pielou and Shannon-Wiener Indices.

More weed biomass and nutrient accumulations were detected in the PK treatment. Weed biomass under the NP, NPK and N1/2PK treatments was 26.4, 26.2, and 26.1%, respectively, of that under the PK treatment. Similarly, the N, P and K accumulations, as a proportion of that in the PK treatment, were 25, 14, and 13%, respectively in the NP treatment, 36, 19, and 24%, respectively in the NPK treatment and 25, 12, and 18%, respectively in the N1/2PK treatment (Table 4). These results indicated that fewer resources were captured by the weed community in the treatments with N and P, especially in the N1/2PK treatment. If a single weed species has a shoot biomass to total weed biomass ratio > 10%, it is regarded as a dominant weed species. More than two dominant weed species were found under the control, PK and NK treatments, with four species present in the PK treatment. Only one dominant weed species (*G. aparine*) was present in the balanced fertilization treatment. Thus, the results also indicated that balanced fertilization could not only maintain a stable crop yield, but also greatly reduce weed productivity.

Lower solar radiation under high soil fertility conditions resulted in fewer weed species compared with the control and PK treatments. However, the number of weed species was only 73% and 64% in the NP and NPK treatments, respectively and their number in the N1/2PK treatment was the same as in the PK treatment (Table 4). These results demonstrated that *G. aparine* is sensitive to soil P, and reducing the P application rate indirectly improved the growth of other weed species. Blackshaw and Brandt [41] also reported that the competitiveness of the high P-responsive species was progressively improved as the P dose increased. Likewise, with a lower P dose in the N1/2PK treatment, its competitive ability would decrease and more opportunities would be available for other species. The N1/2PK treatment could be considered as a balanced fertilization treatment in adjusting the interactions of wheat and weeds in agroecosystems. Although wheat yield in the N1/2PK treatment was 31% less than that in the NPK treatment, its weed species number, Shannon-Wiener and Pielou Indices were significantly higher than those in the NPK treatment. It is noteworthy that a certain amount of *G. aparine* would have a high biodiversity value for invertebrates and is important for seed-eating birds [22]. Thus, applying half the P rate in the NPK treatment could be considered as a better strategy to provide desirable farming benefits and maintain the biodiversity of weed communities. The negative effects on wheat yield could be balanced by the simultaneous positive effects of reduced herbicide usage. Appropriate combinations and rates of fertilization would not only be helpful for keeping desirable crop productivity, but also maintaining the biodiversity of weed communities in an agroecosystem.

Author Contributions

Conceived and designed the experiments: FC DZW. Performed the experiments: CPC KYW RHL YT JFP LLT. Analyzed the data: LLT. Contributed reagents/materials/analysis tools: JX LLT. Wrote the paper: LLT.

References

1. Vollmann J, Wagentristl H, Hartl W (2010) The effects of simulated weed pressure on early maturity soybeans. Eur J Agron 32: 243–248.
2. Odero DC, Mesbah AO, Miller SD, Kniss AR (2011) Interference of redstem filaree (*Erodium cicutarium*) in sugarbeet. Weed Sci 59: 310–313.
3. Marshall EJP, Brown VK, Boatman ND, Lutman PJW, Squire GR, et al. (2003) The role of weeds in supporting biological diversity within crop fields. Weed Res 43: 77–89.
4. Fried G, Petit S, Dessaint F, Reboud X (2009) Arable weed decline in Northern France: Crop edges as refugia for weed conservation. Biol Conserv 142: 238–243.
5. Altieri MA (1999) The ecological role of biodiversity in agroecosystems. Agric Ecosyst Environ 74: 19–31.
6. Van Elsen T (2000) Species diversity as a task for organic agriculture in Europe. Agric Ecosyst Environ 77: 101–109.
7. Marsh SP, Llewellyn RS, Powles SB (2006) Socail costs of herbicide resistance: the case of resistance to glyphosate. International Association of Agricultural Economists in its series 2006 Annual Meeting, August 12–18, 2006, Queensland, Australia with number 25413.
8. Storkey J, Cussans JW (2007) Reconciling the conservation of in-field biodiversity with crop production using a simulation model of weed growth and competition. Agric Ecosyst Environ 122: 173–182.
9. Di Tomaso JM (1995) Approaches for improving crop competitiveness through the manipulation of fertilization strategies. Weed Sci 43: 491–497.
10. Angonin C, Caussanel JP, Meynard JM (1996) Competition between winter wheat and *Veronica hederifolia*: influence of weed density and the amount and timing of nitrogen application. Weed Res 36: 175–187.
11. Blackshaw RE, Semach G, Janzen HH (2002) Fertilizer application method affects nitrogen uptake in weeds and wheat. Weed Sci 50: 634–641.
12. Blackshaw RE, Molnar LJ, Larney FJ (2005) Fertilizer, manure and compost effects on weed growth and competition with winter wheat in western Canada. Crop Prot 24: 971–980.
13. Yin LC, Cai ZC, Zhong WH (2005) Changes in weed composition of winter wheat crops due to long-term fertilization. Agric Ecosyst Environ 107: 181–186.
14. Yin LC, Cai ZC, Zhong WH (2006) Changes in weed community diversity of maize crops due to long-term fertilization. Crop Prot 25: 910–914.
15. Mahn EG (1988) Changes in the structure of weed communities affected by agro-chemicals: what role does nitrogen play? Ecological Bulletins, No. 39, Ecological Implication of Contemporary Agriculture: Proceeding of the 4th European Ecology Symposium, 7–12 September 1986, Wageningen, 71–73.
16. Kleijn D, van der Voort LAC (1997) Conservation headlands for rare arable weeds: the effects of fertilizer application and light penetration on plant growth. Biol Conserv 81: 57–67.
17. Nie J, Yin LC, Liao YL, Zheng SX, Xie J (2009) Weed community composition after 26 years of fertilization of late rice. Weed Sci 57: 256–260.
18. Murphy C, Lemerle D (2006) Continuous cropping systems and weed selection. Euphytica 148: 61–73.
19. Haas H, Streibig JC (1982) Changing patterns of weed distribution as a result of herbicide use and other agronomic factors, in: LeBaron HM, Gressel J (Eds.), Herbicide Resistance in Plants. John Wiley & Sons, New York, pp. 57–79.
20. Pyšek P, Lepš J (1991) Response of a weed community to nitrogen fertilization: a multivariate analysis. J Veg Sci 2: 237–244.
21. Bengtsson J, Ahnström J, Weibull AC (2005) The effects of organic agriculture on biodiversity and abundance: a meta-analysis. J Appl Ecol 42: 261–269.
22. Rassam G, Latifi N, Soltani A, Kamkar B (2011) Impact of crop management on weed species diversity and community composition of winter wheat fields in Iran. Weed Biol Manag 11: 83–90.
23. Tilman D, Wedin D, Knops J (1996) Productivity and sustainability influenced by biodiversity in grassland ecosystems. Nature 379: 718–720.
24. Tilman D, Reich PB, Knops JMH (2006) Biodiversity and ecosystem stability in a decade-long grassland experiment. Letters 441, doi: 10.1038/nature04742.
25. Derksen DA (1996) Weed community ecology: tedious sampling or relevant science? A Canadian perspective. Phytoprotection 77: 29–39.
26. Yoshida S, Forno DA, Cock JH, Gomez KA (1976) Laboratory Manual for Physiological Studies of Rice. International Rice Research Institute, Philippines.
27. Magurran AE (1988) Ecological diversity and its measurements. Princeton, USA: Princeton University Press.
28. Poggio SL (2005) Structure of weed communities occurring in monoculture and intercropping of field pea and barley. Agric Ecosyst Environ 109: 48–58.
29. Forcella F, Benech Arnold RL, Sanchez R, Ghersa CM (2000) Modeling seedling emergence. Field Crop Res 67: 123–139.
30. Tang LL, Wan KY, Cheng CP, Li RH, Wang DZ, et al. (2013) Effect of fertilization patterns on the assemblage of weed communities in an upland winter wheat field. J Plant Ecol doi: 10.1093/jpe/rtt018.
31. Ochoa-Hueso R, Manrique E (2010) Nitrogen fertilization and water supply affect germination and plant establishment of the soil seed bank present in a semi-arid Mediterranean scrubland. Plant Ecol 210: 263–273.
32. Pal RW, Pinke G, Botta-Dukát Z, Campetella G, Bartha S, et al. (2013) Can management intensity be more important than environmental factors? A case study along an extreme elevation gradient from central Italian cereal fields, Plant Biosystems - An international Journal Dealing with all Aspects of Plant Biology: Official Journal of the Societa Botanica Italiana 147: 2, 343–353.
33. Šilc U, Vrbnicanin S, Bozic D, Carni A, Dajic Stevanovic Z (2009) Weed vegetation in the north-western Balkans: diversity and species composition. Weed Res 49: 602–612.
34. Lamberti-Raverot B, Puijalon S (2012) Nutrient enrichment affects the mechanical resistance of aquatic plants. J Exp Bot 63: 6115–6123.
35. Li RH, Qiang S, Qiu DS, Chu QH, Pan GX (2008) Effects of long-term different fertilization regimes on the diversity of weed communities in oilseed rape fields under rice-oilseed rape cropping system. Biodivers Sci 16: 118–125. (In Chinese with English abstract)
36. Fried G, Norton LR, Reboud X (2008) Environmental and management factors determining weed species composition and diversity in France. Agric Ecosyst Environ 128: 68–76.
37. Hanzlik K, Gerowitt B (2011) The importance of climate, site and management on weed vegetation in oilseed rape in Germany. Agric Ecosyst Environ 141: 323–331.
38. Pinke G, Karácsony P, Czúcz B, Botta-Dukát Z, Lengyel A (2012) The influence of environment, management and site context on species composition of summer arable weed vegetation in Hungary. Appl Veg Sci 15: 136–144.
39. Goldberg DE, Miller TE (1990) Effects of different resource additions of species diversity in an annual plant community. Ecology 71: 213–225.
40. Tilman D, Pacala S (1993) The maintenance of species richness in plant communities, in: Ricklefs RE, Schulter D (Eds.), Species Diversity in Ecological Communities. Chicago, USA: University of Chicago Press, 13–25.
41. Blackshaw RE, Brandt RN (2009) Phosphorus fertilizer effects on the composition between wheat and several weed species. Weed Biol Manag 9: 46–53.

5

Bacterial Communities in Malagasy Soils with Differing Levels of Disturbance Affecting Botanical Diversity

Leah C. Blasiak[1], Alex W. Schmidt[2], Honoré Andriamiarinoro[3], Temesgen Mulaw[1], Rado Rasolomampianina[4], Wendy L. Applequist[3], Chris Birkinshaw[3], Félicitée Rejo-Fienena[4], Porter P. Lowry, II[3,5], Thomas M. Schmidt[2], Russell T. Hill[1]*

1 Institute of Marine and Environmental Technology, University of Maryland Center for Environmental Science, Baltimore, Maryland, United States of America, **2** The Center for Microbial Sciences, University of Michigan, Ann Arbor, Michigan, United States of America, **3** Missouri Botanical Garden, St. Louis, Missouri, United States of America, **4** Laboratoire de Microbiologie de Environnement, Centre National de Recherches sur l'Environnement, Antananarivo, Madagascar, **5** Département Systématique et Évolution (UMR 7205), Muséum National d'Histoire Naturelle, Paris, France

Abstract

Madagascar is well-known for the exceptional biodiversity of its macro-flora and fauna, but the biodiversity of Malagasy microbial communities remains relatively unexplored. Understanding patterns of bacterial diversity in soil and their correlations with above-ground botanical diversity could influence conservation planning as well as sampling strategies to maximize access to bacterially derived natural products. We present the first detailed description of Malagasy soil bacterial communities from a targeted 16S rRNA gene survey of greater than 290,000 sequences generated using 454 pyrosequencing. Two sampling plots in each of three forest conservation areas were established to represent different levels of disturbance resulting from human impact through agriculture and selective exploitation of trees, as well as from natural impacts of cyclones. In parallel, we performed an in-depth characterization of the total vascular plant morphospecies richness within each plot. The plots representing different levels of disturbance within each forest did not differ significantly in bacterial diversity or richness. Changes in bacterial community composition were largest between forests rather than between different levels of impact within a forest. The largest difference in bacterial community composition with disturbance was observed at the Vohibe forest conservation area, and this difference was correlated with changes in both vascular plant richness and soil pH. These results provide the first survey of Malagasy soil bacterial diversity and establish a baseline of botanical diversity within important conservation areas.

Editor: Christina A. Kellogg, U.S. Geological Survey, United States of America

Funding: This study was supported by a grant from the National Science Foundation (NSF) Directorate for Biological Sciences Integrative Organismal Systems (IOS-0919728), Fogarty International Center, National Cancer Institute, National Institute of Allergy and Infectious Diseases, National Institute of Mental Health, National Institute on Drug Abuse, National Heart Lung and Blood Institute, National Center for Complementary and Alternative Medicine, Office of Dietary Supplements, National Institute of General Medical Sciences, Biological Sciences Directorate of the NSF, and the Office of Biological and Environmental Research of the U.S. Department of Energy under Cooperative Agreement U01 TW00313 with the International Cooperative Biodiversity Groups. This project was also supported by the National Research Initiative of the Cooperative State Research, Education and Extension Service, USDA, Grant #2008-35621-04732. The funders had no role in study design, data collection and analysis, decision to publish, or preparation of the manuscript.

Competing Interests: The authors have declared that no competing interests exist.

* E-mail: hill@umces.edu

Introduction

While the effects of land-use on the biodiversity of plants and animals are well-studied, much less is known about how disturbance shapes microbial communities. Ecological studies on the effects of disturbance on plant communities have shown that smaller disturbances may increase floral species richness, while larger ones can lead to decreased richness, loss of botanical biodiversity, or even extinction of locally endemic species (reviewed in [1]). Large disturbances such as fires or conversion of native habitats to agriculture may take centuries to completely reverse, and the re-establishment of a primary or secondary plant community often follows a predictable pattern of community succession. Advances in sequencing technology, using deep pyrosequencing of 16S rRNA gene amplicons, now permit interrogation of soil bacterial community composition in sufficient detail to begin to investigate the effects of land-use disturbance on below-ground microbial diversity. Additionally, it is now possible to investigate whether impacts to above-ground plant diversity are paralleled by changes in soil bacterial diversity. Understanding the effects of land-use on soil bacterial diversity may add a new dimension to conservation goals because soil bacteria provide critical ecosystem services and are an important source of biodiversity for drug discovery, and because bacterial communities may be early indicators of ecosystem disturbance [2,3,4].

Recent studies have used multiplexed pyrosequencing of 16S rRNA genes and other non-culture-based methods to identify parameters that correlate with soil bacterial community diversity or composition, but their causal relationships in driving soil bacterial community structure remain unclear. Many studies have now documented the important role of soil pH, which is strongly correlated to soil bacterial diversity (richness and evenness) at local and global scales [5,6,7,8]. Other soil edaphic parameters, including soil nitrogen (N) amendment [9,10] and total N or carbon (C) [6,11] have been found to be significantly associated with shifts in bacterial community composition. The taxonomic composition of soil bacterial communities can also be differentiated by land-use. For example, vegetation type [7], conversion of native vegetation to agriculture [12,13,14], and agriculture management practices [14,15,16] all correlate with changes in taxonomic composition, although these changes often co-vary with changes in soil chemistry. Vegetation type correlates with bacterial community composition locally within a given soil type or as a secondary factor after soil edaphic properties [7,17,18,19]. These correlations with plant diversity may represent indirect interactions (covariance of biogeographic patterns in plant and microbial communities) or direct interactions through influence of plant root exudates, organic matter deposition, and/or specific interactions with plant rhizospheres. For example, bacterial isolation from roots of endemic Malagasy rosewood trees has yielded a wide diversity of root-nodulating bacteria [20].

We hypothesized that loss of plant species richness in Malagasy forests through both human and non-human impacts is correlated with a loss of bacterial diversity and a change in bacterial community composition. Several studies have documented decreases in bacterial diversity or functional diversity with different types of disturbance, including impacts from wildfires, camping, and agriculture [9,21,22,23,24]. However, grassland soils have generally been found to harbor higher bacterial diversity than forest soils, and so disturbance of forests and conversion to grasslands may actually increase soil bacterial diversity and richness [6,13]. The most detailed study thus far on the impact of forest disturbance on microbial communities demonstrated a significant effect of timber harvesting, organic matter removal, and soil compaction on soil bacterial community composition in coniferous forests even 10–15 years after the disturbance and despite replanting with tree seedlings [3].

In this study, we aimed to quantify the changes in soil bacterial communities with differing levels of disturbance to forest plant communities in Madagascar. Madagascar is a biodiversity hotspot with significant conservation value, and the country contains an astonishing botanical diversity including an estimated 13,000–14,000 species of vascular plants of which nearly 90% are endemic to this island nation [25,26,27]. The study sites included three Malagasy forests that have been prioritized for conservation efforts based on work by the Missouri Botanical Garden as part of an International Biodiversity Cooperative Group grant. Within each forest, we established two 0.1 ha plots that were representative of different disturbance levels present at the site as assessed by their plant community structure. Vascular plant species richness and plant community structure, including the number and size of stems, were quantified within each plot, and the soil bacterial communities were investigated using barcoded 454 pyrosequencing of the 16S rRNA gene. This study represents the most detailed characterization to date of plant community structure and soil bacterial diversity in Malagasy forests. The resulting data provide an important initial reference for monitoring the effects of future conservation efforts within the selected forests. The observed patterns of bacterial community composition may also influence the design of sampling strategies to maximize access to bacterial taxonomic and genetic diversity for drug discovery efforts.

Materials and Methods

Site description

This study was conducted within three humid forests (as defined by [28]) located in eastern Madagascar: the forests of Analalava, Pointe à Larrée, and Vohibe (Table 1). All three sites are located in a per-humid climate zone [29], but while Analalava and Vohibe are located on basement rock, Pointe à Larrée is situated on old sand dunes [30]. The loose sand geology at Pointe à Larrée and the proximity of this forest to the sea result in a distinct subclass of humid forest known as littoral forest [28]. The Vohibe Forest is at a higher elevation than the other two sites. Each of the three forests is owned by the Malagasy government, and they are currently the focus of activities that will lead to their being classified and managed as new protected areas within Madagascar's protected area network. Permits for access to sampling sites and for collection of botanical and soil samples were obtained from the Malagasy Ministère de l'Environnement et des Forêts.

A total of six plots of 50 m×20 m were established, with two plots at each of the three sites (Table 1). The two plots within each forest were selected subjectively to maximize the difference between the levels of disturbance, both human and natural, on the vegetation. All sites have been subject to historical and/or recent disturbance from cyclones. At Analalava and Pointe à Larrée, no areas of the forest have been completely protected from human impact; sites were thus selected that represent moderate and high levels of selective timber exploitation. In the Vohibe forest, a site that has been well protected from human disturbance was contrasted with one that was subject to a single cycle of shifting cultivation prior to 2002 and is now largely a monoculture of a tall rhizomatous herb, Madagascar Cardamom [*Aframomum angustifolium* (Sonn.) K. Schum.].

Botanical diversity

The diversity of vascular plants within each 0.1 ha plot was characterized by botanists from the Missouri Botanical Garden (Table 2). Plants were identified by means of nested plots. Within the entire 0.1 ha plot all plant stems with diameter at breast height (DBH) ≥5 cm were identified. Additionally, within five 16 m^2 plots, arranged at regular intervals along the longest transect of the plot, all plants with DBH between 1 cm to 5 cm and all herbs were identified. We did not attempt to identify tree seedlings or non-vascular plants. When possible, plants were identified to a formally named species, but in many cases such precise identification could not be made because the plants were infertile (i.e. neither flowering nor fruiting). In such cases, the plants were classified only to morphospecies: distinct identities defined on the basis of habit and leaf morphology. Plants could rarely be identified with confidence in the field, so voucher herbarium specimens were made and identified later in the laboratory using the botanical literature, mounted reference herbarium specimens available in the herbarium at the Parc Botanique et Zoologique de Tsimbazaza, Antananarivo, and the electronic database of Malagasy flora (http://www.tropicos.org/project/mada). Fertile specimens were preserved at the herbaria of the Missouri Botanical Garden and the Parc Botanique et Zoologique de Tsimbazaza (MO and TAN respectively). The measurements of stem diameters were also used to describe the structure of the vegetation in the plots in terms of stem basal area and stem abundance in various DBH classes.

Soil sampling and chemical analysis

A total of 30 soil cores of 4 cm diameter and 10 cm depth were sampled from the six plots (3 forests ×2 impact levels ×5 soil cores). Five soil cores were taken in 10 m intervals along the central length of each 50×20 m plot. Soil samples were collected from Analalava and Pointe à Larrée in September 2011 and from Vohibe in October 2011, and were transported on ice to the Centre National de Recherches sur l'Environnement (CNRE) laboratory in Antananarivo, Madagascar, and stored at −20°C

Table 1. Description of the 0.1 ha plots.

Site	Forest	Subjective evaluation of disturbance	Georeference (and elevation)	Known disturbance history
A1	Analalava	High	S 17° 42′ 42,3″/E 049° 27′ 18,7″ (49 m)	Subjected to selective exploitation of medium and large trees and the collection of fence posts until 2005 when the forest was protected. Impacted by Cyclone Ivan in 2008 but sustained little damage.
A2	Analalava	Moderate	S 17° 42′ 18,9″/E 049° 27′ 22,9″ (56 m)	As at A1, but with selective exploitation of mainly large trees.
P1	Pointe à Larrée	High	S 16° 47′ 14,5″/E 049° 44′ 27,3″ (15 m)	In past decades and until present this forest is subjected to the selective exploitation of large trees. The forest was impacted by Cyclone Ivan in 2008 with consequent major damage.
P2	Pointe à Larrée	Moderate	S 16° 47′ 45,5″/E 049° 44′ 38,0″ (15 m)	As at P1, but with selective exploitation of large trees now rare.
V1	Vohibe	Very high	S 19° 09′ 45,7″/E 048° 35′ 02,3″ (598 m)	Subjected to one cycle of shifting cultivation (trees cut, left to dry and burnt, and then land cultivated until soil exhausted). The last crops were cultivated on this land in about 2002.
V2	Vohibe	Very low	S 19° 09′ 32,2″/E 048° 34′ 39,1″ (646 m)	In the past possibly subjected to very rare selective exploitation of precious woods and timber.

until processing. Soil cores were homogenized and sieved through a 2 mm mesh. Soil pH, C, N, and C:N ratio (CN) were determined at FOFIFA, the Centre National de la Recherche Appliquée au Développement Rural in Madagascar (Table S1). Briefly, soil pH was measured as a water/soil suspension, total organic carbon (C) was determined according to the ANNE method [31], and the total N by the Kjeldahl method [32].

DNA extraction and pyrosequencing

DNA was extracted from 0.25 g of soil from homogenized soil cores using the PowerSoil DNA isolation kit (Mo Bio, Carlsbad, CA) according to the manufacturer's instructions, except that the DNA yield was increased by pooling the DNA from two DNA extraction kits for each sample. A total of 60 DNA samples were prepared (two technical replicates per soil core ×5 cores per plot ×6 plots). DNA samples were dried at 50°C immediately after extraction at the CNRE laboratory and shipped dry at room temperature to Michigan for sequencing. The V4–V6 region of the 16S rRNA gene was PCR amplified in triplicate using the following barcoded primers: forward primer 518F = CCAGCAGCYGCGGTAAN, forward primer adapter = CCTATCCCCTGTGTGCCTTGGCAGTCTCAG, reverse primer 1046R = CGACRRCCATGCANCACCT, reverse primer adapter = CCATCTCATCCCTGCGTGTCTCCGACTCAG (http://www.hmpdacc.org/doc/16S_Sequencing_SOP_4.2.2.pdf) [33]. The PCR products were then purified using Agencourt® AMPure® XP beads (Beckman Coulter, Brea, CA) and the Life Technologies Dynamag-2 Magnet. DNA concentrations were quantified and samples were combined into an equimolar pool and sequenced using a Roche GS Junior 454 Sequencer. Sequences and associated metadata were deposited in the MG-RAST database and are publically accessible (http://metagenomics.anl.gov/linkin.cgi?project=6283) with accession numbers 4538937.3–4538996.3.

Sequence processing

The V4–V6 region of the 16S rRNA gene was sequenced from 60 soil samples (2 technical replicates ×5 cores per plot ×6 plots). Sequence processing was performed in mothur v1.28.0 [34] and generally followed the recommendations described by Schloss *et al.* [35]. Raw sequences (598,730 reads) were denoised with PyroNoise [36] and then sequences with mismatches to forward primer or barcode sequences, ambiguous bases, and homopolymer stretches greater than 8 bp were removed. Barcodes and primer sequences were removed, short sequences were discarded, and the remaining sequences were trimmed to exactly 250 bp. The denoised and quality filtered sequences were then aligned to the Silva reference alignment (v102) [37]. Preclustering of aligned sequences was employed using an abundance weighted single-linkage preclustering step at ~2% difference (maximum 2 bp difference over 250 bp) [38]. Chimeras were identified and removed using the UChime algorithm [39] and sequences were classified with an 80% bootstrap cutoff to the Ribosomal Database Project (RDP) 16S rRNA reference database Release 9 [40] using the RDP Classifier [41]. Sequences classified as *Archaea*, *Eukaryota*, chloroplasts, mitochondria, and unclassified sequences were removed (<0.01% of sequences), leaving only sequences classified as *Bacteria*. Pairwise distances were calculated on aligned sequences and then sequences were clustered into operational taxonomic units (OTUs) at 3% difference using the average neighbor method. The taxonomy of each OTU was assigned in mothur as the consensus taxonomy for at least 51% of the sequences within the OTU.

Sequence processing was performed with technical replicates separated (60 samples) and then with technical replicates pooled (30 samples). Sequences were subsampled to either 5,800 sequences (technical replicates separated) or 9,800 sequences (technical replicates pooled), to allow comparisons between samples with different sequencing depth. Beta diversity analysis using NMDS ordination of Bray-Curtis distances demonstrated that technical replicates clustered together and thus had similar bacterial communities (Figure S1). The mean Bray-Curtis dissimilarity between pairs of technical replicates was 0.21±0.06 (standard deviation) while the mean dissimilarity between all pairs of samples within each site was 0.39±0.05. This result indicates that the soil core homogenization was sufficient, because replicate soil samples from the same core gave reproducible bacterial communities. Thus, each sequenced soil sample is representative of the entire soil core. Therefore, technical replicates were pooled for all further data analysis. The final quality-filtered and

Table 2. Description of plant community structure by plot.

Site	Subjective evaluation of disturbance	# of morpho-species of vascular plants	Stem basal area (m^2/0.1 ha) by diameter size class			Number woody stems per 0.1 ha by stem diameter size class (cm)						
			≥10 cm	≥5 to <10 cm	≤5 cm	≥1-<5	≥5-<10	≥10-<20	≥20-<30	≥30-<40	≥40-<50	≥50
A1	High	214	1.26	1.01	2.30	1875	272	53	8	2	0	0
A2	Moderate	225	1.99	1.45	3.44	2338	391	96	12	2	0	0
P1	High	103	1.86	0.64	2.50	1300	180	47	14	5	1	0
P2	Moderate	89	2.44	0.76	3.20	1250	213	61	20	5	0	1
V1	Very high	32	0.74	0.17	0.91	163	40	17	4	0	2	0
V2	Very low	215	4.10	0.66	4.75	1225	179	73	23	9	2	3

subsampled dataset contained 294,000 sequences with 30 samples and 9,800 sequences per sample.

Statistical analyses

Comparisons of soil parameters across plots were made using one-way analysis-of-variance (ANOVA) in R [42]. Bacterial diversity measures including the Inverse Simpson's Index and Bray-Curtis dissimilarity matrices were calculated on 3% OTUs in mothur [34]. Relationships between bacterial diversity and soil chemical data were investigated by plotting and with the function cor.test in R. Nonmetric multidimensional scaling (NMDS) ordinations on Bray-Curtis dissimilarities, permuted analysis of variance (PERMANOVA), and permuted analysis of multivariate dispersions were performed using the functions metaMDS, adonis, and betadisper from the vegan package in R [43,44]. Because of the large differences between forests, including elevation, climate, and sampling date, the forests were treated as three separate experiments for pairwise comparisons of disturbance levels by PERMANOVA. Canonical analysis of principal coordinates (CAP) implemented through the capscale function in vegan was used to perform constrained ordination to test for linear relationships between bacterial community composition (Bray-Curtis dissimilarities) and soil chemical parameters within each forest. All permutation tests were performed with 999 permutations. Figures were prepared with R and the package ggplot2 [45].

Results and Discussion

Botanical diversity

The botanical diversity of the six plots is summarized in Table 2. At Analalava and Vohibe, the plots in less disturbed forest had more vascular plant morphospecies than those in more disturbed forest. The reverse was true for the plots at Pointe à Larrée, with the more disturbed plot having higher plant species richness. The moderately disturbed plot at Pointe à Larrée included much lower diversity of vascular plants than the moderately disturbed plot at Analalava or the little disturbed plot at Vohibe. The total stem DBH and stem abundance in various size classes showed a striking structural difference between the little disturbed plot at Vohibe and its very disturbed counterpart. The plots at Analava and Pointe à Larrée also had a generally consistent decrease in stem DBH and stem abundance with disturbance, but the differences between the disturbance levels much less pronounced. While the little disturbed forest plot at Vohibe included more large trees and a much higher total stem basal area than the two plots at Analalava, both of the latter included a morphospecies richness of vascular plants similar to that of the little disturbed plot at Vohibe.

Bacterial community composition

In order to compare the relative abundance of bacterial taxonomic groups observed in this study, 3% OTUs were classified according to the RDP taxonomy and grouped at the phylum level, except for proteobacterial groups, which were further divided by class (Figure 1). The most abundant bacterial group in the Pointe à Larrée and Vohibe plots was *Betaproteobacteria*, representing an average of about 20–30% of sequences within each plot. In contrast, the Analalava plots contained mostly *Acidobacteria* and *Verrucomicrobia*. Other abundant groups with an average of >5% of sequences across all plots were *Actinobacteria*, *Alphaproteobacteria*, *Gammaproteobacteria*, and *Firmicutes*. The bacterial community composition was further analyzed by grouping OTUs at the genus level (Figure 2). Abundant genera across all soil samples included *Burkholderia*, the *Acidobacteria* Group 1, *Rhodococcus*, and *Bacillus*. The functional and ecological roles of the *Acidobacteria* remain poorly characterized, but there is evidence for strong correlations between abundance of acidobacterial groups and pH [8,46], as well as indications that many *Acidobacteria* may be adapted to an oligotrophic lifestyle [47]. Naether *et al.* observed that the Group 1 *Acidobacteria* were dominant in temperate forest soils, while the Group 6 Acidobacteria were most prevalent in grassland soils [46]. The abundance of the betaproteobacterial genus *Burkholderia* observed here is consistent with the recent identification of this genus as an acid tolerant group that may outcompete other taxa in acidic soils such as those in our study [48]. A BLAST analysis of a representative sequence from the dominant *Burkholderia* OTU in our dataset showed that the closest relative was *Burkholderia sabiae* strain STM7319, which clusters within the plant-beneficial-environmental (PBE) group of *Burkholderia* [49]. The next two most abundant *Burkholderia* OTUs in our dataset also clustered with this group rather than the typically pathogenic *Burkholderia cepacia* complex. Members of the PBE clade are often associated with plants and may be capable of nitrogen fixation or otherwise promote plant growth. Amplicon-based community profiling has many well-documented limitations, including DNA extraction and primer bias, which constrain our ability to draw conclusions about absolute abundance of bacterial groups. However, despite any bias, sequencing results are reproducible, so comparisons can be made concerning relative abundance of taxa in samples with consistent DNA extraction and sequencing methods [50].

Bacterial diversity and disturbance

The Inverse Simpson diversity index is a measure of alpha diversity that accounts for the richness and evenness of OTUs observed within a single sample. We observed a trend of lower bacterial diversity in the less disturbed plots (Table 3, Figure S2), although this difference was not significant. We observed no significant difference in the Inverse Simpson index between each of the plots (treating each of the six plots separately) or between each of the forests (combining the two plots in each forest). Comparisons made using richness estimators (number of OTUs, Chao1) and a phylogenetic diversity measure (Faith's PD) gave similar results (data not shown). Comparisons using Smith and Wilson's metric of evenness demonstrated a significant difference at the level of plots ($p = 0.002$) and forests ($p = 0.008$). A Tukey's honestly significant difference (HSD) test showed that the high impact plot at Pointe à Larrée had significantly lower evenness than the low impact plot at Vohibe ($p = 0.003$) and the moderate and high impact plots at Analalava ($p = 0.002$ and 0.04 respectively). However, there was no significant difference in evenness between any of the pairs of plots within each forest ($p > 0.05$).

These results indicate that disturbance from selective timber harvesting or historical cultivation did not significantly affect the soil bacterial diversity, although there was a slight trend towards higher bacterial diversity with disturbance in all forests. We did not detect a significant difference in bacterial diversity or richness even between the Vohibe plots where we observed dramatic differences in plant species richness and vegetation structure.

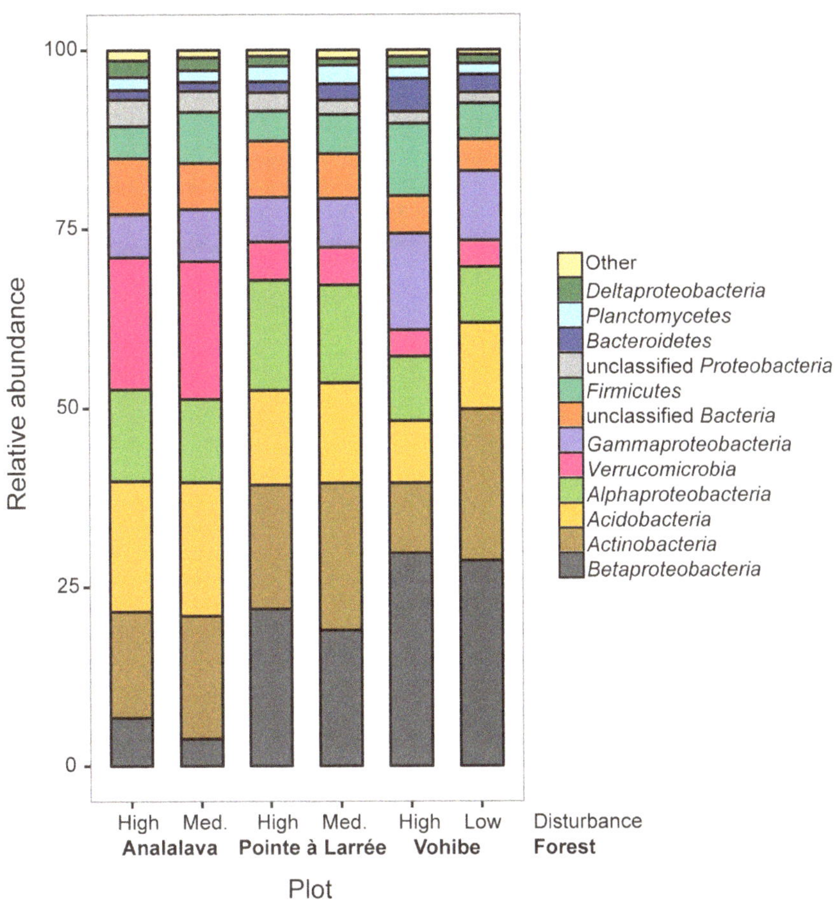

Figure 1. Average relative abundance of dominant bacterial taxa by plot. Sequences in 3% OTUs were classified and grouped at the phylum level, with *Proteobacteria* further divided by class. Taxa representing <1% of the total sequences were grouped as Other.

Figure 2. Log transformed abundance of dominant bacterial genera by sample. Sequences in 3% OTUs were classified and grouped at the genus level. Taxa representing <0.3% of the total sequences were grouped as Other. Darker color indicates higher relative abundance. Asterisks indicate taxa with significant shifts in abundance between disturbance levels at Vohibe (see Table 5).

Table 3. Bacterial diversity statistics[a] calculated at the OTU (3%) level.

Plot	Good's coverage %	# OTUs	Inverse Simpson	Shannon	Smith and Wilson's Evenness
A1	95.8 (0.9)	970 (178)	51.0 (23.3)	5.08 (0.36)	0.34 (0.02)
A2	96.5 (0.5)	819 (78)	37.1 (10.0)	4.77 (0.17)	0.37 (0.01)
P1	94.0 (0.9)	1212 (174)	40.0 (25.1)	5.12 (0.47)	0.29 (0.03)
P2	95.7 (0.9)	972 (199)	34.0 (13.6)	4.88 (0.42)	0.33 (0.04)
V1	94.9 (0.8)	1031 (186)	43.1 (14.7)	4.96 (0.37)	0.33 (0.03)
V2	96.0 (0.8)	838 (120)	29.1 (14.6)	4.64 (0.32)	0.37 (0.02)

[a]Shown are the mean and (sd) of each statistic calculated for 5 sequencing samples per plot.

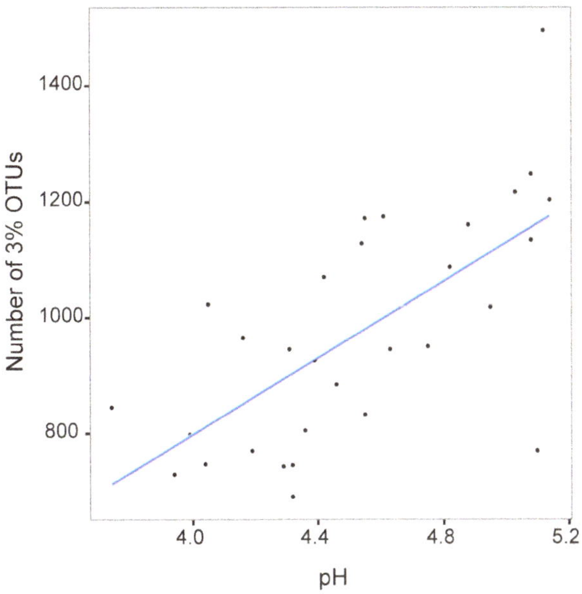

Figure 3. Richness and soil pH. Bacterial community richness is shown as the number of 3% OTUs. Pearson's $r = 0.66$, $p < 0.0001$.

Other studies have found correlations in bacterial diversity with land-use, with agricultural sites having higher diversity than forested sites in tropical soils, but these sites also differed in soil environmental parameters [6,13]. The results of our study more closely resemble those of Suleiman *et al.* in the Brazilian Pampa, where no significant difference in alpha diversity was found between eight-year-old de-forested pasture as compared to surrounding undisturbed forest [12]. While we did not find significant effects of land-use on soil bacterial richness at the resolution afforded by the 16S rRNA gene, there could be finer level changes in the diversity of functional genes or natural product biosynthetic genes [51]. It remains an open question whether bacterial taxonomic diversity (richness) correlates with ecosystem function, and if loss of microbial diversity affects community resilience or function [21,52,53]. Moreover, a recent study found large fluctuations in soil bacterial diversity over time, with temporal variability exceeding that attributed to land-use [53]. More sampling replicates obtained from different seasons might be required to reveal differences in bacterial diversity with disturbance level or botanical diversity.

Bacterial diversity and soil properties

Several studies have documented a strong relationship between bacterial richness and soil pH at both global and local scales [5,6,8]. Although we did not observe site-dependent changes in diversity with pH, we hypothesized that changes in soil properties irrespective of site might be related to bacterial community richness. In fact, when looking at all samples collectively, the number of OTUs per sample was positively correlated with soil pH (Pearson's $r = 0.66$, $p < 0.0001$), which varied from 3.7–5.1 in our study (Figure 3). Measures of diversity, including the Inverse Simpson's Index, were also positively correlated with soil pH across all samples. This relationship was significant but weaker than that observed by Rousk *et al.* at a single site, a gradient of pH 4–8 at the Hoosfield acid strip long-term agricultural research site in Harpenden, UK [8]. We found no significant correlations between richness and the other measured soil parameters including organic C, total N, or C:N ratio.

Bacterial community composition varies with forest more than with disturbance level

An NMDS ordination of Bray-Curtis dissimilarities showed strong clustering of bacterial communities by forest as compared to levels of disturbance (Figure 4). The significance of this clustering was tested using a nested PERMANOVA design (sites nested within forests) with Bray-Curtis dissimilarities calculated on 3% OTUs. We observed significant differences between bacterial community compositions at the three forest sites (pseudo-F = 10.47, $p = 0.001$) and a significant interaction effect between forests and sites (pseudo-F 1.87, $p = 0.009$). Pairwise comparisons

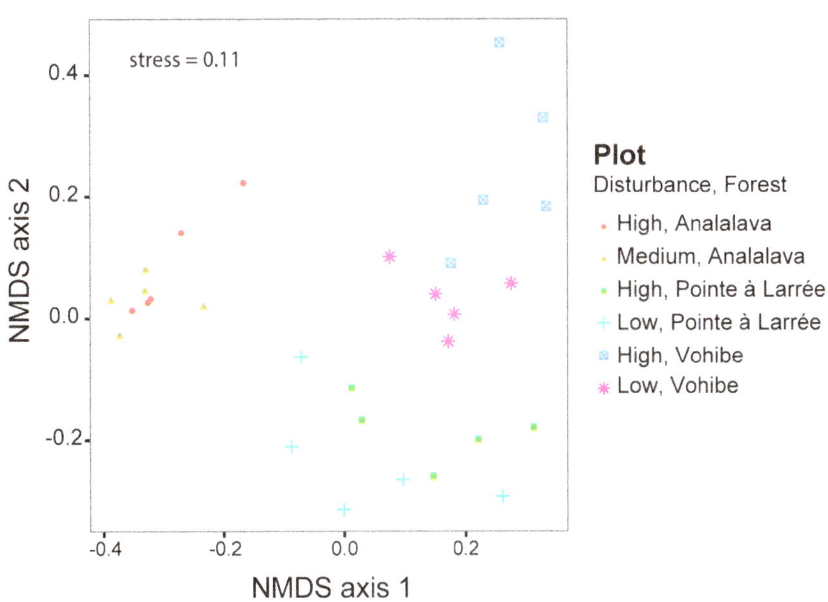

Figure 4. Nonmetric multidimensional scaling (NMDS) plot of pairwise Bray-Curtis dissimilarities between bacterial communities in soil calculated with 3% OTUs.

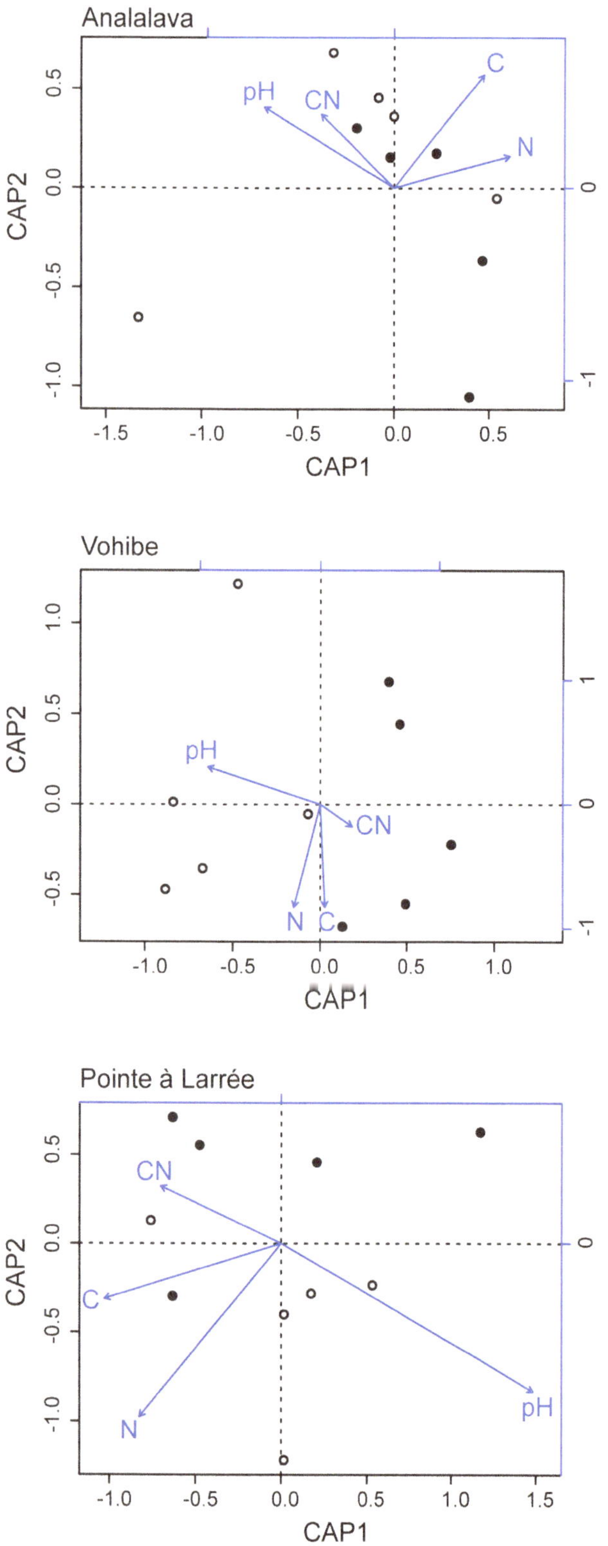

Figure 5. CAP ordination biplots of the bacterial community composition data and soil chemical properties (arrows) within each forest. Closed circles indicate sites from the lower disturbance plot and open circles indicate sites from the higher disturbance plot.

between forests also showed strong differences in the bacterial community structures (A-V: $t = 12.95$; A-P: $t = 11.20$; V-P: $t = 5.44$, by PERMANOVA, Bonferroni corrected $p = 0.003$ for these pairwise comparisons). The samples from Analalava had significantly lower spread than those from Pointe à Larrée, but no other significant differences were detected in the dispersion of samples within forests (pairwise comparisons with betadisper, A-V: $t = 3.80$, $p = 0.20$; A-P: $t = 7.03$, $p = 0.02$; V-P: $t = 0.065$, $p = 2.41$, with Bonferroni correction).

Pairwise PERMANOVAs were then performed between the impact levels within each forest in order to test whether soil bacterial community composition varied with the level of disturbance. Only the bacterial communities at Vohibe varied significantly with disturbance level ($t = 3.0$, $p = 0.007$). The NMDS ordination and smaller PERMANOVA pseudo-F/t statistic indicate that the difference between the bacterial communities of the two sites at Vohibe is less than that observed between forests. Pairwise tests for homogeneity of group variance between the impact levels within each forest were not significant ($p > 0.05$ for all pairwise comparisons). The two Vohibe sites were the most different in terms of impact level (very high vs. very low impact), as indicated by the measured differences in plant morphospecies richness and tree cover (Table 2). It is therefore perhaps not surprising that we were able to observe a difference in bacterial structure between this pair of plots. By contrast, no areas representing a low level of disturbance were available at Analalava and Pointe à Larrée, and so the selected pairs of sites represent moderate and high levels of impact from selective harvesting of trees. To test for differences in soil parameters between plots, we used one-way ANOVAs followed by Tukey's HSD test, and the only significant differences between paired plots within each forest were in pH at Vohibe and in total N at Pointe à Larrée. The pH levels of the soil samples from the high disturbance Vohibe forest were higher compared to those of the low disturbance plot, perhaps resulting from the history of burning followed by shifting cultivation in the disturbed plot. Thus, the observed change in bacterial community structure between sites with different impact levels at Vohibe is correlated with differences in botanical diversity and with a change in soil pH.

Beta diversity and soil properties

We expected that soil samples with more similar soil chemistry would also host more similar bacterial communities. To test this hypothesis, CAP ordinations of the community composition data (relative abundance of 3% OTUs) constrained by the soil chemical data were performed within each forest (Figure 5). Within all forests, pH was significantly correlated with bacterial community structure ($p < 0.05$ for all tests). The total N was only significant within the Pointe à Larrée forest. Total C and CN were not significant within any forest and C tended to covary with N. As expected given the previous unconstrained ordination, pH was significantly aligned with CAP axis 1 and the two disturbance sites at Vohibe also separate along this axis. Past studies have demonstrated associations of bacterial community composition with pH, C, N, and/or C:N ratio, depending on the sample type or treatment [6,11,24]. Fierer *et al.* [9] also observed a correlation between bacterial community phylogenetic distances and high soil N amendment levels in temperate agricultural soils. Strong shifts in soil bacterial community composition have been reported with nitrogen amendment to soil samples and year-long incubation under controlled, laboratory conditions [54]. While pH is strongly correlated with bacterial community composition in the majority of soil bacterial community studies, the importance of other soil edaphic properties may vary depending on the soil type or treatment.

Table 4. Bacterial phylotypes at the Class level with differing abundance between Vohibe high impact (V1) and low impact (V2) sites (Metastats $p<0.05$).

RDP Classification	V1 rel. abundance (%)	V2 rel. abundance (%)	Abundance at higher impact plot
Phylum_Class	mean (sd)	mean (sd)	
*Firmicutes*_unclassified	0.14 (0.04)	0.46 (0.10)	lower
Acidobacteria_Gp2	1.1 (0.2)	2.3 (0.3)	lower
Verrucomicrobia_Opitutae	0.09 (0.02)	0.03 (0.02)	higher
Actinobacteria_Actinobacteria	9.8 (1.6)	21.2 (4.5)	lower

Differentially abundant groups between plots at Vohibe

The program Metastats [55] was used to determine which bacterial groups were differentially abundant between the Vohibe plots and therefore contributed to the significant difference in bacterial community composition between these plots. At the OTU level, this analysis detected greater than 200 bacterial groups with significant changes in abundance (data not shown). In order to relate these changes to classified bacterial taxa, the Metastats analysis was performed with OTUs grouped at the class and genus levels of the RDP taxonomy (Table 4 and Table 5). Differences in abundance of bacterial taxa were considered

Table 5. Bacterial phylotypes at the Genus level with differing abundance between Vohibe high impact (V1) and low impact (V2) sites (Metastats $p<0.05$).

RDP Classification	V1 rel. abundance (%)	V2 rel. abundance (%)	Abundance at higher impact plot
Class_Genus	**mean (sd)**	**mean (sd)**	
Actinobacteria_Solirubrobacter	0.00008 (0.00002)	0.00029 (0.00004)	lower
*Deltaproteobacteria*_unclassified	0.0037 (0.0004)	0.0016 (0.0004)	higher
Gammaproteobacteria_Pseudomonas	0.05 (0.01)	0.006 (0.003)	higher
*Actinobacteria*_unclassified	0.0023 (0.0005)	0.0045 (0.0003)	lower
Alphaproteobacteria_Bradyrhizobium	0.010 (0.001)	0.0044 (0.0006)	higher
*Betaproteobacteria*_unclassified	0.061 (0.006)	0.026 (0.009)	higher
*Firmicutes*_unclassified	0.001 (0.003)	0.0045 (0.0009)	lower
*Alphaproteobacteria*_unclassified	0.0039 (0.0006)	0.0014 (0.0004)	higher
Betaproteobacteria_Cupriavidus	0.014 (0.003)	0.003 (0.002)	higher
Acidobacteria Gp2_Gp2	0.011 (0.002)	0.023 (0.003)	lower
*Alphaproteobacteria*_unclassified	0.013 (0.001)	0.021 (0.003)	lower
Actinobacteria_Streptacidiphilus	0.010 (0.005)	0.036 (0.008)	lower
Planctomycetacia_Gemmata	0.00008 (0.00004)	0.0006 (0.0002)	lower
Gammaproteobacteria_Coxiella	0.00014 (0.00005)	0	higher
*Gammaproteobacteria*_unclassified	0.0007 (0.0002)	0.00002 (0.00002)	higher
Actinobacteria_Nocardioides	0.0018 (0.0007)	0.00002 (0.00002)	higher
Caldilineae_Caldilinea	0.00006 (0.00002)	0	higher
Betaproteobacteria_Georgfuchsia	0.00006 (0.00002)	0	higher
Alphaproteobacteria_Sphingomonas	0.012 (0.004)	0.0008 (0.0002)	higher
Sphingobacteria_Ferruginibacter	0.0010 (0.0004)	0.0024 (0.0005)	lower
*Anaerolineae*_unclassified	0.00010 (0.00005)	0	higher
Actinobacteria_Kitasatospora	0.007 (0.003)	0.05 (0.02)	lower
Acidobacteria Gp22_Gp22	0.00008 (0.00004)	0	higher
Alphaproteobacteria_Hyphomicrobium	0.00008 (0.00004)	0	higher
Gammaproteobacteria_Salmonella	0.00008 (0.00004)	0	higher
Planctomycetacia_Planctomyces	0.00016 (0.00005)	0.00004 (0.00002)	higher
*Actinobacteria*_unclassified	0.010 (0.002)	0.018 (0.004)	lower

significant with $p<0.05$. At the class level, the relative abundance of bacterial groups *Acidobacteria* Group 2, unclassified *Firmicutes*, and *Actinobacteria* were all lower in the high impact Vohibe plot as compared to the low impact plot (Table 4). Conversely, the abundance of class *Opitutae* within the phylum *Verrucomicrobia* was higher in the high impact plot compared to the low impact plot samples from Vohibe. Changes in acidobacterial groups with pH have been well documented. For example, Rousk *et al.* [8] also observed a decrease in the relative abundance *Acidobacteria* Gp 2 with increasing pH in temperate agricultural soil. In the same study, no consistent trend in the abundance of *Actinobacteria* was observed from pH 4–8, but within the relevant pH range of our samples (~4–5), Rousk *et al.* did observe a decrease in actinobacterial abundance. At a larger geographic scale, Fierer *et al.* [9] found that the abundance of the phylum *Acidobacteria* decreased with increasing pH, while the relative abundance of *Actinobacteria* increased [56]. However, the increasing abundance of *Actinobacteria* was observed mainly between pH 5–7. *Verrucomicrobia* were not dominant groups in these studies, perhaps because of primer bias, and so differential abundances were not reported.

When the shifts in the bacterial community composition at Vohibe were examined in greater detail at the genus level, we also observed higher abundance of the genus *Bradyrhizobium* and of an unclassified group within the *Oxalobacteraceae* in the high impact plot; both of these clades contain nitrogen-fixing members (Table 5). Interestingly, several bacterial genera known for their roles as human and animal pathogens had significantly higher abundance with disturbance, including *Pseudomonas*, *Coxiella*, and *Salmonella*. This shift might reflect the history of shifting cultivation in the higher impact site at Vohibe. Corroborating the results at the class level, the abundance of several actinobacterial genera, including *Solirubrobacter*, *Streptacidiphilus*, and *Kitasatospora* as well as unclassified *Actinobacteria*, were lower in the high impact plot (Table 5 and Figure S3). Only one actinobacterial genus, the *Nocardioides*, was observed in significantly higher abundance in the higher impact plot.

Members of the phylum *Actinobacteria* are well known for their role in degrading complex organic matter in soils and for their potential as producers of bioactive natural products. The significant decrease in the relative abundance of *Actinobacteria* amplicons in the disturbed plot at Vohibe indicates that sampling across disturbance levels could improve the sampled biodiversity for this natural product discovery target phylum. Similarily, Zabinski & Gannon [22] showed that disturbance from recreational camping decreased the percentage of culturable bacteria with *Actinomycete*-like morphology in subalpine soils. However, the sequencing survey performed in this study measures relative abundance of amplicons and cannot distinguish between true shifts in the numbers of *Actinobacteria* and changes in abundance due to shifts in other groups. The results from our work suggest that disturbance from shifting cultivation at Vohibe could decrease the proportion of *Actinobacteria,* and we plan future experiments to test this hypothesis directly using comparisons of culturable bacterial isolates, direct sequencing, and characterization of the diversity of natural product biosynthetic genes.

Conclusions

Contrary to our initial hypothesis, we found no significant differences in soil bacterial diversity with different disturbance levels, although there was a non-significant trend towards higher richness with disturbance. Comparison of technical replicates from each bulk soil core demonstrated that the observed bacterial community composition within samples from the same homogenized core were reproducible. We detected a significant change in soil bacterial community composition between the low and high disturbance plots in Vohibe forest, but not at Analalava or Pointe à Larrée where there was less difference between disturbance levels in the plots sampled. The Vohibe sites exhibited a large decrease in woody plant morphospecies richness and a decrease in tree cover in the high disturbance plot compared to relatively intact, native forest in the low disturbance plot. The change in bacterial community composition with disturbance at Vohibe was also correlated with an increase in soil pH, which may have resulted from historical burning during shifting cultivation. The study design did not allow us to differentiate between co-varying parameters and so future experiments will be required to distinguish direct and indirect effects. Future work might also look for shifts in microbial functional genes that could be masked by this OTU level taxonomic analysis. Our study focused on the soil bacterial diversity, but changes in botanical diversity and land use would also be expected to influence archaeal and fungal taxa that are known to serve important roles within soil microbial communities. The relative changes in bacterial community composition between sites were much greater than between disturbance levels, suggesting that widespread geographic sampling across forests would maximize access to bacterial diversity rather than collecting multiple samples from different disturbance levels within a given forest.

Supporting Information

Figure S1 Nonmetric multidimensional scaling (NMDS) plot of pairwise Bray-Curtis distances between soil samples calculated with 3% OTUs. Technical replicates (two DNA samples from the same homogenized bulk soil core) are labeled with the same number.

Figure S2 Mean bacterial alpha diversity in each plot as measured by the Inverse Simpson's index of diversity. Errors bars represent the standard error of the mean for five soil samples.

Figure S3 Average relative abundance of actinobacterial genera by plot. Sequences in 3% OTUs were classified and grouped at the genus level. Taxa representing <0.1% of the total sequences were grouped as Other.

Table S1 Chemical data for homogenized soil cores.

Acknowledgments

We thank David Kingston for his leadership as Principal Investigator of the International Cooperative Biodiversity Group "Biodiversity Conservation & Drug Discovery in Madagascar". We would also like to thank the site-based Conservation Facilitators at the three study sites for their help during the fieldwork: Anselme Tilahimena at Analalava, Adolphe Lehavana at Pointe à Larrée, and Fortunat Rakotoarivony at Vohibe. This is IMET contribution No. 13-113 and UMCES contribution No. 4827.

Author Contributions

Conceived and designed the experiments: LCB TM RR FRF WLA CB PPL TMS RTH. Performed the experiments: HA TM RR CB AWS. Analyzed the data: LCB AWS CB WLA. Contributed reagents/materials/analysis tools: AWS TMS. Wrote the paper: LCB AWS CB PPL TMS RTH.

References

1. Keddy P (2005) Putting the plants back into plant ecology: Six pragmatic models for understanding and conserving plant diversity. Annals of Botany 96: 177–189.
2. Griffiths BS, Philippot L (2013) Insights into the resistance and resilience of the soil microbial community. Fems Microbiology Reviews 37: 112–129.
3. Hartmann M, Howes CG, VanInsberghe D, Yu H, Bachar D, et al. (2012) Significant and persistent impact of timber harvesting on soil microbial communities in Northern coniferous forests. ISME J 6: 2199–2218.
4. Daniel R (2004) The soil metagenome—a rich resource for the discovery of novel natural products. Curr Opin Biotechnol 15: 199–204.
5. Lauber CL, Hamady M, Knight R, Fierer N (2009) Pyrosequencing-based assessment of soil pH as a predictor of soil bacterial community structure at the continental scale. Appl Environ Microbiol 75: 5111–5120.
6. Tripathi BM, Kim M, Singh D, Lee-Cruz L, Lai-Hoe A, et al. (2012) Tropical soil bacterial communities in Malaysia: pH dominates in the equatorial tropics too. Microb Ecol 64: 474–484.
7. Griffiths RI, Thomson BC, James P, Bell T, Bailey M, et al. (2011) The bacterial biogeography of British soils. Environ Microbiol 13: 1642–1654.
8. Rousk J, Baath E, Brookes PC, Lauber CL, Lozupone C, et al. (2010) Soil bacterial and fungal communities across a pH gradient in an arable soil. ISME J 4: 1340–1351.
9. Fierer N, Lauber CL, Ramirez KS, Zaneveld J, Bradford MA, et al. (2012) Comparative metagenomic, phylogenetic and physiological analyses of soil microbial communities across nitrogen gradients. ISME J 6: 1007–1017.
10. Ramirez KS, Lauber CL, Knight R, Bradford MA, Fierer N (2010) Consistent effects of nitrogen fertilization on soil bacterial communities in contrasting systems. Ecology 91: 3463–3470; discussion 3503–3414.
11. Nemergut DR, Cleveland CC, Wieder WR, Washenberger CL, Townsend AR (2010) Plot-scale manipulations of organic matter inputs to soils correlate with shifts in microbial community composition in a lowland tropical rain forest. Soil Biology & Biochemistry 42: 2153–2160.
12. Suleiman AK, Manoeli L, Boldo JT, Pereira MG, Roesch LF (2013) Shifts in soil bacterial community after eight years of land-use change. Syst Appl Microbiol 36: 137–144.
13. Rodrigues JL, Pellizari VH, Mueller R, Baek K, Jesus Eda C, et al. (2013) Conversion of the Amazon rainforest to agriculture results in biotic homogenization of soil bacterial communities. Proc Natl Acad Sci U S A 110: 988–993.
14. Buckley DH, Schmidt TM (2003) Diversity and dynamics of microbial communities in soils from agro-ecosystems. Environ Microbiol 5: 441–452.
15. Berthrong ST, Buckley DH, Drinkwater LE (2013) Agricultural management and labile carbon additions affect soil microbial community structure and interact with carbon and nitrogen cycling. Microb Ecol 66: 158–170.
16. Wu T, Chellemi DO, Graham JH, Martin KJ, Rosskopf EN (2008) Comparison of soil bacterial communities under diverse agricultural land management and crop production practices. Microb Ecol 55: 293–310.
17. Mitchell RJ, Hester AJ, Campbell CD, Chapman SJ, Cameron CM, et al. (2010) Is vegetation composition or soil chemistry the best predictor of the soil microbial community? Plant and Soil 333: 417–430.
18. Andrew DR, Fitak RR, Munguia-Vega A, Racolta A, Martinson VG, et al. (2012) Abiotic factors shape microbial diversity in Sonoran Desert soils. Appl Environ Microbiol 78: 7527–7537.
19. Liliensiek AK, Thakuria D, Clipson N (2012) Influences of plant species composition, fertilisation and *Lolium perenne* ingression on soil microbial community structure in three Irish grasslands. Microbial Ecology 63: 509–521.
20. Rasolomampianina R, Bailly X, Fetiarison R, Rabevohitra R, Bena G, et al. (2005) Nitrogen-fixing nodules from rose wood legume trees (*Dalbergia* spp.) endemic to Madagascar host seven different genera belonging to *alpha*- and *beta-Proteobacteria*. Molecular Ecology 14: 4135–4146.
21. Levine UY, Teal TK, Robertson GP, Schmidt TM (2011) Agriculture's impact on microbial diversity and associated fluxes of carbon dioxide and methane. ISME J 5: 1683–1691.
22. Zabinski CA, Gannon JE (1997) Effects of recreational impacts on soil microbial communities. Environ Manage 21: 233–238.
23. Lupwayi NZ, Rice WA, Clayton GW (1998) Soil microbial diversity and community structure under wheat as influenced by tillage and crop rotation. Soil Biology & Biochemistry 30: 1733–1741.
24. Ferrenberg S, O'Neill SP, Knelman JE, Todd B, Duggan S, et al. (2013) Changes in assembly processes in soil bacterial communities following a wildfire disturbance. ISME J 7: 1102–1111.
25. Myers N, Mittermeier RA, Mittermeier CG, da Fonseca GA, Kent J (2000) Biodiversity hotspots for conservation priorities. Nature 403: 853–858.
26. Callmander MW, Phillipson PB, Schatz GE, Andriambololonera S, Rabarimanarivo M, et al. (2011) The endemic and non-endemic vascular flora of Madagascar updated. Plant Ecology and Evolution 144: 121–125.
27. Phillipson PB, Schatz GE, Lowry PP II, Labat JN (2006) A catalogue of the vascular plants of Madagascar. In: Ghanzanfar SA, Beentje HJ, editors. Taxonomy and ecology of african plants: their conservation and sustainable use. Addis Ababa, Ethiopia. 613–627.
28. Moat P, Smith P (2007) Atlas of the vegetation of Madagascar. Kew: Royal Botanic Gardens.
29. Cornet A (1974) Essai de cartographie bioclimatique à Madagascar. Notice Explicative 55. Paris: ORSTOM. 1–28.
30. Du Puy DJ, Moat J, editors (1996) A refined classification of the primary vegetation of Madagascar based on underlying geology: using GIS to map its distribution and assess its status. Paris: ORSTOM. 205–218.
31. Aubert G (1978) Methode d'analyse des sols. Marseille.
32. Bremner JM, Mulvaney CS (1982) Regular Kjeldahl Method. In: Page AL, Miller RH, Keeney DR, editors. Methods of Soil Analysis Part 2: Chemical and Microbiological Properties. 2nd ed. Madison: American Society of Agronomy, Inc. 610–614.
33. Mitreva M, and the Human Microbiome Project Consortium (2012) Structure, function and diversity of the healthy human microbiome. Nature 486: 207–214.
34. Schloss PD, Westcott SL, Ryabin T, Hall JR, Hartmann M, et al. (2009) Introducing mothur: open-source, platform-independent, community-supported software for describing and comparing microbial communities. Appl Environ Microbiol 75: 7537–7541.
35. Schloss PD, Gevers D, Westcott SL (2011) Reducing the effects of PCR amplification and sequencing artifacts on 16S rRNA-based studies. PLoS One 6: e27310.
36. Quince C, Lanzen A, Davenport RJ, Turnbaugh PJ (2011) Removing noise from pyrosequenced amplicons. BMC Bioinformatics 12: 38.
37. Quast C, Pruesse E, Yilmaz P, Gerken J, Schweer T, et al. (2013) The SILVA ribosomal RNA gene database project: improved data processing and web-based tools. Nucleic Acids Res 41: D590–596.
38. Huse SM, Welch DM, Morrison HG, Sogin ML (2010) Ironing out the wrinkles in the rare biosphere through improved OTU clustering. Environ Microbiol 12: 1889–1898.
39. Edgar RC, Haas BJ, Clemente JC, Quince C, Knight R (2011) UCHIME improves sensitivity and speed of chimera detection. Bioinformatics 27: 2194–2200.
40. Cole JR, Wang Q, Cardenas E, Fish J, Chai B, et al. (2009) The Ribosomal Database Project: improved alignments and new tools for rRNA analysis. Nucleic Acids Res 37: D141–145.
41. Wang Q, Garrity GM, Tiedje JM, Cole JR (2007) Naive Bayesian classifier for rapid assignment of rRNA sequences into the new bacterial taxonomy. Appl Environ Microbiol 73: 5261–5267.
42. Team RC (2012) R: A language and environment for statistical computing. Vienna, Austria: R Foundation for Statistical Computing.
43. Warton DI, Wright ST, Wang Y (2012) Distance-based multivariate analyses confound location and dispersion effects. Methods in Ecology and Evolution 3: 89–101.
44. Oksanen J, Blanchet FG, Kindt R, Legendre P, Minchin PR, et al. (2013) vegan: Community Ecology Package. R package version 2.0-7 ed.
45. Wickham H (2009) ggplot2: elegant graphics for data analysis: Springer New York.
46. Naether A, Foesel BU, Naegele V, Wust PK, Weinert J, et al. (2012) Environmental factors affect Acidobacterial communities below the subgroup level in grassland and forest soils. Appl Environ Microbiol 78: 7398–7406.
47. Fierer N, Bradford MA, Jackson RB (2007) Toward an ecological classification of soil bacteria. Ecology 88: 1354–1364.
48. Stopnisek N, Bodenhausen N, Frey B, Fierer N, Eberl L, et al. (2013) Genus-wide acid tolerance accounts for the biogeographical distribution of soil *Burkholderia* populations. Environ Microbiol. doi: 10.1111/1462-2920.12211.
49. Suarez-Moreno ZR, Caballero-Mellado J, Coutinho BG, Mendonca-Previato L, James EK, et al. (2012) Common features of environmental and potentially beneficial plant-associated *Burkholderia*. Microb Ecol 63: 249–266.
50. Pilloni G, Granitsiotis MS, Engel M, Lueders T (2012) Testing the limits of 454 pyrotag sequencing: reproducibility, quantitative assessment and comparison to T-RFLP fingerprinting of aquifer microbes. PLoS One 7: e40467.
51. Reddy BV, Kallifidas D, Kim JH, Charlop-Powers Z, Feng Z, et al. (2012) Natural product biosynthetic gene diversity in geographically distinct soil microbiomes. Appl Environ Microbiol 78: 3744–3752.
52. Philippot L, Spor A, Henault C, Bru D, Bizouard F, et al. (2013) Loss in microbial diversity affects nitrogen cycling in soil. ISME J 7: 1609–1619.
53. Lauber CL, Ramirez KS, Aanderud Z, Lennon J, Fierer N (2013) Temporal variability in soil microbial communities across land-use types. ISME J 7: 1641–1650.
54. Ramirez KS, Craine JM, Fierer N (2012) Consistent effects of nitrogen amendments on soil microbial communities and processes across biomes. Global Change Biology 18: 1918–1927.
55. White JR, Nagarajan N, Pop M (2009) Statistical methods for detecting differentially abundant features in clinical metagenomic samples. PLoS Comput Biol 5: e1000352.
56. Fierer N, Jackson RB (2006) The diversity and biogeography of soil bacterial communities. Proc Natl Acad Sci U S A 103: 626–631.

6

Persistence and Dissipation of Chlorpyrifos in Brassica Chinensis, Lettuce, Celery, Asparagus Lettuce, Eggplant, and Pepper in a Greenhouse

Meng-Xiao Lu[1,2], Wayne W. Jiang[3], Jia-Lei Wang[1], Qiu Jian[4], Yan Shen[2], Xian-Jin Liu[1,2], Xiang-Yang Yu[1,2]*

1 Pesticide Biology and Ecology Research Center, Nanjing, Jiangsu, China, **2** Key Laboratory of Food Safety Monitoring and Management of Ministry of Agriculture, Nanjing, Jiangsu, China, **3** Department of Entomology, Michigan State University, East Lansing, Michigan, United States of America, **4** Institute for the Control of Agrochemicals, Ministry of Agriculture, Beijing, China

Abstract

The residue behavior of chlorpyrifos, which is one of the extensively used insecticides all around the world, in six vegetable crops was assessed under greenhouse conditions. Each of the vegetables was subjected to a foliar treatment with chlorpyrifos. Two analytical methods were developed using gas chromatography equipped with a micro-ECD detector (LOQ = 0.05 mg kg^{-1}) and liquid chromatography with a tandem mass spectrometry (LOQ = 0.01 mg kg^{-1}). The initial foliar deposited concentration of chlorpyrifos (mg kg^{-1}) on the six vegetables followed the increasing order of brassica chinensis<lettuce<celery<asparagus lettuce<eggplant <pepper. The initial deposition of chlorpyrifos showed differences among the six selected vegetable plants, ranging from 16.5±0.9 mg kg^{-1} (brassica chinensis) to 74.0±5.9 mg kg^{-1} (pepper plant). At pre-harvest interval 21 days, the chlorpyrifos residues in edible parts of the crops were <0.01 (eggplant fruit), < 0.01 (pepper fruit), 0.56 (lettuce), 0.97 (brassica chinensis), 1.47 (asparagus lettuce), and 3.50 mg kg^{-1} (celery), respectively. The half-lives of chlorpyrifos were found to be 7.79 (soil), 2.64 (pepper plants), 3.90 (asparagus lettuce), 3.92 (lettuce), 5.81 (brassica chinensis), 3.00 (eggplant plant), and 5.45 days (celery), respectively. The dissipation of chlorpyrifos in soil and the six selected plants was different, indicating that the persistence of chlorpyrifos residues strongly depends upon leaf characteristics of the selected vegetables.

Editor: Youjun Zhang, Institute of Vegetables and Flowers, Chinese Academy of Agricultural Science, China

Funding: This work was financially supported by the Independent Innovation Fund of Agricultural Sciences in Jiangsu Province (cx (12) 3090) and the Natural Science Foundation of China (31071719). The funders had no role in study design, data collection and analysis, decision to publish, or preparation of the manuscript.

Competing Interests: The authors have declared that no competing interests exist.

* Email: yu98190@gmail.com

Introduction

Chlorpyrifos [*O,O*-diethyl *O*-(3,5,6-trichloro-2-pyridyl) phosphorothioate] is an organophosphorous insecticide, acaricide, and nematicide used to control a broad spectrum of foliage and soil-born insect pests on a variety of food and feed crops [1–2]. It is ranked as one of the most extensively used insecticides all around the world. In China, since the use of several highly-toxic organophosphorous insecticides was banned in 2006, chlorpyrifos has been recommended as one of the alternative insecticides and broadly used in agriculture. Extensive use of chlorpyrifos has led to a potential risk of residues in various crops. Chlorpyrifos is of great environmental concerns due to its widespread use in the past several decades and its potential toxic effects on human health. Thus, the degradation study of chlorpyrifos has become increasing important in recent years [3–5]. In a market monitoring study conducted between 2007 and 2010, chlorpyrifos was detected in approximately 22.8% of 2082 samples of 17 vegetable commodities collected from Zhejiang Province, China with a highest residue of 3.47 mg kg-1. The residue levels in 1.4% of vegetable samples were found to be higher than the maximum residue limits (MRLs) of China [3].

Although most pesticides are effective to control pests in agricultural industry, inappropriate uses of pesticides may lead to public concerns on food safety and human health [5–10], environmental contamination [11–12], insect resistance and resurgence [12–14], etc. In China, the use of agrochemicals is critical to provide food supplies for its growing population with its limited arable land. Ideally, control the harmful organism efficiently, having no or minimum pesticide residues in the harvested crops, or at least lower than the statutory MRLs [3,15]. Therefore, the field dissipation studies of pesticide persistence in foods and pesticide residue behavior is of particular importance in order to find out which pesticide application strategies are efficient to control insect pests while leaving minimum residues [16,17]. There are many factors that influence the dissipation behavior of pesticides in plants, including the climate conditions (temperature, humidity, light intensity, etc) [18], the crop species [19–22] the nature of the chemicals, the formulations, and the application methods [23–24]. Due to the difference in the extension of the foliar, which resulted in the different initial pesticide deposition, and/or the difference in the metabolism system of different crops, dissipation of a pesticide on various crops may be markedly different [25–26]. A dissipation study of pesticides on leafy vegetables showed that among the tested vegetables, spinach and amaranth could incur higher pesticide deposition and that half-lives ($t_{1/2}$ = 1.37–5.17 days) of chlorpyrifos were from on different leafy vegetables [22]. For chiral pesticide malathion, the calculated $t_{1/2}$ values of the enantiomers were relatively short, ranging from 0.83 to 1.43 days in five plants. The degradation of the two enantiomers in Chinese cabbage (brassica chinesis), rape, and sugar beet was highly selective, while non-enantioselectivity was found in paddy rice and wheat [20–21]. For better understanding the possible residue risk of a pesticide, dissipation studies for different crop species in the specific growing conditions are necessary to test if the established application strategies are suitable.

It is concluded that agrochemicals enter into plants via two major pathways, which are either via foliar treatment - foliar deposition followed by entering into the inner parts of the crops, or via soil treatment - root uptake from the soil [27–28]. For the foliar pesticide applications, the agrochemicals are deposited directly on foliar surfaces of the crops and the excess agrochemicals will precipitate in the soil. The resulting deposition fractions are determined predominantly by crop species, growth stage of crop, pesticide formulation, and spraying technology [29]. Generally, optimization of chlorpyrifos uses is foliar application with minimizing losses of the applied pesticides from plants to soil. Spray loss of chlorpyrifos may lead to the soil environment pollution and extend its residue duration in plants. The chlorpyrifos residues were found to persist in soils, as the half-lives ranged generally between 50 and 120 days [30–32]. It was reported in the literature that the chlorpyrifos residues were found in soils for over one year after the applications. Pesticide persistence in soils may depend on the formulation, rate of application, soil type, climate, and other conditions [33–37]. Plant rhizosphere plays an important role in the degradation of pesticides in soils [21,36].

In the present study, six vegetables were selected, including two fruit vegetables (pepper and eggplant) and 4 whole plant edibles leafy vegetables (brassica chinensis, lettuce, celery, and asparagus lettuce). They were treated with chlorpyrifos by foliar application under the controlled conditions in a greenhouse. The initial deposition of chlorpyrifos on the crop foliar was analyzed by GC and on the fruits by LC/MS/MS and the dynamics of pesticide residues in the plants and rhizosphere soils were to be monitored. The work was to evaluate the persistence and dissipation behavior of chlorpyrifos in different vegetables and soil. The results would help to provide understanding of the residue characteristics of chlorpyrifos in vegetables, and help to guide proper and safe use of pesticides on vegetables to ensure food safety.

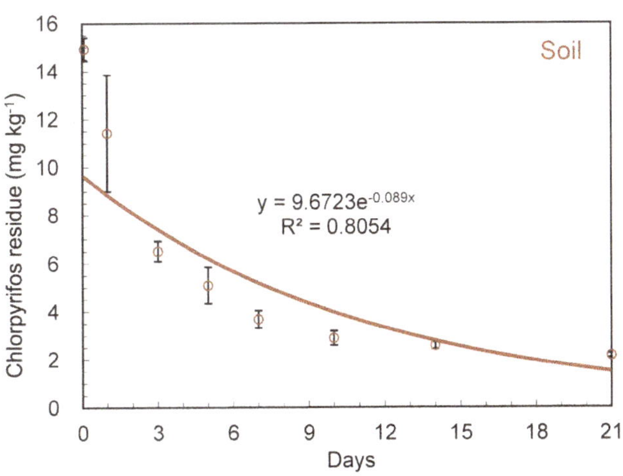

Figure 1. Dissipation dynamic of chlorpyrifos in soil.

Materials and Methods

Instruments and Reagents

Gas chromatography was an Agilent GC 7890 (Agilent Technologies, Santa Clara, CA, USA) equipped with a micro-ECD detector and an analytical column HP-5ms J&W Ultra Inert capillary column (30 m length×0.25 mm I.D.×0.25 µm film thickness, Agilent Technologies, USA). LC/MS/MS (Agilent Technologies, USA) contained a 1200 SL HPLC system coupled to an Agilent G6410A triple quadrupole mass spectrometer. The column was an Agilent ZORBAX SB-C18 (2.1×150 mm, 5 µm) analytical column. Chlorpyrifos-EC 40% (Hubei Xian Long Chemical Co., LTD) was purchased from a local pesticide store. The analytical reference substance, chlorpyrifos (certified analytical standard, 99.7%) was purchased from National Standard Company (Tianjin, China). Two stock solutions (1000 mg L^{-1}) were prepared by dissolving the chlorpyrifos standard (100 mg) in 100 mL of acetone (for GC) and acetonitrile (for LC/MS/MS), respectively. Working solutions were prepared by diluting the stock solution or a working solution using the organic solvents (acetone for GC and acetonitrile for LC/MS/MS). Acetone, n-hexane, acetonitrile, sodium chloride, and anhydrous sodium sulfate were of analytical grades, purchased from Kermel Chemical Reagent Co., Ltd (Tianjin, China). The 0.22 µm SCAA-104 membranes and 500 mg florisil SPE cartridges were purchased from Anpel Scientific Instrument Co., Ltd (Shanghai, China).

The six selected vegetables were brassica chinensis (*Brassicachinensis L.*), lettuce (*Lactuca sativa* spp), pepper (*Capsicum annuum* spp), eggplant (*Solanum melongena* L), celery (*Apium graveolens*), and asparagus lettuce (*Asparagus Lettuce* spp). The pepper and eggplant seeds were grown in a nursery tray before being transplanted to the field at the 2–3 leaf stage seedlings, and the other four vegetables were directly sowed in the field. The row spacing and inter-plant spacing were set to be 30 cm for the vegetables, except for celery which was sowed in rows with row spacing of 30 cm.

Experiment design

Field experiments were conducted in a controlled environment in a greenhouse at the Experiment Station of Jiangsu Academy of Agricultural Science (JAAS), Nanjing, China from September to November 2013. JAAS permitted the study in its greenhouse and this study did not use protected area of land or sea, neither with relevant protected wildlife. Since the work was completed in the greenhouse, no specific permission was required. This study did not involve endangered or protected species. The soil was of sandy loam texture dried and its contents contained 30% of sand, 53% silt, 15% clay, and 2% organic matter. There were four 30 m^2 trial plots to be selected for each vegetable, i.e., three replicates and one control. Between the plots there was a buffer strip of 0.5 m in width. Chlorpyrifos-EC 40% was applied by foliar spraying at a rate of 0.97 kg a.i./ha on October 21, 2013. At the time of spraying, brassica chinensis, lettuce, and asparagus lettuce were at the stage when the leaves overspread; celery was 15–16 cm of height and the leaves scattered; pepper and eggplant were at the stages of 30–40 cm of height and the leaves of the adjacent plants touched each other and overlapped. In this study, the plants were

Table 1. Dynamic equations, correlation coefficients and half-lives of chlorpyrifos in soil and six vegetable plants (foliar).

Matrix	Dynamic equation	Correlation coefficient (R^2)	Half-life (days)
Soil	$C_t = 9.672\ e^{-0.089t}$	0.8054	7.79
Brassica chinensis	$C_t = 11.74\ e^{-0.1192t}$	0.9618	5.81
Lettuce	$C_t = 17.30\ e^{-0.1769t}$	0.9359	3.92
Celery	$C_t = 40.62\ e^{-0.1271t}$	0.9648	5.45
Asparagus lettuce	$C_t = 31.97\ e^{-0.1775t}$	0.8850	3.90
Pepper	$C_t = 40.23\ e^{-0.2622t}$	0.9315	2.64
Eggplant	$C_t = 48.28\ e^{-0.2307t}$	0.9756	3.00

allowed to the leaves to overlap to cover most of the surface of soil and thus the loss of pesticides sprayed was minimum. The temperature inside the greenhouse was controlled between 16–26°C. Three representative samples of whole plant and rhizosphere soil were collected at 0 (2 h after application), 1, 3, 5, 7, 10, 14, and 21 days intervals after pesticides application. Pepper and eggplant fruit samples were collected on 0, 3, 7, 14, and 21 days.

Extraction and purification

All samples (plants and fruits) were homogenized using a Philips blender (Shanghai, China) and the ground samples were stored in a freezer (at −20°C) until analysis.

Plant samples. The extraction of chlorpyrifos residues from plants and fruits was carried out by the procedure as follows. Five (5.0) g of the homogenized sample was weighed into a 50 mL Teflon centrifuge tube. The extraction solvent (acetonitrile, 10 mL) was added. The samples were then mixed thoroughly for 1 min with a vortex mixer, followed by high-speed homogenizing for about 2 min. After addition of 2 g of sodium chloride, the samples were vortexed immediately for 1 min and centrifuged for 5 min at 5000 rpm. An aliquot of 1 mL of the supernatant was transferred into a 10 mL glass test tube, and then evaporated just to dryness under a stream of nitrogen (40°C). The residue was dissolved in 1 mL of hexane and then subjected to Florisil SPE column clean-up. The SPE column was pre-conditioned by rinsing it with 5 mL of hexane. The extraction was added to the SPE column followed by eluting with 10 mL of a mixture containing acetone and n-hexane (9:1, v/v). The eluate was evaporated to dryness. The residues were redissolved with acetone to 1 mL. The final extract was filtered through a 0.2 μm SCAA-104 membrane followed by GC analysis.

Fruit samples. The method of sample extraction was a modification of a reference method [38]. Fruit sample (5.0 g) was weighted into a polypropylene centrifuge tube. An extraction solvent (10 mL of acetonitrile) was added. The sample was homogenized for 1 min using the homogenizer. The homogenizer probe was rinsed with a portion of 5 mL of the extraction solvent. The extracts were combined. The sample was centrifuged at 5000 rpm for 5 min. Transferred 2 mL of supernatant into a 15 mL centrifuge tube containing PSA (100 mg), ODS-C18 (100 mg) and florisil (100 mg). The sample was vortexed for 1 min and centrifuged at 5000 rpm for 2 min. After centrifugation, the supernatant was filtered using a 0.22 μm nylon filter into an autosampler vial for LC/MS/MS analysis.

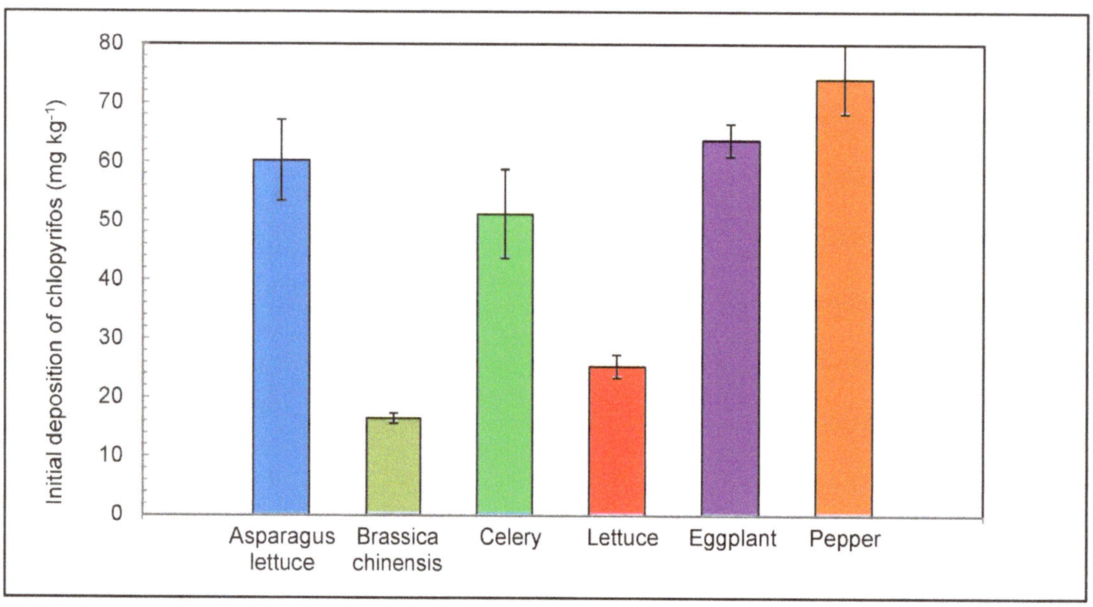

Figure 2. Initial depositions of chlorpyrifos on the six vegetable plants.

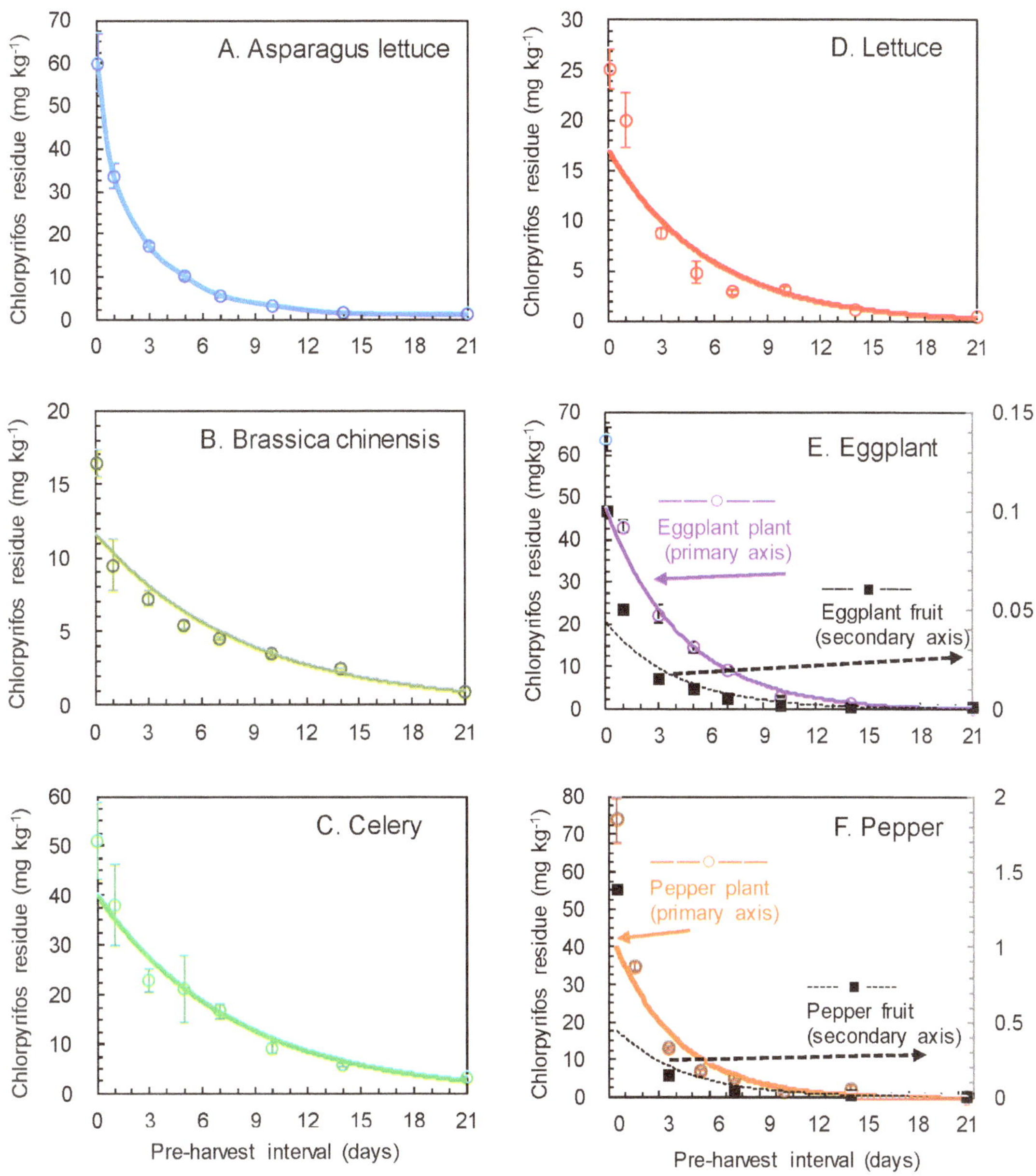

Figure 3. Dissipation dynamic of chlorpyrifos in vegetables and fruits. A. Asparagus lettuce, B. Brassica chinensis, C. Celery, D. Lettuce, E. Eggplant (eggplant plant - solid curve with marker ○ and primary axis; eggplant fruit – dashed curve with marker ■ and secondary axis), and F. Pepper (pepper plant - solid curve with marker ○ and primary axis; pepper fruit – dashed curve with marker ■ and secondary axis).

Soil. Five (5.0) g of soil was weighed into a 50 mL Teflon centrifuge tube and 5 g of sodium chloride was added. The contents were thoroughly mixed. Then, 20 mL of acetonitrile was added. The mixture was vortexed for 1 min, ultrasonically extracted at 40–45°C for 30 min, shaken on a rotary shaker for 2 h, and centrifuged at 5000 rpm for 5 min. Then 1 mL of the supernatant was transferred into a centrifuge tube and dried under a stream of nitrogen (40°C). The chlorpyrifos residues were redissolved in 1 mL of acetone followed by GC analysis.

GC and LC/MS/MS instrument analyses

GC. The conditions for the analysis were: detector temperature, 280°C; injector temperature 270°C; oven temperature program starting at 120°C, 3.67 min at 120–230°C (ramp 30°C min^{-1}), 5 min at 230°C, 2 min at 230–270°C (ramp 20°C min^{-1}), 2 min at 270°C; carrier gas, N_2 at 1 mL/min; injection volume 1.0 μL, in a splitless mode. A linear calibration curve was used and the calibration range was 0.01–5 mg kg^{-1}. Under these conditions chlorpyrifos retention times were approximately 7.52 min. The software was Agilent ChemStation Rev. B04.03 software for instrument control, data acquisition and processing.

LC/MS/MS. The instrument analysis method was an adaption of reference methods [38]. The HPLC conditions for the analysis were: mobile phase A: water containing 0.1% of formic acid (v/v); mobile phase B: acetonitrile containing 0.1% of formic acid (v/v); flow rate 0.4 mL min^{-1}; injection volume 10 μL; mobile phase gradients of binary pump: 10% B (0–1.50 min), 95% B (1.51–4.50 min), 95% B (4.51–6.00 min), and

Table 2. Comparison of chlorpyrifos residues in edible parts of brassica chinensis, lettuce, celery, asparagus lettuce, eggplant and pepper with Maximum Residue Limits (MRLs).*[1]

Commodity	Crop group*[2]	Residue (mg kg⁻¹)		MRL (mg kg⁻¹)			
		PHI 7 days	PHI 21 days	CAC*[3]	China*[4]	USA*[5]	EU*[6]
Edible plants:							
Brassica chinensis	5B	4.55	0.97	1.0	0.1	1.0	0.5
Lettuce	4A	3.07	0.56	*[7]	0.1	*[7]	0.05
Celery	4B	16.9	3.50	*[7]	0.05	*[7]	0.05
Asparagus Lettuce	4B	5.96	1.47	*[7]	0.1	*[7]	*[7]
Edible fruits:							
Eggplant (fruit)	8–10B 8–10C	0.01	<0.01	*[7]	*[7]	*[7]	0.5
Pepper (fruit)	8–10B	0.03	<0.01	2.0 (sweet pepper)	*[7]	1.0	1.0

*[1]Please note that this work intended to study pesticide persistence instead of for MRL establishment. The comparison described in above the table is used to study the persistence and dissipation of chlorpyrifos in the selected crops. The high residues above were due to the high application rate and different formulations.

*[2]Code of Federal Regulations Title 40 Part 180.41 Crop group table (40 CFR 180.41). Group 4A/4B Leafy vegetables (except brassica vegetables), 5B Brassica leafy vegetable, 8–10B 8–10C Fruiting vegetable group.

*[3]CODEX Alimentarius: List of standards: http://www.codexalimentarius.org/standards/list-of-standards/en/?provide = standards&orderField = fullReference&sort = asc&num1 = CAC/MRL.

*[4]Chinese National Standards (GB2763-2012): Maximum Residue Limits for Pesticides in Foods.

*[5]The United States Tolerances and Exemptions for Pesticide Chemical Residues in Food: http://www.ecfr.gov/cgi-bin/text-idx?SID = e33dfa87fab3f5ddecad25dafa7028ea&node = 40:25.0.1.1.27.3.19.113&rgn = div8.

*[6]Pesticide EU-MRLs (Regulation (EC) No 396/2005, MRLs updated on 28/01/2014) http://ec.europa.eu/sanco_pesticides/public/index.cfm?event = substance.resultat&s = 1.

*[7]MRLs not currently established or registrations canceled.

10% B (6.01–7.00 min). The MS/MS conditions were: gas temperature 350°C; gas flow 10 L/min; nebulizer pressure 45 psi; and capillary voltage 4000 V. In order to achieve the highest sensitivity, the fragment, voltage and the collision energy were optimized. MRM (chlorpyrifos) transitions were 350.1>198 (quantitation, collision energy, CE, 20 V) and 350.1>97 (identification, CE 30 V). The retention time was 3.89 min. The software was Agilent MassHunter software for instrument control, data acquisition and processing.

Data analysis

The degradation rate constant and half-life were calculated using a first-order rate equation:

$$C_t = C_0 e^{-kt}$$

where C_t and C_0 represent the concentrations of the chlorpyrifos residues at the day t and day 0 (2 h), respectively, and k is the degradation rate constant. The half-life ($t_{1/2}$) is defined as the time required for the pesticide residue level to fall to the half of the initial residue level of day 0 (i.e., C_0) and was calculated using the following equation:

$$t_{1/2} = (ln\, 2)/k$$

Results and Discussion

Method validation

GC method. Quantification was accomplished by using the standard curve constructed by plotting analyte concentrations against peak areas. Good linearity was achieved with the chlorpyrifos concentration and the correlation coefficient was 0.9995. Recoveries of chlorpyrifos at different fortification levels, i.e., 0.05, 1, and 10 mg kg^{-1}, were determined in three replicates for validation of the method. The recoveries of chlorpyrifos in the soil were 72.5%–89.6% with the relative standard deviation (RSD) 2.1%–7.2% and $R^2 = 0.9846$. For the six vegetable plants, the recoveries were from 79.3%–97.0% with RSD 3.0%–15% and $R^2 = 0.9910$. The limit of quantification (LOQ) was defined as the lowest fortification concentration whose signal-to-noise (S/N) ratio was equal to or greater than 10 and thus the LOQ of the GC analysis was 0.05 mg kg^{-1}. The sample extracts in which the chlorpyrifos residues were greater than 10 mg kg^{-1} were diluted 10 times prior to the evaporation/cleanup steps. And the diluted samples were re-analyzed to ensure the residues were in the acceptable ranges.

LC/MS/MS method. Three fortification levels, i.e., 0.01, 0.1, and 2 mg kg^{-1}, were analyzed. For pepper fruits, the average recoveries of chlorpyrifos residue in the soil were 87.1%–106% with the relative standard deviation (RSD) from 1.1% to 7.4% and $R^2 = 0.9991$. For eggplant fruits, the average recoveries were from 86.1% to 97.0%, with RSD 2.9%–5.9% and R2 = 0.9978. The LOQ of the LC/MS/MS analysis was set to be 0.01 mg kg^{-1}.

Soil and whole plant samples of the six selected crops were analyzed using GC (LOQ = 0.05 mg kg^{-1}). The fruit samples of pepper and eggplant were analyzed by LC/MS/MS (LOQ = 0.01 mg kg^{-1}). These methods were capable of conducting the analyses in this study.

Degradation of chlorpyrifos

Soil. The rhizosphere soil samples were collected at different intervals after chlorpyrifos was applied. The chlorpyrifos residues in soil were analyzed by GC. The concentration of chlorpyrifos in the soil decreased over time (Figure 1). The average initial deposition of chlopyrifos was 14.9±0.5 mg kg^{-1} (i.e., 2 h, day 0) and the final residue was 2.2±0.1 mg kg^{-1} on day 21. The first-order kinetic equation of chlorpyrifos dissipation is $C_t = 4.84e^{-0.089t}$ (Table 1) with correlation coefficient $R^2 = 0.8054$ and the half-life $t_{1/2} = 7.79$ days. The half-lives of chlorpyrifos in soil were in the range of 3–7 days reported by Singh et al. [30]. Singh et al. observed that chlorpyrifos persisted in a low pH soil, i.e., less than 3% of the pesticide had degraded after 10 days and more than 50% of chlorpyrifos was dissipated at a higher pH soil (pH 8.5). Chai et al. reported that the half-lives in humid tropical soils from Malaysia were typically 7–120 days [32]. However, Chai, et. al. also reported that some half-lives were 257 days in the soils containing less soil microbial populations [32]. In the literature, it was reported long environmental dissipation half-lives of chlorpyrifos, i.e., up to 4 years, depending on application rate, ecosystem, and pertinent environments [33]. Since chlorpyrifos presented low water solubility and a higher log K_{ow}, it had a strong tendency to sorb to organic matter and soil. Stability and effectiveness had made chlorpyrifos one of the most popular pesticides worldwide but on the other side its persistence had raised environmental concerns [34].

Initial Deposition. After foliar application, the whole plants of the six selected crops were collected. The chlorpyrifos residues in plants were analyzed by GC. The initial depositions of chlorpyrifos in the six selected plants are compared in Figure 2. As can be seen in Figure 2, the initial depositions (2 h, day 0) on the six plants were in an increasing order: 16.5±0.87 mg kg^{-1} (brassica chinensis), 25.3±2.0 mg kg^{-1} (lettuce), 51.2±7.6 mg kg^{-1} (celery), 60.3±6.8 mg kg^{-1} (asparagus lettuce) < 63.7±2.8 mg kg^{-1} (eggplant) <74.0±5.9 mg kg^{-1} (pepper), respectively.

It is presumed that the initially deposited chlorpyrifos amount mainly depended upon the surface area of the foliar which the pesticide was sprayed on in spite of the leaf characteristics of the plants, such as leaf roughness, content of cuticular waxes, etc. which were assumed to contribute little to the initial depositions. Therefore, the concentration of the initial deposition is directly proportional to the foliar area and inversely proportional to the biomass of the whole plant. The foliages of pepper plant, eggplant plant, celery, and asparagus lettuce were overlapping to maximize the effective foliar surface area. The shorter plants (lettuce and brassica chinensis) had an area of uncovered soils between rows. As a result, lettuce and brassica chinensis had the lowest initial depositions. Since it was lightest in weight, the pepper plant had over all the largest initial deposition of chlorpyrifos.

Dissipation. The dynamic equations of chlorpyrifos degradation are given in Table 1. The curves of dissipation in the six plants and two fruits are described in Figure 3. As can be seen in Figure 3, the dynamics curves demonstrated that the chlorpyrifos residues dissipated significantly in the first a few days and persisted in the crops for extended period of time. For example, at preharvest interval (PHI) 21 days, the chlorpyrifos residues in the six plants decreased to 1.47±0.22 mg kg^{-1} ((A) asparagus lettuce), 0.97±0.03 mg kg^{-1} ((B) brassica chinensis), 3.50±0.27 mg kg^{-1} ((C) celery), 0.56±0.06 mg kg^{-1} ((D) lettuce), 0.53±0.06 mg kg^{-1} ((E) eggplant), and 0.15±0.01 mg kg^{-1} ((F) pepper), respectively. As can be seen in Table 1, the half-lives (from low to high) were found to be: 0.91 days (pepper plants) <3.92 days (lettuce) <3.92

days (asparagus lettuce) <5.82 days (brassica chinensis) <3.00 days (eggplant plants) <5.46 days (celery).

The possible mechanisms are thought to be due to the difference in the activities of pesticide degradation enzymes and/or pesticide degradation endophytes among these plants [25]. It is interesting to observe that the half-lives of the crop plants seemed to be a reverse order of the initial deposition. Pepper plants had the highest initial deposition but shortest half-live (0.92 day) while brassica chinensis had the lowest initial deposition but longest half-live (5.82 days). The difference in the calculated half-life values indicated that different vegetables had different degradation rates. One of the key factors could be photodegradation of chlorpyrifos [37]. The pepper plant had a greater effective foliar area which led to the highest pesticide position, and exposed to the ultraviolet lights. Also, after application, the leaf characteristics may affect how the pesticide would be retained on the surfaces of the leaves and then be penetrating into the plant tissues such as leaf surface roughness [39–40] and the content of water repellent cuticular waxes [41–44].

Edible parts of the vegetables. The edible parts are edible leaves and stems of brassica chinensis, lettuce, celery, and asparagus, and fruits of eggplant and pepper. The residue in pepper and eggplant fruits were analyzed by LC/MS/MS. The comparison of the residue data are given in Table 2. A number of existing Maximum Residue Limits (MRLs) and crop grouping are also included in Table 2. The higher residues of chlorpyrifos in the plants were due to high initial depositions. It was observed that the residues of foliages were significantly higher than those in the edible fruits (eggplant and pepper fruits), i.e., pepper had 0.03 mg kg^{-1} and <0.01 mg kg^{-1} of chlorpyrifos residues and eggplant had <0.10 mg kg^{-1} at PHI 7 days and PHI 21 days, respectively (Table 2). In Figure 3 (E and F), The half-lives were calculated to be 2.84 days (pepper fruits) and 3.15 days (eggplant fruits), respectively. It was observed that same residues exceeded the MRLs. However, it should be noted that the purpose of this work was to study the residues of chlorpyrifos change in the crops. Therefore, higher application rates were used in order to monitor such changes.

Because of chlorpyrifos' high hydrophobicity (high K_{ow} value), the pesticide would readily enter into the inter parts from the surfaces resulting in high residue levels. A portion of the residues may be transferred from leaves to the growing fruits. For eggplant and pepper fruits, the difference in chlorpyrifos residues mainly depended upon the composition of the surface waxes from pepper and eggplant [41,44]. Bauer et. al. [44] reported that the bell pepper contained 39% of fraction 1 (mainly C20–C35 of alkanes and aldehydes) and 61% of fraction 2 (15 various triterpenes) while the eggplant cultivars had 77% of fraction 1 and 23% of fraction 2. Fraction 2 consisted of 15 triterpenes, including α- and β-amyrin, lupeol, glutinol, 3β-friedelanol, friedelin, taraxerol, taraxasterol, δ-amyrin, germanicol, multi-florenol, ω-taraxasterol, isomultiflor-enol, isobauerenol and bauerenol, as well as n-alkanoic acids 2-hydroxy-alkanoic acids [44]. These chemicals would enhance the dissipation of chlorpyrifos.

The comparisons of the chlorpyrifos residue data listed in Table 2 and presented in Figure 3 indicated that chlorpyrifos was relatively stable and persisted in the crops, especially leafy crops. The rate of degradation of pesticide residue is affected by environmental conditions, nature of the pesticide, application rate, formulation, and plant species, etc. [45]. The vegetables selected in this study are minor crops representing a range of various species and it was intended to promote the process. For the crop grouping, celery may be a good reprehensive species of leafy vegetable (group 4B, Table 2) with consideration of initial deposition and dynamic data.

Conclusions and Implications

The study investigated the residue behavior of chlorpyrifos in six vegetables in the greenhouse. The results of chlorpyrifos of initial depositions on different vegetables detected after pesticide application showed differences among the six selected crops. The half-lives of chlorpyrifos in the six vegetables were different indicating that different vegetables had different capacities for metabolizing chlorpyrifos.

Author Contributions

Conceived and designed the experiments: XYY MXL XJL. Performed the experiments: MXL. Analyzed the data: MXL WWJ QJ XYY. Contributed reagents/materials/analysis tools: MXL JLW YS. Contributed to the writing of the manuscript: WWJ MXL.

References

1. FAO (2000) Pesticide Residues in Food, 2000: Report of the Joint Meeting of the FAO Panel of Experts on Pesticide Residues in Food and the Environment and the WHO Core Assessment Group on Pesticide Residues, Geneva, Switzerland, 20–29 September 2000. 45–59 p.
2. Lemus R, Abdelghani A (2000) Chlorpyrifos: an unwelcome pesticide in our homes. Rev Env Health 15: 421–433.
3. Yuan Y, Chen C, Zheng C, Wang X, Yang G, et al. (2014) Residue of chlorpyrifos and cypermethrin in vegetables and probabilistic exposure assessment for consumers in Zhejiang Province, China. Food Control 36: 63–68.
4. Gao Y, Chen S, Hu M, Hu Q, Luo J, et al. (2012) Purification and Characterization of a Novel Chlorpyrifos Hydrolase from *Cladosporium cladosporioides* Hu-01. PLoS ONE 7(6): e38137.
5. Wentzell J, Cassar M, Kretzschmar D (2014) Organophosphate-Induced changes in the PKA regulatory function of Swiss cheese/NTE lead to behavioral deficits and neurodegeneration. PLoS ONE 9(2): e87526.
6. Rauh VA, Perera FP, Horton MK, Whyatt RM, Bansal R, et al. (2012) Brain anomalies in children exposed prenatally to a common organophosphate pesticide. PNAS 109: 7871–7876.
7. Janssens L, Stoks R (2013) Fitness Effects of Chlorpyrifos in the Damselfly *Enallagma cyathigerum* Strongly Depend upon Temperature and Food Level and Can Bridge Metamorphosis. PLoS ONE 8(6): e68107.
8. Canesi L, Negri A, Barmo C, Banni M, Gallo G, et al. (2011) The Organophosphate Chlorpyrifos Interferes with the Responses to 17β-Estradiol in the Digestive Gland of the Marine Mussel *Mytilus galloprovincialis*. PLoS ONE 6(5): e19803.
9. Sasikala C, Jiwal S, Rout P, Ramya M (2012) Biodegradation of chlorpyrifos by bacterial consortium isolated from agriculture soil. World J Microbio Biotech, 28(3), p1301.
10. Trunnelle KJ, Bennett DH, Tulve NS, Clifton MS, Davis MD, et al. (2014) Urinary pyrethroid and chlorpyrifos metabolite concentrations in northern California families and their relationship to indoor residential insecticide levels, Part of the study of use of products and exposure related behavior (SUPERB). Environ. Sci. Technol., 48 (3): 1931–1939.
11. Watts M (2012) Chlorpyrifos as a Possible Global Persistent Organic Pollutant. Pesticide Network North America, Oakland, CA, USA. Available: http://www.ipen.org/cop6/wp-content/uploads/2013/04/Chlorpyrifos_as_POP_final.pdf. Accessed on March 17, 2014.
12. Popp J, Peto K, Nagy J (2013) Pesticide productivity and food security. A review. Agron Sustain Dev 33: 243–255.
13. Zhang NN, Liu CF, Yang F, Dong SL, Han ZJ (2012) Resistance mechanisms to chlorpyrifos and F392W mutation frequencies in the acetylcholine esterase ace1 allele of field populations of the tobacco whitefly, Bemisia tabaci in China. J Insect Sci, Vol 12 Article 41.
14. Ouyang Y, Chueca P, Scott SJ, Montez GH, Grafton-Cardwell EE (2010) Chlorpyrifos Bioassay and Resistance Monitoring of San Joaquin Valley California Citricola Scale Populations. J Econ Ent 103(4): 1400–1404.
15. Mouron P, Heijne B, Naef A, Strassemeyer J, Hayer F, et al. (2012) Sustainability assessment of crop protection systems: SustainOS methodology and its application for apple orchards. Agri Syst 113: 1–15.
16. MacLachlan J, Hamilton D (2010) Estimation methods for maximum residue limits for pesticides. Reg Toxicol Pharmacol, 58: 208–218.

17. Malhat F, Kamel E, Saber A, Hassan E, Youssef A, et al. (2013) Residues and dissipation of kresoxim methyl in apple under field condition. Food Chem 140: 371–374.
18. Garau VL, Angioni A, Aguilera Del Real A, Russo MT, Cabras P (2002) Disappearance of azoxystrobin, cyprodinil, and fludioxonil on tomato in a greenhouse. J Agri Food Chem 50: 1929–1932.
19. Cabras P, Meloni M, Manca MR, Pirisi FM, Cabitza F, et al. (1988) Pesticide residues in lettuce. 1. Influence of the cultivar. J Agri Food Chem: 36: 92–95.
20. Wang M, Zhang Q, Cong L, Yin W, Wang M (2014) Enantioselective degradation of metalaxyl in cucumber, cabbage, spinach and pakchoi. Chemosphere 95: 241–256.
21. Sun H, Xu J, Yang S, Liu G, Dai S (2004) Plant uptake of aldicarb from contaminated soil and its enhanced degradation in the rhizosphere. Chemosphere. 54: 569–574.
22. Fan S, Zhang F, Deng K, Yu C, Liu S, et al. (2013) Spinach or amaranth contains highest residue of metalaxyl, fluazifop-p-butyl, chlorpyrifos, and lambda-cyhalothrin on six leaf vegetables upon open field application. J Agri Food Chem 61: 2039–2044.
23. Montemurro N, Grieco F, Lacertosa G, Visconti A (2002) Chlorpyrifos decline curves and residue levels from different commercial formulations applied to oranges. J Agri Food Chem 50: 5975–5980.
24. Cabras P, Meloni M, Gennari M, Cabitza F, ubeddu M (1989) Pesticide residues in lettuce. 2. Influence of formulations. J Agri Food Chem 37: 1405–1407.
25. Xia XJ, Zhang Y, Wu JX, Wang JT, Zhou YH, et al. (2009) Brassinosteroids promote metabolism of pesticides in cucumber. J Agri Food Chem 57: 8406–8413.
26. Itoiz ES, Fantke P, Juraske R, Kounina A, Vallejo AA (2012) Deposition and residues of zaoxystrobin and imidacloprid on greenhouse lettuce with impliacations of human consumption. Chemosphere 89: 1034–1041.
27. Collins C, Fryer M, Grosso A (2006) Plant uptake of non-ionic organic chemicals. Environ Sci Tech 40: 45–52.
28. Juraske R, Castells F, Vijay A, Muñoz P, Antón A (2009) Uptake and persistence of pesticides in plants: Measurements and model estimates for imidacloprid after foliar and soil application. J Hazard Mater 165: 683–689.
29. Hauschild M (2000) Estimating pesticide emissions for LCA of agricultural products. In: Weidema BP, Meeusen MJG. (Eds.), Agricultural Data for Life Cycle Assessments. Agricultural Economics Research Institute, The Hague, 64–79 p.
30. Singh BK, Walker A, Wright DJ (2006) Bioremedial potential of fenamiphos and chlorpyrifos degrading isolates: influence of different environmental conditions. Soil Biol Biochem 38: 682–93.
31. Chen S, Liu C, Peng C, Liu H, Hu M, et al. (2012) Biodegradation of Chlorpyrifos and Its Hydrolysis Product 3,5,6-Trichloro-2-Pyridinol by a New Fungal Strain *Cladosporium cladosporioides* Hu-01. PLoS ONE 7(10): e47205.
32. Chai LK, Wong MH, Hansen HCB (2013) Degradation of chlorpyrifos in humid tropical soils. J Environ Manage 125: 28–32.
33. Gebremariam SY, Beutel MW, Yonge DR, Flury M, Harsh JB (2012) Adsorption and Desorption of Chlorpyrifos to Soils and Sediments. Rev Environ Contam Tox 215: 123–175.
34. Kamrin MA (1997) Pesticide profiles toxicity, environmental impact, and fate. Lewis publishers: Boca Raton, FL 147–152 p.
35. US EPA (1999) Reregistration eligibility science chapter for chlorpyrifos fate and environmental risk assessment chapter. US EPA, office of prevention, pesticides and toxic substances, office of pesticide programs, environmental fate and effects division, US government printing office: Washington, DC.
36. Fang C, Radosevich M, Fuhrmann JJ (2001) Atrazine and phenanthrene degradation in grass rhizosphere soil. Soil Biol Biochem 33: 671–678.
37. Nieto LM, Hodaifa G, Vives SR, Casares JAG, Casanov MS (2009) Photodegradation of phytosanitary molecules present in virgin olive oil. J Photoch Photobio A 203: 1–6.
38. Liang Y, Wang W, Shen Y, Liu Y, Liu XJ (2012) Dynamics and residues of chlorpyrifos and dichlorvos in cucumber grown in greenhouse. Food Control 26: 231–234.
39. Gaskin RE, Steele KD, Foster WA (2005) Characterizing plant surfaces for spray adhesion and retention. New Zealand Plant Protection 58: 1790–183.
40. Hunche M, Bringe K, Schmitz-Eiberger M, Noga G (2006) Leaf surface characteristics of apple seedlings, bean seedlings and Kohlrabi plans and their impact on the retention and rainfastness of mancozeb. Pest Manage Sci 62: 839–847.
41. Bargel H, Koch K, Cerman Z, Neinhuis C (2006) Structure-function relationships of the plant cuticle and cuticular waxes – a smart material? Funct Plant Biol. 33: 893–910.
42. Wagner P, Furstner R, Barthlott W, Neinhuis C (2003) Quantitative assessment to the structural basis of water repellency in natural and technical surfaces. J Exp Bot 54: 1295–1303.
43. Malhat F, Badawy HMA, Barakat DA, Saber AN (2014) Residues, dissipation and safety evaluation of chromafenozide in strawberry under open field conditions. Food Chem 152: 18–22.
44. Bauer S, Schulte E, Their H-P (2005) Composition of the surface waxes from bell pepper and eggplant. Eur Food Res Technol 220: 5–10.
45. Fantke P, Juraske R (2013) Variability of Pesticide Dissipation Half-Lives in Plants. Environ Sci Tech 47: 3548–3562.

7

Dryland Soil Hydrological Processes and Their Impacts on the Nitrogen Balance in a Soil-Maize System of a Freeze-Thawing Agricultural Area

Wei Ouyang[1]*, Siyang Chen[2], Guanqing Cai[1], Fanghua Hao[1]

1 School of Environment, State Key Laboratory of Water Environment Simulation, Beijing Normal University, Beijing, China, **2** Marine Monitoring and Forecasting Center of Zhejiang, Hangzhou, China

Abstract

Understanding the fates of soil hydrological processes and nitrogen (N) is essential for optimizing the water and N in a dryland crop system with the goal of obtaining a maximum yield. Few investigations have addressed the dynamics of dryland N and its association with the soil hydrological process in a freeze-thawing agricultural area. With the daily monitoring of soil water content and acquisition rates at 15, 30, 60 and 90 cm depths, the soil hydrological process with the influence of rainfall was identified. The temporal-vertical soil water storage analysis indicated the local *albic* soil texture provided a stable soil water condition for maize growth with the rainfall as the only water source. Soil storage water averages at 0–20, 20–40 and 40–60 cm were observed to be 490.2, 593.8, and 358 m^3 ha^{-1}, respectively, during the growing season. The evapo-transpiration (ET), rainfall, and water loss analysis demonstrated that these factors increased in same temporal pattern and provided necessary water conditions for maize growth in a short period. The dry weight and N concentration of maize organs (root, leaf, stem, tassel, and grain) demonstrated the N accumulation increased to a peak in the maturity period and that grain had the most N. The maximum N accumulative rate reached about 500 mg $m^{-2}d^{-1}$ in leaves and grain. Over the entire growing season, the soil nitrate N decreased by amounts ranging from 48.9 kg N ha^{-1} to 65.3 kg N ha^{-1} over the 90 cm profile and the loss of ammonia-N ranged from 9.79 to 12.69 kg N ha^{-1}. With soil water loss and N balance calculation, the N usage efficiency (*NUE*) over the 0–90 cm soil profile was 43%. The soil hydrological process due to special *soil* texture and the temporal features of rainfall determined the maize growth in the freeze-thawing agricultural area.

Editor: Guoping Zhang, Zhejiang University, China

Funding: We are grateful for assistance with the data requirements of the Bawujiu Farm in Heilongjiang Province. The research discussed in this paper benefited from financial support provided by the National Natural Science Foundation of China (Grant No. 41371018, 51121003), the Supporting Program of the "Twelfth Five-year" Plan for Sci & Tech Research of China (2012BAD15B05), and The Fundamental Research Funds for the Central Universities. The funders had no role in study design, data collection and analysis, decision to publish, or preparation of the manuscript.

Competing Interests: The authors have declared that no competing interests exist.

* Email: wei@itc.nl

Introduction

During the agricultural tillage management, the maximum crop production and minimum diffuse nitrogen (N) loading are the priority issues that need to be considered at the same time [1][2]. It is essential to understand the fates of soil water (SW) and N in agricultural systems in order to attain higher crop yields and N usage efficiency [3]. In developing countries, there is much political and commercial pressure for controlling the N application with the sacrifice of the crop harvest [4]. The dryland agriculture in the Sanjian Plain, Northeast China, is a key food base and the most water limited agricultural zone in China [5]. In this freeze-thawing area, the soil N and water efficiency are the keys for dryland tillage sustainability due to the short growing season. However, there are few reports about the dryland soil hydrological process and N use efficiency in freeze-thawing agricultural areas [6].

The soil nutrient N is a major limiting factor for crop growth and production, which is directly related to the N fertilisation and soil N background level [7]. The soil microbial community, which affects the soil N cycle [8], is less active in this freeze-thawing area. The application of chemical N is a basic agricultural practice in worldwide, which causes excessive discharge of N to the aquatic environment [9]. The European Union agricultural landscape contributes about 55% of the diffuse pollution for water eutrophication. This is a major emerging environmental issue in the developing counties [10]. The soil eco-hydrological process is an essential channel for N transport in the soil-crop system.

Plant organs have different allocation rates with N during growth, and the canopy green area has higher N absorption than the other organs during the jointing period [11]. The N accumulations in the upper leaves of the canopy also follow the eco-hydrological pattern in farmland [12]. Some models have been developed to simulate the N absorption in different organs during crop growth with the impacts of the solar radiation, rainfall, temperature, soil hydrology and N concentration [13]. With the temporal pattern of N accumulation in crop organs, the N cycle and efficiency can also be analyzed. The diffuse N discharge from agricultural systems depends on diverse factors, including the climatic, soil properties and agronomic features [14][15]. The tillage practice, soil hydrological process and N usage efficiency are the main potential factors to consider when the goal is to reduce the agricultural diffuse N pollution [16]. The dryland maize farmland in a freeze-thawing agricultural area presents different N discharge patterns due to the special hydrological process [17].

The soil hydrological process is the bridge for diffuse N loss from soil to water bodies [18]. The nitrate-N is the dominant type of diffuse N in the soil profile and it can move swiftly downstream flow with the rainfall through the soil pathways [19]. The soil hydrological process is also closely related to the soil texture, which is the medium for water movement and crop growth. In the freeze-thawing dryland, the soil has short-term, transitional phase and rainfall is the only water source for N movement with crop growth [20]. The soil pattern in the study area is meadow *albic* soil (*Albic Luvisols*) and has a 20–30 cm depth impermeable stratum. The *albic* soil provides a stable soil water (SW) condition in the crop root zone and interfaces with the soil hydrological process. With the consideration of high N efficiency, it is necessary to analyze the special pattern of soil hydrology and its impact on the N efficiency [21]. Evaluating the response of N loss to the combination of rainfall and the soil hydrological process can help to identify the interactions of water with N and reduce diffuse N pollution.

The general objective is to maximize the N efficiency in tillage practices. Typically, of about 20%–70% of N is lost from soil-crop systems and this loss can cause other environmental issues [22]. The quality of soil properties will decline under the combination of intensive application of N fertilizer and low N efficiency; this in turn, will also deliver some risk to water [23]. There are several methods to improve the N efficiency, including accurate fertilizer application and irrigation [24][25]. The dryland soil hydrological process in a freeze-thawing agricultural area presents a special pattern and also affects the N efficiency [26]. This study aims (1) to identify the temporal-vertical dynamics of dryland SW within an *albic* soil layer; (2) to express the dry weight and N accumulation of maize organs in dryland of a freeze-thawing agricultural area; (3) to explore the variation of soil loss, potential N loss, and N use efficiency (*NUE*) in dryland during the growing season.

Materials and Methods

2.1 Ethics statement

The field experiment was conducted in the farmland of Bawujiu Farm, which has been rented by Wang Dongli for 50 years. He is also the contact person for future permission. No specific permissions were required for these locations. The field studies did not involve endangered or protected species. This study did not involve vertebrate. The specific location of our study is 133° 50′–134° 33′ E, 47° 18′47°50′ N.

2.2 Study area description

The case study area was conducted on a farm in the north-eastern part of China. The farm has a long history and on the east is adjoined to Russia (Fig. 1). The observation site is located in the southern part of the Farm and the elevation is 48 m. The only water source for dryland tillage is precipitation and maize is the dominant crop. This area the temperate continental monsoon climate has an average annual precipitation of 588 mm and an average yearly temperature of 2.94 °C [27]. From October to the following April, the temperature is below zero and, as a result, the soil freeze-thawing process affects the tillage process. The crop growing season is from May to October.

2.3 Field monitoring and soil sample collection

In order to identify soil physical-chemical properties, soil samples at four depths (0–15, 15–30, 30–60 and 60–90 cm) were collected in 2010. The samples from three points chosen at random in each layer were mixed into a composite soil sample and placed in plastic bags for lab measurements. The soil samples were collected before sowing and after harvest. Soil particle size distributions were measured by laser diffraction after the removal of organic residue (MasterSizer S, Malvern Instruments, Malvern, UK). The soil's total nitrogen (TN) and organic carbon (OC) concentrations were measured with a CHN Elemental Analyser (Euro VectorS.P.A EA3000, 136 Milan, Italy) [28]. The ammonia N in rainfall was analyzed with the aid of Nessler's reagent and the nitrate-N concentration was determined by the Ultraviolet Spectro Photometric Method. The soil's pH was determined with a pH meter (METTLER TOLEDO, Switzerland) after mixing the fresh soil with deionised water (1:2.5, w/v) [24]. The soil N stock at four depths was calculated with the following equation,

$$Stock_i = BD_i * H_i * C_i * 10^5 \qquad (1)$$

where $Stock_i$ is the soil N storage at layer i (kg/m^2), BD_i is the soil bulk density at layer i (g/cm^3), H_i is the soil depth at layer i (cm), and C_i is the soil N concentration at layer i (mg/kg).

For the soil hydrological process monitoring, the SW collector head (suction cup: Teflon and quartz; OD 21 mm×L 95 mm; porosity: 2 μm; conductivity: 3.31×10^{-6} mm sec^{-1}, PRENART, Danmark) was set at 15, 30, 60 and 90 cm depths. It was covered with a mixed solution of quartz powder and stirred, which can cause a 0.05 MPa vacuum pressure and in direct contact with the soil. The collector tail was connected with PVC pipe and the water samples were collected in bottles by a portable manual vacuum pump. The soil water storage was calculated with soil volumetric water content at different depths. The mean soil water storage at each layer was defined as the average of SW in the upper and lower soil layers. The evapo-transpiration in dryland is calculated with the Bowen ratio and energy balance (BREB). The SW balance was calculated with the following equation [29],

$$SL = R - I_r - ET - \Delta S \qquad (2)$$

where SL is the soil water loss (mm), R is the precipitation (mm), ET is the evapo-transpiration (mm), I_r is the canopy interception (mm), and ΔS is the soil water storage change (mm).

The local meteorological characteristics were monitored with a ZENO station (Coastal, Seattle, WA, USA). Automated soil temperature sensors (Thermistor, Coastal, Seattle, WA, USA) and soil volumetric water content sensors (TDR type, Coastal, Seattle, WA, USA) were placed in each of the four different soil layers. The monitoring sensors were repeated at double at horizontal 0.5 m space.

2.4 Crop sample collection and measurement

The local dryland is intensively cultivated with the aid of mechanized equipment and the maize (*Zea mays L.*) planting density is 75,000 plants ha^{-1}. The detailed maize planting date, harvest date, relevant tillage and fertilization practices are listed in Table 1. The maize samples in 2011 were collected monthly and the sample in 2010 were only collected in October due to the limited experimental conditions. The detailed process of sample collection and monitoring of crop features (height (CH), leaf area index (LAI) and root depth (RD)) are listed in a previously published paper [24]. The maize samples were taken to the lab

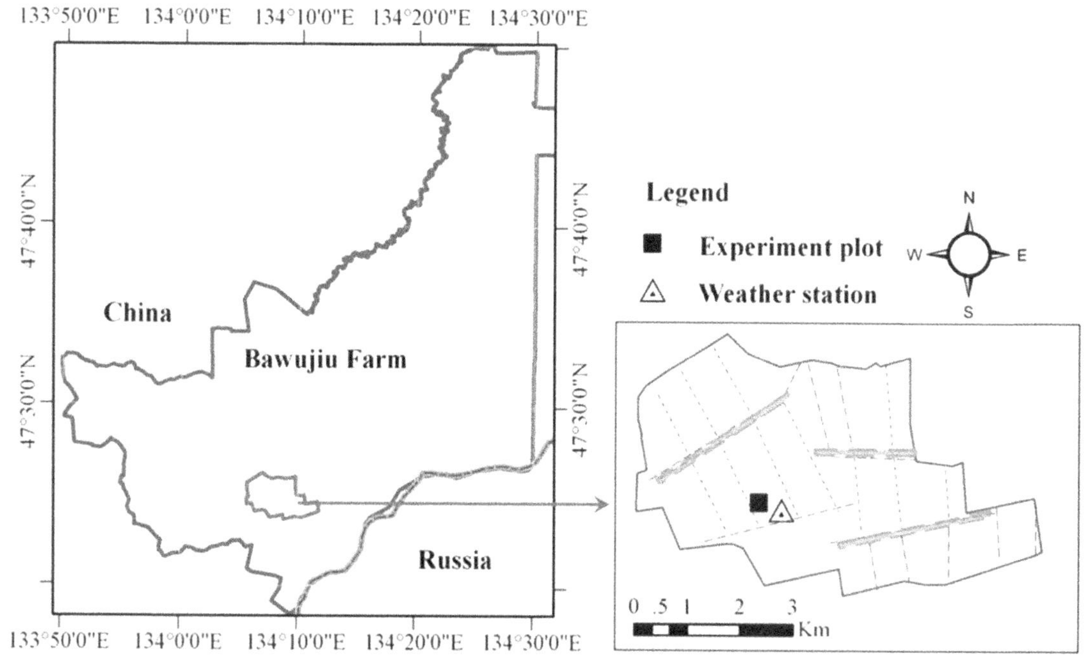

Figure 1. Location of the experimental dryland site in Northeast of China.

immediately and the root, leaves, stems, tassels, and grains were separated. The samples were washed clean and then dried for 48 h at 65°C. The dry weight of organ samples was determined after fixing the temperature for 30 min at 105°C. The dried maize samples were crushed and sieved for crop organ N content (%) analysis (Vario EL, Elementar Co. Ltd., Germany). The N uptake amount by crop was calculated based on crop dry weight and plant N content (%).

2.5 Nitrogen supply and loss calculation

The N use efficiency (*NUE*) in a crop-soil system is defined as the ratio of N uptake (N_c) by a crop to N supply (N_s) in a system [30]. The N_s was defined as the total available N to the crop, which included fertilizer N (N_f), mineralized N (N_m), initial mineralized N in soil ($N_{min\ initial}$), N fixed by soil (N_x) and N deposition (N_d) from rainfall. The N_s is calculated as:

$$N_s = N_f + N_{\min\ initial} + N_m + N_x + N_d \tag{3}$$

The field N_m is calculated with the crop-soil N equation [31]:

$$N_m + (N_{\min\ initial} - N_{\min\ final}) = (N_l + N_c + N_{gl}) - (N_f + N_d + N_x) \tag{4}$$

where N_m is the mineralized N, $N_{min\ initial}$ is the initial mineralized N concentration in soil, $N_{min\ final}$ is the final mineralized N concentration in soil, N_l is the soil loss N, N_{gl} is the N loss in gas. N_c is the N crop absorption, N_x is the N fixed by soil, and N_d is the N deposition.

Table 1. Dryland tillage practices and fertilization information.

Crop year	Planting date	Harvest date	Tillage prior to planting	Main fertilisation (kg ha^{-1})
2010	8/Jun.	9/Oct.	Tilling	525
2011	30/May.	5/Oct.		525
Detailed fertilisation data and amounts * (kg ha^{-1})				
2010	2011	N	P	K
8/Jun.	30/May.	71.25	58.65	27.23
27/Jun.	20/Jun.	34.50	13.80	10.02
5/Jul.	1/Jul.	29.45		

*Based on computation of the N, P, and potassium (K) elements.

The mineralized N is calculated with the following equation,

$$N_m = (N_l + N_c) - (N_f + N_d) - (N_{\min initial} - N_{\min final}) \quad (5)$$

The potential soil water N loss (N_l) is defined as the total loss of nitrate and ammonia, which was estimated with the N concentration in SW and the water volume in dryland [32]. Based on the SW loss and the N concentration in SW at different depths, the soil N loss by water was calculated.

2.6 Data analyses

Data analysis was performed with Sigma plot 10.0 and SPSS 16.0 software. The differences in dry weight and their percentages on crop organs (including root, leaf, stem, tassel, grain) in different growing stages were tested with ANOVA. The difference in the crop N uptake allocation proportion, the N accumulation in different organs, and the mineralized N concentration in different soil layers were statistically analyzed by ANOVA. The multiple comparisons were analyzed with the Duncan method.

Results

3.1 Temporal-spatial distribution of soil water in dryland

With the soil-maize water monitoring system, the daily precipitation, SW acquisition rate, and SW content in the entire growing period was determined (Fig. 2). The monthly precipitation in the years 2010 and 2011 displayed a similar pattern, except for the lower value in September 2010. The precipitation occurred mainly between July and September, which was also the prime period for maize growth. With the light precipitation in April and May, the maize seed was sowed in moist soil. The precipitation in August and September of 2010 was 64 mm and 120 mm less than in the corresponding months of 2011. The cumulative precipitation from April to September in 2011 was 575 mm, which was 141 mm more than that in 2010. The precipitation was the only water source for the dryland in this freeze-thawing agricultural area, which directly impacted the SW acquisition rate. The SW could be collected only after the rainfall and the vertical difference in the SW acquisition rate was slight (around 0.002 $cm^3\ s^{-1}$). In the entire monitoring period, the biggest acquisition rate occurred at the 90 cm depth after the rainfall in early September. The direct reason for it was that only smaller amount of the water can move deeply in the vertical direction after the absorption by roots.

The SW content increased in a stable manner during the growth period and also coincided with the accumulative precipitation. In all of June, the SW content at the 30 cm depth was the highest, the reason for this was that uptake of water in growing roots resulted in water accumulation in the rhizosphere. Due to the 73.6 mm rainfall on July 3, the SW content at the 30 cm depth increased by 20% to 0.30 $cm^3\ cm^{-3}$. In the next two days, the SW content was transported in the vertical direction and it increased to 0.30 $cm^3\ cm^{-3}$ in the deep layer. This vertical distribution also confirmed the retention role of roots in the early growing season. The surface SW content increased significantly after August due to the precipitation and then was transported to the deep soil layers. Later, the impact of the crop roots decreased and the soil texture was the main factor for SW transport in September.

In order to further analyse SW content fluctuation in the temporal-vertical dimensions, a statistical analysis was performed on the data from the soil layers during the growing season (30 May–30 September) (Table 2). The SW content averages at the 15 cm and 60 cm depths were very close with values of 0.245 and 0.230 $cm^3\ cm^{-3}$, respectively. The mean SW content at the 30 cm and 90 cm depths were higher than at the other two layers with values of 0.272 $cm^3\ cm^{-3}$ and 0.280 $cm^3\ cm^{-3}$, respectively. The close relationship between the SW acquisition rate and the SW content that was shown in Fig. 2, was also confirmed by the statistical results. The SW acquisition rate had the same vertical pattern as the SW content and the higher values appeared at the 30 and 90 cm depths. Before maize seeding (May 30), the SW could be collected when the SW content reached about 0.25 $cm^3\ cm^{-3}$. After maize seeding, the SW could be collected when it reached a content of over 0.30 $cm^3\ cm^{-3}$. The SW acquisition rate was greatly and primarily influenced by rainfall.

3.2 Temporal variation of soil water storage

Based on the monitoring of the SW characteristics, the soil volumetric water content (SVW) and soil water storage (SWS) in the top 30 and 90 cm were calculated, respectively (Fig. 3). The SVW and SWS of the 0–90 cm and 0–30 cm depth showed similar temporal trends. In the vertical dimension, the SVW of the same two depths (0–30 cm and 0–90 cm) had similar trends in most periods, except in the first month which revealed a difference due to thawing. The thawing process from the surface to the deep layer and the tillage in preparation for sowing in the top soil decreased the SVW at 0–30 cm depth. There was almost no difference in the SVW between the two depths from mid-May to August. Before the crop harvest period in September, the deeper soil had a higher SVW content than the surface layer due to less root activity and strong precipitation. The SWS of the 0–30 cm depth had a seasonal variation and had a more direct relationship with precipitation events. Before the rainfall on 1 July, the SWS remained at a stable level from mid-May to the end of June. The SWS had a periodic decrease when no rainfall occurred in late July. The SWS reached its summit with the large rainfall in the later of August.

The SWS was calculated with the vertical variation in the SVW. The statistical characteristics of SWS and their proportions at each depth are listed in Table 3. The soil water storage averages at 0–20, 20–40 and 40–60 cm were 490.2, 593.8, and 358 $m^3\ ha^{-1}$. The vertical differences indicated the middle level has a higher soil water content as a consequence of soil texture and crop root zone influences. Because precipitation was the only water source for dryland, the SWS at the three layers in the dry period was smaller than the average for the entire growing season, which also revealed the impact of maize ET on SWS. The SWS in the vertical dimension had the same trend in the entire period and the middle layer had the largest value (548.9 $m^3\ ha^{-1}$). The differences in SWS between the two periods become greater when going from the surface to the deeper layer, an observation derived from the rainfall contribution to the SWS in deeper layer.

Based on daily monitoring data, the accumulative patterns of evapo-transpiration (ET), rainfall, and water loss in the growing season were compared (Fig. 4). The ET was at a low level in the first month of the growing period and then assumed a stable increasing rate beyond the 50th day after sowing. The accumulated SW loss was consistent with the fluctuation in the rainfall. The ET and rainfall stayed with the SW loss over the first 22 days. The accumulated rainfall amount increased by 79.8 mm; a large increase occurred on the 35th day when the ET and water loss increased slightly. The rainfall increased intensively after 80 days, which also caused an increase in the SW loss. The accumulated amounts of rainfall increased from 269.3 mm after 80 days to 440.9 mm after 120 days, and the SW loss increased from 47.9 mm after 75 days to 126.3 mm after 120 days. The temporal pattern demonstrated that the SW loss and ET of dryland were strongly correlated with the rainfall. The ET was also a maize

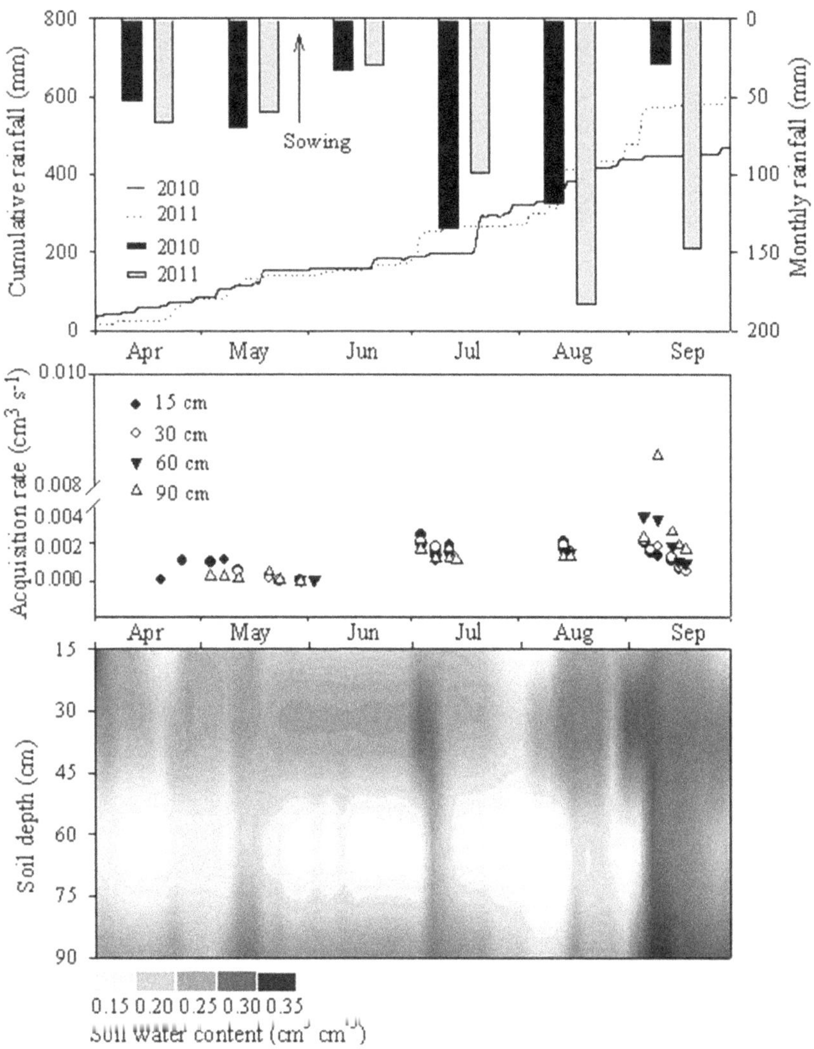

Figure 2. Daily precipitation in 2010 and 2011, and acquisition rate, and soil hydrological pattern of dryland in 2011.

growth feature and its similar pattern with rainfall also proved it was the dominant factor for crop growth.

3.3 Variation of crop biomass in the growing period

The dry weight of the different maize organs (root, leaf, stem, tassel, and grain) in growing stages were analyzed (Fig. 5). Also shown in the Fig. 5 are the percentages of the total dry weight comprised by each organ in each of four growing seasons. The dry organ weight varied greatly from the seedling stage to maturity, a development period which is relatively short in this freeze-thawing area. The temporal pattern of dry weight in 2011 revealed the changes that occurred over four growing sections. The dry weight of grains increased significantly in the grain filling and maturity sections. The dry weight percentage did not change as intensively as the weight, and it was also found that the grains comprised nearly half of the maize weight in the maturity period. With the N concentration in dried organs, the N absorption amount was quantified along with the organ dry weight. Due to logistical limitations, the crop organ sample in 2010 was only collected in the maturity period. The grain dry weights were larger in 2010 than in 2011. Comparing the dry weights of the organs in the maturity stage of the two years, the dry weights of the roots and tassel were significantly but their percentages differed little.

3.4 Nitrogen uptake by crop and accumulation

The temporal N dynamics of crop uptake was then analyzed with the samples obtained from the maize organs (Fig. 6). As the maize grew, the N uptake in leaf first increased and then decreased with the maximum occurring in the jointing period. From grain filling to maturity, N uptake by leaf diminished insignificantly with a relative flat trend compared with that from the jointing to grain filling stages. As maize grew N uptake in stem decreased significantly decrease from 1 to 2 with a probable transfer of much N to tassel. The difference of N uptake by stem in the grain filling and maturity stages was insignificant, but the N uptake by grain increased significantly ($P<0.001$). The N uptake by leaf decreased insignificantly, which indicated that a large amount of N was transferred to tassel in the early time period and then transferred from tassel to grain (N uptakes by tassel in grain filling and maturity were 8.12 g kg^{-1} and 3.57 g kg^{-1}, respectively). There were insignificant differences in N uptakes of stem and leaf during the maturity stage in 2010 and 2011. The N uptake by grain in 2010 was 9.1% more than that in 2011.

Table 2. Statistical analysis of soil water content and acquisition rate at four depths during maize growing period.

Index / Depth (cm)	Soil water content (cm^3 cm^{-3})				Soil water acquisition rate (cm^3 s^{-1})			
	15	30	60	90	15	30	60	90
Minimum	0.210	0.235	0.205	0.240	0.00008	0.00005	0.00009	0.00003
Maximum	0.308	0.333	0.306	0.342	0.00241	0.00208	0.00324	0.00856
Mean	0.245	0.272	0.230	0.280	0.00125	0.00114	0.00140	0.00167
Std. Error	0.018	0.020	0.024	0.024	0.00077	0.00078	0.00091	0.00220
Coefficient of variance	0.074	0.072	0.106	0.088	0.615	0.685	0.648	1.320

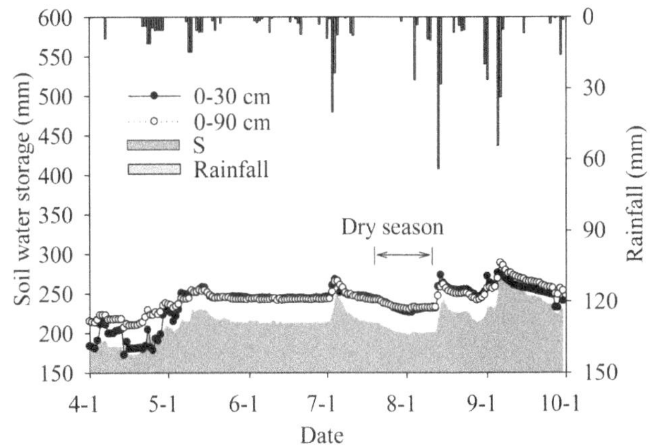

Figure 3. Temporal patterns of daily rainfall, soil water storage (SWS) and volumetric water content at 30 and 90 cm depths.

The accumulated N in different organs gradually increased in the early stages and reached a summit in the maturity period. From the jointing to grain filling period, the N in stem increased about 4.23 times and the amount in leaf increased less than five times. The N amounts in crop organs were not only related to the N uptake efficiency, but they also depended on the organ dry weight. According to the analysis in the maturity period, grain accumulated most of N, which occurred in a short time section. The total N amount of maize organs in 2010 was 309.9 kg ha^{-1}, which was larger than in 2011; the difference in grain was the dominant factor. The differences of N accumulation in leaf and stem during the maturity stages in the years 2010 and 2011 were insignificant.

Seasonal dynamics of N accumulative rates in organs during the growth period were analyzed (Fig. 7). From June to August, the N accumulative rates of stem and leaf increased rapidly and then declined. When the stem and leaf accumulated less N, the absorption efficiency in grain increased to 383 mg $m^{-2}d^{-1}$. The leaf was the most effective organ in uptaking N and the accumulation rate was about 500 mg $m^{-2}d^{-1}$.

3.5 Variations of soil N stocks in the growing season

The N concentrations at four depths before sowing and after harvest were analyzed, which demonstrated the temporal-vertical soil N dynamics. The N (nitrate and ammonia) concentrations at different depths were first analyzed after the growing season (Table 5). Before the sowing, the ammonia-N stock at four depths ranged from 18.8 to 21.3 kg N ha^{-1} with no occurrence of a significant difference. After the harvest, the deeper layer experienced more loss than the surface layer. The middle two layers had close values and the 12.5 kg N ha^{-1} in top layer was the largest value. The nitrate-N concentration in soil was much larger than the ammonia-N. At the beginning of the growing season, the nitrate-N concentration was more than 121 kg ha^{-1} at all depths and had slight differences at 60–90 cm depth. The nitrate-N concentration decreased to 90 kg ha^{-1} after the harvest. The upper two layers had similar patterns and the concentrations ranged between 88–90 kg ha^{-1}. There was no significant difference in the concentration in the deeper two layers and their concentrations ranged from 63.2 to 68.0 kg ha^{-1}. The nitrate-N was more variable than ammonia-N and had a larger stock gap.

Table 3. Dryland soil water storage and its proportion at three depths.

Depth/cm	Item \ Duration	Entire growth season (30 May.–30 Sep.)	Dry season (20 Jul.–12 Aug.)
0–20	SWS/m^3 ha^{-1}	490.2±34.6	441.1±19.5
	Proportion/%	23.4±1.04	23.3±0.4
0–40	SWS/m^3 ha^{-1}	1084±62.9	990.0±55.9
	Proportion/%	51.8±2.6	52.0±0.76
0–60	SWS/m^3 ha^{-1}	1442±111	1303.0±70.6
	Proportion/%	68.8±1.6	68.5±0.55

3.6 Nutrient loss and supply in soil-crop system

Based on field water monitoring and SW content in the 0–90 cm profile, the soil water loss (SL) and the leached N were calculated (Table 6). The rainfall and evapo-transpiration (ET) were calculated in the four stages of the maize growing season. The SL had a closer correlation with rainfall than with ET. The rainfall was nearly 100 mm in the jointing stage, but the SL was 11.5 mm due to the intensive ET. Moving from the grain filling stage to the maturity stage, the ET increased and more SL occurred under the higher rainfall event. The leached N was also calculated with the N concentration in soil water. The nitrate-N was the dominant N lost from the 90 cm soil profile, especially in the grain filling and maturity stages. The leached ammonia-N had similar temporal patterns and the largest amount was 0.18 kg ha^{-1}.

Based on the N monitoring of rainfall and the survey of N fertilization, the N in rainfall and fertilizer was calculated. With the SWS and N concentration in SW in 0–90 cm profile, the *NUE* in the maize-soil system was calculated (Table 7). The N input from rainfall and fertilizer in the entire maize growing season was 16.5 and 120 kg ha^{-1}, respectively. The soil N supply in the entire growth period was about 641 kg ha^{-1}. With the nitrate and ammonia concentrations in SW and water volume, the N loss was estimated at 15.6 kg ha^{-1}. Based on the N balance, it was calculated that the *NUE* was 43%. On the other hand, the excessive N from dryland was an important potential source for N discharge to the environment.

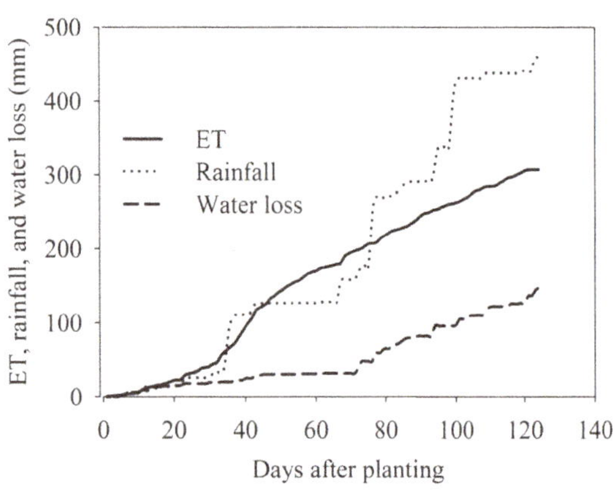

Figure 4. Cumulative evapo-transpiration (ET), rainfall, and water loss throughout maize growth period.

Discussion

4.1 The dryland soil hydrological process in a freeze-thawing area

Rainfall is the only water source in dryland and it has a direct impact on SW content and storage [33]. The field monitoring done in four stages as part of our study indicated that the temporal patterns of SW content and acquisition rate closely followed the rainfall amount (Fig. 2). The soil roughness in a cultivated area is increased by tillage, which enhances the vertical transport of SW [34]. The SL was 146.7 mm (32% of rainfall in the growing season). However, the vertical distribution of SW content and soil water storage demonstrate that the water has a short vertical movement range. The SW can be sampled only after the rainfall events, but the acquisition rates showed slight vertical differences. These two findings both demonstrated that the vertical movement of water is slow even with large volumes and the speed can affect the range.

Compared with the long growing periods in other dryland agricultural areas, the local maize growing season is short and has a higher efficiency in the use of water and N [35]. A proper SW condition in the root zone is essential for crop growth in the maize seeding and grain filling stages. The SWS can remain at a stable level (about 1000 m^3 ha^{-1} of 40 cm profile) even in the dry season without precipitation (Fig. 3). All of these factors indicate that the *albic* soil texture can provide a positive water condition for crop growth [24]. The SW also controls surficial N movements in the dryland and provides a strong signal for crop absorption [36]. Crop transpiration will increase when the maize is near its maturity period, and the precipitation normally increases and meets this need during the same period (Fig. 4). The combined conditions of special soil texture and precipitation patterns directly determine the dryland soil hydrological pattern, which is the critical factor for maize growth in this freeze-thawing area.

4.2 Crop growth and N accumulation differences in plant organs

The quantification and assessment of the organ biomass in crops can help in understanding the impacts of SW and N on crop growth. This information is also necessary for assessing N leaching [37]. The crop growth is measured by the organ dry weight (Fig. 5), which indicates the crop growth rate is higher than in other drylands under warming temperatures. The yearly precipitation in 2011 was more than in 2000 (Fig. 2), but did not lead to the higher grain dry weight in 2011 (Fig. 5). The difference proves that a higher maize yield is based on a combination of climate, soil and tillage. The grain filling period in this freeze-thawing area occurs in August, at a time when temperature and precipitation are at

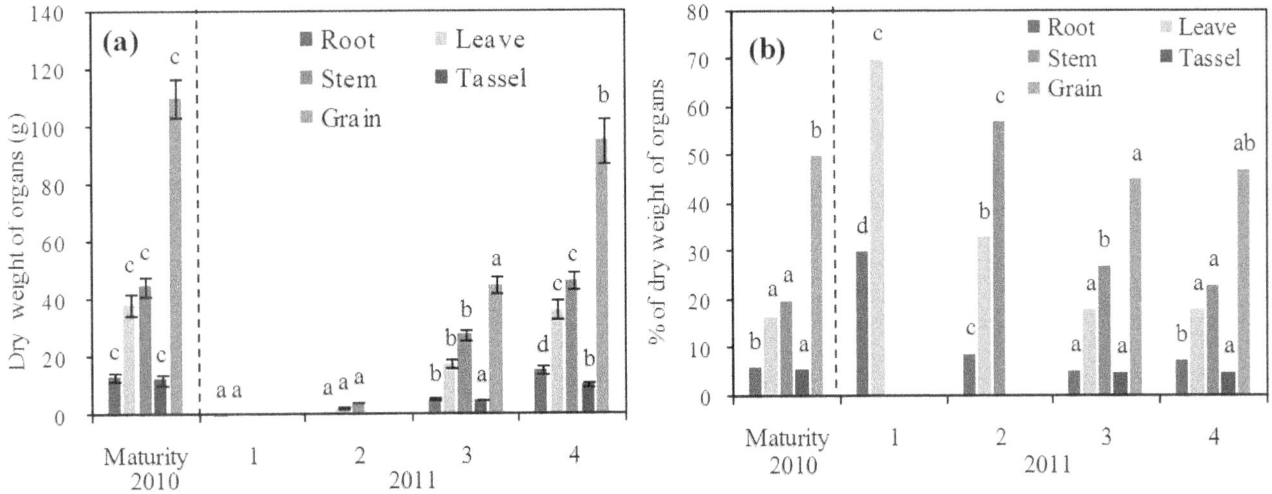

Figure 5. Dry weights (a) and percentage of maize organ weights (b) at different growth stages (Same letters in same organ between different growth stages indicate no significant difference. 1-Seedling, 2-Jointing, 3-Grain filling, 4-Maturity).

their highest levels. The active ET facilitates the transport of water and N from soil to crop, which finally causes a higher grain filling performance [38].

The absorbed N by maize organs in the entire growth period had different allocation rates (Fig. 7), which is determined by the photosynthesis of the organs and the eco-hydrological process. The crop accumulative rate in its canopy, stem and grain varied significantly, which means the N absorption capability for tissue production of dedicated organs shifts due to the N flux from soil to above ground biomass [39]. The soil N concentration in the vertical direction proved that the subsurface has a significant variance due to the root uptake. The accumulation rates of different organs over time also follow the typical sigmoid curve [40]. The temporal patterns of N in maize organs also indicate the N absorption is not only related to the soil N concentration, but that there is a complicated systemic feedback signal within the soil-crop system [41]. The SW availability, ET and N concentration are the key signaling factors. As the soil N is transported in crops with water, there are plenty of field observations and models to identify the growth stress reaction to the SW and N conditions [42][43]. This information provides detailed guidelines for N and water adjustment in the soil-crop system.

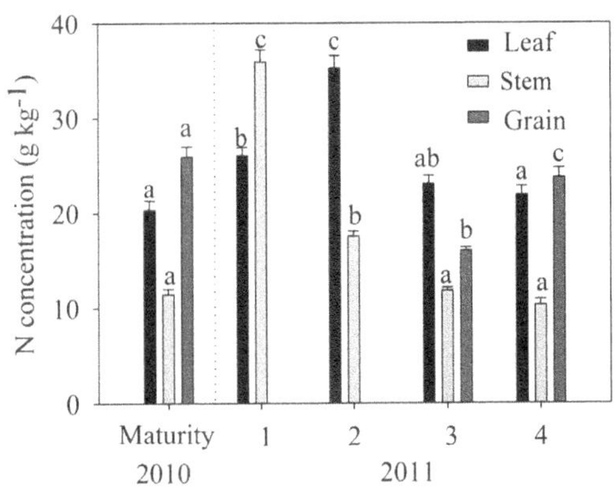

Figure 6. Seasonal dynamics of N concentration in maize organs (Same letters in same crop organs indicates no significant change between different growth stages. 1-Seedling, 2-Jointing, 3-Grain filling, 4-Maturity).

4.3 Implication for N leach control and soil N optimization

The N uptake efficiency of a crop is the main component for *NUE* and the temporal patterns in Table 4 indicated that the highest amount of grain uptake of N occurs in the graining filling and maturity periods. The maize uptake of N will increase until the optimum levels of water and N in soils are reached [44]. Overlaying this information with the leached N loading, it was found that the intensive N uptake and loss occurred at same time. The strong ET and rainfall in that time period were the reason for the N transport in the crop-soil system [45]. The soil N gap analysis revealed that the observed N gap was similar in the vertical direction, which also demonstrates that the soil N can move among the soil depths with hydrological processes and maintain the N concentration balances [3]. The temporal N analysis in crop organs, soil and leached water explain the maize growth in this freeze-thawing area, which also provides a guideline for N fertilization while, also considering SW conditions.

The N pollution to groundwater due to excessive N fertilizer application is an important environmental concern [46]. The

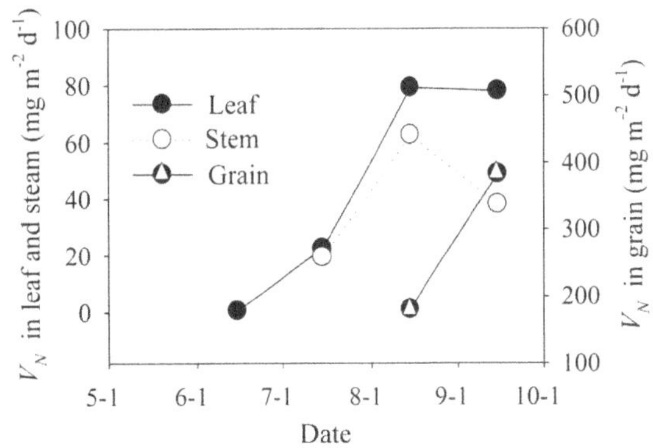

Figure 7. Seasonal dynamics of N accumulative rates (VN) in maize organs.

Table 4. Temporal pattern of N amount (kg ha^{-1}) of maize organs during growing season*.

Year	Growth stage	Leaf	Stem	Grain
2010	Mature	58.6f	38.3ef	213de
2011	Seedling	0.14a		
	Jointing	6.85c	5.81a	
	Grain filling	30.6d	24.6d	54.1b
	Maturity	60.0f	36.0e	169c

*Same lowercase letter in one column indicats no significant difference (P<0.001); Same uppercase letter in one row indicates no significant difference (P<0.001).

Table 5. Mineral N stocks (kg ha^{-1}) and variations at each depth of growing season.

Depth (cm)	Before sowing		After harvest		Stock gap	
	NO_3^--N	NH_4^+-N	NO_3^--N	NH_4^+-N	NO_3^--N	NH_4^+-N
0–15	147d	21.3b	90.1d	12.5c	−48.9c	−8.64bc
15–30	144d	21.2b	88.1d	8.12b	−53.8cd	−11.2c
30–60	130cd	19.3b	68.0c	9.02b	−65.3e	−9.79c
60–90	121c	18.8b	63.2c	5.74a	−56.8d	−12.69c

Same letters in one column indicates no significant difference (P<0.001).

Table 6. Amounts of soil water loss (SL) and N leached below maize root zone (90 cm).

Growth stage	Period	R	ET	SL	leached N (kg ha^{-1})	
		(mm)			NO_3^--N	NH_4^+-N
Seedling	30 May to 29 Jun.	29.8	44.6	19.6	0.03	0.01
Jointing	30 Jun. to 25 Jul.	96.5	119.0	11.5	3.36	0.12
Grain filling	26 Jul. to 23 Aug.	165	67.3	48.6	5.53	0.18
Maturity	24 Aug. to 30 Sep.	168	76.6	67.5	6.26	0.10

yearly *NUE* is about 43% in this area, which is a relatively high value in comparison with similar dryland agriculture areas [47]. The N leached from farmland mainly depends on the difference between the N obtained by the soil-water-crop system and the N uptake by the crop [48]. As Table 6 shows, the leached N is mainly related to the loss of SW, which is the gap between rainfall and ET. The rainfall in August 2011 was above the long term monthly value, which caused the highest SW and N loss during the growing season. Under the normal climatic conditions, the *NUE* in this area can reach 50%. The nitrate-N is the dominant form of leached N to the groundwater, which is a direct consequence of fertilization and soil nitrification [49]. Models have been developed to optimize the fertilizer application to farmland based on a consideration of the soil hydrological process, which can help to maximize the crop yield with minimal discharge to the environment [50]. The model with local characteristic parameters,

Table 7. The N balance and usage efficiency (*NUE*) in crop-soil system during the maize growth period.

Rainfall	Fertilizer	Initial soil N		Final soil N		N loss		Soil supply N	*NUE*
(kg N ha^{-1})		(kg N ha^{-1})		(kg N ha^{-1})		(kg N ha^{-1})		(kg N ha^{-1})	(kg kg^{-1})
		NO_3^--N	NH_4^+-N	NO_3^--N	NH_4^+-N	NO_3^--N	NH_4^+-N		
16.5	120	542.9±29.76	80.6±4.29	309.4±18.4	35.4±4.37	15.2	0.41	641±14.4	0.43±0.03

should be applied in the tillage practice to obtain an improved *NUE*.

Conclusion

The results of this study have shown that combined conditions of soil texture, soil hydrology and precipitation were basic factors for maize growth in the freeze-thawing area in Northeast China. The rainfall significantly affected the pattern of SW content and the allocation of N in the maize organs. Precipitation was the only water source for the dryland, but the variation in SW acquisition rates and SW content remained at a stable level due to the special *albic soil* texture. The dry weight and N accumulation of maize organs in the entire growing period proved that the soil hydrological condition was also enhanced.

The N loss was dependent on both the hydrological process and the N in the water, which was then decided by the rainfall water depth and the applied N. The NO_3-N was the dominant leachate, which was closely correlated with the amount of applied N fertilizer. The *NUE* was about 43%, which was also related to the initial soil N and maize uptake. In order to improve the *NUE* and decrease the amounts of leached nitrogen, it would be better to improve the fertilizer application amounts while giving consideration to rainfall patterns and N accumulation rates.

Author Contributions

Conceived and designed the experiments: WOY. Performed the experiments: SYC GQC. Analyzed the data: WOY SYC. Contributed reagents/materials/analysis tools: FHH. Wrote the paper: WOY.

References

1. Buckley C, Carney P (2013) The potential to reduce the risk of diffuse pollution from agriculture while improving economic performance at farm level. Environ Sci Policy 25(1): 118–126.
2. Quemada M, Baranski M, Nobel-de Lange MNJ, Vallejo A, Cooper JM (2013) Meta-analysis of strategies to control nitrate leaching in irrigated agricultural systems and their effects on crop yield. Agric Ecosyst Environ, 174(15): 1–10.
3. Ferrant S, Oehler F, Durand P, Ruiz L, Salmon-Monviola J, et al. (2011) Understanding nitrogen transfer dynamics in a small agricultural catchment: Comparison of a distributed (TNT2) and a semi distributed (SWAT) modeling approaches. J Hydrol 406(1–2): 1–15.
4. Panda RK, Behera S (2003) Non-point source pollution of water resources: Problems and perspectives. J Food Agric Environ 1(3–4): 308–311.
5. Wang XB, Dai KA, Wang Y, Zhang XM, Zhao QS, et al. (2010) Nutrient management adaptation for dryland maize yields and water use efficiency to long-term rainfall variability in China. Agric Water Manage 97(9): 1344–1350.
6. Yang YG, Xiao HL, Wei YP, Zhao LJ, Zou SB, et al. (2011) Hydrologic processes in the different landscape zones of Mafengou River basin in the alpine cold region during the melting period. J Hydrol 409(1–2): 149–156.
7. Grzebisz W (2013) Crop response to magnesium fertilization as affected by nitrogen supply. Plant Soil 368(1–2): 23–39.
8. Matejek B, Huber C, Dannenmann M, Kohlpaintner M, Gasche R, et al. (2010) Microbial N turnover processes in three forest soil layers following clear cutting of an N saturated mature spruce stand. Plant Soil 337(1–2): 93–110.
9. Moreau P, Ruiz L, Vertes F, Baratte C, Delaby L, et al. (2013) CASIMOD'N: An agro-hydrological distributed model of catchment-scale nitrogen dynamics integrating farming system decisions. Agric Syst 118: 41–51.
10. Kersebaum KC, Steidl J, Bauer O, Piorr HP (2003) Modelling scenarios to assess the effects of different agricultural management and land use options to reduce diffuse nitrogen pollution into the river Elbe. Phys Chem Earth 28: 537–545.
11. Hirel B, Le Gouis J, Ney B, Gallais A (2007) The challenge of improving nitrogen use efficiency in crop plants: towards a more central role for genetic variability and quantitative genetics within integrated approaches. J Exp Bot 58: 2369–2387.
12. Pask AJD, Sylvester-Bradley R, Jamieson PD, Foulkes MJ (2012) Quantifying how winter wheat crops accumulate and use nitrogen reserves during growth. Field Crop Res 126(14): 104–118.
13. Bonato O, Schulthess F, Baumgärtner J (1999) Simulation model for maize crop growth based on acquisition and allocation processes for carbohydrate and nitrogen. Ecol Modell 124(1): 11–28.
14. Binder DL, Dobermann A, Sander DH, Cassman KG (2002) Biosolids as nitrogen source for irrigated maize and rainfed sorghum. Soil Sci Soc Am J 66(2): 531–543.
15. Favaretto N, Norton LD, Johnston CT, Bigham J, Sperrin M (2012) Nitrogen and Phosphorus Leaching as Affected by Gypsum Amendment and Exchangeable Calcium and Magnesium. Soil Sci Soc Am J 76(2): 575–585.
16. Lam QD, Schmalz B, Fohrer N (2012) Assessing the spatial and temporal variations of water quality in lowland areas, Northern Germany. J Hydrol 438–439(17): 137–147.
17. Friend AD, Stevens AK, Knox RG, Cannell MGR (1997) A process-based, terrestrial biosphere model of ecosystem dynamics (Hybrid v3.0). Ecol Modell 95(2–3): 249–287.
18. Dorioz JM, Ferhi A (1994) Non-point pollution and management of agricultural areas: Phosphorus and nitrogen transfer in an agricultural watershed. Water Res 28(2): 395–410.
19. Keeney DR, Follett RF (1991) Managing Nitrogen for Groundwater Quality and Farm Profitability, SSSA, Madison, WI. 1–7.
20. Kim DG, Vargas R, Bond-Lamberty B, Turetsky MR (2012) Effects of soil rewetting and thawing on soil gas fluxes: a review of current literature and suggestions for future research. Biogeoscience 9(7): 2459–2483.
21. Lewis DR, McGechan MB, McTaggart IP (2003) Simulating field-scale nitrogen management scenarios involving fertiliser and slurry applications. Agr Syst 76(1): 159–180.
22. Dawson JC, Huggins DR, Jones SS (2008) Characterizing nitrogen use efficiency in natural and agricultural ecosystems to improve the performance of cereal crops in low-input and organic agricultural systems. Field Crop Res 107: 89–101.
23. Schroder JL, Zhang HL, Girma K, Raun WR, Penn CJ, et al. (2011) Soil Acidification from Long-Term Use of Nitrogen Fertilizers on Winter Wheat. Soil Sci Soc Am J 75(3): 957–964.
24. Caliskan S, Ozkaya I, Caliskan ME, Arslan M (2008) The effects of nitrogen and iron fertilization on growth, yield and fertilizer use efficiency of soybean in a Mediterranean-type soil. Field Crop Res 108(2): 126–132.
25. Razzaghi F, Plauborg F, Jacobsen SE, Jensen CR, Andersen MN (2012) Effect of nitrogen and water availability of three soil types on yield, radiation use efficiency and evapotranspiration in field-grown quinoa. Agric Water Manage 109: 20–29.
26. Hao FH, Chen SY, Ouyang W (2013) Temporal rainfall patterns with water partitioning impacts on maize yield in a freeze-thaw zone. J Hydrol 486(12): 412–419.
27. Ouyang W, Huang HB, Hao FH, Shan YS, Guo BB (2012) Evaluating spatial interaction of soil property with non point source pollution at watershed scale: the phosphorus indicator in Northeast China. Sci Total Environ 432, 412–421.
28. Jackson JE (1991) A user's guide to principal components. Wiley-Interscience, New York.
29. Moroizumi T, Hamada H, Sukchan S, Ikemoto M (2009) Soil water content and water balance in rainfed fields in Northeast Thailand. Agric Water Manage 96(1): 160–166.
30. Huggins DR, Pan L (1993) Nitrogen efficiency component analysis: an evaluation of cropping system differences in productivity. Agronomy J 85: 898–905.
31. Meisinger JJ, Randall GW (1991) Estimating nitrogen budgets for soil-crop systems. In: Follet, R.F., et al. (Eds.), Managing Nitrogen for Groundwater Quality and Farm Profitability. SSSA, Madison,WI, USA, 85–124.
32. Vázquez N, Pardo A, Suso ML, Quemada M (2006) Drainage and nitrate leaching under processing tomato growth with drip irrigation and plastic mulching. Agric Ecosyst Environ 112(4): 313–323.
33. Tanveer SK, Wen XX, Lu XL, Zhang JL, Liao YC (2013) Tillage, Mulch and N Fertilizer Affect Emissions of CO2 under the Rain Fed Condition. PLOS ONE: 8(9): e72140.
34. Dörner J, Horn R (2009) Direction-dependent behaviour of hydraulic and mechanical properties in structured soils under conventional and conservation tillage. Soil Till Res 102(2): 225–232.
35. Murungu FS, Chiduza C, Muchaonyerwa P, Mnkeni PNS (2011) Mulch effects on soil moisture and nitrogen, weed growth and irrigated maize productivity in a warm-temperate climate of South Africa. Soil Till Res 112(1): 58–65.
36. Gheysari M, Mirlatifi SM, Homaee M, Asadi ME, Hoogenboom G (2009) Nitrate leaching in a silage maize field under different irrigation and nitrogen fertilizer rates. Agric Water Manage 96(6): 946–954.
37. Wegehenkel M, Mirschel W (2006) Crop growth, soil water and nitrogen balance simulation on three experimental field plots using the Opus model—A case study. Ecol Modell 190(1-2): 116–132.
38. Katerji N, Campi P, Mastrorilli M (2013) Productivity, evapotranspiration, and water use efficiency of corn and tomato crops simulated by AquaCrop under contrasting water stress conditions in the Mediterranean region. Agric Water Manage 130: 14–26.
39. Strullu L, Cadoux S, Preudhomme M, Jeuffroy MH, Beaudoin N (2011) Biomass production and nitrogen accumulation and remobilisation by Miscanthus × giganteus as influenced by nitrogen stocks in belowground organs. Field Crop Res 121(3): 381–391.

40. Shoji S, Gandeza AT, Kimura K (1991) Simulation of crop response to polyolefin-coated urea: II. Nitrogen uptake by corn. Soil Sci Soc Am J 55(5): 1468–1473.
41. Nacry P, Bouguyon E, Gojon A (2013) Nitrogen acquisition by roots: physiological and developmental mechanisms ensuring plant adaptation to a fluctuating resource. Plant Soil, 370(1–2): 1–29
42. Kovács GJ (2005) Modelling of adaptation processes of crops to water and nitrogen stress. Phys Chem Earth, Parts A/B/C 30(1–3): 209–216.
43. Overman AR, Scholtz RV (2013) Accumulation of Biomass and Mineral Elements with Calendar Time by Cotton: Application of the Expanded Growth Model. PLOS ONE 8(9): e72810.
44. Kerbiriou PJ, Stomph TJ, Van Der Putten PEL, Van Bueren ETL, Struik PC (2013) Shoot growth, root growth and resource capture under limiting water and N supply for two cultivars of lettuce (Lactuca sativa L.). Plant Soil 371(1–2): 281–297.
45. Asseng S, Turner NC, Keating BA (2001) Analysis of water- and nitrogen-use efficiency of wheat in a Mediterranean climate. Plant Soil 233(1): 127–143.
46. Fang QX, Yu Q, Wang EL, Chen YH, Zhang GL, et al. (2006) Soil nitrate accumulation, leaching and crop nitrogen use as influenced by fertilization and irrigation in an intensive wheat-maize double cropping system in the North China Plain. Plant Soil 284(1–2): 335–350.
47. Ouyang W, Wei XF, Hao FH (2013) Long-term soil nutrient dynamics comparison under smallholding land and farmland policy in northeast of China. Sci Total Environ 450–451(15): 129–139.
48. Hall JA, Bobe G, Hunter JK, Vorachek WR, Stewart WC, et al. (2013) Effect of Feeding Selenium-Fertilized Alfalfa Hay on Performance of Weaned Beef Calves. PLOS ONE 8(3): e58188.
49. Lteif A, Whalen JK, Bradley RL, Camire C (2010) Nitrogen transformations revealed by isotope dilution in an organically fertilized hybrid poplar plantation. Plant Soil 333(1–2): 105–116.
50. Sierra J, Brisson N, Ripoche D, Noel C (2003) Application of the STICS crop model to predict nitrogen availability and nitrate transport in a tropical acid soil cropped with maize. Plant Soil 256(2): 333–345.

8

Soil Aggregates and Associated Organic Matter under Conventional Tillage, No-Tillage, and Forest Succession after Three Decades

Scott Devine[1], Daniel Markewitz[1]*, Paul Hendrix[2], David Coleman[2]

1 Warnell School of Forestry and Natural Resources, The University of Georgia, Athens, Georgia, United States of America, **2** Odum School of Ecology, The University of Georgia, Athens, Georgia, United States of America

Abstract

Impacts of land use on soil organic C (SOC) are of interest relative to SOC sequestration and soil sustainability. The role of aggregate stability in SOC storage under contrasting land uses has been of particular interest relative to conventional tillage (CT) and no-till (NT) agriculture. This study compares soil structure and SOC fractions at the 30-yr-old Horseshoe Bend Agroecosystem Experiment (HSB). This research is unique in comparing NT and CT with adjacent land concurrently undergoing forest succession (FS) and in sampling to depths (15–28 cm) previously not studied at HSB. A soil moving experiment (SME) was also undertaken to monitor 1-yr changes in SOC and aggregation. After 30 years, enhanced aggregate stability under NT compared to CT was limited to a depth of 5 cm, while enhanced aggregate stability under FS compared to CT occurred to a depth of 28 cm and FS exceeded NT from 5–28 cm. Increases in SOC concentrations generally followed the increases in stability, except that no differences in SOC concentration were observed from 15–28 cm despite greater aggregate stability. Land use differences in SOC were explained equally by differences in particulate organic carbon (POC) and in silt-clay associated fine C. Enhanced structural stability of the SME soil was observed under FS and was linked to an increase of 1 Mg SOC ha^{-1} in 0–5 cm, of which 90% could be attributed to a POC increase. The crushing of macroaggregates in the SME soil also induced a 10% reduction in SOC over 1 yr that occurred under all three land uses from 5–15 cm. The majority of this loss was in the fine C fraction. NT and FS ecosystems had greater aggregation and carbon storage at the soil surface but only FS increased aggregation below the surface, although in the absence of increased carbon storage.

Editor: Mary O'Connor, University of British Columbia, Canada

Funding: This research was funded with internal support for the Warnell School of Forestry and Natural Resources, the Odum School of Ecology, and the University of Georgia. The funders had no role in study design, data collection and analysis, decision to publish, or preparation of the manuscript.

Competing Interests: The authors have declared that no competing interests exist.

* E-mail: dmarke@uga.edu

Introduction

Understanding the mechanisms for soil carbon storage and stabilization relative to land management is increasingly relevant as soil organic carbon (SOC) is known to be a major pool of global C and SOC is critical to sustaining soil productivity. Since physical protection of SOC within stable soil aggregates is considered to be one of the major SOC stabilization mechanisms, [1–3], the effect of land management on aggregate stability is accepted as a key factor in determining SOC levels [4]. There have been many comparisons of soil aggregates under conventional tillage relative to no-tillage agriculture but relatively few that have incorporated other land uses such as abandoned agricultural fields or forest succession, particularly over a decades timescale.

The three-decade agroecosystem experiment at Horseshoe Bend (HSB), Athens, GA, USA, has long focused on SOC differences in side-by-side plots of conventional tillage agriculture (CT) and no-till agriculture (NT). This research has demonstrated increased SOC in the 0–5 cm layer of NT [5] and some role for aggregation in SOC protection [6]. More recent work at HSB has incorporated adjacent plots of forest succession (FS), which has demonstrated a similar increase in 0–5 cm SOC relative to CT as well as an increase in particulate organic carbon (POC) in 5–15 and 15–30 cm subsurface soil layers [7]. Although previous CT-NT comparisons have been reported from HSB this study extends analyses to subsurface soil layers (15–28 cm) previously not studied at HSB for aggregation. These subsurface soils are recognized to affect the carbon balance of CT-NT studies [8]. Furthermore, few studies have compared CT and NT with other land uses but here we simultaneously investigate the role of aggregation in forest succession to evaluate the robust nature of aggregates for SOC stabilization. As such there is an opportunity to understand if physical stabilization through aggregation is functioning similarly in 0–5 cm soils in NT and FS, and if aggregation in subsurface layers of FS is stabilizing POC at rates greater than under agroecosystems.

Under these contrasting land uses (CT-NT-FS) soil structure can both directly and indirectly affect SOC stabilization. Narrow pores can directly exclude extracellular enzymes, microbes, and soil micro-fauna. Pore size also indirectly affects decomposition by regulating fluxes that influence microbial activity, such as oxygen, heat, solutes, and water. This explains how intra-aggregate pores, which are finer (and perhaps more tortuous) than inter-aggregate pores [9], can slow decomposition within aggregates. The abundant research on the relationship between soil aggregation and SOC has produced several contrasting models of how physical protection of SOC occurs.

The aggregate hierarchy model proposes that soil structure arises from aggregation of fine mineral particles into microaggregates (53 μm to 250 μm) continuing to aggregate to macroaggregates (>250 μm) with increasing dependence on more transient, organic binding agents as the scale increases [10]. This model suggests that SOC concentration increases with increasing aggregate size and that cultivation exposes these binding agents to microbial attack with subsequent loss of SOC [11].

Oades [12] later modified this theory hypothesizing that microaggregates formed around POC (53–2000 μm) when enmeshed within stable macroaggregates. Thus, soils lacking mechanical disturbance such as NT or FS result in decreased macroaggregate turnover and higher SOC due to microaggregate formation and stabilization of fine POC (53–250 μm) [13,14]. Loss of SOC in cultivated soils is explained by shorter macroaggregate turnover times and a reduced opportunity for formation of microaggregates around a fine POC core [3].

More recently, it has been hypothesized that increased pore connectivity in soils that are not tilled allows greater movement of dissolved organic carbon (DOC) from larger pores surrounding aggregates into intra-aggregate pore spaces [15]. In tilled soils, disruption of the pore network effectively isolates substantial portions of the intra-aggregate pore space from replenishment of DOC by root exudates, decomposition of POC, or litter leachate laden with DOC. The focus on DOC contrasts with that on POC as the fraction of SOC most susceptible to management and soil structural disruption [16–22].

Given the previous research at HSB the objective of this study was to quantify SOC in various soil aggregate fractions under the different land uses to better understand the role of the aforementioned mechanisms in SOC stabilization [23,24]. It was hypothesized that fractionation of SOC into POC and fine C (presumably adsorbed DOC) offers a means to address differences in the proposed mechanisms of carbon storage. In particular, it was hypothesized that if macro- and micro-aggregate protection of POC is the predominate mechanism increasing SOC then aggregate associated POC contents should be greater in land uses with higher SOC content. Conversely, if micro-scale fluxes of DOC are important to attaining maximum SOC storage capacity, then this should be recovered as fine C (<53 μm) in systems with significantly higher levels of SOC.

In addition to the POC and fine C fractionation, a soil moving experiment (SME) was established to determine the extent to which destructured soil can re-develop macroaggregate stability in the field under the contrasting land uses. This SME better represents land use attributes as opposed to laboratory incubations [5,6]. It was hypothesized that differential recovery rates of macroaggregates in SME soil under the contrasting field conditions would reflect a greater abundance of organic binding agents as indexed by soil microbial biomass (i.e., bacteria and fungi). We hypothesized these rates would be greatest in forests due to an expectation of greater fungal abundance.

Methods

Site history

The Horseshoe Bend (HSB) agroecosystem experiment (33°57′N, 83°23′W) is a long-term comparison of conventional tillage (CT) and no-tillage (NT) treatments begun in 1978. The site is under the management of the Odum School of Ecology at The University of Georgia so no special permission was required for the current research and no endangered or protected species were involved in or negatively impacted by this research.

Based on a series of aerial photos, HSB had been cleared entirely of trees except for a narrow strip along the Oconee River at least since 1938 [7]. The site was used for pasture and forage production until 1965 when the Institute of Ecology acquired HSB for research [25]. The area underwent secondary succession in 1966 except for the future agroecosystem site, which was tilled to plant a crop of millet for a separate, earlier study [26]. Afterwards, the site went fallow and was used for a prescribed burn study and a N fertilization study [27,28]. In 1978, the study area was divided into equal numbers of CT and NT 0.1 ha plots (n = 4 for each type), and the old-field vegetation was mowed [25].

Site description

Horseshoe Bend is so named because it was established in a bend on the banks of the Oconee River and as such the site is very flat (<1% slope). The soils on this river terrace have previously been identified as clayey, kaolinitic, thermic, Rhodic Kanhapludults [29] but have recently been reclassified as a fine-loamy, kaolinitic, thermic, Typic Kanhapludalf based on new particle-size analysis data in the control section and measurements of base saturation (Table 1) that are >98% in all three treatments at 140 cm (i.e., 125 cm below the top of the kandic horizon). Average rainfall from 1945 to 2003 in Athens was 125 cm with a mean annual temperature of 16.4°C

Experimental design

When the site was originally sub-divided in 1978 the CT-NT treatments were assigned randomly in a completely randomized design [25]. Since the agroecosystem plots were in fallow from 1966, the current NT plots have not been tilled for forty-two years. Tillage has been performed twice annually in most years with moldboard plowing to a depth of 15 cm, followed by disking. In 1981, each agricultural plot was split into different winter crop treatments with winter rye (*Secale cereale*) grown on half of each plot and N-fixing crimson clover (*Trifolium incarnatum*) grown on the other half until 1984 and then again from 1989 to 2007 [25,30]. From 1978–1998, summer crops have included sorghum (*Sorghum bicolor*), soybeans (*Glycine max*), corn (*Zea mays*), and kenaf (*Hibiscus cannabinus*). In 1999 to 2007, a second split-plot was assigned to the agroecosystem experiment for a comparison between summer crops of non-Bt and Bt-cotton (*Gossypium hirsutum* L.) [31]. Soil sampling in the current study sought to estimate an average effect of NT and CT on the measured properties by sampling equally among the winter cover split-plots before compositing; the effects of transgenic cotton on soil C and N were considered negligible, since earlier work showed no difference in decomposition rates between non-Bt and Bt-cotton residues [32].

In 2007, four secondary, hardwood forest plots bordering the tillage experiment were established. These areas represent a third treatment, referred to herein as forest succession (FS). Most likely, these afforested plots were not tilled in 1966, which means they essentially underwent a pasture to forest conversion. All FS plots are on the same terrace as the agroecosystem plots and were clear of forest from 1938 to 1967, as observed in the aerial photographs [7].

Aggregate size fractions and stability

To quantify aggregate stability, four samples were obtained per plot at the end of the summer growing season in October 2007 using a 5 cm diameter corer attached to a slide-hammer to a depth of 28 cm. Each core was divided into 0–5, 5–15, and 15–28 cm

Table 1. Soil chemical attributes (mean ±1SD) for the Horseshoe Bend Agroecosystem Experiment, Athens, GA.

Treatment	Depth	pHs[1]	C	N	P[2]	K	Ca	Mg	ECEC[3]	Bulk Density
	cm		--------g/kg--------		µg/g	------------------------cmol$_c$/kg------------------------				g/cm^3
Conventional	0–5	5.16±0.04	9.73±0.75	0.75±0.06	4.1±0.6	0.36±0.04	2.13±0.22	0.54±0.03	3.04±0.27	1.45±0.09
Till	5–15	4.90±0.04	6.99±0.58	0.58±0.05	2.2±0.4	0.25±0.03	1.93±0.13	0.49±0.02	2.72±0.18	1.61±0.03
	15–30	5.42±0.08	2.93±0.35	0.27±0.03	1.7±0.1	0.18±0.01	1.92±0.05	0.58±0.03	2.69±0.07	1.68±0.04
	30–50	5.69±0.06	1.84±0.12	0.20±0.02	0.7±0.2	0.12±0.01	1.89±0.07	0.58±0.01	2.58±0.06	1.64±0.03
	50–100	6.00±0.09	1.16±0.05	0.14±0.01	0.4±0.2	0.08±0.01	1.82±0.10	0.41±0.06	2.31±0.15	1.61±0.03
	100–150	6.05±0.06	0.85±0.04	0.10±0.00	0.4±0.1	0.04±0.00	1.87±0.05	0.38±0.07	2.29±0.11	
	150–200	5.98±0.09	0.54±0.02	0.06±0.00	0.4±0.0	0.03±0.01	1.20±0.07	0.33±0.09	1.56±0.12	
No Till	0–5	5.12±0.14	17.93±0.42	1.53±0.04	11.7±2.2	0.34±0.04	4.06±0.35	0.90±0.06	5.32±0.43	1.32±0.02
	5–15	5.09±0.12	6.92±0.31	0.54±0.03	4.3±1.2	0.27±0.05	2.31±0.13	0.57±0.03	3.18±0.17	1.50±0.06
	15–30	5.45±0.04	3.45±0.25	0.30±0.01	1.2±0.2	0.19±0.03	2.00±0.02	0.67±0.04	2.86±0.05	1.55±0.05
	30–50	5.74±0.06	2.14±0.13	0.21±0.01	0.5±0.1	0.14±0.02	1.97±0.10	0.69±0.04	2.80±0.10	1.60±0.05
	50–100	5.87±0.08	1.19±0.08	0.14±0.01	0.4±0.0	0.08±0.01	2.04±0.07	0.43±0.04	2.54±0.07	1.67±0.04
	100–150	5.93±0.05	0.98±0.07	0.11±0.00	0.3±0.0	0.04±0.01	1.99±0.10	0.30±0.07	2.32±0.06	
	150–200	5.93±0.05	0.64±0.06	0.07±0.01	0.3±0.0	0.03±0.00	1.48±0.15	0.27±0.11	1.77±0.19	
Forest	0–5	4.33±0.15	22.4±1.44	1.55±0.11	1.5±0.7	0.23±0.05	2.42±0.51	0.81±0.13	3.97±0.53	1.06±0.05
Succession	5–15	4.04±0.10	8.02±0.34	0.60±0.04	1.0±0.4	0.11±0.03	0.85±0.33	0.38±0.07	2.34±0.20	1.38±0.02
	15–30	4.38±0.06	3.52±0.18	0.30±0.01	0.5±0.1	0.06±0.01	1.34±0.19	0.49±0.02	2.44±0.14	1.65±0.02
	30–50	4.76±0.12	2.11±0.07	0.19±0.01	0.4±0.0	0.04±0.01	1.62±0.20	0.37±0.01	2.23±0.16	1.57±0.06
	50–100	5.19±0.14	1.40±0.21	0.14±0.01	0.4±0.1	0.03±0.00	1.46±0.09	0.37±0.04	1.89±0.08	1.63±0.04
	100–150	5.50 ±0.16	0.81 ±0.06	0.08±0.01	0.4±0.1	0.02±0.00	1.04±0.11	0.46±0.11	1.54±0.16	
	150–200	5.30 ±0.21	0.61 ±0.07	0.06±0.01	0.3±0.1	0.03±0.00	0.61±0.13	0.55±0.15	1.20±0.26	

N = 4 per treatment type. Soils were collected in 2009.
[1]Salt pH in 0.01 M $CaCl_2$.
[2]Mehlich I extractable P and cations.
[3]Effective cation exchange capacity by sum of cations method.

and then composited by depth. Composite samples were passed through a 10 mm mesh by gently breaking the cores along planes of weakness before air-drying. These air-dried samples were dry-sieved by hand through 6.3, 4, 2, and 1 mm screens to obtain dry aggregate size classes. A separate subsample of the <1 mm dry aggregates was sieved on a 0.25 mm screen. The objective of dry-sieving was to determine the initial aggregate size distribution before wet-sieving [33]. Sub-samples of 50 g were then weighed out by re-combining all the dry aggregate size classes according to their proportional mass for the bulk soil. This ensured that each sub-sample accurately represented the bulk soil dry aggregate size distribution [34]. These 50 g sub-samples were then re-wetted via capillary action followed by wet-sieving with a modified Yoder apparatus [35], which permits complete recovery of all soil from each sample. The method obtains four wet aggregate size classes: (i) >2000 µm; (ii) 250–2000 µm; (iii) 53–250 µm; and (iv) <53 µm. Wet aggregate mean-weighted diameters (MWD) were calculated by multiplying the proportion of soil in each of these four aggregate size classes by the mid-point of the size class. The calculation of the dry MWD used the same size divisions as the wet MWD except no <53 µm fraction was included, since dry sieving material through a 53 µm sieve is not practical. The aggregate stability index was then calculated by dividing the wet MWD by the dry MWD; an index of 1 represents perfect structural stability.

Soil organic C (SOC) fractionation and determination

Soil sub-samples were pulverized in a Spex 8200 ball-mill grinder (Metuchen, NJ) and total SOC and N were determined by dry combustion on a CE Elantech NC 1110 analyzer (Lakewood, NJ). Earlier mineralogical work on subsamples showed no evidence of carbonates so no pre-treatments were performed (Devine, unpublished data, 2008). To obtain particulate organic carbon (POC) and fine-C fractions, a 3–5 g subsample of each aggregate size fraction was dispersed by adding 15–25 mL of 0.5% Na-hexametaphosphate and shaking in 50 mL centrifuge tubes for 16 h on a reciprocal shaker at 200 rpm. The resulting soil slurry was washed through a 53 µm sieve using 0.5 L of DI water to fractionate the soil into: (1) sand+POC and (2) silt+clay+fine-C. A rubber policeman was used to aid in dispersion of stable soil aggregates on top of the 53 µm sieve when necessary. After drying both fractions in a forced-air oven at 60°C, fine-C and POC fractions were pulverized and C and N determined. A mass correction was made for the amount of Na-HMP recovered in the fine fraction prior to estimation of fine-C and N concentrations.

Since the <53 µm fraction does not contain sand, total SOC and POC are calculated on a sand-free basis for each aggregate size fraction in order to make comparisons among all the aggregate size classes [36]:

$$\frac{g\_POC}{kg\_sandfree\_aggregate} = \left(\frac{g\_POC}{kg\_sand}\right) * \left(\frac{kg\_sand}{kg\_silt + clay}\right)$$

$$\frac{g\_SOC}{kg\_sandfree\_aggregate} = (\frac{g\_SOC}{kg\_soil}) * (\frac{kg\_soil}{kg\_silt + clay})$$

Soil moving experiment

A soil moving experiment (SME) was carried out to determine the extent to which destructured soil re-develops aggregate stability among the different land uses. Soil moving or transplant experiments have been carried out to investigate soil processes including C accumulation or N cycling [37,38]. In this study, destructured soil was utilized, as opposed to intact cores, which is similar to the use of sand or sieved soil used in root ingrowth core experiments [39]. First, CT soil from 0–15 cm was collected from all four plots. CT soil was utilized since previous research at HSB has measured this soil to have the least SOC and aggregate stability. Sole use of CT soil, however, also assumes that there is no inherent internal difference between destructured CT, NT, or FS soil and that differences during the experiment would result only from external differences due to land use. After collection soil was dried at 60°C in a forced-air oven, and crushed with a rolling pin to destroy larger macroaggregates and to pass a 1 mm sieve, referred to henceforth as "initial soil." A 10 cm diameter auger was used to remove soil to 20 cm depth from four locations in each CT, NT, and FS plot at the end of July 2007. A PVC pipe of an equal diameter was placed as an insert in each hole to prevent the side walls from collapsing during filling. Each hole was then filled with the initial soil and uniformly tamped with a capped PVC pipe to a bulk density of approximately 1.4 g soil cm^{-3}. Soil packing was tested with success in the laboratory by filling the same PVC pipe insert to the desired bulk density. The PVC insert was removed after packing, and the edge of the SME core marked with flags and aluminum nails. At the end of August 2008, the same 5 cm diameter corer previously used to sample aggregates was used to extract a smaller soil core within the 10 cm diameter SME core. All cores were located (some with the aid of a metal detector), divided into 0–5 and 5–15 cm depth classes, and composited by plot. In one FS plot, only two cores were sampled due to obvious mixing with native soil that had occurred from animal activity; in three other plots (two CT and one FS), only three cores were sampled due to obvious animal activity.

The same procedure as described above was used to measure aggregate stability, total SOC, and POC in the two depth classes. Microbial biomass C was also determined on field moist, replicate subsamples following the 24 hour chloroform fumigation—K_2SO_4 extraction procedure [40]. Extracts were analyzed for non-purgable organic carbon (NPOC) (Shimadzu TOC 5000, Columbia, Maryland, USA). Microbial biomass C (MBC) was estimated following Wu et al. [41]:

$$MBC = (NPOC_{fumigated} - NPOC_{non-fumigated}) * 2.22$$

Statistical analysis

All statistical analyses were completed with SAS 9.3.1 (SAS, Cary, NC). All data were checked for normality using Shapiro-Wilk and Kolmogorov-Smirnoff but no transformations were deemed necessary. Differences in soil properties were tested with a one-way ANOVA by each depth class and aggregate size combination with land use as the main effect (n = 4). Equality of variance among treatment classes was ensured with Levine's test. The same analysis was performed by each depth class and land use combination with aggregate size as the main effect. For the soil moving experiment where starting values of soil properties were known, changes in total SOC, POC, and fine-C were analyzed using a repeated measures ANOVA by depth class to investigate the effect of time and the land-use *x* time interaction. When the main effects were significant, post-hoc tests were performed using the LSMEANS statement in PROC GLM to evaluate the significance of mean separations with p-values adjusted by Tukey's honestly significant difference ($\alpha = 0.05$).

Results

Aggregate size distributions and stability

Aggregate stability decreased significantly from FS and NT to CT in the upper 5 cm (Figure 1; see also Table S1). In the 5–28 cm layers FS soil was significantly more stable than both NT and CT (Figure 1). This response can also be seen as: (i) a higher percentage of wet-sieved, large macroaggregates (>2000 μm) in the FS and NT soils from 0–5 cm compared to CT, (ii) more large macroaggregates in FS from 15–28 cm compared to both CT and NT, and (iii) a higher proportion of fines (<53 μm) in 0–15 cm under CT compared to both NT and FS (Figure 2; see also Table S2).

Land use effects on aggregate associated C

There were significant effects of land use on C fraction concentrations in all aggregate size classes from 0–5 cm and the majority of size classes from 5–15 cm. From 0–5 cm, NT and FS aggregate size fractions were significantly elevated for SOC and fine fractions compared to CT. In POC, increasing C from CT>NT>FS was evident in all aggregate sizes but significant at 0.05 for only the >2000 μm (Figure 3; see also Table S3, part 1). From 5–15 cm, there were no significant differences between CT and NT for any of the size class and C fraction combinations, while the two largest aggregate size classes were significantly elevated in FS with respect to NT for SOC and all three classes were greater for POC (Figure 4; see also Table S4, part 1). FS also exceeded CT but only in the small and micro aggregate fractions for POC. There were no significant differences for any of the size class and C fraction combinations from 15–28 cm (Figure 5; see also Table S5, part 1).

Overall, the majority of C at HSB under all land uses is contained in the water-stable macroaggregate fraction to a depth of at least 28 cm (Figure 6; see also Table S6). By summing the C contribution from each aggregate size fraction to calculate a C concentration for each depth class by treatment combination, differences in POC from 0–5 cm explained 34% of the total SOC difference between NT and CT, 43% of the total SOC difference between FS and CT, and 51% of the total SOC difference between FS and NT. From 5–15 cm, differences in POC explained 56% of the difference between FS and CT and 47% of the difference between FS and NT.

Aggregate associated C within a land use

In terms of significant differences among the aggregate size classes within a land use and depth class, the most consistent result was that the <53 μm aggregate fraction had lower total SOC concentrations when expressed on a sand-free basis (Figures 3–5; see also Table S3, part 2, Table S4, part 2, and Table S5, part2). Under FS from 0–15 cm, there were no other significant differences among the size classes >53 μm (Figures 3 and 4) and there were no significant differences in POC among the aggregate size classes for any depth (Figures 3–5).

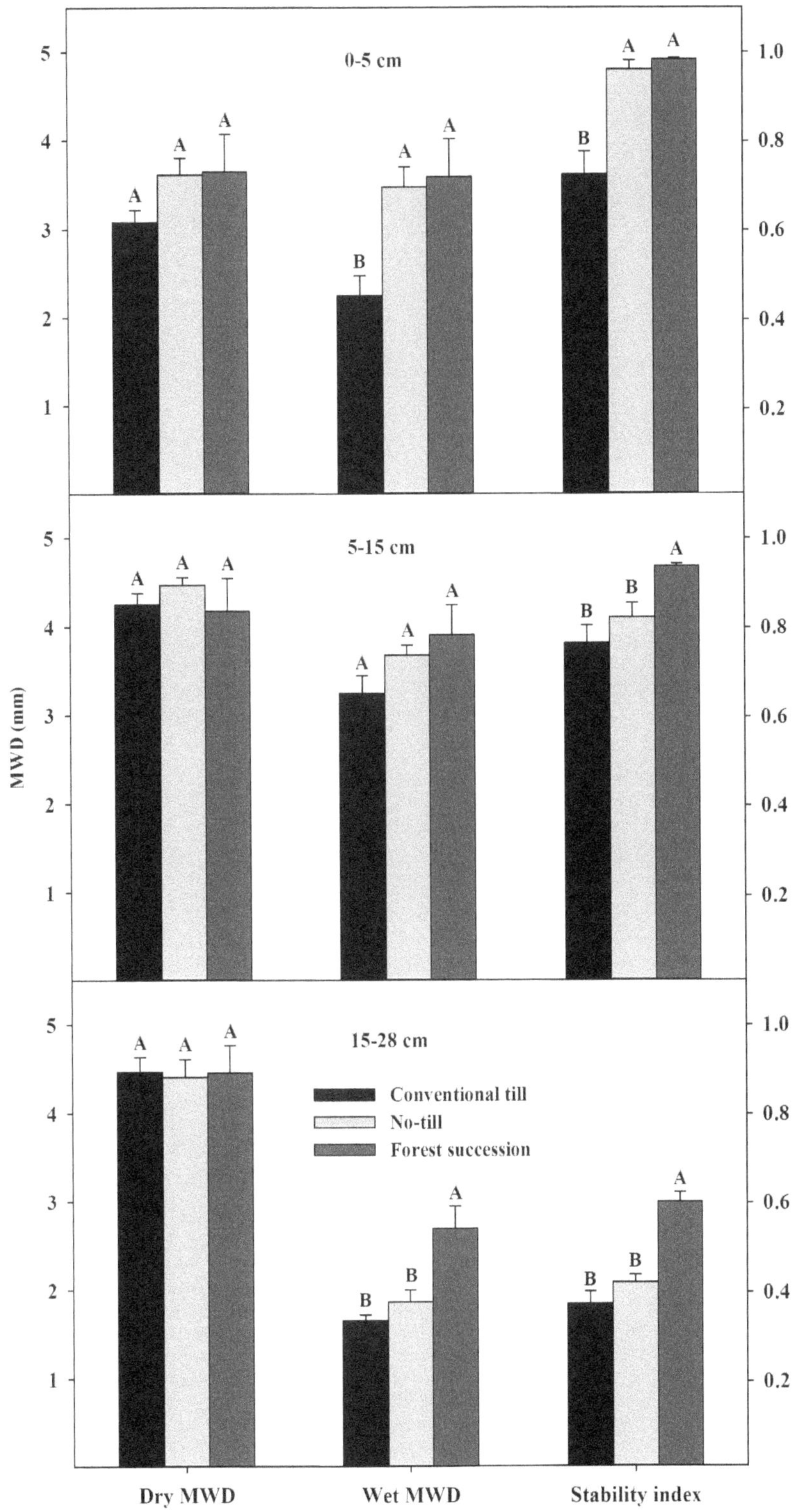

Figure 1. Dry-sieved and wet-sieved aggregates' mean weight diameter (MWD) (mean ± 1 SE) at the Horseshoe Bend Agroecosystem Experiment (n = 4), Athens, GA, USA, October 2007. The aggregate stability index is calculated by dividing the Wet MWD by the Dry MWD. Different letters indicate statistical significance between land uses within a depth class based on Tukey's HSD with a = 0.05. See also Table S1

Under NT, in the upper 15 cm both the large macroaggregates and the <53 μm fractions were SOC depleted relative to the small macroaggregates (250–2000 μm) and microaggregates (Figures 3 and 4). This was similar under CT from 0–5 cm where the small macroaggregates (250–2000 μm) were significantly elevated in total SOC and POC compared to the other size fractions

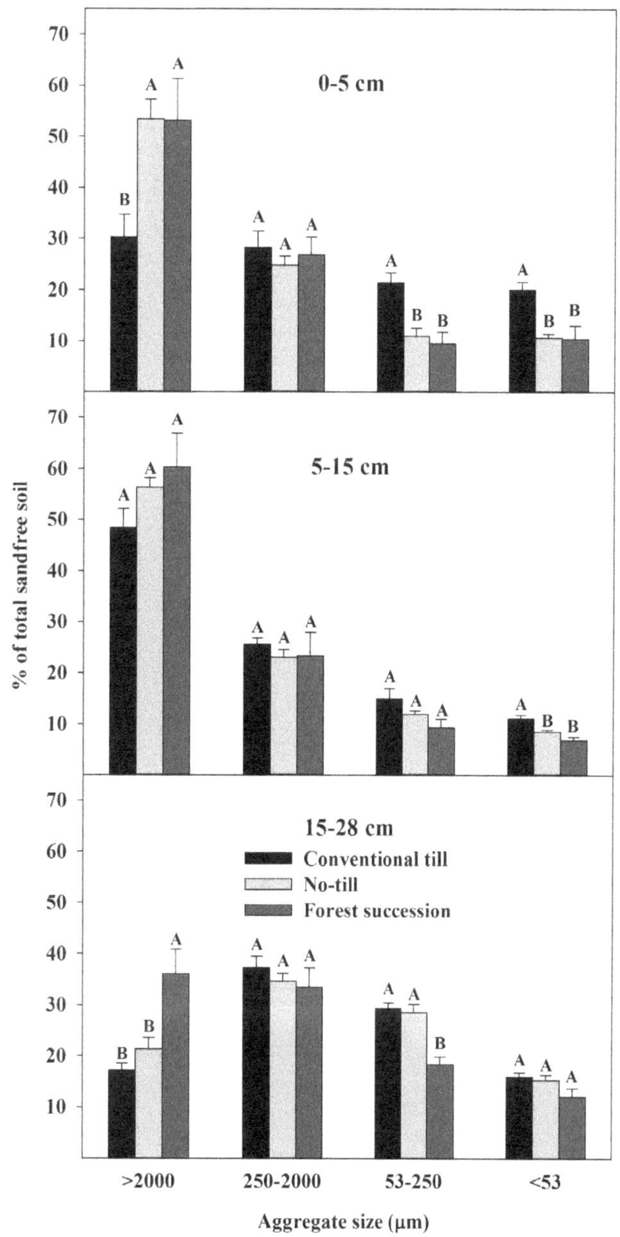

Figure 2. Distribution of water-stable aggregates on a sand-free soil basis (mean ±1 SE) at the Horseshoe Bend Agroecosystem Experiment (n = 4), Athens, GA, USA, October 2007. Different letters within an aggregate size class and depth class represent a significant difference between land uses (α = 0.05). See also Table S2.

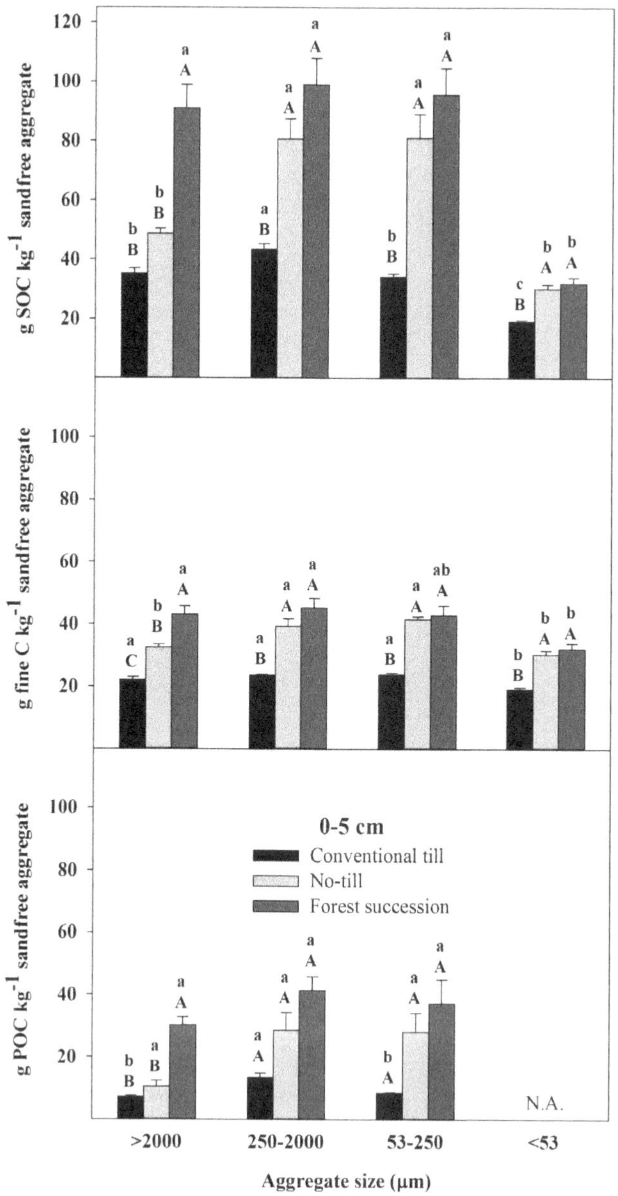

Figure 3. Total soil organic C (SOC), fine C (<53 μm), and particulate organic C (POC) (53–2000 μm) concentrations on a sand-free basis from 0–5 cm in four different aggregate size fractions (mean ±1 SE) in the Horseshoe Bend Agroecosystem Experiment (n = 4) Athens, GA, USA, October 2007. Different uppercase letters within an aggregate size class and C fraction represent a significant difference between land uses. Different lowercase letters within a C fraction and a land use represent a significant difference among aggregates size classes (α = 0.05). See also Table S3, part 1 and Table S3, part 2.

(Figure 3). Under CT from 5–15 cm, both the small macroaggregates and microaggregates were elevated in total SOC and POC compared to the large macroaggregates but only the small macroaggregates were significantly elevated compared to the large macroaggregates and only for POC (Figure 4). From 15–28 cm, the large macroaggregates were the most carbon rich under all land uses, having significantly greater total SOC and fine C concentrations compared to the <250 μm size fractions (Figure 5).

Recovery of POC and fine C

Summation of POC and fine C compared to direct analysis of total SOC revealed that recoveries of POC and fine C were, on average, 86% from 0–5 cm, 88% from 5–15 cm, and 92% from 15–28 cm. A precipitate formed on the glass beakers used to dry the fine C slurry and was more difficult to remove in the beakers containing 0–5 cm samples, suggesting this may have been the source of the <100% recovery of C.

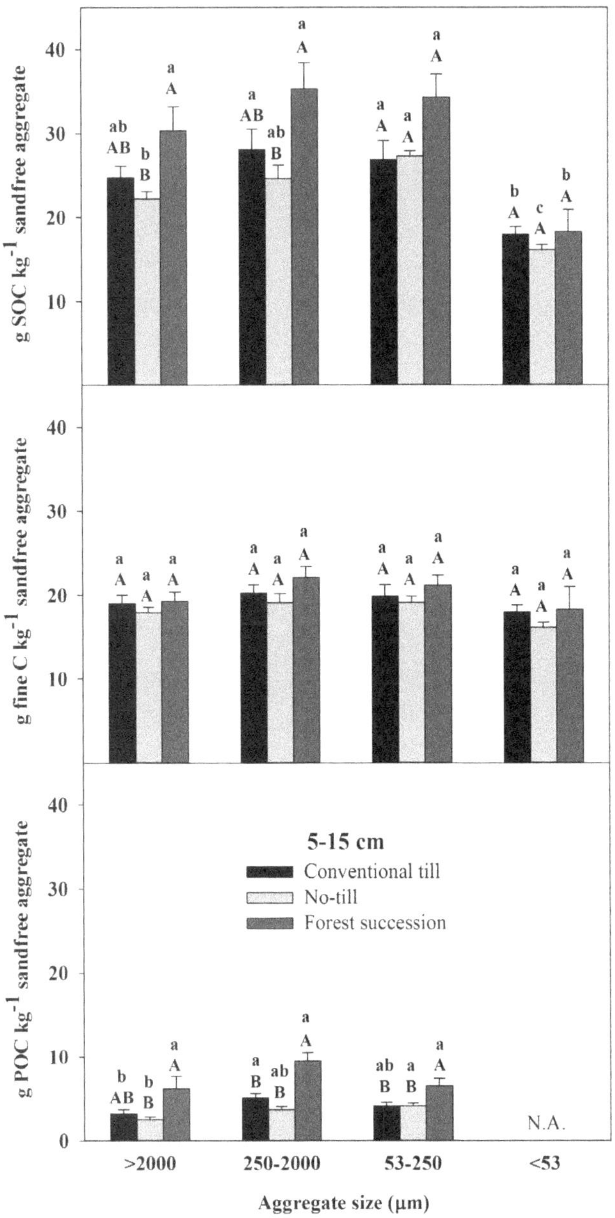

Figure 4. Total soil organic C (SOC), fine C (<53 μm), and particulate organic C (POC) (53–2000 μm) concentrations on a sand-free basis from 5–15 cm in four different aggregate size fractions (mean ±1 SE) in the Horseshoe Bend Agroecosystem Experiment (n = 4) Athens, GA, USA, October 2007. Different uppercase letters within an aggregate size class and C fraction represent a significant difference between land uses. Different lowercase letters within a C fraction and a land use represent a significant difference among aggregates size classes ($\alpha = 0.05$). See also Table S4, part 1 and Table S4, part 2.

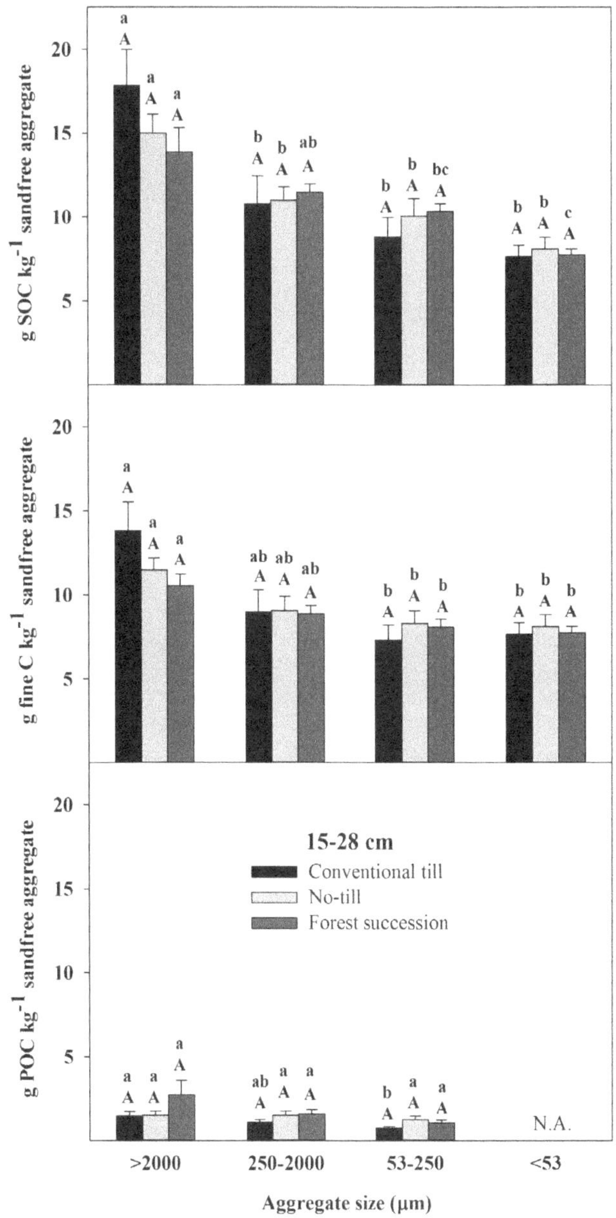

Figure 5. Total soil organic C (SOC), fine C (<53 μm), and particulate organic C (POC) (53–2000 μm) concentrations on a sand-free basis from 15–28 cm in four different aggregate size fractions (mean ±1 SE) in the Horseshoe Bend Agroecosystem Experiment (n = 4) Athens, GA, USA, October 2007. Different uppercase letters within an aggregate size class and C fraction represent a significant difference between land uses. Different lowercase letters within a C fraction and a land use represent a significant difference among aggregates size classes ($\alpha = 0.05$). See also Table S5, part 1 and Table S5, part 2.

Soil moving experiment (SME): Structural recovery, C change, and microbial biomass

The SME soil under FS had a significantly greater stability index compared to both CT and NT at 0–5 and 5–15 cm after one year (Table 2; see also Table S7). There were no significant differences in aggregate stability between CT and NT at either depth in the SME (Table 2). Overall, aggregate stability tended to be greater from 0–5 cm compared to 5–15 cm under NT and FS. This depth stratification did not occur under CT (Table 2).

Repeated measures analysis showed a significant time x treatment interaction effect from 0–5 cm for total SOC ($p = 0.03$) and POC ($p = 0.0004$) but not fine C ($p = 0.50$), suggesting a significant departure from initial values of SOC and POC was dependent upon land use. The overall effect of time from 0–5 cm was significant for POC ($p = 0.0007$), which increased after one year. The effect of time was not significant

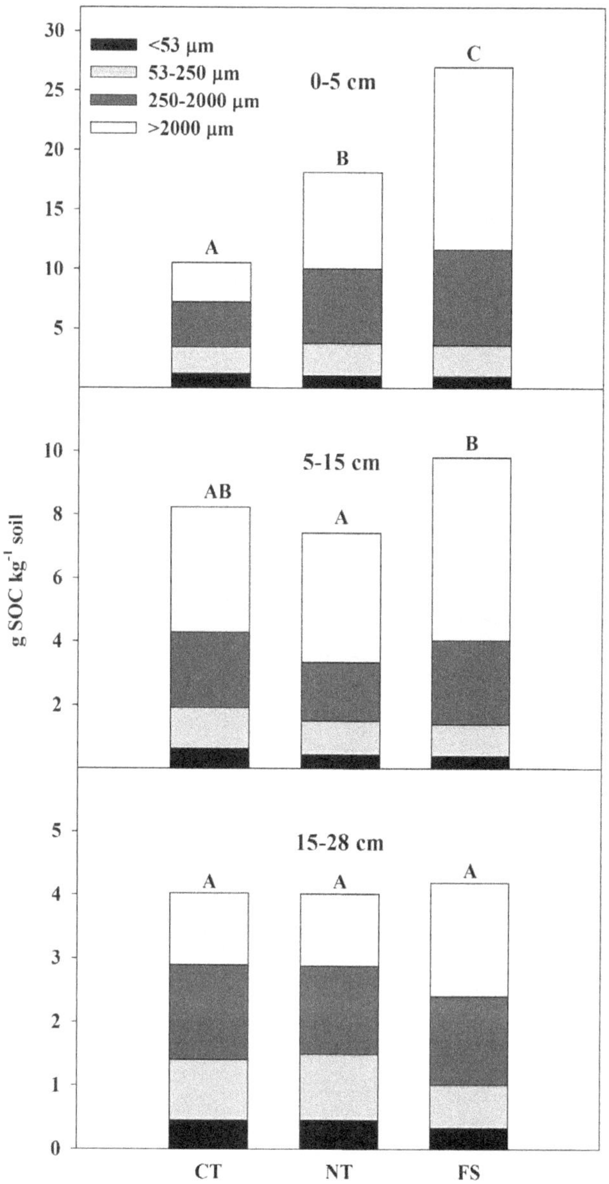

Figure 6. Contributions of four different aggregate size fractions to total soil organic C (SOC) concentrations. Different letters indicate a significant difference between land uses in a depth class ($\alpha = 0.05$). Means of the sums of the four fractions are shown (n = 4) from Horseshoe Bend, Athens, GA, USA, October 2007. See also Table S6.

for SOC ($p = 0.27$) or fine C ($p = 0.35$) for 0–5 cm, which means there was no clear directional change for these components compared to initial values across all land uses (Table 3: see also S8). A test of mean separation showed that total SOC from 0–5 cm was significantly elevated in the FS soil compared to CT ($p = 0.08$) and compared to NT ($p = 0.03$) (Table 3). POC was also significantly elevated from 0–5 cm in the FS soil compared to both CT and NT ($p < 0.002$) (Table 3; see also Table S8). From 5–15 cm, declines in C were observed after one year. The effect of time was significant for SOC ($p < 0.0001$), POC ($p = 0.0004$), and fine C ($p < 0.0001$), meaning a significant decrease in these components from the initial values had occurred (Table 3). The time x treatment interaction was significant for POC at $p = 0.099$

Table 2. Aggregate stability (mean ±1 SD) at two depths at the end of a soil moving experiment (n = 4).

Depth	*Land use*	*Dry MWD*[†]	*Wet MWD*	*Stability index*
cm		-----------mm-----------		Wet/Dry
0–5	Conventional Till	4.3 ± 0.4^a	1.4 ± 0.2^a	0.33 ± 0.06^a
	No-Till	4.5 ± 0.3^a	1.9 ± 0.3^a	0.43 ± 0.09^a
	Forest Succession	3.9 ± 0.3^a	3.1 ± 0.2^b	0.79 ± 0.03^b
5–15	Conventional Till	4.8 ± 0.6^a	1.2 ± 0.2^a	0.26 ± 0.08^a
	No-Till	4.8 ± 0.6^a	1.0 ± 0.2^a	0.22 ± 0.05^a
	Forest Succession	3.6 ± 0.5^b	2.1 ± 0.3^b	0.60 ± 0.10^b

Conventional till soil was initially dried and crushed to pass a 1 mm sieve and then installed under three different land uses for a period of one year at Horseshoe Bend, Athens, GA, USA, July 2007–August 2008.
Different letters indicate statistical significance among treatment means within a depth class based on Tukey's HSD with $\alpha = 0.05$. See also Table S7.
[†]Mean-weighted diameter.

possibly reflecting a lower net loss of POC from 5–15 cm under FS (Table 3). The interaction was not significant for SOC ($p = 0.85$) or fine C ($p = 0.89$), since there was a more uniform decrease in these components from the initial value under all the land uses.

SME microbial biomass carbon (MBC), like aggregate stability and total SOC from 0–5 cm, was elevated in FS relative to the agroecosystem soils, though not significantly ($p > 0.27$; Table 3). Overall, there were no significant differences among the land uses from 0–5 cm in terms of MBC concentrations. CT MBC were significantly elevated compared to NT from 5–15 cm ($p = 0.02$). For all three land uses, MBC tended to be higher from 0–5 cm compared to 5–15 cm (Table 3).

Discussion

Changes in soil structure and SOC after 30 years of different management

Earlier work from HSB reported that the three decades of divergent land use have led to 20% greater SOC content in the FS

Table 3. Total soil organic C (SOC), particulate organic C (POC) and microbial biomass C (MBC) (mean ±1 SD) at two depths at the end of a soil moving experiment (n = 4).

Land use	*Depth*	*SOC*	*POC*	*Fine C*	*MBC*
	-cm-	---------------g kg^{-1} soil--------------			--mg kg^{-1}--
Initial	0–5	9.2 ± 0.2	1.8 ± 0.1	7.5 ± 0.1	NA
Conventional till		9.1 ± 0.3^a	2.0 ± 0.2^a	7.2 ± 0.2^a	397 ± 27^a
No-till		8.7 ± 0.2^a	1.7 ± 0.1^a	7.0 ± 0.2^a	366 ± 30^a
Forest succession		10.9 ± 1.7^b	3.3 ± 0.6^b	7.6 ± 1.2^a	422 ± 72^a
Initial	5–15	9.2 ± 0.2	1.8 ± 0.1	7.5 ± 0.1	NA
Conventional till		8.3 ± 0.2^a	1.5 ± 0.1^a	6.8 ± 0.2^a	317 ± 25^a
No-till		8.3 ± 0.2^a	1.4 ± 0.1^a	6.8 ± 0.3^a	274 ± 13^b
Forest succession		8.4 ± 0.6^a	1.7 ± 0.2^a	6.7 ± 0.5^a	305 ± 16^{ab}

Conventional till soil was initially dried and crushed to pass a 1 mm sieve and then installed under the three different land uses for a period of one year at Horseshoe Bend, Athens, GA, USA, July 2007–August 2008.
Different letters indicate statistical significance between treatment means within a depth class based on Tukey's HSD with $\alpha = 0.05$. See also Table S8.

(61.5±4.3 Mg ha^{-1}) and NT (60.0±4.5 Mg ha^{-1}) soil profiles (0–2 m) relative to CT (51.9±4.9 Mg ha^{-1}) [7]. SOC content differences in the 0–5 cm layer accounted for 56% of the 0–2 m mean content difference between FS and CT and 66% of the 0–2 m content difference between NT and CT. These increases in SOC at the surface under FS and NT are linked to enhanced aggregate stability.

The increase in structural stability under NT compared to CT was limited to a depth of 5 cm, while the FS soil demonstrated significantly greater stability compared to CT to a depth of 28 cm and compared to NT from 5–28 cm (Figure 1). Previous work at HSB on aggregate size fractions showed that all aggregates >106 μm from 0–5 cm were significantly more stable in NT compared to CT and from 5–15 cm the large macroaggregates (>2000 μm) were more stable in NT [23]. In the current study, the difference in aggregate stability between NT and FS from 5–15 cm is evident in the aggregate stability metric (Figure 1). The stability metric had lower coefficients of variation within a land-use than did either the dry or wet mean-weighted diameter, indicating this metric may be more sensitive to the effects of land use on soil structure than analysis of aggregate size distributions produced from one methodology (i.e., wet or dry sieving).

Significant differences in SOC were evident among the land uses in different aggregate size fractions. Both NT and FS generally had elevated concentrations in all fractions compared to CT from 0–5 cm with increases in POC and fine C each explaining 40–50% of concentration differences (Figure 3). With respect to our initial hypothesis this relatively equal contribution indicates that both stabilization of POC within microaggregates formed in stable macroaggregates [13,42], and flux of DOC into unsaturated micropores [15] might play a role in soil carbon accumulation. The effect of tillage could disrupt either of these processes and afforestation did not favor either process in the 0–5 cm surface soil.

The FS soil also had elevated concentrations of SOC from 5–15 cm compared to NT for macro- and microaggregates and compared to CT for microaggregates with the increases in POC explaining 35–60% of the concentration differences (Figure 4). For CT and NT there were no differences in SOC concentrations among the aggregate size fractions from 5–15 cm, which has been found previously at HSB [6,23,24,29]. Below 15 cm differences in aggregate stability were evident but significant differences in SOC fractions among the aggregate size fractions were not evident (Figure 5). This demonstrates that enhanced aggregate stability is not always linked to increased stabilization and storage of SOC.

Within the FS soil from 0–15 cm, only the wet-sieved fine fraction (<53 μm) had a significantly lower C concentration than the other fractions within each depth class (Figures 3 and 4), highlighting the importance of POC to increasing total SOC under afforestation. In both agroecosystems, large macroaggregates have significantly lower proportions of POC compared to small macroaggregates at all depths (Table 4; see also Table S9) and most of the difference in SOC concentration at 0–5 cm and 5–15 cm is accounted for by differences in macroaggregate C (Figure 6). As such, most of the C difference in the agroecosystem compared to the FS soil occurs at the macroaggregate level.

In previous research at HSB, Beare et al. [5] observed the highest C concentrations in the large NT microaggregates (106–250 μm) compared to the macroaggregates from 0–5 cm and speculated that this was due to formation of microaggregates around decomposing residues in stable macroaggregates, followed by their release upon macroaggregate breakdown, as originally formulated by Oades [12]. Work that followed this at HSB also found the highest C concentrations in a broader 53–250 μm microaggregate class from 0–5 cm in NT [6], but this result was not found more recently [24] or in the present work, in which small macroaggregates were found to have similar C concentrations compared to microaggregates (Figure 3). These contrasting results at the same study site may be due to methodological differences in aggregate sieving. Beare et al. [5] used tension-wetted field-moist soil before wet-sieving and Bossuyt et al. [6] used capillary-wetted field moist soil before wet-sieving, whereas Arce Flores and Coleman [24] and the present work air-dried soils for storage and then capillary-wetted soils before wet-sieving. Beare and Bruce [43] clearly demonstrated that aggregate size distributions depend upon both pre-treatment and sieving methodologies.

Finally, aggregate hierarchy theory, which suggests that inter-microaggregate organic matter should increase C concentration of stable macroaggregates compared to microaggregates, is only evident in the 15–28 cm depth class (Figures 3–5). In all the 0–15 cm soils, large macroaggregates (>2000 μm) were relatively depleted in total SOC compared to the small macroaggregates (250–2000 μm) and microaggregates (53–250 μm). The method for isolating water-stable aggregates used repeatedly at HSB may not be forceful enough to isolate large macro-aggregates of higher C concentration in the upper 15 cm. If a more forceful fractionation procedure was used, such as sonication or slaking, then perhaps low C macroaggregates would be dispersed and their particles transferred to smaller fractions. Previous research has also shown that wet-sieving of pre-wetted soil does not support aggregate hierarchy theory, as was done in this study, while slaking dry soil does [11]. Also, aggregate hierarchy has been shown to be less apparent in soils dominated by 1:1 clays with high oxide contents like at HSB compared to soils dominated by 2:1 clays with low oxide contents [44,45].

Soil moving experiment

Significant differences in aggregate stability were evident after moving crushed and sieved (1 mm) soil from the CT plots back to each land use for both the 0–5 and 5–15 cm depth classes after one year (Table 2), suggesting that water-stable macroaggregates can be quickly formed under the different land uses but most quickly under FS, a result consistent with our initial hypothesis.

Similar rapid changes in soil structure were observed in a soil moving experiment in the tropics where larger blocks of soil were transferred from forest to pasture and vice-versa to study the effects of an invasive earthworm species [46]. The effects of different drying-wetting cycles, root densities, and levels of microbial activity are also known to affect soil structural development over time periods of days to a few years [47,48], so the rapid, differential development of aggregate stability under different land uses is consistent with previous and current observations.

Since forest soils are fungal-dominated systems, any observed difference in aggregate stability was hypothesized to result from more rapid binding by fungal hyphae under FS but no significant differences were observed in MBC between land uses (Table 3). The chloroform-incubation method for MBC is related to direct counts of fungal and bacterial cells [49] but the method has failed to efficiently extract fungal biomass from some forest soils [50]. On the other hand, previous work at HSB has shown a sensitivity of the chloroform-incubation method to treatments since both MBC and fungal hyphae lengths significantly decreased after application of fungicide to a mixed meadow that had been recently plowed at HSB [51]. Furthermore, the importance of fungal hyphae in stabilizing aggregates was demonstrated in the agroecosystems by treating sub-plots with fungicides and observing declines in

Table 4. The proportion of particulate organic carbon (POC) within aggregate fractions and whole soil (mean ±1 SD).

Land use	*Depth*	*Large macroaggregates*	*Small macroaggregates*	*Microaggregates*	*Whole soil*[1]
	--cm--	---% POC---			
CT	0–5	20.2±1.3[a]	31.0±4.2[a]	24.7±0.6[a]	22.6±0.9[a]
NT		21.3±6.3[a]	34.8±7.9[a]	33.7±7.3[a]	26.9±6.6[a]
FS		33.8±5.7[b]	41.8±4.6[a]	38.1±9.8[b]	35.5±5.1[b]
CT	5–15	13.0±2.4[ab]	18.4±1.9[a]	15.3±1.2[ab]	13.8±1.8[a]
NT		11.4±2.6[b]	14.9±2.6[a]	15.0±2.2[a]	12.1±2.3[a]
FS		19.7±6.4[a]	26.9±4.7[b]	18.8±2.0[b]	20.4±5.1[b]
CT	15–28	8.2±0.9[a]	10.4±1.2[a]	8.8±0.8[a]	8.2±0.5[a]
NT		10.1±2.0[a]	13.6±2.8[a]	12.5±1.9[a]	10.8±1.2[a]
FS		20.0±13.6[a]	13.7±3.5[a]	10.3±2.7[a]	15.1±6.7[a]

Horseshoe Bend, Athens, GA, USA, October 2007.
Different letters indicate statistical significance between treatment means within a depth class and aggregate class based on Tukey's HSD with $\alpha = 0.05$. See also Table S9.
[1]Calculated by summation of individual fractions.

aggregate size distributions [29]. As such, further efforts will be needed to conclusively demonstrate the role of fungal hyphae in aggregate formation in FS at HSB.

Along with aggregate stability, total SOC also increased significantly from 0–5 cm under FS from 9.2 g SOC kg^{-1} soil to 10.8 g SOC kg^{-1} soil after one year. This increase in SOC could, however, be explained by mixing of a relatively small portion of the *in situ* FS soil SOC (9–22%) with the SME soil. In contrast, however, the significant increase in POC from 1.8 g POC kg^{-1} soil to 3.3 g POC kg^{-1} soil, which explains 90% of the increase in SOC, cannot be explained by *in situ* soil mixing since POC in 0–5 cm *in situ* forest soil accounts for only 35% of total SOC. The observed increase, which is equal to an increased storage of approximately 1 Mg C ha^{-1}, could result partly from leaf litter inputs under FS that are estimated to be 2 Mg C ha^{-1} yr^{-1} (Markewtiz, 2009, unpublished data) or annual fine root production that has been reported to range globally from 0.3–8.2 Mg C ha^{-1} yr^{-1} [52]. Under a simulated no-till experiment with a Montana Hapludoll 14C labeled SOC in aggregates was derived predominantly from fine root litter [42,53].

Finally, decreases in SOC compared to the initial concentration from 5–15 cm under all three land uses showed that C mineralization exceeded new C inputs in this sub-surface layer. This enhanced decomposition was possibly triggered by the loss of macroaggregate structure from crushing the soil before its installation. Interestingly, decreases in POC concentrations explained less than half (11–36%) of the loss in total SOC under all three land uses, suggesting that fine C was more susceptible to decomposition after structural disruption. This reduction occurred even though the soil used in the moving experiment was CT soil in which macroaggregate protection of SOC had already been impacted during 30 years of continuous moldboard plowing.

Conclusions

Changes in aggregate size distributions after 30 years of differing land management were only clearly evident to a depth of 5 cm between conventional tillage (CT) and no-tillage (NT) systems. When adjacent land undergoing forest succession (FS) was added to the comparison, structural changes were evident to 28 cm. Changes in SOC concentration and composition occurred along with changes in structural stability to a depth of 15 cm, consistent with a reduced capacity for tilled soil to physically protect organic matter from decomposition. Although differences in stability were evident from 15–28 cm, no significant difference in SOC concentration was observed among the land uses, indicating that increased aggregate stability is not always linked to an increase in SOC storage. SOC results lend support to two different models of how SOC becomes physically protected in systems lacking mechanical disturbance, one that highlights the importance of particulate organic carbon (POC) and the other dissolved organic carbon (DOC). Since POC contributed only about half of the differences in SOC between land uses, it is possible that micro-scale fluxes of DOC are also playing an important role in maintaining higher fine-C concentrations in soils with greater pore connectivity, such as the NT and FS soils in this study. The importance of macroaggregate stability for the protection of both POC and fine C was also demonstrated by the destruction of large macroaggregates in a soil moving experiment in which losses of C from 5–15 cm were observed under all land uses after 1 yr. The SME experiment also demonstrated that the recovery of water-stable macroaggregates can be rapid, since significant increases in stability under FS were observed in the experiment to 15 cm. Finally, a gain of 1 Mg C ha^{-1} was detected in the 0–5 cm soil layer under FS during the SME, 90% of which could be accounted for by a gain in POC. The stabilization of this POC may have been enhanced by recovery of water-stable aggregates under FS.

Supporting Information

Table S1 ANOVA results for Figure 1. ANOVA table reports tests of significance among Land Uses (conventional tillage, no tillage, forest succession) by aggregate attribute (Dry mean weighted diameter, Wet mean weighted diameter, Aggregate Stability {wet/dry}) and soil depth (0–5, 5–15, 15–28 cm).

Table S2 ANOVA results for Figure 2. ANOVA table reports tests of significance among Land Uses (conventional tillage, no tillage, forest succession) by aggregate size fraction (>2000, 250–2000, 53–250, and <53 μm) and soil depth (0–5, 5–15, 15–28 cm).

Table S3 1. ANOVA results for Figure 3 in soil depth 0–5 cm for land use. ANOVA table reports tests of significance among land uses (CT, NT, and FS) within a carbon fraction (Soil organic carbon, particulate organic carbon, fine carbon) and size class (>2000, 250–2000, 53–250, and <53 µm). 2. ANOVA results for Figure 3 in soil depth 0–5 cm for aggregate size class. ANOVA table reports tests of significance among size classes (2000, 250–2000, 53–250 and <53 µm) within a carbon fraction and land use. Note for POC the test is only for three size classes (2000, 250–2000, and 53–250 µm).

Table S4 1. ANOVA results for Figure 4 in soil depth 5–15 cm for land use. ANOVA table reports tests of significance among land uses (CT, NT, and FS) within a carbon fraction and size class. 2. ANOVA results for Figure 4 in soil depth 5–15 cm for aggregate size class. ANOVA table reports tests of significance among size classes (2000, 250–2000, 53–250 and <53 µm) within a carbon fraction and land use.

Table S5 1. ANOVA results for Figure 5 in soil depth 15–28 cm for land use. ANOVA table reports tests of significance among land uses (CT, NT, and FS) within a carbon fraction and size class. 2. ANOVA results for Figure 5 in soil depth 15–28 cm for aggregate size class. ANOVA table reports tests of significance among size classes (2000, 250–2000, 53–250 and <53 µm) within a carbon fraction and land use.

Table S6 ANOVA results for Figure 6. ANOVA table reports tests of significance among Land Uses (conventional tillage, no tillage, forest succession) by soil depth (0–5, 5–15, 15–28 cm) for the sum of C contents in all aggregate fractions.

Table S7 ANOVA results for Table 2. ANOVA table reports tests of significance among Land Uses (conventional tillage, no tillage, forest succession) by aggregate attribute (Dry mean weighted diameter, Wet mean weighted diameter, Aggregate Stability {wet/dry}) and soil depth (0–5, 5–15 cm).

Table S8 ANOVA results for Table 3. ANOVA table reports tests of significance among Land Uses (conventional tillage, no tillage, forest succession) by soil carbon component (soil organic carbon, particulate organic carbon, fine carbon, microbial biomass carbon) and soil depth (0–5, 5–15 cm).

Table S9 ANOVA results for Table 4. ANOVA table reports tests of significance among Land Uses (conventional tillage, no tillage, forest succession) by aggregate size fraction (>2000, 250–2000, 53–250, and <53 µm) and soil depth (0–5, 5–15, 15–28 cm).

Acknowledgments

We are indebted to the legacy of researchers who initiated and maintained the agroecosystem experiment at Horseshoe Bend. This includes but is not limited to: Drs. Gary Barrett, Eugene Odum, Dac Crossley, Benjamin Stinner, Peter Groffman, Michael Beare, and Shuijin Hu. Dr. David Radcliffe offered a critical review of the manuscript. Luke Worsham and Patrick Bussell provided valuable laboratory assistance. The UGA Graduate School and UGA Warnell School of Forestry and Natural Resources provided research support.

Author Contributions

Conceived and designed the experiments: SD DM PH DC. Performed the experiments: SD DM. Analyzed the data: SD DM. Contributed reagents/materials/analysis tools: DM PH DC. Wrote the paper: SD DM PH DC. Historical data: PH DC.

References

1. Mikutta R, Kleber M, Torn MS, Jahn R (2006) Stabilization of soil organic matter: association with minerals or chemical recalcitrance? Biogeochemistry 77: 25–56.
2. von Lutzow M, Kogel-Knabner I, Ekschmitt K, Matzner E, Guggenberger G, et al. (2006) Stabilization of organic matter in temperate soils: mechanisms and their relevance under different soil conditions - a review. European Journal of Soil Science 57: 426–445.
3. Six J, Conant RT, Paul EA, Paustian K (2002) Stabilization mechanisms of soil organic matter: Implications for C-saturation of soils. Plant and Soil 241: 155–176.
4. Six J, Elliott ET, Paustian K, Doran JW (1998) Aggregation and soil organic matter accumulation in cultivated and native grassland soils. Soil Science Society of America Journal 62: 1367–1377.
5. Beare MH, Cabrera ML, Hendrix PF, Coleman DC (1994) Aggregate-protected and unprotected organic-matter pools in conventional-tillage and no-tillage soils. Soil Science Society of America Journal 58: 787–795.
6. Bossuyt H, Six J, Hendrix PF (2002) Aggregate-protected carbon in no-tillage and conventional tillage agroecosystems using carbon-14 labeled plant residue. Soil Science Society of America Journal 66: 1965–1973.
7. Devine S, Markewitz D, Hendrix P, Coleman DC (2011) Soil carbon change through two meters during forest succession alongside a 30-yr agroecosystem experiment. Forest Science 57: 36–50.
8. Baker JM, Ochsner TE, Venterea RT, Griffis TJ (2007) Tillage and soil carbon sequestration – what do we really know? Agriculture, Ecosystems and Environment 118: 1–5.
9. Horn R, Taubner H, Wuttke M, Baumgartl T (1994) Soil physical-properties related to soil-structure. Soil & Tillage Research 30: 187–216.
10. Tisdall JM, Oades JM (1982) Organic matter and water-stable aggregates in soils. Journal of Soil Science 33: 141–163.
11. Elliott ET (1986) Aggregate structure and carbon, nitrogen, and phosphorus in native and cultivated soils. Soil Science Society of America Journal 50: 627–633.
12. Oades JM (1984) Soil organic-matter and structural stability -Mechanisms and implications for management. Plant and Soil 76: 319–337.
13. Six J, Elliott ET, Paustian K (2000) Soil macroaggregate turnover and microaggregate formation: a mechanism for C sequestration under no-tillage agriculture. Soil Biology & Biochemistry 32: 2099–2103.
14. Six J, Callewaert P, Lenders S, De Gryze S, Morris SJ, et al. (2002) Measuring and understanding carbon storage in afforested soils by physical fractionation. Soil Science Society of America Journal 66: 1981–1987.
15. Smucker AJM, Park EJ, Dorner J, Horn R (2007) Soil micropore development and contributions to soluble carbon transport within macroaggregates. Vadose Zone Journal 6: 282–290.
16. Chan KY (2001) Soil particulate organic carbon under different land use and management. Soil Use and Management 17: 217–221.
17. Tiessen H, Stewart JWB (1983) Particle-size fractions and their use in studies of soil organic matter: II. cultivation effects on organic-matter composition in size fractions Soil Science Society of America Journal 47: 509–514.
18. Causarano HJ, Franzluebbers AJ, Shaw JN, Reeves DW, Raper RL, et al. (2008) Soil organic carbon fractions and aggregation in the Southern Piedmont and Coastal Plain. Soil Science Society of America Journal 72: 221–230.
19. Franzluebbers AJ, Arshad MA (1997) Particulate organic carbon content and potential mineralization as affected by tillage and texture. Soil Science Society of America Journal 61: 1382–1386.
20. Skjemstad JO, Swift RS, McGowan JA (2006) Comparison of the particulate organic carbon and permanganate oxidation methods for estimating labile soil organic carbon. Australian Journal of Soil Research 44: 255–263.
21. Wander MM, Traina SJ, Stinner BR, Peters SE (1994) Organic and conventional management effects on biologically-active soil organic-matter pools Soil Science Society of America Journal 58: 1130–1139.
22. Cambardella CA, Elliott ET (1992) Particulate soil organic-matter changes across a grassland cultivation sequence. Soil Science Society of America Journal 56: 777–783.
23. Beare MH, Hendrix PF, Coleman DC (1994) Water-stable aggregates and organic-matter fractions in conventional-tillage and no-tillage soils. Soil Science Society of America Journal 58: 777–786.
24. Arce Flores S, Coleman DC (2006) Comparing water-stable aggregate distributions, organic matter fractions, and carbon turnover using ^{13}C natural

abundance in conventional and no tillage soils. Master's Thesis. Athens: Institute of Ecology, University of Georgia.
25. Hendrix PF (1997) Long-term patterns of plant production and soil carbon dynamics in a Georgia Piedmont agroecosystem. In: Paul EA, Paustian K, Elliott ET, Cole CV, editors. Soil Organic Matter in Temperate Agroecosystems: Long-Term Experiments in North America. Boca Raton, FL: CRC Press, Inc. pp. 235–245.
26. Barrett GW (1968) Effects of an acute insecticide stress on a semi-enclosed grassland ecosystem. Ecology 49: 1019–1035.
27. Bakelaar RG, Odum EP (1978) Community and population level responses to fertilization in an old-field ecosystem. Ecology 59: 660–665.
28. Odum EP, Pomeroy SE, Dickinson JC III, Hutcheson K (1973) The effect of a late winter litter burn on the composition, productivity, and diversity of a 4-year old fallow-field in Georgia. Proc Annu Tall Timbers Fire Ecol Conf: 399–412.
29. Beare MH, Hus S, Coleman DC, Hendrix PF (1997) Influences of mycelial fungi on soil aggregation and organic matter storage in conventional and no-tillage soils. Applied Soil Ecology 5: 211–219.
30. Coleman DC, Hunter MD, Hendrix PF, Crossley DAJ, Arce-Flores S, et al. (2009) Long-term consequences of biological and biogeochemical changes in the Horseshoe Bend LTREB agroecosystem, Athens, Ga. In: Bohlen P, editor. Agroecosystem management for ecological, economic and social sustainability. New York: Taylor and Francis.
31. Coleman D, Hunter M, Hendrix P, Crossley D, Simmons B, et al. (2006) Long-term consequences of biochemical and biogeochemical changes in the Horseshoe Bend agroecosystem, Athens, GA. European Journal of Soil Biology 42: S79–S84.
32. Lachnicht SL, Hendrix PF, Potter RL, Coleman DC, Crossley DA (2004) Winter decomposition of transgenic cotton residue in conventional-till and no-till systems. Applied Soil Ecology 27: 135–142.
33. Nimmo JR, Perkins KS (2002) Aggregate stability and size distribution. In: Dane JH, Topp GC, editors. Methods of Soil Analysis: Part 4–Physical Methods. Madison, Wisconsin, , USA: Soil Science Society of America, Inc. pp. 317–328.
34. Kemper WD, Chepil WS (1965) Size distribution of aggregates. In: Black CA, editor. Methods of soil analysis. Madison, WI: ASA. pp. 499–510.
35. Yoder RE (1936) A direct method of aggregate analysis of soils and a study of the physical nature of erosion losses. Journal of the American Society of Agronomy 28: 337–351.
36. Elliott ET, Palm CA, Reuss DE, Monz CA (1991) Organic-matter contained in soil aggregates from a tropical chronosequence - correction for sand and light fraction. Agriculture Ecosystems & Environment 34: 443–451.
37. Balser TC, Firestone MK (2005) Linking microbial community composition and soil processes in a California annual grassland and mixed-conifer forest. Biogeochemistry 73: 395–415.
38. Mack MC, D'Antonio CM (2003) Exotic grasses alter controls over soil nitrogen dynamics in a Hawaiian woodland. Ecological Applications 13: 154–166.
39. Neill C (1992) Comparison of soil coring and ingrowth methods for measuring belowground production. Ecology Letters 73: 1918–1921.
40. Vance ED, Brookes PC, Jenkinson DS (1987) An extraction method for measuring soil microbial biomass. Soil Biology & Biochemistry 19: 703–707.
41. Wu J, Joergensen RG, Pommerening B, Chaussod R, Brookes PC (1990) Measurement of soil microbial biomass C by fumigation extraction - an automated procedure Soil Biology & Biochemistry 22: 1167–1169.
42. Gale WJ, Cambradella CA, Bailey TB (2000) Root-derived carbon and the formation and stabilization of aggregates. Soil Science Society American Journal 64: 201–207.
43. Beare MH, Bruce RR (1993) A comparison of methods for measuring water-stable aggregates - implications for determining environmental-effects on soil structure. Geoderma 56: 87–104.
44. Six J, Paustian K, Elliott ET, Combrink C (2000) Soil structure and organic matter: I. Distribution of aggregate-size classes and aggregate-associated carbon. Soil Science Society of America Journal 64: 681–689.
45. Oades JM, Waters AG (1991) Aggregate hierarchy in soils. Australian Journal of Soil Research 29: 815–828.
46. Barros E, Curmi P, Hallaire V, Chauvel A, Lavelle P (2001) The role of macrofauna in the transformation and reversibility of soil structure of an oxisol in the process of forest to pasture conversion. Geoderma 100: 193–213.
47. Horn R, Dexter AR (1989) Dynamics of soil aggregation in an irrigated desert loess. Soil & Tillage Research 13: 253–266.
48. Feeney DS, Crawford JW, Daniell T, Hallett PD, Nunan N, et al. (2006) Three-dimensional microorganization of the soil-root-microbe system. Microbial Ecology 52: 151–158.
49. Jenkinson DS, Powlson DS, Wedderburn RWM (1976) Effects of biocidal treatments on metabolism in soil. 3. Relationship between soil biovolume, measured by optical microscopy, and flush of decomposition caused by fumigation Soil Biology & Biochemistry 8: 189–202.
50. Ingham ER, Griffiths RP, Cromack K, Entry JA (1991) Comparison of direct vs fumigation incubation microbial biomass estimates from ectomycorrhizal mat and non-mat soils. Soil Biology & Biochemistry 23: 465–471.
51. Hu S, Coleman DC, Hendrix PF, Beare MH (1995) Biotic manipulation effects on soil carbohydrates and microbial biomass in a cultivated soil. Soil Biology & Biochemistry 27: 1127–1135.
52. Nadelhoffer KJ, Raich JW (1992) Fine root production estimates and belowground carbon allocation in forest ecosystems. Ecology 73: 1139–1147.
53. Gale WJ, Cambradella CA (2000) Carbon dynamics of surface residue- and root-derived organic matter under simulated no-till. Soil Science Society American Journal 64: 190–195.

9

Dynamic Succession of Soil Bacterial Community during Continuous Cropping of Peanut (*Arachis hypogaea* L.)

Mingna Chen, Xiao Li, Qingli Yang, Xiaoyuan Chi, Lijuan Pan, Na Chen, Zhen Yang, Tong Wang, Mian Wang, Shanlin Yu*

Shandong peanut research institute, Qingdao, China

Abstract

Plant health and soil fertility are affected by plant–microbial interactions in soils. Peanut is an important oil crop worldwide and shows considerable adaptability, but growth and yield are negatively affected by continuous cropping. In this study, 16S rRNA gene clone library analyses were used to study the succession of soil bacterial communities under continuous peanut cultivation. Six libraries were constructed for peanut over three continuous cropping cycles and during its seedling and pod-maturing growth stages. Cluster analyses indicated that soil bacterial assemblages obtained from the same peanut cropping cycle were similar, regardless of growth period. The diversity of bacterial sequences identified in each growth stage library of the three peanut cropping cycles was high and these sequences were affiliated with 21 bacterial groups. Eight phyla: *Acidobacteria, Actinobacteria, Bacteroidetes, Chloroflexi, Gemmatimonadetes, Planctomycetes, Proteobacteria* and *Verrucomicrobia* were dominant. The related bacterial phylotypes dynamic changed during continuous cropping progress of peanut. This study demonstrated that the bacterial populations especially the beneficial populations were positively selected. The simplification of the beneficial microbial communities such as the phylotypes of *Alteromonadales, Burkholderiales, Flavobacteriales, Pseudomonadales, Rhizobiales* and *Rhodospirillales* could be important factors contributing to the decline in peanut yield under continuous cropping. The microbial phylotypes that did not successively changed with continuous cropping, such as populations related to *Rhizobiales* and *Rhodospirillales*, could potentially resist stress due to continuous cropping and deserve attention. In addition, some phylotypes, such as *Acidobacteriales*, *Chromatiales* and *Gemmatimonadales*, showed a contrary tendency, their abundance or diversity increased with continuous peanut cropping progress. Some bacterial phylotypes including *Acidobacteriales, Burkholderiales, Bdellovibrionales,* and so on, also were affected by plant age.

Editor: Gabriele Berg, Graz University of Technology (TU Graz), Austria

Funding: This study was supported by grants from China Agriculture Research System (C ARS-14), the National Natural Science Foundation of China (31000728; 31100205; 31200211), the Natural Science Fund of Shandong Province (ZR2011CQ 036; ZR2012586 CQ031), the Promotive Research Fund for Young and Middle-aged Scientists of Shandong Province (BS2010NY023), Qingdao Municipal Science and Technology Plan Project (11-2-4-9-(3)-jic; 11-2-3-26-nsh; 12-1-4-11-(2)-jch). The funders had no role in study design, data collection and analysis, decision to publish, or preparation of the manuscript.

Competing Interests: The authors have declared that no competing interests exist.

* Email: yshanlin1956@163.com

Introduction

The bacterial populations in the root related environment contribute to plant health by mediating nutrient acquisition, by causing soil borne diseases or by controlling the growth or activity of plant pathogens [1,2,3]. The interactions between soil, bacteria and plants in the root related environment determine soil fertility and sustainability and plant quality [3,4]. The structural and functional diversity of soil bacterial communities are influenced by several biotic and abiotic factors, such as climate and season, soil type and structure, treatment or management of the soil, plant species and development stage [5,6,7,8,9,10]. In agriculture, soil type, crop selection and cropping sequence have been shown to be the key determining factors on soil microbial communities [6,11,12]. On the other hand, microbial community characteristics have been shown to be efficient bio-indicators of soil condition and land management success due to their quick response to environmental changes, although soil productivity and sustainability depends on a complex of physical, chemical and biological properties [13,14]. Because of the interactions among bacteria, soil and plants, bacterial community succession has been shown to be a potential indicator of environmental change [13,14,15] and active management of soil bacterial communities, such as the manipulation of microbial populations, could be a promising approach to improving soil conditions, developing natural suppression of soil borne diseases and improving crop productivity [4,16]. However, the specific populations, characteristics and relationships between plants and soil bacterial communities are known relatively little.

Peanut, an important oil and economic crop worldwide, is very adaptable to climatic conditions and grows in many different environments between latitudes 40°N and 40°S [17]. However, its productivity is negatively affected by continuous cropping. The yield of other main world crops, such as wheat, rice, corn and soybean, also decline under continuous cropping [18,19,20]. It has been reported that the negative influence of continuous cropping on soil productivity has a significant effect on soil microbial biomass and community structure [21,22]. Previously, we investigated the succession characteristics of soil eukaryotic microorganisms under continuous peanut cropping and found that soil eukaryotic microorganisms showed significant dynamic changes during continuous cropping, especially with regards to the pathogenic and beneficial fungi selected [23]. However, there is no comprehensive knowledge of specific soil bacterial inhabitants, characteristics, succession pattern and how soil bacteria interact with continuous peanut cropping.

This investigation is an attempt to study the soil bacterial community characteristics of peanut and bacterial succession under continuous peanut cultivation by comparing the soil bacterial community structure when peanut was grown over three continuous cropping cycles and its representative growth stages. This was achieved by using cultivation-independent 16S rRNA gene clone library analyses. The aims of this study were to determine the phylogenetic affiliations of the most common and predominant populations of soil bacteria at different peanut growth stages under a continuous cropping regime as well as identify the diversity and succession patterns of soil bacteria during continuous cropping progress.

Materials and Methods

Ethics statement

No specific permits were required for the described field studies. No specific permissions were required for these locations and activities. The location is not privately-owned or protected in any way and the field studies did not involve endangered or protected species.

Management description and soil sampling

The study was conducted between 2008 and 2010 at Shandong Peanut Research Institute in Qingdao, China (36.10°N, 120.41°E). Peanuts were planted in May each year and harvested in October. Pot culture experiments were undertaken and fourty Huayu 25 seedlings (a pot a seedling) were planted, each year. The soil consisted of 28.5% clay, 41% loam, and 30.5% sand, with no known history of peanut cropping. In order to reduce interference from external factors on the soil bacterial community: The same peanut variety was employed during all the cropping cycles; two drought sheds were built, the plants were irrigated with sterile water and no fertilizer was applied during the growing seasons and fallow periods; Samples were collected at the same time of the three planting cycles, and at 5–10 cm soil depths. Two different peanut growth stages in the three cycles were selected: the seedling stage (twenty days after planting), and the pod-maturing stage (harvest time, about 129 days after planting). The soils were sampled using a soil probe (1.5 cm diameter) at 5–10 cm soil depths and five replicate test plants were used for each stage. The samples were stored at −80°C and were used for bacterial community analysis.

Genomic DNA extraction and 16S rRNA gene library construction

DNA was extracted from the pooled material for each sample with Ultra-Clean soil DNA kits (MoBio, USA). Bacteria-specific primers, 27F and 1387R primers [24], were used for 16S rRNA gene amplification. The reaction was performed for 25 cycles using the 'Easy-A High-Fidelity' PCR master mix from Stratagene (USA). Products from six PCRs were pooled, purified, using a Gel Extraction kit (Omega, USA), and ligated into pMD19-T simple vectors (Takara, Japan). The hybrid vectors were cloned into *Escherichia coli* TOP10. Clone libraries were constructed by random selection of approximately 100 white colonies from a single plate for each library. All clone libraries were named on the basis of the crop cycle and the plant growth stage of soil sampling. S1 and M1 were libraries constructed respectively for the seedling stage and pod-maturing stage in the first crop cycle; S2 and M2 were libraries constructed respectively for the two stages in the second cycle and S3 and M3 were libraries constructed respectively for the two stages in the third cycle.

16S rRNA gene sequencing and data analysis

Cloned 16S rRNA gene fragments were re-amplified by using primers RV-M (5′-GAGCGGATAACAATTTCACACAGG-3′) and M13-D (5′-AGGGTTTTCCCAGTCACGACG-3′) that flanked the insertion site of the cloning vector. Two primers, V-27 and V-1387 [25] spaned both the pMD19-T simple cloning vector RV-M annealing end and one of the 16S rRNA gene insert ends. The primers were used separately with primer M13-D for PCR determination of the orientations of the inserts. Primer M13-D or RV-M was chosen based on insert orientation and all positive clones were sequenced to improve data accuracy and integrity.

Operational taxonomic units (OTUs) in each library were defined by using the Mothur program at the 3% distance level [26]. Chimeric sequences were identified using Greengenes, a chimera-checked 16S rRNA gene database [27]. Sampling efficiency and the sequence phylotype diversity of the libraries were analyzed using the rarefaction curve method and the aRarefactWin program [25]. Similarities between the soil bacterial assemblages at different peanut growth stages during different crop cycles were established by performing cluster analyses using the software package SPSS 11.5 (SPSS, United States). The frequency (F) for the OTUs was calculated as follows: $F = (m/N) \times 100$, where m is the clone number of an OTU in a library and N is the total number of clones in the same library. Log-transformed frequency data were used to minimize the influence of preferential PCR amplification [25,28]. The relative abundance and occurrence of the bacterial orders or phylotypes detected in the six 16S rRNA gene libraries were visualized as a heatmap using Genesis program 1.7. Sample clustering and bacterial phylotype clustering were analyzed by using complete linkage clustering method.

The nearest phylogenetic neighbor sequence for each OTU sequence was determined in the GenBank database using the BLASTN program [29]. Sequences were aligned using ClustalX v.1.81.9 [30]. Phylogenetic trees were constructed using the DNADIST and NEIGHBOR programs in the PHYLIP package (version 3.65c) [31] and bootstrap support was evaluated using 100 replicates.

Nucleotide sequence accession numbers

The 16S rRNA gene sequences determined have been deposited in the GenBank database under accession numbers KF182746–KF183313.

Results

Statistical and comparative analyses of the six libraries

Six 16S rRNA gene libraries were constructed over three continuous cropping cycles and over two peanut growth stages for each cycle. A total of 618 clones were screened and 576 of them contained a proper insert. A total of 351 OTUs were identified, eight of which (eight clones) were probably chimeras, based on the chimera-check results. Rarefaction analysis indicated the high OTU phylotype diversity of the soil bacteria (Figure S1).

The OTU number ranged from 69 to 82 in the six libraries. The clone libraries obtained from the same cycle, but from different growth stages, typically shared 15 to 17 common OTUs, accounting for 20.0 to 23.2% of the OTUs in each library (Table 1). The dynamics of the bacterial community structure were further analyzed by performing cluster analyses. The hierarchical (Figure 1) and k-means (Figure S2) as well as heatmap sample (Figure 2) clustering results were consistent and showed that the libraries obtained from different growing stages, but from the same cropping cycle, were grouped together. The results demonstrated that the soil bacterial populations changed when peanut was grown continually, regardless of growth stage.

16S rRNA gene sequence analyses

All the 343 valid OTUs were used for sequence analysis, most (74.9%) of which showed high levels of similarity (≥97% identity) with known sequences in the GenBank database and 12.9% were moderately related (≤94% identity) to known sequences. Phylogenetic analysis showed that 342 OTUs were affiliated with the domain, *Bacteria*, and the remaining OTU was related to the chloroplast 16S rRNA gene from *Daviesia angulata*. There were 21 groups related to *Bacteria* identified in the libraries (Figure 3). Eight phyla: *Acidobacteria, Actinobacteria, Bacteroidetes, Chloroflexi, Gemmatimonadetes, Planctomycetes, Proteobacteria* and *Verrucomicrobia* were predominant, accounting for 93.0% of the 568 clones and 91.0% of the 343 valid OTUs (Figure 4). The phylum, *Proteobacteria*, dominated in all the libraries, accounting for 45.6% of the 568 clones, and showed the greatest diversity, accounting for 37.9% of the 343 OTUs. The phylum, *Acidobacteria* was relatively abundant, accounting for 15.3% of the obtained valid clones and was highly diverse, accounting for 14.0% of the OTUs analyzed. The other six phyla sequences accounted for 3.4 to 7.2% of the obtained clones and 4.7 to 8.5% of the analyzed OTUs (Figure 4). The bacterial orders or phylotypes affiliated with these phyla showed dynamic succession with continuous cropping of peanut (Figure 2).

***1) Proteobacteria* sequences.** One hundred and thirty OTUs (45.6% of the clones) were related to *Proteobacteria*, which predominated in all the libraries and showed the greatest diversity. Four classes: *α-proteobacteria*, *β-proteobacteria*, *δ-proteobacteria* and *γ-proteobacteria*, were affiliated with the *Proteobacteria* phylum (Figure 3). The *α-proteobacteria* sequences were most abundant and diverse, accounting for 40.9% of the *Proteobacteria* clones and for 39.2% of the corresponding OTUs. The *γ-proteobacteria* and *β-proteobacteria* sequences were relatively abundant, accounting for 30.5 and 18.9% of the *Proteobacteria* clones, respectively, and were relatively diverse, accounting for 23.8 and 20.7% of the corresponding OTUs, respectively. The *δ-proteobacteria* sequences accounted for 9.7% of the *Proteobacteria* clones and for 16.2% of the corresponding OTUs. All four class sequences showed considerable dynamic population changes.

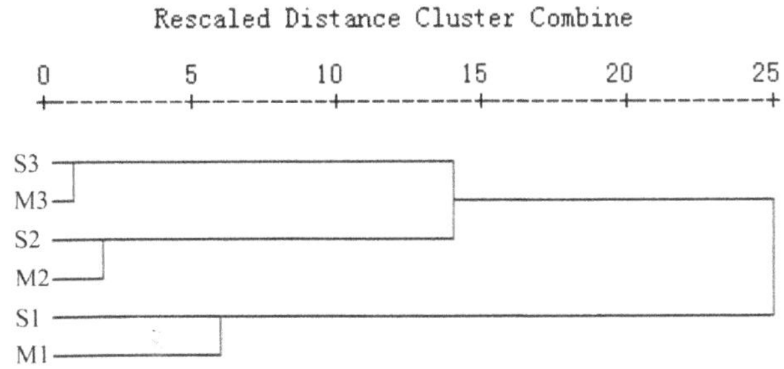

Figure 1. Dendrogram based on a hierarchical clustering analysis of the 16S rRNA gene clone libraries, constructed using the squared Euclidean distance similarity and Ward linkage procedures.

The *α-proteobacteria* sequences were abundant and diverse in all clone libraries and the clone abundance accounted for 16.8, 16.6 and 22.8%, respectively, of the first, second and third cycle library clones. Five orders or subdivisions were identified as being related to *α-proteobacteria*, including the orders of *Caulobacterales, Rhizobiales, Rhodospirillales, Sphingomonadales*, and an uncultured *α-proteobacteria* subdivision (Figure S3). These subdivisions showed complex dynamic changes with continuous peanut cropping (Figure 2, Figure 5). The *Rhizobiales* and *Sphingomonadales* sequences were the most abundant *α-proteobacteria* members in the libraries, accounting for 34.9 and 39.6%, respectively, of the 106 *α-proteobacteria* clones. The sequences affiliated with *Rhizobiales* were diverse, accounting for 47.1% of the 51 *α-proteobacteria* OTUs. Their abundance slightly decreased and then increased during continuous cropping, accounting for 7.1, 5.4, 4.9, 5.6, 6.7 and 10.6%, of the clones identified in the six libraries, respectively. Their clone diversity showed a similar tendency, accounting for 8.7, 6.3, 3.7, 6.3, 8.0 and 9.3%, respectively, of the OTUs identified in the six libraries. Eleven OTUs were related to *Sphingomonadales* and the clone abundance accounted for 5.2, 8.3 and 8.7%, respectively, of the clones obtained in the three cycle libraries. Seven OTUs were related to *Rhodospirillales* and accounted for 3.1, 1.6 and 3.3%, respectively, of the clones obtained in the three cycle libraries. Only two OTUs were related to the *Caulobacterales* order and these sequences appeared in all the second and third cycle libraries, but

Table 1. Analysis of the clone libraries constructed over three continuous cropping cycles and representative growing stages of peanut.

Library	No. of clones	No. of OTUs	No. of unique OTUs	No. of OTUs shared by libraries from the same cropping cycle	% shared OTUs
S1	99	69	58	16	23.2
M1	92	80	69		20.0
S2	103	82	67	17	20.7
M2	90	79	71		21.5
S3	90	75	64	15	20.0
M3	94	75	62		20.0

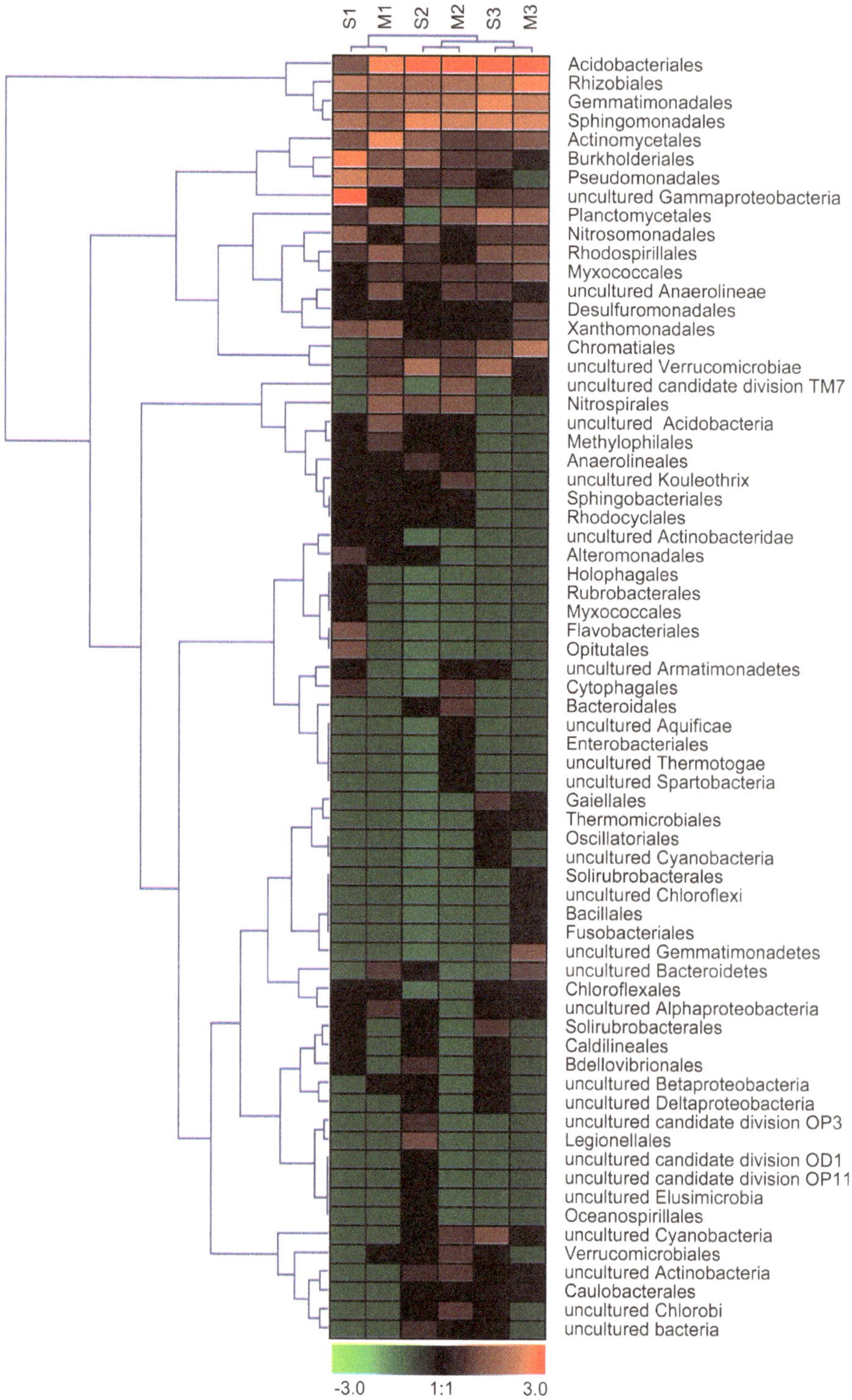

Figure 2. Heatmap analyses of the bacterial orders or phylotypes detected in the six 16S rRNA gene libraries. The color code indicates relative abundance, ranging from green (low abundance) to black to red (high abundance).

were not found in any first cycle libraries. The OTU related to *Myxococcales* was only identified in library S1.

Thirty-one OTUs were affiliated with *γ-proteobacteria* and the related clones accounted for 22.5% of the clones in the first cycle libraries, but only 9.8, and 9.2% of the clones in the second and third cycle libraries, respectively. These sequences were highly diverse and were affiliated with eight orders or subdivisions: *Alteromonadales*, *Chromatiales*, *Enterobacteriales*, *Legionellales*, *Oceanospirillales*, *Pseudomonadales*, *Xanthomonadales* and an uncultured *γ-proteobacteria* subdivision (Figure S4). These subdivision sequences showed dynamic population changes with continuous cropping (Figure 2, Figure 5). The clones affiliated with *Chromatiales*,

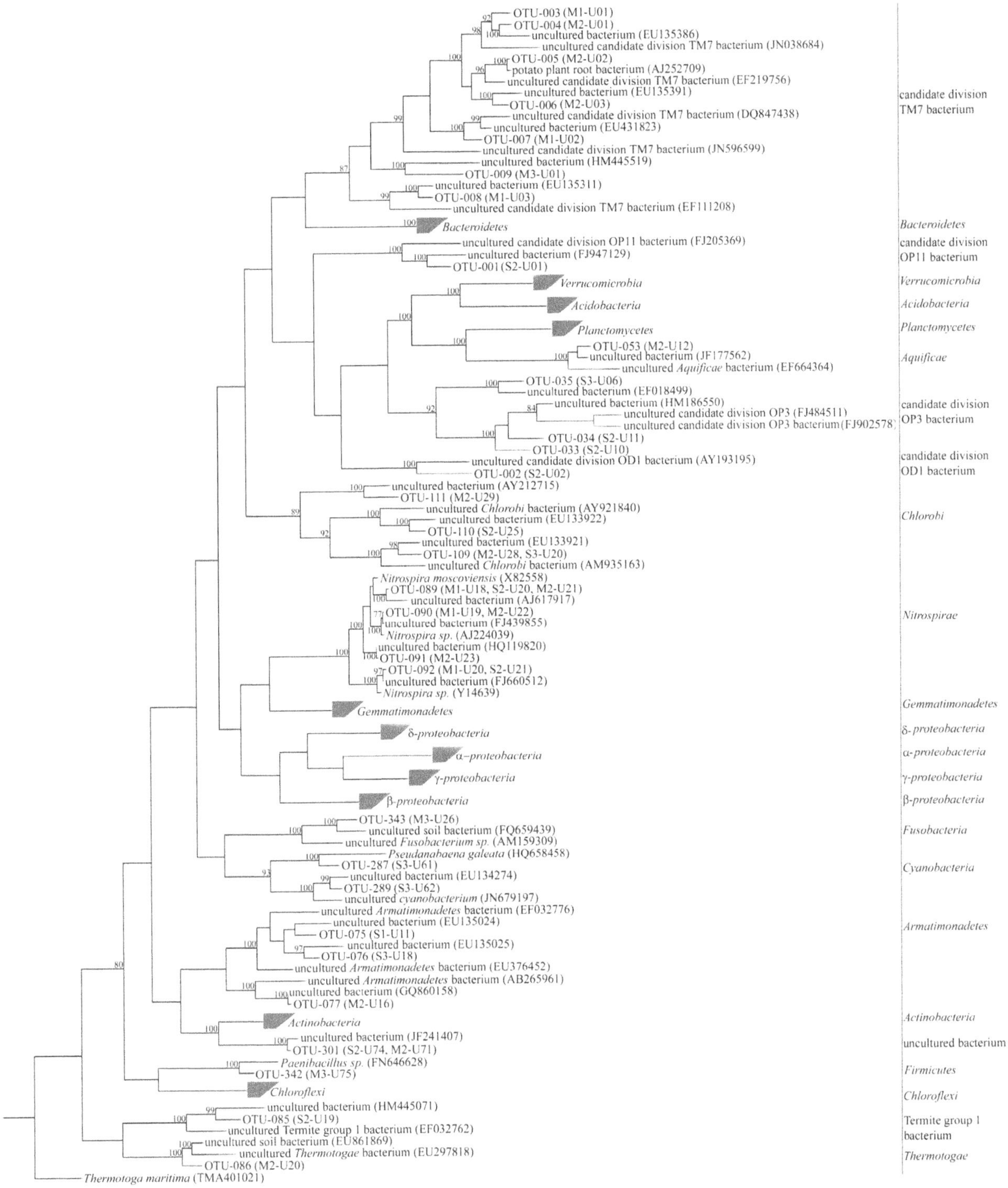

Figure 3. Phylogenetic tree of the bacterial sequences recovered from the six libraries, constructed using the neighbor-joining method with the Kimura two-parameter model for nucleotide change. The predominant groups which were just denoted the phylogenetic positions in this tree, presented in other separate phylogenetic trees. The libraries OTUs occurred were labeled. Scale bar, 0.1 substitutions per nucleotide position. Bootstrap values (100 replicates) above 70% are indicated at the nodes. The tree was rooted using the sequence related to *Thermotoga maritima* (TMA401021).

Pseudomonadales, *Xanthomonadales* and the uncultured *γ-proteobacteria* subdivision were relatively abundant and common in the analyzed libraries, accounting for 19.0, 22.8, 15.2 and 31.6%, respectively, of the 79 *γ-proteobacteria* clones. Six OTUs were affiliated with *Chromatiales* and appeared from the pod-maturing stage of the first cycle. Their abundance increased with continuous cropping and accounted for 1.0, 2.1 and 4.9% of the clones identified in the three cycle libraries, respectively. The sequences affiliated with

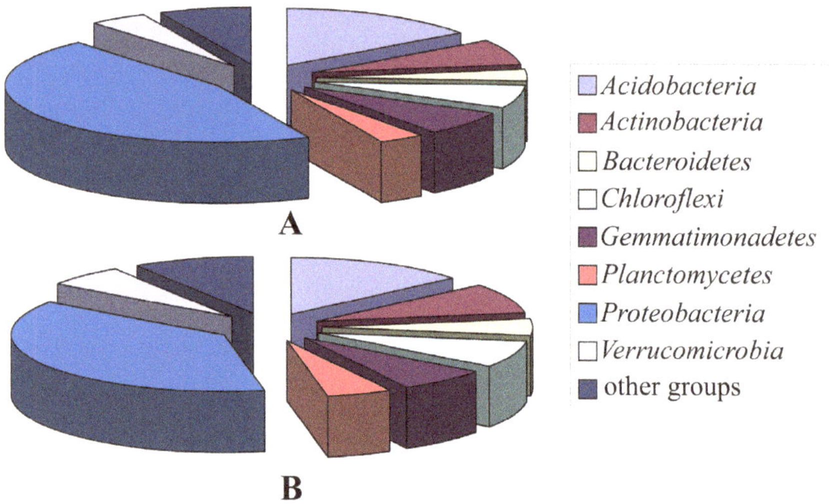

Figure 4. The abundance and diversity analyses of the bacterial groups identified in the six libraries. A shows the clone percentage of these groups and B shows the OTU percentage.

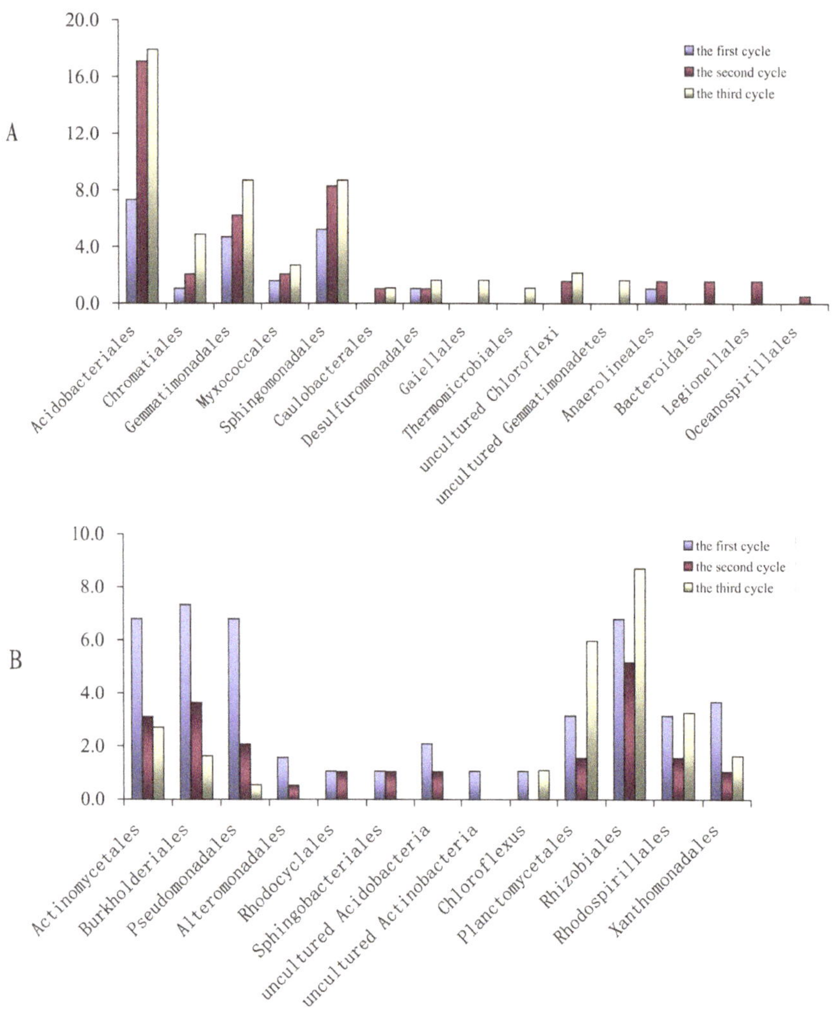

Figure 5. The clone abundance analyses of the bacterial orders or phylotypes which showed succession changes in populations with continuous cropping cycles. (The clone abundance was calculated as: $m1/M \times 100$, where $m1$ is the number of related clones detected in the libraries for the same cropping cycle and M is the total number of clones in the same libraries.)

Pseudomonadales were seen in five libraries, but not in library M3. The clone abundance and diversity decreased with continuous cropping and accounted for 6.8, 2.1 and 0.5%, respectively, of the clones obtained in the three cycle libraries. The OTU number was 4, 3 and 1 in the three cycle libraries, respectively. Seven OTUs were related to *Xanthomonadales* and were common in all six libraries, accounting for 3.7, 1.0 and 1.6%, respectively, of the clones obtained in the three cycle libraries. The uncultured γ-*proteobacteria* sequences in the seedling stage library of the first cycle were most abundant, accounting for 17.2% of the clones identified in library S1 and decreased considerably from the pod-maturing stage of the first cycle, accounting for no more than 3% of the corresponding clones in the later five libraries. In addition, the clone abundance in the seedling stage library was higher than in the pod-maturing stage library for the same cycle (Figure 6) and the clone diversity showed a similar tendency, accounting for 4.4, 1.3, 3.6, 0, 2.7 and 1.3% respectively of the OTUs identified in libraries: S1, M1, S2, M2, S3 and M3. Only three OTUs were affiliated with *Alteromonadales* and were seen only in the S1, M1 and S2 libraries, but did not occur in the other three libraries during the later stages. Sequences related with the other three orders were rare and only seen in the second cycle libraries.

Twenty-seven OTUs were affiliated with β-*proteobacteria* and the clone abundance decreased significantly with continuous cropping, accounting for 13.1, 8.3 and 4.3% of the clones detected in the three cycle libraries, respectively. These sequences were associated with orders or subdivisions of *Burkholderiales*, *Methylophilales*, *Nitrosomonadales*, *Rhodocyclales* and an uncultured β-*proteobacteria* subdivision with 96–98% similarity (Figure S5). The sequences related to *Burkholderiale*, which was the dominant β-*proteobacteria* phylotype in the libraries, accounted for 55.6% of the 27 OTUs and 49.0% of the 49 related clones. The clone abundance decreased significantly in the same growth stage libraries with continuous cropping and accounted for 11.1, 4.9 and 2.2% of the clones detected in the seedling stage libraries of the three cycles, respectively, and 3.3, 2.2 and 1.1% of the clones in the three pod-maturing stage libraries, respectively (Figure 5, Figure 6). The clone diversity showed a similar tendency, accounting for 10.1, 4.9 and 2.7% of the OTUs identified in the three seedling stage libraries, respectively, and 3.8, 2.5, and 1.3% of the OTUs identified in the three pod-maturing stage libraries, respectively. The analyses also showed that the clone abundance and diversity in the seedling stage libraries were higher than in the pod-maturing stage libraries of the same cycle. The Sequence related to the orders of *Methylophilales* and *Rhodocyclales* occurred in all the first two cycle libraries, but were not seen in any third cycle libraries (Figure 5). The *Nitrosomonadales* and uncultured β-*proteobacteria* sequences showed no significant changes over the three cycle libraries.

The abundance of δ-*proteobacteria* sequences was low and moderately increased over the three cycle libraries, accounting for 3.1, 4.7 and 5.4% of the corresponding library clones, respectively. These sequences were affiliated with the orders: *Bdellovibrionales*, *Desulfuromonadales*, *Myxococcales* and an uncultured δ-*proteobacteria* subdivision (Figure S6). The clone diversity and abundance of the *Desulfuromonadales* and *Myxococcales* sequences showed a slightly increasing tendency during continuous cropping (Figure 5). The OTU number for the *Desulfuromonadales* clones was 1, 2 and 3 in the three cycle libraries, respectively, and the related clones accounted for 1.0, 1.0 and 1.6%, respectively, of the clones obtained in the three cycle libraries. The OTU number of *Myxococcales* clones respectively was 3, 4 and 5 and the related clones accouted for 1.6, 2.1 and 2.7% of the clones in the three cycles, respectively. Sequences affiliated with *Bdellovibrionales* (3 OTUs) occurred in all three cycle libraries, but only occurred in the seedling stage libraries. The uncultured δ-*proteobacteria* sequences (2 OTUs) were found in the second and third cycle libraries and only occurred in the seedling stage libraries (Figure 6).

***2) Acidobacteria* sequences.** Forty-eight OTUs (15.3% of the total clones) were affiliated with the *Acidobacteria* phylum. Two classes of *Acidobacteria* and *Holophagae* were identified as being related to *Acidobacteria* in the analyzed libraries. Forty two OTUs were affiliated with the *Acidobacteria* class, which was the predominant phylotype of the *Acidobacteria* phylum and accounted for 92.0% of the corresponding clones. Only one OTU (1 clone) was related to *Holophagae*; the remaining five OTUs were related to uncultured *Acidobacteria* (Figure S7). The three subgroups all showed dynamic changes with continuous cropping (Figure 5). All *Acidobacteria* class sequences were affiliated with the *Acidobacteriales* orde and the abundance significantly increased in the same growth stage library of the second cycle, but there were no significant differences between the same growth stage libraries of the second

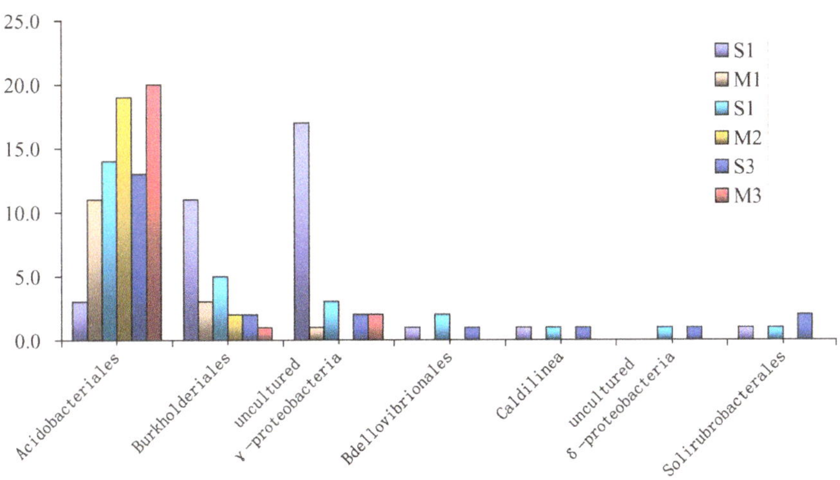

Figure 6. The clone abundance analyses of the bacterial orders or phylotypes which showed succession changes in populations with peanut growth stage. (The clone abundance was calculated as: ($n1/N \times 100$, where $n1$ is the number of related clones detected in a library and N is the total number of clones in the same library.)

and third cycles. The related clones accounted for 3.0, 13.6 and 14.4%, respectively, of the clones identified in the seeding stage libraries of each cycle and for 12.0, 21.1 and 21.3% of the clones identified in the pod-maturing stage libraries, respectively (Figure 6). Clone diversity also showed a similar tendency. There were 3, 8 and 8 OTUs identified in the seeding stage libraries of each cycle, respectively, and there were 8, 14 and 17 OTUs identified in the pod-maturing stage libraries, respectively. The analysis also indicated that clone abundance and diversity both increased during the pod-maturing stage compared to the seedling stage of the same cropping cycle. The *Holophagae* sequence, which was affiliated with uncultured *Holophaga sp.* with 96% similarity, was only seen in library S1. The five OTUs related to uncultured *Acidobacteria* occurred in all the four first and second cycle libraries, but not in any third cycle libraries.

3) Sequences of other relatively abundant groups. The *Actinobacteria* sequences were relatively abundant and diverse, accounting for 7.2% of the total clones and 8.5% of the analyzed OTUs. These sequences were affiliated with two classes of *Actinobacteridae, Rubrobacteridae* and uncultured *Actinobacteria* (Figure S8). Phylotypes affiliated with these three classes or subgroups showed dynamic changes over the three cycle libraries (Figure 5). The sequences related to *Actinobacteridae* were most abundant and diverse, accounting for 63.0% of the *Actinobacteria* clones and 69.0% of the corresponding OTUs. The *Actinomycetales* orders and an uncultured *Actinobacteridae* subdivision were identified in libraries affiliated with the *Actinobacteridae* class (Figure S8). The *Actinomycetales* clone abundance decreased in the second cycle libraries, but there were no significant differences between the second and third cycle libraries, which accounted for 7.9, 3.1 and 3.3%, respectively, of the clones in the three cycle libraries. Clone diversity showed a similar tendency and the OTU numbers were 10, 5 and 5, respectively, in the three cycle libraries. The OTU related to uncultured *Actinobacteridae* only occurred in the two first cycle libraries, S1 and M1. Three orders: *Rubrobacterales, Solirubrobacterales* and *Gaiellales* were identified as being related to the *Rubrobacteridae* class (Figure S8). Four OTUs were affiliated with the *Solirubrobacterales* order and were seen in all three seedling stage libraries and the pod-maturing stage library of the third cropping cycle, but did not occur in the earlier two pod-maturing stage libraries (Figure 6). The OTU related to *Rubrobacterales* only occurred in library S1, but the OTU related to *Gaiellales* only occurred in the two third cycle libraries, S3 and M3. Three OTUs (14.6% of the *Actinobacteria* clones) were affiliated with uncultured *Actinobacteria* and these sequences were seen in all the four second and third cycle libraries, but did not occur in any first cycle libraries.

The sequences related to the *Chloroflexi* phylum accounted for 6.5% of the 568 clones and for 7.9% of the 343 valid OTUs. There were 19 OTUs being affiliated with classes of *Anaerolineae, Caldilineae, Thermomicrobia* and the remaining 8 OTUs were related to uncultured *Chloroflexi* sequences and the genus, *Kouleothrix*, which has not been fully described in the GenBank database (Figure S9). Ten OTUs (45.5% of the *Chloroflexi* clones) were affiliated with the *Anaerolineae* class and these sequences were related to *Anaerolineales* and uncultured *Anaerolineae* sequences. The *Anaerolineales* sequences were only seen in the four first and second cycle libraries (Figure 5), and the uncultured *Anaerolineae* sequences did not show any dynamic tendencies over the three cycle libraries. Six OTUs (seven clones) were affiliated with the *Chloroflexales* order, which belonged to the *Caldilineae* class, and these sequences were related to two genera of *Chloroflexus* and *Caldilinea*. The *Chloroflexus* genus sequences were seen in the four first and third cycle libraries, but did not occur in any second cycle libraries, and the *Caldilinea* genus sequences were only seen in three of the libraries representing the seedling stage of each cycle (Figure 5 and 6). Two OTUs (two clones) were affiliated with the *Thermomicrobiales* order, which belonged to the *Thermomicrobia* class, and were only seen in the two third cycle libraries (Figure 5). The *Kouleothrix* sequences (four OTUs) were seen in all the first and second cycle libraries, but not in any third cycle libraries. In contrast, the uncultured *Chloroflexi* sequences (five OTUs) were only seen in the four second and third cycle libraries (Figure 5).

The sequences related to the *Gemmatimonadetes* phylum accounted for 7.0% of both total clones, and valid OTUs. Twenty-two OTUs (94.9% of *Gemmatimonadetes* clones) were affiliated with the *Gemmatimonadetes* class and the other two OTUs were related to uncultured *Gemmatimonadetes* (Figure S10). All the sequences affiliated with the *Gemmatimonadetes* class were related to the *Gemmatimonadales* order and their abundance showed an increasing tendency during continuous cropping, accounting for 4.7, 6.2 and 8.7% respectively of the three cycle libraries (Figure 5). Clone diversity also showed a similar tendency. There were 9, 10 and 12 OTUs, respectively, in the three cycle libraries. The sequences related to uncultured *Gemmatimonadetes* were only seen in library M3 (Figure 6).

The sequences affiliated with the *Bacteroidetes, Planctomycetes* and *Verrucomicrobia* phyla were relatively abundant and diverse. These sequences accounted for 3.4, 3.5 and 4.4%, respectively, of the total clones and accounted for 4.7, 5.0, and 6.1%, respectively, of the total valid OTUs. Fourteen OTUs related to *Bacteroidetes* were affiliated with four classes: *Bacteroidia, Cytophagia, Flavobacteriia* and *Sphingobacteriia*, and were, respectively, related to the *Bacteroidales, Cytophagales, Flavobacteriales* and *Sphingobacteriales* orders (Figure S11). The remaining five OTUs were affiliated with uncultured *Bacteroidetes*. The analysis indicated that sequences related to the *Sphingobacteriales* order were only seen in the four first and second cycle libraries, but did not occur in any third cycle libraries (Figure 5). Sequences related to the *Bacteroidales* only were identified in the two second cycle libraries and sequences affiliated with *Flavobacteriales* only were seen in library S1. The other order or subdivision sequences didn't show any significant tendencies over the three cycle libraries. All the sequences affiliated with the *Planctomycetes* phylum were related to the *Planctomycetales* order, which belonged to the *Planctomycetia* class (Figure S12). Clone abundance decreased in the second cycle libraries and increased significantly in the third cycle libraries, accounting for 3.1, 1.6 and 6.0%, respectively, of the clones identified in the three cycle libraries (Figure 5). Clone diversity showed a similar tendency and there were 5, 3 and 10 OTUs identified in the three cycle libraries, respectively. The sequences affiliated with the *Verrucomicrobia* phylum were related to the *Opitutae, Spartobacteria, Verrucomicrobiae* and uncultured *Verrucomicrobia* classes (Figure S13). The sequences related to the *Opitutae* and *Verrucomicrobiae* classes were affiliated with the *Opitutales* and *Verrucomicrobiales* orders, respectively. Sequences affiliated with the *Spartobacteria* class were not described. The analysis indicated that all orders or subdivisions affiliated with *Verrucomicrobia* did not show any significant dynamic change tendencies with continuous cropping.

Discussion

In this study, 16S rRNA gene clone library analyses were undertaken to study soil bacterial succession under continuous peanut cultivation and warrant comparison with the previous study of soil eukaryotic microorganism succession [23]. The high diversity of soil bacteria in each growth stage library of the three peanut cropping cycles was shown based on the sequence analyses.

A total of 343 OTUs were identified in the six libraries and these sequences were affiliated with 21 bacterial groups. Eight phyla: *Acidobacteria, Actinobacteria, Bacteroidetes, Chloroflexi, Gemmatimonadetes, Planctomycetes, Proteobacteria* and *Verrucomicrobia* were predominant, accounting for 93.0% of the 568 clones and 91.0% of the 343 valid OTUs. The bacterial communities identified in our libraries were much more diverse than the populations found by Sudini et al. [11] in a different peanut-cropping sequence soil, but the eight abundant bacterial groups identified in our libraries were also all identified by Sudini et al. [11] and the *Proteobacteria* and *Acidobacteria* phyla were common predominant members. There have been few reports about specific soil bacterial phylotype succession with continuous cropping of plants, but, taking into account previous reports on the root associated soil microbial communities around clover, bean or alfalfa [12], there is a suggestion that plant species could have a relatively greater impact than soil type and plant age in determining the soil microbial community structure. Consequently, the analyses of bacterial succession in this study can provide general information about the relationships between the soil microbial community and continuous peanut cropping.

It is known that soil bacterial populations are influenced by a wide range of factors. Soil type, plant species and cropping patterns are the factors that most affect the bacterial community structure in soil [6,11,12]. In order to reduce interference from other factors and truly reflect the relationship between soil bacterial succession and continuous cropping with peanut, several measures were used, as described in the Materials and Methods. Cluster analyses demonstrated that the soil bacterial assemblages obtained from the same cropping cycle were similar, regardless of growth stage. Heatmap analyses and phylogenetic analyses also showed dynamic changes in bacterial populations with continuous peanut cropping. This was consistent with the earlier study into soil eukaryotic microorganisms [23]. These results demonstrated that the soil microbial communities were indeed affected by continuous cropping of peanut. It has been reported that soil microbial biomass and structure were also significantly influenced by continuous cropping with cucumber, maize, rice and pea [21,22,32,33]. These findings indicated that successional change in soil microbial communities with continuous cropping could be a common feature.

Analyses of sequences obtained in the six libraries showed that the successional trends were complex and can be mainly divided into four types: trends showing a continuous increase or decrease; a trend where the population increases and then decreases or a trend where the population deceases and then increases.

Data analyses showed that the abundance and/or diversity of clones affiliated with *Acidobacteriales, Chromatiales, Gemmatimonadales, Myxococcales* and *Sphingomonadales* were relatively abundant in the libraries and all successively increased with continuous peanut cropping. In addition, some populations that were not abundant in the libraries, including the *Caulobacterales, Desulfuromonadales, Gaiellales, Thermomicrobiales* phylotypes and two uncultured bacterial phylotypes: uncultured *Chloroflexi* and uncultured *Gemmatimonadetes*, also showed an successively increasing tendency with continuous peanut cropping.

Other phylotypes showed the opposite tendency with continuous peanut cropping. Their abundance and/or diversity successively decreased during continuous cropping. These phylotypes, including: *Actinomycetales, Burkholderiales* and *Pseudomonadales*, were relatively abundant in the libraries. Additionally, many non-abundant phylotypes identified in our libraries, including: *Alteromonadales, Methylophilales, Rhodocyclales, Sphingobacteriales* and two uncultured phylotypes (uncultured *Acidobacteria* and uncultured *Actinobacteridae*) also showed a successively decreasing tendency with continuous cropping.

There were another two different dynamic tendencies during continuous peanut cropping. The abundance and/or diversity of *Anaerolineales* and *Kouleothrix* increased with continuous cropping in the first and second cycles but disappeared in the third cycle libraries. The phylotypes of *Bacteroidales, Legionellales* and *Oceanospirillales* appeared in the second cycle libraries, but not occurred in any of the first and third cycle libraries. All these phylotypes were not abundant in the libraries. In contrast, the abundance and/or diversity of some phylotypes decreased with continuous cropping in the first and second cycles but increased in the third cycle libraries. These phylotypes were *Planctomycetales, Rhizobiales, Rhodospirillales, Xanthomonadales* and the *Chloroflexus* genus which belonged to *Chloroflexales*.

The complex dynamics of the bacterial populations identified in this study raises many questions, such as why did these populations change over time and what roles did the changing bacterial population play during continuous cropping?

Bacterial populations in soil that are affiliated with the *Alteromonadales, Burkholderiales, Flavobacteriales, Pseudomonadales, Rhizobiales* and *Rhodospirillales* orders have been reported as being beneficial to plant growth [3,4,34,35]. These populations promoted plant growth by being involved in nitrogen fixation, phosphorus solubilization and in the biocontrol of plant pathogens [4,34]. Interestingly, the abundance and/or diversity of these populations identified in our libraries all showed a decreasing tendency with continuous cropping. Most of these populations showed a successively decreasing tendency and the populations related to *Rhizobiales* and *Rhodospirillales* also showed a decreasing tendency in the first two continuous cropping cycles, but increased in the third cycle libraries. It has been reported that bacteria can quickly respond to environmental changes [36] and PGPR (plant growth promoting rhizobacteria) can have positive effects on plant vigor and productivity, especially under stress conditions [34,37,38]. So the populations associated with *Rhizobiales* and *Rhodospirillales* could be quick responders under continuous cropping stress. Based on these analyses, it can be seen that the beneficial bacterial population showed a simplified tendency with continuous cropping, but the succession models were non-single.

The *Pseudomonales* are considered a major group of rhizobacteria that may have the potential to biologically control plant pathogens [34,39]. The abundance and diversity of the sequences of this phylotype identified in the libraries decreased successively with continuous peanut cropping. It was reported that *Fusarium* genus was major pathogens of plant root rot, and *Pseudomonas* was major antagonists [40,41,42]. Interestingly, contrary to the decrease tendency of *Pseudomonales*, sequences affiliated with *Fusarium* showed increasing tendency with continuous cropping of peanut [23,43]. The succession analyses of soil eukaryotic microorganisms in the earlier study demonstrated that the pathogenic and beneficial fungi were positively selected over time and showed, respectively, increased and simplified tendencies with continuous cropping [23]. These analyses, along with previous studies [26,33,44], demonstrated that the trend towards an increase in microbial pathogens and simplification of the beneficial microbial community could be important factors contributing to the decline in peanut growth and yield over many years of continuous cropping.

Plant–microbial interactions in soil are the determinants of plant health and soil fertility. However, due to the high diversity and heterogeneity of soil bacteria, the interactions between bacteria, soil and plants, especially for specific bacteria, are not well understood. Many of the bacterial phylotypes identified in our libraries showed dynamic change during continuous cropping, but only a few have been reported as being able to promote plant growth and the functions of the other phylotypes have not been identified yet. Generally, PGPRs function in the following ways: they synthesize phytohormones, fix nitrogen, solubilize phosphorus, facilitate the uptake of certain nutrients from soil, increase plant tolerance to extreme environmental conditions and lessen or prevent plant diseases [4,34,35]. In contrast, other rhizobacteria suppress plant growth by acting as a carbon drain and causing disease, galls and tumors on plants [45,46]. These reports, combined with the bacterial succession dynamic analyses in this study, could provide the basis for further study into the interactions between soil bacteria and continuous peanut cropping and the decline in peanut growth and yield under continuous cropping.

In addition, the analyses indicated that some bacterial populations also were affected by the growth stage of peanut. The abundance and diversity of sequences affiliated with the *Acidobacteriales* order in the pod-maturing stage library were higher than in the seedling stage library of the same cropping cycle. Oppositely, the clone abundance and diversity affiliated with *Burkholderiales* and uncultured *γ-proteobacteria* in the seedling stage libraries were higher than in the pod-maturing stage libraries of the same cycle. In addition, sequences affiliated with *Bdellovibrionales* and the *Caldilinea* genus, which belong to *Chloroflexales*, *Solirubrobacterales*, and uncultured *δ-proteobacteria* were only seen in the seedling stage libraries of the three cycles, but did not occur in any pod-maturing stage libraries. Based on the analyses, it can be seen that these populations may be related to the growth and development of peanut. It was reported that the rhizobacterial communities were minor affected by plant development [12,47]. Interestingly, the eukaryotic microorganisms did not show any obvious correlation with peanut development in the earlier study [23]. There needs to be further study into the interaction mechanisms between microbial communities and the growth or development of peanut.

In this study and in a previous study [23], the succession dynamics of soil microbial communities with continuous peanut cropping has been comprehensively analyzed. Future studies will investigate the interactions between the microbial populations that showed a change over time and the mechanisms behind the interactions between microbes and peanut under continuous cropping.

Supporting Information

Figure S1 Rarefaction curves for the 16S rRNA gene libraries constructed from each of the soil samples.

Figure S2 Cluster membership based on a k-means clustering analysis of the 16S rRNA gene clone libraries (Case number 1–6 were library S1, M1, S2, M2, S3 and M3, respectively).

Figure S3 Phylogenetic tree representing affiliations of the 16S rRNA gene sequences related to the *α-proteobacteria* class.

Figure S4 Phylogenetic tree representing affiliations of the 16S rRNA gene sequences related to the *γ-proteobacteria* class.

Figure S5 Phylogenetic tree representing affiliations of the 16S rRNA gene sequences related to the *β-proteobacteria* class.

Figure S6 Phylogenetic tree representing affiliations of the 16S rRNA gene sequences related to the *δ-proteobacteria* class.

Figure S7 Phylogenetic tree representing affiliations of the 16S rRNA gene sequences related to the *Acidobacteria* phylum.

Figure S8 Phylogenetic tree representing affiliations of the 16S rRNA gene sequences related to the *Actinobacteria* phylum.

Figure S9 Phylogenetic tree representing affiliations of the 16S rRNA gene sequences related to the *Chloroflexi* phylum.

Figure S10 Phylogenetic tree representing affiliations of the 16S rRNA gene sequences related to the *Gemmatimonadetes* phylum.

Figure S11 Phylogenetic tree representing affiliations of the 16S rRNA gene sequences related to the *Bacteroidetes* phylum.

Figure S12 Phylogenetic tree representing affiliations of the 16S rRNA gene sequences related to the *Planctomycetes* phylum.

Figure S13 Phylogenetic tree representing affiliations of the 16S rRNA gene sequences related to the *Verrucomicrobia* phylum.

Author Contributions

Conceived and designed the experiments: SY. Performed the experiments: MC XL XC. Analyzed the data: QY. Contributed reagents/materials/analysis tools: LP NC ZY TW MW. Wrote the paper: MC.

References

1. Mazzola M (2004) Assessment and management of soil microbial community structure for disease suppression. Annu Rev Phytopathol 42: 35–59.
2. Raaijmakers JM, Mazzola M (2012) Diversity and natural functions of antibiotics produced by beneficial and plant pathogenic bacteria. Annu Rev Phytopathol 50: 403–424.
3. Fabra A, Castro S, Taurian T, Angelini J, Ibañez F, et al. (2010) Interaction among *Arachis hypogaea* L. (peanut) and beneficial soil microorganisms: how much is it known? Crit Rev Microbiol 36(3): 179–194.
4. Hayat R, Ali S, Amara U, Khalid R, Ahmed I (2010) Soil beneficial bacteria and their role in plant growth promotion: a review. Ann Microbiol 60: 579–598.
5. Berg G, Smalla K (2009) Plant species and soil type cooperatively shape the structure and function of microbial communities in the rhizosphere. FEMS Microbiol Ecol 68: 1–13.
6. Garbeva P, van Veen JA, van Elsas JD (2004) Microbial diversity in soil: selection of microbial populations by plant and soil type and implications for disease suppressiveness. Annu Rev Phytopathol 42: 243–270.
7. Rasche F, Velvis H, Zachow C, Berg G, van Elsas JD, et al. (2006) Impact of transgenic potatoes expressing antibacterial agents on bacterial endophytes is comparable to effects of soil, wildtype potatoes, vegetation stage and pathogen exposure. Can J Microbiol 43: 555–566.

8. Jangid K, Williams MA, Franzluebbers AJ, Sanderlin JS, Reeves JH, et al. (2008) Relative impacts of land-use, management intensity and fertilization upon soil microbial community structure in agricultural systems. Soil Biol Biochem 40: 2843–2853.
9. Caesar T, Lenssen AW, Caesar AJ, Sainju UM, Gaskin JF (2010) Effects of tillage on microbial populations associated to soil aggregation in dryland spring wheat system. Eur J Soil Biol 46: 119–127.
10. Guan X, Wang J, Zhao H, Wang J, Luo X, et al. (2013) Soil bacterial communities shaped by geochemical factors and land use in a less-explored area, Tibetan Plateau. BMC Genomics 14: 820.
11. Sudini H, Liles MR, Arias CR, Bowen KL, Huettel RN (2011) Exploring soil bacterial communities in different peanut-cropping sequences using multiple molecular approaches. Phytopathology 101: 819–827.
12. Wieland G, Neumann R, Backhaus H (2001) Variation of microbial communities in soil, rhizosphere, and rhizoplane in response to crop species, soil type, and crop development. Appl Environ Microbiol 67: 5849–5854.
13. Avidano L, Gamalero E, Cossa GP, Carraro E (2005) Characterization of soil health in an Italian polluted site by using microorganisms as bioindicators. Appl Soil Ecol 30: 21–33.
14. Nannipieri P, Ascher J, Ceccherini MT, Landi L, Pietramellara G, et al. (2003) Microbial diversity and soil functions. Eur J Soil Sci 54: 655–670.
15. Welbaum GE, Sturz AV, Dong Z, Nowak J (2004) Managing soil microorganisms to improve productivity of agro-ecosystems. Crit Rev Plant Sci 23: 175–193.
16. Glick BR (2010) Using soilbacteria to facilitate phytoremediation. Biotechnol Adv 28: 367–374.
17. Sharma KK, Bhatnagar-Mathur P (2006) Peanut (*Arachis hypogaea* L.). Methods Mol Biol 343: 347–358.
18. Dobermann A, Dawe D, Roetter RP, Cassman KG (2000) Reversal of rice yield decline in a long-term continuous cropping experiment. Agron J 92: 633–643.
19. Johnson NC, Copeland PJ, Crookston RK, Pfleger FL (1992) Mycorrhizae: possible explanation for yield decline with continuous corn and soybean. Agron J 84: 387–390.
20. Lithourgidis AS, Damalas CA, Gagianas AA (2006) Long-term yield patterns for continuous winter wheat cropping in northern Greece. Eur J Agron 25: 208–214.
21. Nayyar A, Hamel C, Lafond G, Gossen BD, Hanson KG, et al. (2009) Soil microbial quality associated with yield reduction in continuous-pea. Appl Soil Ecol 43: 115–121.
22. Yao HY, Jiao XD, Wu FZ (2006) Effects of continuous cucumber cropping and alternative rotations under protected cultivation on soil microbial community diversity. Plant Soil 284: 195–203.
23. Chen MN, Li X, Yang QL, Chi XY, Pan LJ, et al. (2012) Soil eukaryotic microorganism succession as affected by continuous cropping of peanut - pathogenic and beneficial fungi were selected. Plos One 7: e40659.
24. Dang H, Lovell CR (2000) Bacterial primary colonization and early succession on surfaces in marine waters as determined by amplified rRNA gene restriction analysis and sequence analysis of 16S rRNA genes. Appl Environ Microbiol 66(2): 467–475.
25. Dang HY, Li TG, Chen MN, Huang GQ (2008) Cross-ocean distribution of *Rhodobacterales* bacteria as primary surface colonizers in temperate coastal marine waters. Appl Environ Microbiol 74: 52–60.
26. Liu J, Wang H, Yang H, Zhang Y, Wang J, et al. (2013) Composition-based classification of short metagenomic sequences elucidates the landscapes of taxonomic and functional enrichment of microorganisms. Nucleic Acids Res 41(1): e3.
27. DeSantis TZ, Hugenholtz P, Larsen N, Rojas M, Brodie EL, et al. (2006) Greengenes, a chimera-checked 16S rRNA gene database and workbench compatible with ARB. Appl Environ Microbiol 72: 5069–72.
28. Von Wintzingerode F, Göbel UB, Stackebrandt E (1997) Determination of microbial diversity in environmental samples: pitfalls of PCR-based rRNA analysis. FEMS Microbiol Rev 21: 213–229.
29. Altschul SF, Madden TL, Schäffer AA, Zhang J, Zhang Z, et al. (1997) Gapped BLAST and PSI-BLAST: a new generation of protein database search programs. Nucleic Acids Res 25: 3389–3402.
30. Thompson JD, Gibson TJ, Plewniak F, Jeanmougin F, Higgins DG (1997) The CLUSTAL_X windows interface: flexible strategies for multiple sequence alignment aided by quality analysis tools. Nucleic Acids Res 25: 4876–4882.
31. Felsenstein J (1989) PHYLIP–phylogeny inference package (version 3.2). Cladistics 5: 164–166.
32. Kiharaa J, Martiusb C, Bationod A, Thuitaa M, Lesueure D, et al. (2012) Soil aggregation and total diversity of bacteria and fungi in various tillage systems of sub-humid and semi-arid Kenya. Appl Soil Ecol 58: 12–20.
33. Reichardt W, Mascarina G, Padre B, Doll J (1997) Microbial communities of continuously cropped, irrigated rice fields. Appl Environ Microbiol 63: 233–238.
34. Saharan BS, Nehra V (2011) Plant Growth Promoting Rhizobacteria: A Critical Review. Life Sci Med Res LSMR-21: 1–30.
35. Bhattacharyya PN, DK (2012) Plant growth-promoting rhizobacteria (PGPR): emergence in agriculture. World J Microbiol Biotechnol 28: 1327–1350.
36. Kandeler E, Mosier AR, Morgan JA, Milchunas DG, King JY, et al. (2006) Response of soil microbial biomass and enzyme activities to the transient elevation of carbon dioxide in a semi-arid grassland. Soil Biol Biochem 38: 2448–2460.
37. Liddycoat SM, Greenberg BM, Wolyn DJ (2009) The effect of plant growth-promoting rhizobacteria on asparagus seedlings and germinating seeds subjected to water stress under greenhouse conditions. Can J Microbiol 55: 388–394.
38. Ehteshami SMR, Aghaalikhani M, Khavazi K, Chaichi MR (2007) Effect of phosphate solubilizing microorganisms on quantitative and qualitative characteristics of maize (*Zea mays* L.) under water deficit stress. Pak J Biol Sci 10: 3585–3591.
39. Pliego C, Ramos C, de Vicente A, Cazorla FM (2011) Screening for candidate bacterial biocontrol agents against soilborne fungal plant pathogens. Plant Soil 340: 505–520.
40. Elmer WH, Joseph JP (2011) Effect of biochar amendments on mycorrhizal associations and *Fusarium* crown and root rot of asparagus in replant soils. Plant Dis 95(8): 8960–966.
41. Li YG, Ma FM (2012) Antagonistic mechanism of *Fusarium oxysporum* of soybean root rot by *Bacillus subtilis*. Applied Mechanics and Materials 108: 127–131.
42. Almario J, Moënne-Loccoz Y, Muller D (2013) Monitoring of the relation between 2,4-diacetylphloroglucinol-producing *Pseudomonas* and *Thielaviopsis basicola* populations by real-time PCR in tobacco black root-rot suppressive and conducive soils. Soil Biol Biochem 57: 144–155.
43. Li XG, Ding CF, Zhang TL, Wang XX (2014) Fungal pathogen accumulation at the expense of plant-beneficial fungi as a consequence of consecutive peanut monoculturing. Soil Biol Biochem 72: 11–18.
44. Larkin RP, Honeycutt CW, Griffin TS, Olanya OM, Halloran JM, et al. (2011) Effects of different potato cropping system approaches and water management on soilborne diseases and soil microbial communities. Phytopathology 101, 58–67.
45. Spaepen S, Vanderleyden J, Okon Y (2009) Growth-promoting actions of rhizobacteria. Adv Bot Res 51: 283–320.
46. Raaijmakers JM, Paulitz TC, Steinberg C, Alabouvette C, Moënne-Loccoz Y (2009) The rhizosphere: a playground and battlefield for soilborne pathogens and beneficial microorganisms. Plant Soil 321: 341–361.
47. Picard C, di Cello F, Ventura M, Fani R, Guckert A (2000) Frequency and biodiversity of 2, 4-diacetylphloroglucinol producing bacteria isolated from the maize rhizosphere at different stages of plant growth. Appl Environ Microbiol 66: 948–955.

10

Effects of Biochar on Soil Microbial Biomass after Four Years of Consecutive Application in the North China Plain

Qing-zhong Zhang[1]*, Feike A. Dijkstra[2], Xing-ren Liu[1], Yi-ding Wang[1], Jian Huang[1], Ning Lu[1]

1 Key Laboratory of Agricultural Environment, Ministry of Agriculture, Sino-Australian Joint Laboratory For Sustainable Agro-Ecosystems, Institute of Environment and Sustainable Development in Agriculture, Chinese Academy of Agricultural Sciences, Beijing, China, **2** Centre for Carbon, Water and Food, Department of Environmental Sciences, The University of Sydney, Camden, New South Wales, Australia

Abstract

The long term effect of biochar application on soil microbial biomass is not well understood. We measured soil microbial biomass carbon (MBC) and nitrogen (MBN) in a field experiment during a winter wheat growing season after four consecutive years of no (CK), 4.5 (B4.5) and 9.0 t biochar ha^{-1} yr^{-1} (B9.0) applied. For comparison, a treatment with wheat straw residue incorporation (SR) was also included. Results showed that biochar application increased soil MBC significantly compared to the CK treatment, and that the effect size increased with biochar application rate. The B9.0 treatment showed the same effect on MBC as the SR treatment. Treatments effects on soil MBN were less strong than for MBC. The microbial biomass C:N ratio was significantly increased by biochar. Biochar might decrease the fraction of biomass N mineralized (K_N), which would make the soil MBN for biochar treatments underestimated, and microbial biomass C:N ratios overestimated. Seasonal fluctuation in MBC was less for biochar amended soils than for CK and SR treatments, suggesting that biochar induced a less extreme environment for microorganisms throughout the season. There was a significant positive correlation between MBC and soil water content (SWC), but there was no significant correlation between MBC and soil temperature. Biochar amendments may therefore reduce temporal variability in environmental conditions for microbial growth in this system thereby reducing temporal fluctuations in C and N dynamics.

Editor: Andrew C. Singer, NERC Centre for Ecology & Hydrology, United Kingdom

Funding: This work was supported by the S & T Innovation Program of Chinese Academy of Agricultural Sciences. The funders had no role in study design, data collection and analysis, decision to publish, or preparation of the manuscript.

Competing Interests: The authors have declared that no competing interests exist.

* Email: ecologyouth@126.com

Introduction

Biochar is the product of the thermal degradation of organic materials in the absence of air (pyrolysis) and is distinguished from charcoal by its use as a soil amendment [1]. Application of biochar has been proposed as a novel approach to improve soil fertility, increase soil carbon sequestration and mitigate the greenhouse effect [2]. Nevertheless, the turnover of soil organic matter and biochar and their interaction in soils remain poorly understood. Positive and negative priming effects on native organic carbon mineralization in biochar-amended soils have been reported [3–6], and the priming direction is thought to be controlled by the biochar type (determined by production conditions and sources used) and soil pH. Moreover, the presence of soil organic matter can stimulate the mineralization of the more labile components of biochar over the short term, but over the long term, the biochar-soil interaction may enhance soil C storage via processes of organic matter sorption to biochar and physical protection [3,5,6]. There is strong evidence that priming effect on soil organic matter decomposition relies on microbial biomass [7]. Research on soil microbial biomass dynamics with biochar application will help the understanding of priming effect on soil organic matter and biochar decomposition. Biochar addition can increase soil microbial biomass, and may also affect the soil biological community composition, which in turn will affect nutrient cycling, plant growth, and greenhouse gas emission, as well as soil organic carbon mineralization mentioned above [1].

However, the present knowledge on soil microbial biomass dynamics due to biochar application is mainly based on the comparison between biochar application and no biochar treatment without considering crop residue return practice, and lacks long-term field experimental data [8–12]. Our field experiment with consecutive biochar amendments and crop residue return lasted nearly 4 years in a winter wheat-maize relay cropping system until October 2010, with total inputs of biochar ranging from 18.0 to 36.0 t ha^{-1}. The aim of this paper is to examine effects of biochar and wheat straw residue incorporation on the temporal and spatial variation in microbial biomass carbon (MBC) and microbial biomass nitrogen (MBN) measured within a wheat growing season under different soil depths after four years of consecutive application.

Materials and Methods

Site description

The field experiment was conducted at an experimental station (36°58′N, 117°59′E, elevation 17 m) for ecological and sustainability research in Huantai County, Shandong Province, China, and was begun in 2007 [13]. This site has a warm, temperate, continental monsoon climate with a mean annual temperature of 12.4°C. The mean annual precipitation was 600 mm, with most rainfall occurring in June, July, and August. The soil is classified to Fluvic Cambisol according to the USDA system. The soil was a sandy loam, and the proportion of sand, silt, and clay particles in the top 20 cm of soil was 70.8, 26.9, and 2.3%, respectively. The soil bulk density of the top 20 cm of soil before the biochar amendment was 1.52 g cm^{-3}, and the soil organic matter (SOM) content was 15 g kg^{-1} of soil. The soil pH was 8.1 before experiment, and was not significantly changed due to biochar amendment [13].

Biochar

Milled biochar was purchased from a local company (Jinfu Biochar Company, Liaoning Province), with a density of 0.297 g cm^{-3} and a pH of 8.2 [14]. The biochar was made by incomplete self-combustion of crushed corncob in an open-top concrete tank with an igniting apparatus at the bottom for about 24 hours at about 360°C. The incomplete self-combustion process did not need an external energy source. The biochar contained 65.7% carbon and 0.909% N (analyzed with an elemental analysis apparatus, Flash EA 2000, Thermo Electron Corporation, Italy). Available phosphorus content was 0.08% (extracted with 0.5 M $NaHCO_3$ at a pH of 8.5, and analyzed with a colorimetric method), and available potassium content was 1.60% (extracted with 2.0 M HNO_3, and analyzed with a flare photometer, FP640, Cany, China). Ash content of the biochar was 72.0% (determined by dry combustion in a muffle furnace at 550°C for 2 h).

Experimental design

The experimental crops were winter wheat (*Triticum aestivum* L.) and maize (*Zea mays* L.) in relay cropping. Generally, the maize was sown in early June and matured in late September. The winter wheat was then sown in early October and harvested in early June of the next year. The field experiment was a randomized complete block design and each of the experimental plots was 6 m×6 m = 36 m^2, four treatments (CK, B4.5, B9.0, SR) with three replications. The experiment included a control treatment with no biochar addition (CK), two biochar treatments with 4.5 and 9.0 t ha^{-1} yr^{-1} biochar applied (B4.5 and B9.0, respectively), and a treatment where all wheat and maize straw residue produced in the plot was returned to the field and incorporated into the soil after harvest by a straw returning machine (SR), which allowed us to compare biochar vs. fresh litter effects on microbial biomass. Based on the estimated total amount of aboveground crop residues of about 15 t ha^{-1} yr^{-1} for local normal croplands and the empirical value of about 30% in weight of crop residues left as biochar from the local biochar company, biochar amount can be obtained from the field crop residues is equivalent to that used in the B4.5 treatment.

The biochar was distributed equally to each crop (half applied before sowing of wheat and the other half before sowing of maize). Inorganic basal N fertilizers as urea of 200 kg N ha^{-1} yr^{-1} (165 kg N ha^{-1} yr^{-1} was used before 2009), P as superphosphate of 52.5 kg P_2O_5 ha^{-1} yr^{-1}, and K as potassium sulfate of 37.5 kg K_2O ha^{-1} yr^{-1} were applied in all treatments. Before 2009, all fertilizers were used as base fertilizer. From 2009, half of nitrogen fertilizer was applied as base fertilizer, and the other half was applied as topdressing. Biochar and the base fertilizers were broadcast on the soil surface and incorporated into the soil by rotary tillage to depth of 15 cm before seeding.

Soil sampling

Soil samples were collected during the winter wheat growing period (from October 2010 to June 2011) divided into five functional stages: (1) wheat post-seeding stage; (2) before freezing stage, (3) reviving stage; (4) shooting stage, and (5) harvest stage. Soil samples were collected at depths of 0–5, 5–10, 10–20, and 20–30 cm in each experimental plot. The soil samples were collected from five points randomly, and mixed into one sample, each mixed soil sample was divided into two parts. One part of the soil sample was determined for soil water content (SWC) and the other part was prepared for microbial analysis. All samples were immediately stored in sealed plastic bags in a cooler and transported to laboratory and stored in refrigerator at 4°C. All microbiological determinations were performed within one week of sampling.

Measurement and monitoring

Microbial biomass carbon (MBC) and nitrogen (MBN) in soil were determined by fumigation extraction method [15,16], and the value of 0.45 was used for both the fraction of biomass C mineralized (K_C) and the fraction of biomass N mineralized (K_N). Soil samples were thoroughly mixed and ground to pass through a 2-mm sieve, and then the soil moisture was adjusted to about 40% water holding capacity. We fumigated 20.0 g (dry weight equivalent) of soil with ethanol-free chloroform for 24 h. Both fumigated and non-fumigated soils were extracted with 80 ml of 0.5 M K_2SO_4 by shaking for 30 min on a reciprocating shaker at 40 cycles per minute and then filtered (soil:water = 1:4). The TOC analyzer (Multi N/C 2100, Jena, Germany) was used to determine the C and N in the extracts.

SWC was determined gravimetrically by oven-drying at 105°C for 48 h. Soil temperature at 5-cm depth in each replication was monitored hourly by a temperature probe (tolerance: ±0.2°C over the 0–70°C range; temperature measurement range: −50 to +70°C; model 109, Campbell Scientific, Logan, UT, USA) connected to a datalogger (model U12-006, Jimuduoli, Beijing, China) from 22 October 2010, except during harvesting and sowing (June 10–July 2) of each year.

Data analyses

We used repeated measures ANOVA to test for main effects of treatment (between-subjects factor), soil depth (within-subjects factor) and date (within-subjects factor), and their interactive effects on MBC, MBN and microbial biomass C:N ratio. When necessary, data were log-transformed to reduce heteroscedasticity and improve assumptions of normality. We used Pearson's test to determine whether there was a significant correlation between microbial and environmental soil properties (moisture and temperature) and Fisher's least significant difference (LSD) test to determine the significant difference in coefficient of seasonal variability (CV) of MBC and MBN between different treatments. All statistical analyses were performed with JMP (version 4.0.4; SAS Institute, Cary, NC, USA).

Results

Microbial biomass C (MBC)

Variability of MBC under different rates of biochar and straw residue addition was large (Fig. 1). Treatment effects on MBC were significant (Table 1), and on average, largest in the B9.0 treatment, followed by the SR treatment, and lowest in CK. There were significant treatment*date and treatment*date*depth interactive effects ($P=0.02$ and $P<0.0001$ respectively, Table 1). The greatest MBC was found at depth of 10–20 cm treated with SR on 20 October of 2010 (post-seeding stage), followed by the same treatment in 5 November of 2010 (before freezing stage) among all treatments and both soil layers during the whole experimental period. At depth of 0–5 and 5–10 cm, the MBC content in the treatments of B9.0 and SR in general showed the largest increases compared to CK. The MBC in the B9.0 treatment increased by 118% to 763%, while the MBC in the SR treatment was 2% lower to 722% higher compared to CK, depending on the time of year (Fig. 1a, b). The MBC at 0–5 and 5–10 cm in the B4.5 treatment also increased by 6% to 246%, depending on the time of year.

The MBC in the B9.0 treatment increased by 45% to 294% at soil depths of 10–20 and 20–30 cm compared to CK (Fig. 1c, d). The MBC in the SR treatment decreased by 62% during the shooting stage (25 April) at 20–30 cm soil depth, but showed some of the largest increases (between 50% and 408%) at other times at this depth. The MBC in the B4.5 treatment at 20–30 cm soil depth increased in most of the stages, but decreased at 10–20 cm soil depth by 19%, 40%, 6%, and 10% in the winter wheat post-seeding stage (20 October 2010), before freezing stage (5 November 2010), reviving stage (26 March 2011), and harvest stage (5 June 2011), respectively.

Table 2 presents the coefficients of seasonal variation of the MBC under different treatments. The CV of CK varied between 20% and 38%, and the overall CV was 25%. The overall CV of B4.5 and B9.0 treatments was 11% and 16%, respectively. The CVs were significantly smaller than that of CK. But the CV of SR treatment varied from 40% to 63%, which was significantly higher than that of CK. The CV of the surface soil tended to be higher than that of the subsoil in the same treatment, but not for the SR treatment.

Microbial biomass N (MBN)

As with MBC, there was large variability in MBN among the different biochar and SR treatments, dates, and depths (Fig. 2, Table 1). However, unlike the MBC results, across date and depth the largest MBN was observed in the SR treatment (on average 35, 106, and 31% higher than the CK, B4.5 and B9.0 treatment respectively). The MBN in the B4.5 treatment decreased at most dates and soil depths, and across all dates compared to the CK treatment. Across date, the MBN in the B4.5 treatment decreased by 40, 47, 3, and 30% at 0–5, 5–10, 10–20, and 20–30 cm soil depth. The MBN in the B9.0 treatment often increased at 0–5, 5–10, and 20–30 cm soil depth during the stages of wheat post-seeding stage, before freezing stage, and reviving stage compared to the CK treatment. Large MBN pools were particularly observed in the SR treatment at 10–20 and 20–30 cm soil depth. At the

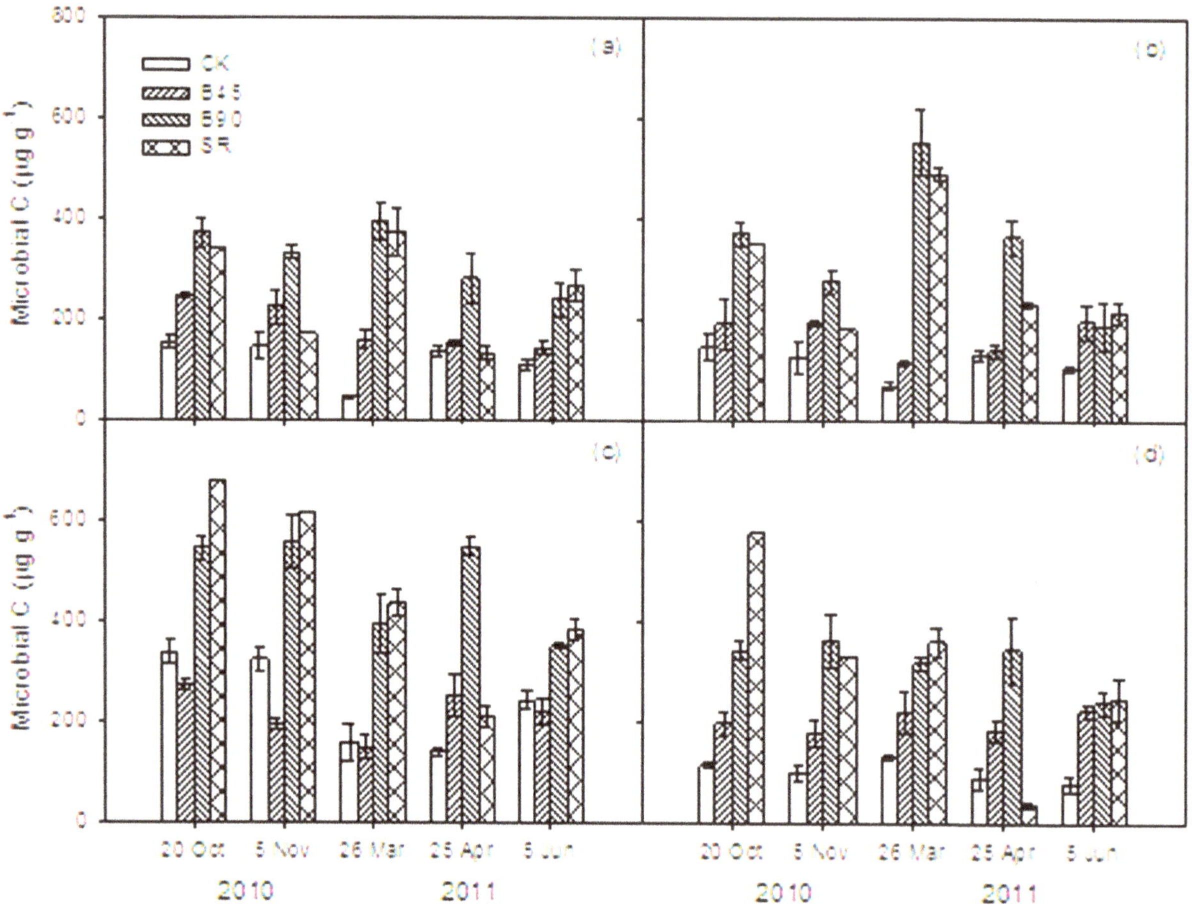

Figure 1. Microbial biomass carbon (MBC) under different treatments (CK: control, B4.5: 4.5 t ha^{-1} yr^{-1} biochar addition, B9.0: 9.0 t ha^{-1} yr^{-1} biochar addition, SR: incorporation of wheat straw) and different soil depths (a: 0–5, b: 5–10, c: 10–20, and d: 20–30 cm) during winter wheat season. Vertical bars represent the standard errors for means of each treatment ($n=3$).

Table 1. Repeated Measures ANOVA P values.

Variable	Microbial C	Microbial N	Microbial C:N
Treatm	0.0003	<0.0001	<0.0001
Date	<0.0001	<0.0001	0.002
Depth	<0.0001	0.0001	<0.0001
Treatm*date	0.02	<0.0001	0.0005
Treatm*depth	0.12	<0.0001	0.0004
Date*depth	<0.0001	0.0001	0.005
Treatm*date*depth	<0.0001	0.0003	0.0007

shooting stage, the MBN of the four treatments were all relatively low.

The CV of MBN in the CK treatment varied from 24% to 44%, and the overall CV was 32% (Table 3). The overall CV of B4.5 was 19%, which was significantly lower than that of other treatments. The overall CV of B9.0 and SR treatments were 49% and 65% respectively, which were significantly larger than that of CK. From this, it can be seen that addition of 4.5 t ha^{-1} of biochar decreased the temporal and spatial fluctuation of MBN, but that a higher biochar addition increased the fluctuation of MBN.

Microbial biomass C:N ratio

Variability in microbial biomass C:N ratio was large among the different treatments. We observed significant treatment effects on microbial biomass C:N ratio (Table 1, Fig. 3), with largest ratios in the B9.0 treatment (on average 181% greater than CK) and B4.5 treatment (on average 93% higher than CK), suggesting that the increase in MBC with biochar addition was larger compared to the increase in MBN. The greatest microbial biomass C:N ratio ratios were found at 10–20 cm and 20–30 cm soil depth in the B9.0 treatment.

Discussion

Soil microbial biomass

The change in MBC reflects the process of microbial growth, death and organic matter degradation. Our results showed that MBC increased with biochar amendment compared to CK, which suggested that microbial growth could be accelerated by biochar addition. Reported biochar effects on soil MBC are quite inconsistent. Several studies found that there was no significant effect of biochar amendment on soil MBC [8,11]. Dempster et al. [12] found that MBC significantly decreased with biochar addition while MBN was unaltered in a coarse textured soil, and others observed the same positive effects of biochar addition on microbial biomass as ours [1,17–19]. Moreover, a positive linear relationship between microbial biomass and biochar concentration was also observed in a highly weathered soil [20]. Biochar type was thought as the driving parameter for any effects on soil microbial biomass, community, and activity [9,21].

Compared to wheat straw residue return, after 4 years of annual application, biochar increased MBC in 0–30 cm soil when applied at a high rate (9.0 t ha^{-1} yr^{-1}, $p<0.05$), but decreased MBC when applied at a low rate (4.5 t ha^{-1} yr^{-1}, $p<0.01$). The amount of C added in the B4.5 treatment most likely did not induce a similar increase in MBC as in the SR treatment, while the amount of C added in the B9.0 treatment did.

The MBN at 0–30 cm soil depth significantly decreased in the B4.5 treatment, but increased in the B9.0 treatment compared to CK ($p<0.01$), whereas the MBN at 0–30 cm soil depth in both biochar treatments were lower compared to SR treatment ($p<0.05$). Zavalloni et al. [8] found that biochar amendment at a rate of 5% had no significant influence on soil MBN in an incubation experiment. The increase in MBC and the decrease in MBN for biochar treatments indicate that biochar in soil acted as a carbon source rather than a nitrogen source for soil microbes. In return, this could have consequences for N cycling (e.g., increased microbial N immobilization with biochar addition). Our previous study in the field showed that biochar addition decreased soil available N by 7–10% compared to CK ($p>0.05$), but increased total N by 14–21% ($p<0.05$) in the 0–15 cm soil [22].

In the relatively unfertile and coarse-textured soil of our study, microbial biomass was probably limited by both a suitable growth environment and by C availability [17]. The labile fraction of biochar has been shown to stimulate microbial activity and abundance in some cases. Sorption of comparatively polar organic matter and nutrients could support energy for microorganisms, while macro-and micropores of biochar, which hold air and water, likely support microorganisms' livable habitat [1].Our results

Table 2. Coefficients of seasonal variation of MBC under different treatments (%).

Treatment	0–5 cm	5–10 cm	10–20 cm	20–30 cm	0–30 cm
CK	37.0ab	25.4ab	38.1ab	20.3b	25.0b
B4.5	25.7bc	22.9b	22.6bc	10.4c	11.2c
B9.0	19.4c	34.2ab	20.5c	15.3b	15.6c
SR	40.6ab	43.1a	40.4a	63.3a	41.6a

Note: Different small letters indicate significant differences among treatments (CK: control, B4.5: 4.5 t ha^{-1} yr^{-1} biochar addition, B9.0: 9.0 t ha^{-1} yr^{-1} biochar addition, SR: incorporation of wheat straw) at the $P<0.05$ level (L.S.D.).

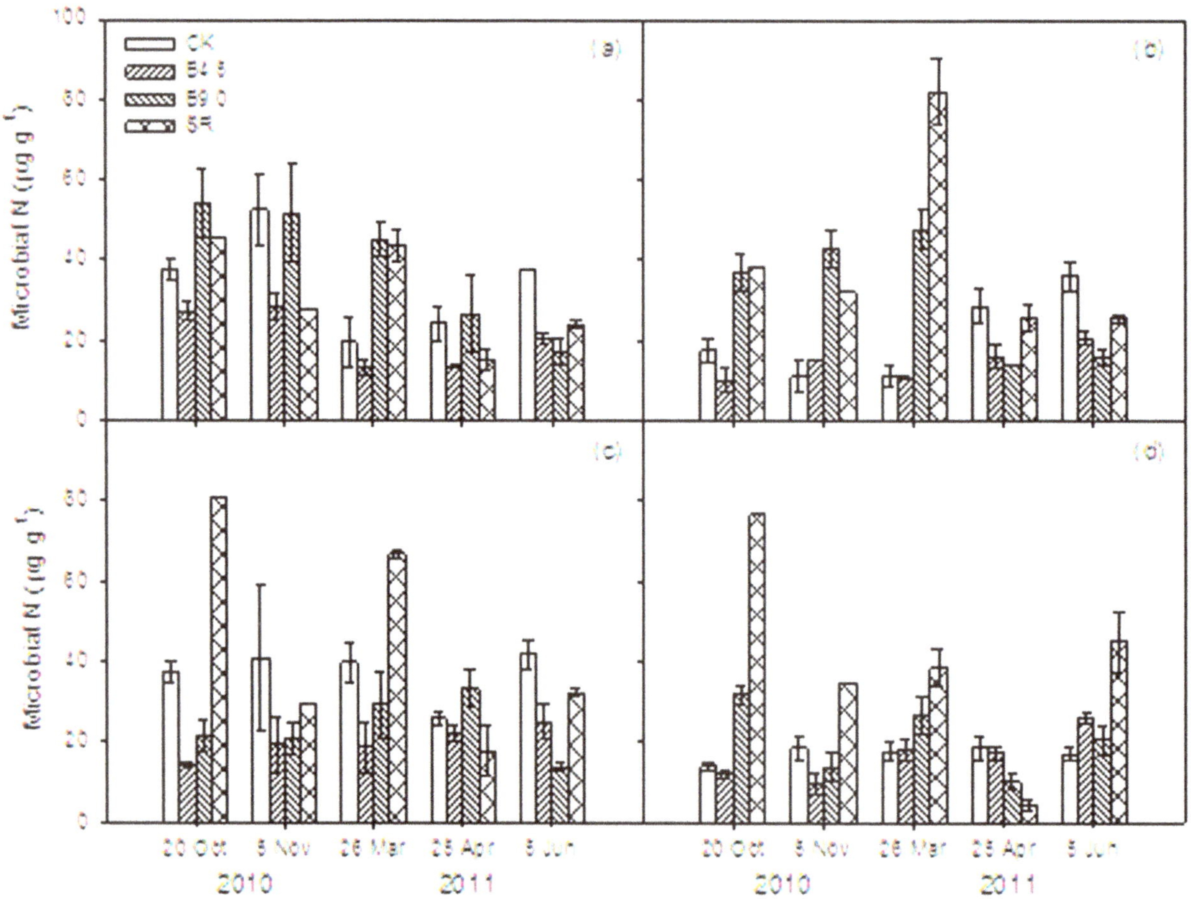

Figure 2. Microbial biomass nitrogen (MBN) under different treatments (CK: control, B4.5: 4.5 t ha^{-1} yr^{-1} biochar addition, B9.0: 9.0 t ha^{-1} yr^{-1} biochar addition, SR: incorporation of wheat straw) and different soil depths (a: 0–5, b: 5–10, c: 10–20, and d: 20–30 cm) during winter wheat season. Vertical bars represent the standard errors for means of each treatment ($n = 3$).

suggest that biochar supplied livable habitat for microorganisms stimulating microorganisms which can use the carbon source from the labile fraction of biochar.

Soil microbial biomass C:N

Changes in the microbial biomass C:N ratio can reflect changes in the relative availability of C and N to microbes, but could also reflect changes in the microbial community structure. In our study, the microbial biomass C:N ratio significantly increased with increasing biochar application and the microbial biomass C:N ratio more than doubled in the B9.0 treatment compared to the control (Fig. 3). A pot experiment showed that there was no significant effect on the microbial biomass C:N ratio of biochar addition at a rate of 25 t ha^{-1} [12]. Nicolardot et al. [23] determined that there was a significantly positive correlation between soil microbial biomass C:N ratio and the C:N ratio of the added organic matter. In contrast, Kushwaha et al. [24] reported a decrease in microbial biomass C:N ratio after straw return, while Kallenbach and Grandy [25] found no effect on microbial biomass C:N ratio after application of organic carbon, suggesting that changes in microbial biomass C:N ratio cannot solely be explained by organic amendments with relatively high C content. Several studies observed a change in soil microbial structure and abundance with biochar application [1]. In our study the

Table 3. Coefficients of seasonal variation of MBN under different treatments (%).

Treatment	0–5 cm	5–10 cm	10–20 cm	20–30 cm	0–30 cm
CK	38.1ab	43.6ab	23.7b	31.4a	31.7b
B4.5	34.9b	23.9b	6.9c	34.9c	18.6c
B9.0	49.5a	52.5a	43.7ab	58.8b	49.1ab
SR	42.0ab	55.3a	66.9a	93.6a	65.4a

Note: Different small letters indicate significant differences among treatments (CK: control, B4.5: 4.5 t ha^{-1} yr^{-1} biochar addition, B9.0: 9.0 t ha^{-1} yr^{-1} biochar addition, SR: incorporation of wheat straw) at the $P<0.05$ level (L.S.D.).

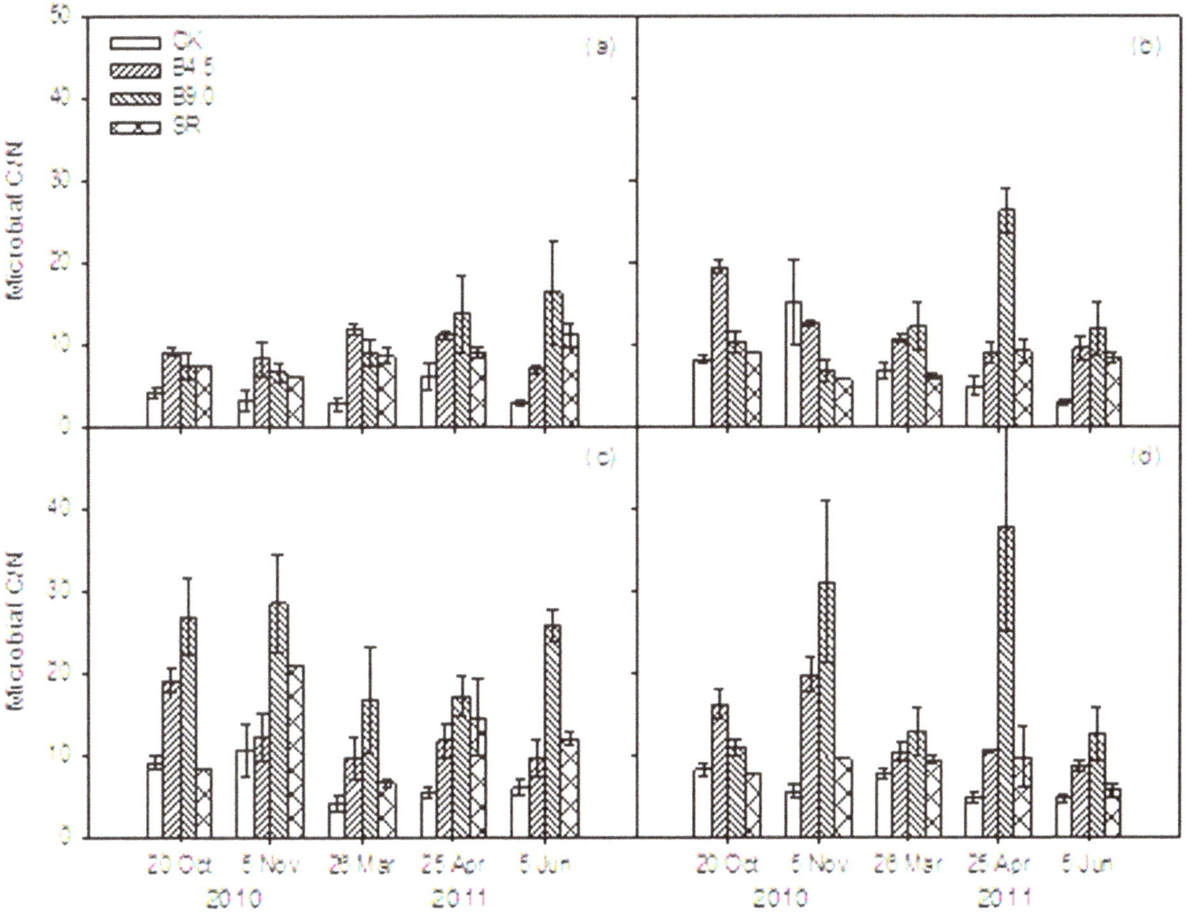

Figure 3. The ratio of soil microbial biomass carbon (MBC) to soil microbial biomass nitrogen (MBC/MBN) under different treatments (CK: control, B4.5: 4.5 t ha^{-1} yr^{-1} biochar addition, B9.0: 9.0 t ha^{-1} yr^{-1} biochar addition, SR: incorporation of wheat straw) and different soil depths (a: 0–5, b: 5–10, c: 10–20, and d: 20–30 cm) during winter wheat season. Vertical bars represent the standard errors for means of each treatment ($n = 3$).

increased microbial C:N ratio with biochar suggests increased microbial N limitation, while it is unclear if the increased microbial C:N ratio was also due to changes in microbial community structure. Moreover, biochar application may alter the release of MBN or the fraction of biomass N mineralized (K_N). The value of K_N had a wide range from 0.28–0.81 [26–28], and it was difficult to be measured precisely. The high values of microbial C:N ratio for biochar treatments especially at 10–20 and 20–30 cm soil depths might be due to the same value of K_N used for all treatments herein. It is worthy of further study on the effect of biochar on the fraction of biomass N mineralized in the fumigation-extraction method.

Relationship between MBC and environmental factors

Data analysis showed that there was a significant positive correlation between SWC and MBC (**Fig. 4**, $R^2 = 0.172$, $P < 0.001$). The positive relationship between SWC and MBC suggests that SWC may have limited microbial activity at the experimental site. Biochar addition can increase soil water content [29,30]. Therefore, biochar application could influence soil microbial biomass via variation in soil water content.

However, the increases in MBC with biochar addition did not always coincide with an increase in SWC. For instance, MBC increased sharply in the reviving stage (March 2011) in the B9.0 treatment while SWC did not. Moreover, our study showed that biochar application showed a limited effect on soil water content [13,14], which also did not coincide with the higher MBC in the biochar treatments than in the CK treatment.

There was no significant correlation between MBC and soil temperature, and our result was supported by other studies [31,32]. For instance, Contin et al. [32] found that MBC did not change significantly at different incubation temperatures in arable and grassland soils. But we should point out that the spike in MBC at 5–10 cm depth of soil on March 26 may be related to the soil temperature increase that occurred during the same time.

The response of microbial biomass to seasonal changes is important in regulating microbial turnover, which may in turn influence nutrient availability and ultimately plant nutrition [33]. Seasonal fluctuations in MBC in the biochar treatments were smaller than in the CK and SR treatments (Table 2), whereas, seasonal fluctuations in MBN showed no obvious differences among the CK, B9.0 and SR treatments, and only the seasonal fluctuations in the B4.5 treatment was lower than in the other treatments. Biochar may have reduced extreme environmental conditions for microbial growth, which could have important implications for C and N dynamics, and crop yield [34,35].

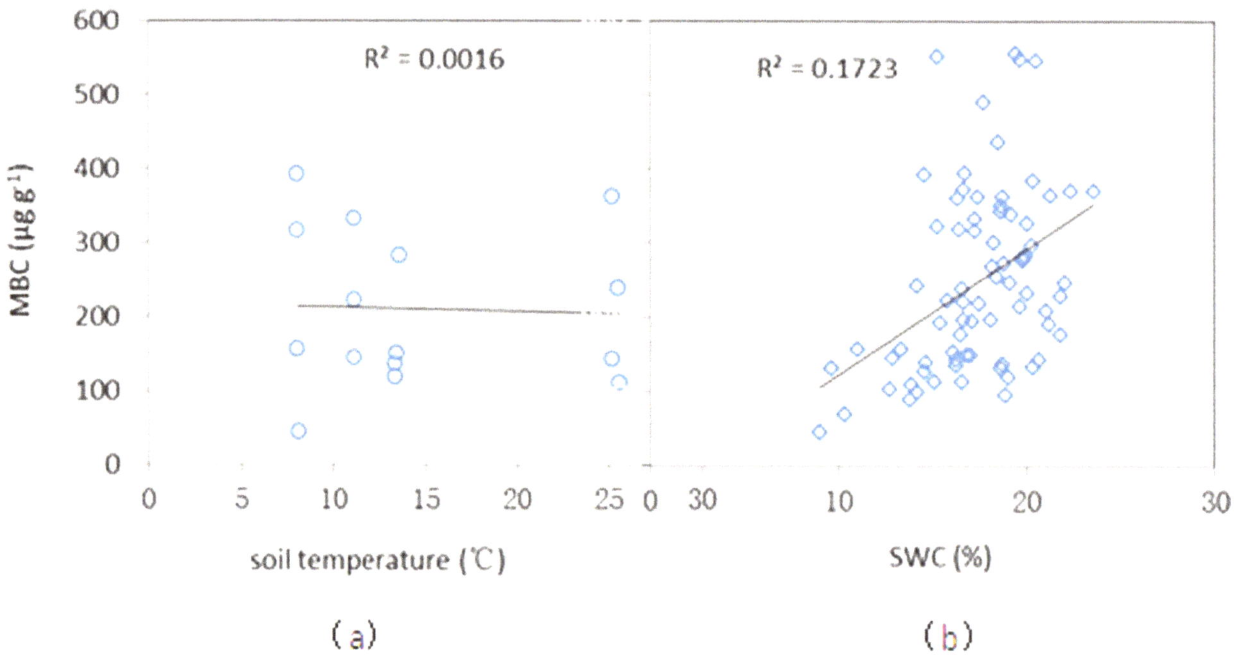

Figure 4. The correlations between soil microbial biomass carbon (MBC) and soil temperature (a), and between MBC and soil water content (SWC, b).

Conclusion

After 4 consecutive years of application, biochar increased soil MBC significantly compared to CK. The biochar application at a rate of 9.0 t ha^{-1} yr^{-1} reached and even exceeded the effect of wheat residue return, whereas smaller differences in soil MBN were found among treatments. Soil MBC for biochar treatments showed less seasonal fluctuation compared to CK and SR treatments, suggesting that biochar provided a more suitable habitat for soil microorganisms. Biochar treatment showed the highest value of soil microbial biomass C:N ratio, and one possible reason might be that biochar could decease the fraction of biomass N mineralized.

Acknowledgments

This work was supported by the S&T Innovation Program of Chinese Academy of Agricultural Sciences.

Author Contributions

Conceived and designed the experiments: QZZ. Performed the experiments: JH NL QZZ. Analyzed the data: QZZ FAD XRL YDW. Contributed reagents/materials/analysis tools: QZZ FAD YDW XRL. Contributed to the writing of the manuscript: QZZ FAD XRL.

References

1. Lehmann J, Rillig MC, Thies J, Masiello CA, Hockaday WC, et al. (2011) Biochar effects on soil biota – A review. Soil Biology and Biochemistry 43: 1812–1836.
2. Lehmann J (2007) A handful of carbon. Nature 447: 143–144.
3. Zimmerman AR, Gao B, Ahn M-Y (2011) Positive and negative carbon mineralization priming effects among a variety of biochar-amended soils. Soil Biology and Biochemistry 43: 1169–1179.
4. Luo Y, Durenkamp M, De Nobili M, Lin Q, Brookes PC (2011) Short term soil priming effects and the mineralisation of biochar following its incorporation to soils of different pH. Soil Biology and Biochemistry 43: 2304–2314.
5. Cross A, Sohi SP (2011) The priming potential of biochar products in relation to labile carbon contents and soil organic matter status. Soil Biology and Biochemistry 43: 2127–2134.
6. Keith A, Singh B, Singh BP (2011) Interactive priming of biochar and labile organic matter mineralization in a smectite-rich soil. Environmental Science & Technology 45: 9611–9618.
7. Thiessen S, Gleixner G, Wutzler T, Reichstein M (2013) Both priming and temperature sensitivity of soil organic matter decomposition depend on microbial biomass – An incubation study. Soil Biology & Biochemistry 57: 739–748.
8. Zavalloni C, Alberti G, Biasiol S, Vedove GD, Fornasier F, et al. (2011) Microbial mineralization of biochar and wheat straw mixture in soil: A short-term study. Applied Soil Ecology 50: 45–51.
9. Steinbeiss S, Gleixner G, Antonietti M (2009) Effect of biochar amendment on soil carbon balance and soil microbial activity. Soil Biology and Biochemistry 41: 1301–1310.
10. Luo Y, Durenkamp M, De Nobili M, Lin Q, Devonshire BJ, et al. (2013) Microbial biomass growth, following incorporation of biochars produced at 350°C or 700°C, in a silty-clay loam soil of high and low pH. Soil Biology and Biochemistry 57: 513–523.
11. Castaldi S, Riondino M, Baronti S, Esposito FR, Marzaioli R, et al. (2011) Impact of biochar application to a Mediterranean wheat crop on soil microbial activity and greenhouse gas fluxes. Chemosphere 85: 1464–1471.
12. Dempster DN, Gleeson DB, Solaiman ZM, Jones DL, Murphy DV (2012) Decreased soil microbial biomass and nitrogen mineralisation with Eucalyptus biochar addition to a coarse textured soil. Plant and Soil 354: 311–324.
13. Zhang QZ, Wang XH, Du ZL, Liu XR, Wang YD (2013) Impact of biochar on nitrate accumulation in an alkaline soil. Soil Research 51: 521–528.
14. Zhang QZ, Wang YD, Wu YF, Wang XH, Du ZL, et al. (2013) Effects of biochar ammendment on soil thermal conductivity, reflectance, and temperature. soil Science Society of America Journal 77: 1478–1487.
15. Brookes PC, Landman A, Pruden G, Jenkinson DS (1985) Chloroform fumigation and the release of soil nitrogen: A rapid direct extraction method to measure microbial biomass nitrogen in soil. Soil Biology and Biochemistry 17: 837–842.
16. Wu J, Joergensen RG, Pommerening B, Chaussod R, Brookes PC (1990) Measurement of soil microbial biomass C by fumigation-extraction—an automated procedure. Soil Biology and Biochemistry 22: 1167–1169.
17. Kolb SE, Fermanich KJ, Dornbush ME (2008) Effect of charcoal quantity on microbial biomass and activity in temperate soils. Soil Science Society of America Journal 73: 1173–1181.

18. O' Neill B, Grossman J, Tsai MT, Gomes JE, Lehmann J, et al. (2009) Bacterial community composition in Brazilian Anthrosols and adjacent soils characterized using culturing and molecular identification. Microbial Ecology 58: 23–35.
19. Liang B, Lehmann J, Sohi SP, Thies JE, O'Neill B, et al. (2010) Black carbon affects the cycling of non-black carbon in soil. Organic Geochemistry 41: 206–213.
20. Steiner C, Das KC, Garcia M, Förster B, Zech W (2008) Charcoal and smoke extract stimulate the soil microbial community in a highly weathered xanthic Ferralsol. Pedobiologia 51: 359–366.
21. Lehmann J, Joseph S (2009) Biochar for Environmental Management: An Introduction. In: Lehmann J, Joseph S, editors. Biochar for Environmental Management: Science and Technology. London: Earthscan. pp. 1–12.
22. Guo W, Chen HX, Zhang QZ, Wang YD (2011) Effects of biochar application on total nitrogen and alkali-hydrolyzable nitrogen content in the topsoil of the high-yield cropland in north China Plain. Ecology and Environmental Sciences 20: 425–428.
23. Nicolardot B, Recous S, Mary B (2001) Simulation of C and N mineralisation during crop residue decomposition: a simple dynamic model based on the C: N ratio of the residues. Plant and Soil 228: 83–103.
24. Kushwaha CP, Tripathi SK, Singh KP (2000) Variations in soil microbial biomass and N availability due to residue and tillage management in a dryland rice agroecosystem. Soil and Tillage Research 56: 153–166.
25. Kallenbach C, Grandy AS (2011) Controls over soil microbial biomass responses to carbon amendments in agricultural systems: A meta-analysis. Agriculture, Ecosystems and Environment 144: 241–252.
26. Joergensen RG (1996) The fumigation-extraction method to estimate soil microbial biomass: Calibration of the kEC value. Soil Biology and Biochemistry 28: 25–31.
27. Greenfield LG (1995) Release of microbial cell N during chloroform fumigation. Soil Biology and Biochemistry 27: 1235–1236.
28. Sparling G, Zhu C (1993) Evaluation and calibration of biochemical methods to measure microbial biomass C and N in soils from western australia. Soil Biology and Biochemistry 25: 1793–1801.
29. Chen Y, Shinogi Y, Taira M (2010) Influence of biochar use on sugarcane growth, soil parameters, and groundwater quality. Australian Journal of Soil Research 48: 526–530.
30. Piccolo A, Pietramellara G, Mbagwu JSC (1996) Effects of coal derived humic substances on water retention and structural stability of Mediterranean soils. Soil Use and Management 12: 209–213.
31. Joergensen RG, Brookes PC, Jenkinson DS (1990) Survival of the soil microbial biomass at elevated temperatures. Soil Biology and Biochemistry 22: 1129–1136.
32. Contin M, Corcimaru S, De Nobili M, Brookes PC (2000) Temperature changes and the ATP concentration of the soil microbial biomass. Soil Biology and Biochemistry 32: 1219–1225.
33. Wardle DA (1992) A comparative assessment of factors which influence microbial biomass carbon and nitrogen levels in soil. Biological Reviews of the Cambridge Philosophical Society 67: 321–358.
34. Kumar K, Goh KM (1999) Crop residues and management practices: effects of soil quality, soil nitrogen dynamics, crop yield, and nitrogen recovery. Advances In Agronomy 68: 197–319.
35. Lou Y, Li Z, Zhang T, Liang Y (2004) CO_2 emissions from subtropical arable soils of China. Soil Biology & Biochemistry 36: 1835–1842.

11

Effects of Winter Cover Crops Residue Returning on Soil Enzyme Activities and Soil Microbial Community in Double-Cropping Rice Fields

Tang Hai-Ming[1]*, Xiao Xiao-Ping[1]*, Tang Wen-Guang[1], Lin Ye-Chun[2], Wang Ke[1], Yang Guang-Li[1]

1 Hunan Soil and Fertilizer Institute, Changsha, PR China, **2** Guizhou Academy of Tobacco Science, Guiyang, PR China

Abstract

Residue management in cropping systems is useful to improve soil quality. However, the studies on the effects of residue management on the enzyme activities and microbial community of soils in South China are few. Therefore, the effects of incorporating winter cover crop residue with a double-cropping rice (*Oryza sativa* L.) system on soil enzyme activities and microbial community in Southern China fields were studied. The experiment has conducted at the experimental station of the Institute of Soil and Fertilizer Research, Hunan Academy of Agricultural Science, China since winter 2004. Four winter cropping systems were used: rice–rice–ryegrass (*Lolium multiflorum* L.) (R-R-Ry), rice–rice–Chinese milk vetch (*Astragalus sinicus* L.) (R-R-Mv), rice–rice–rape (*Brassica napus* L.) (R-R-Ra) and rice–rice with winter fallow (R-R-Fa). The result indicated that the enzyme activities in the R-R-Ry, R-R-Mv and R-R-Ra systems were significantly higher ($P<0.05$) than in the R-R-Fa system during the early and late rice season. The β-glucosidase activities reached peak values at the tillering stage after residue application, and alkaline phosphatase activities reached peak values at the booting stage after residue application, respectively, the activities of β-glucosidase and alkaline phosphatase gradually decreased after this. Arylsulfatase activities reached peak values at the maturity stage. Arylamidase activities reached peak values at the maturity stage. The numbers of aerobic bacteria, actinomycete and fungus of residue treatments were significantly higher ($P<0.05$) than that the R-R-Ra system. However, the number of anaerobic bacteria under the R-R-Ry and R-R-Mv systems was significantly lower ($P<0.05$) than that under the R-R-Fa system during early rice and late rice growth stage. Thus, incorporation of winter cover crops into rotations may increase enzyme activities and microbial community in soil and therefore improve soil quality.

Editor: Hauke Smidt, Wageningen University, Netherlands

Funding: This study was supported by the Hunan Provincial Natural Science Foundation of China (No. 12JJ4022), and the Public Research Funds Projects of Agriculture, Ministry of Agriculture of the P.R. China (No. 201103001). The funders had no role in study design, data collection and analysis, decision to publish, or preparation of the manuscript.

Competing Interests: The authors have declared that no competing interests exist.

* Email: hntfsxxping@163.com (XXP); tanghaiming66@163.com (THM)

Introduction

Enzymes play an important role in the cycling of nutrients in nature, and soil enzyme activity can be used as an index of soil microbial activity and fertility [1]. Soil enzymes are involved in energy transfer, and consequently affect environmental quality and crop productivity [2,3]. The number of enzyme production, and the activity and the stability of free and adsorbed enzymes are controlled by environmental conditions and ecological interactions. Enzymes may respond to changes in soil management more quickly than other soil variables and therefore might be useful as early indicators of biological changes. Enzyme activity profiles reflect soil functional diversity, which is influenced by the genetic diversity of soil microorganisms, plants and animals and is closely related to environmental factors and ecological interactions [4]. Among the different enzymes in soils, arylamidase, alkaline phosphatase, β-glucosidase and arylsulfatase are important for the transformation of plant nutrients. β-glucosidase catalyzes glucose formation and is an important enzyme in the terrestrial carbon cycle, glucose being an important energy source for microbial biomass [3]. Phosphatases play an important role in transforming organic phosphorus into inorganic forms that are suitable for plants. Arylamidase catalyzes one of the most important reactions in N mineralization: it releases amino acids that are used as substrates for amidohydrolases from soil organic matter. Arylsulfatase is an extracellular enzyme that catalyses the hydrolysis of organic sulphate esters, releasing SO_2 that can be used by plants.

Soil microbes play an important role in ecosystem function and may act as filters or valves that regulate the intra-system cycling of soil nutrients [5]. Soil microbes play an important role in the maintenance of soil fertility (i.e. nutrient cycling and organic matter decomposition). The ability of microbes to maintain soil fertility and regulate nutrient cycling may be largely dependent on the composition of soil microorganisms communities [6,7], and changes in microbial diversity or community structure could have dramatic impacts on ecosystem processes [8].

Winter cover crops, which are grown during an otherwise fallow period, are a possible means of improving nutrient dynamics in the surface layer of intensively managed cropping systems. Hermawan and Bomke [9] suggested that growing winter cover crops such as annual ryegrass may protect aggregate breakdown during winter and result in a better soil structure after spring tillage, as opposed to leaving soil bare. Other potential benefits of winter cover crops are the prevention of nitrate leaching [10]; weed infestation [11]; and improvement of soil water retention, soil organic matter content and microbial activity [12]. Recycling of crop residues has been suggested to improve overall soil conditions, reduce the requirement for N fertilizers and support sustainable rice (*Oryza sativa* L.) productivity. Chinese milk vetch (*Astragalus sinicus* L.), ryegrass (*Lolium multiflorum* L.) and rape (*Brassica napus* L.) are the main winter cover crops in China.

In recent years, many studies have shown that the enzyme and microbial activities of soil are affected by soil tillage and residue management [13], application of fertilizer and organic matter [14,15], crop rotations [16], and field management [17,18]. However, relatively few studies on soil enzyme activities and soil microbes in double-rice cropping system with winter cover crop rotations have been conducted in southern China.

Our relatively rare study was conducted in a double-cropping paddy field, the winter cover crops tested were ryegrass (*Lolium multiflorum* L.), Chinese milk vetch (*Astragalus sinicus* L.), and rape (*Brassica napus* L.), and a control plot with weed and no cover crop was also used. Usually, enzymes and microbial activities affected by management practices were used as indicators of ecological stability. The objectives of this research were to determine the effects of different winter cover crops residues returning on soil enzyme activities and soil microbial community in a double-cropping rice system.

Materials and Methods

Sites

The experiment has conducted at the experimental station of the Institute of Soil and Fertilizer Research, Hunan Academy of Agricultural Science, China (28°11′58″ N, 113°04′47″ E) since winter 2004. The typical cropping system in this area is double-cropping rice. The soil type is a Fe–accumuli–Stagnic Anthrosol derived from Quaternary red clay (clay loam). The characteristics of the surface soil (0–20 cm) are: pH 5.40, soil organic carbon (SOC) 13.30 g kg^{-1}, total nitrogen 1.46 g kg^{-1}, available N 154.5 mg kg^{-1}, total phosphorous 0.81 g kg^{-1}, available P 39.2 mg kg^{-1}, total potassium 13.0 g kg^{-1}, and available potassium 57.0 mg kg^{-1} [19]. All these values were tested before the experiment in 2004. This region has the subtropical monsoonal humid climate, with a long hot period and short cold period. The average annual precipitation is approximately 1500 mm and the annual mean temperature is 17.1°C, and the annual frost-free period is approximately from 270 days to 310 days. The daily precipitation and daily mean temperature data during the 2010 early and late rice growing season is presented in Fig. 1.

Experimental design and soil sampling

The cropping system was the winter cover crops were planted after double cropping rice harvested every year. The study was continuously conducted for 7 years after straw recycling of the winter cover crop. The four cropping systems used were rice–rice–ryegrass (R-R-Ry), rice–rice–Chinese milk vetch (R-R-Mv), rice–rice–rape (R-R-Ra) and rice–rice with winter fallow (R-R-Fa). A randomized block design was adopted in plots, with three replications. The plot area was 1.1 m^2 (1 m×1.1 m). Before sowing the winter cover crops, 75.0 kg ha^{-1} N (163.0 kg urea) and 45.0 kg ha^{-1} P_2O_5 (375.0 kg diammonium phosphate) were applied as basal fertilizer. The winter cover crops planted in winter 2004 were cut and incorporated into soil by means of a disk plow (conventional tillage) in the following spring. When winter cover crops were harvested, the straw residue of ryegrass, Chinese milk vetch, and rape straw residue were weighted and returned at the same quantity of 22500.0 kg ha^{-1} to the soil surface. And the plots were plowed once to a depth of 20 cm using a moldboard plow on the fifteen day before rice seedling transplanting.

One-month-old seedlings were transplanted at a density of 150,000 plants ha^{-1} (one seed per 16 cm×16 cm), with 2–3 plants per hill. For both early and late rice, basal fertilizer was applied at the rate of 150.0 N kg ha^{-1} as urea (60%, basal; 40%, top-dressed at the tillering stage), 75.0 kg ha^{-1} P_2O_5 as diammonium phosphate and 120.0 kg ha^{-1} K_2O as potassium sulfate.

Data were collected during the early and late rice growth period in 2010. Soil samples in the ploughed layer (0–20 cm) were collected from the centre of four hills of rice plants by using a drill at different rice growth periods.

Laboratory analyses

The soil samples were passed through a 2-mm sieve and kept moist in a refrigerator at 4°C until use. Arylamidase (EC 3.4.11.2) activity was assayed by incubating 1.0 g moist soil with 3.0 mL of 0.1 M THAM buffer (pH 8.0) and 1.0 mL of an 8.0 mM solution of L-leucine β-naphthylamide hydrochloride [20]. Alkaline phosphatase (EC 3.1.3.1), β-glucosidase (EC 3.2.1.21) and arylsulfatase (EC 3.1.6.1) activities were determined as described by Tabatabai [3], and the activity was reported as μg *p*-nitrophenol g^{-1} h^{-1}. All the measurements were repeated for three times.

Colony forming units (CFUs) of soil anaerobic bacteria, aerobic bacteria, actinomycetes, and fungi were enumerated by a 10-fold dilution plate technique. And the number of aerobic bacteria were identified by spreading 100 μl of diluted sample on LB agar medium. The medium for actinomycetes contained 1% (w/v) soluble starch, 0.2% (w/v) $(NH_4)_2SO_4$, 0.1% (w/v) K_2HPO_4, 0.1% (w/v) $MgSO_4{\cdot}7H_2O$, 0.7% (w/v) NaCl, 0.3% (w/v) $CaCO_3$, and 2% (w/v) agar. The number of CFUs of actinomycetes formed on this medium was determined by probing colonies that developed with a dissecting needle: if the colony remained as a discrete, small mass, it was considered to be an actinomycete. The number of CFUs of fungi was estimated on Martin's agar medium containing 1.25 g streptomycin l^{-1} and 33 mg rose bengal l^{-1}. Three replicates of the inoculated agar plates were incubated at 28°C for 3 d for bacteria, 5 d for fungi, and 7 d for actinomycetes, after which colonies were counted [17]. The number of aerobic bacteria was determined according to the method described by Oliveira et al. [21]. The number of anaerobic bacteria was determined according to the method described by Min et al. [22].

Statistical analysis

All data were expressed as mean ± standard error. The data were analyzed as a randomized complete block, using the PROC ANOVA procedure of SAS [23]. Mean values were compared using the least significant difference (LSD) test, and a probability value of 0.05 was considered to indicate statistical significance.

Results

Dynamics of β-glucosidase activity during the rice growth period

In the early rice season, the activity of β-glucosidase in soils was significantly affected by the residue management practice. The R-R-Ry had the highest β-glucosidase activity and the R-R-Fa had the lowest activity (Fig. 2), and the activity decreased in the following order: R-R-Ry>R-R-Ra≈R-R-Mv>R-R-Fa. In the late rice season, the highest activity was the treatment that the rape was used as residues, and the activity trend was as follows: R-R-Ra>R-R-Ry≈R-R-Mv>R-R-Fa.

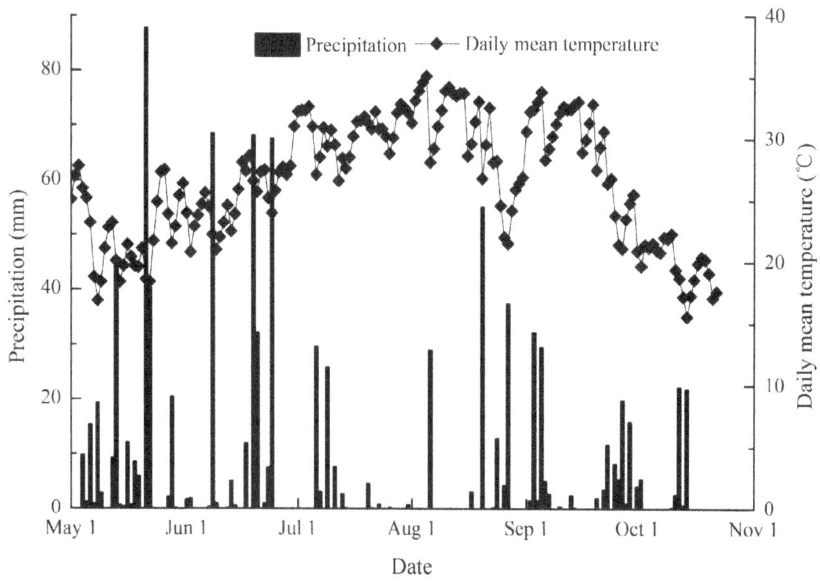

Figure 1. Daily precipitation and daily mean temperature at the study site during the experimental period.

Dynamics of alkaline phosphatase activity during the rice growth period

With the addition of the winter cover crop residues, alkaline phosphatase activity in the R-R-Ry, R-R-Mv, and R-R-Ra soils was higher than that in R-R-Fa (Fig. 3). In other words, the alkaline phosphatase activity was enhanced by the application of cover crop residues in the rice growth season. The alkaline phosphatase activity of soils in the early rice season was higher than that in the late rice season. In the early and late rice seasons, alkaline phosphatase activity decreased as follows: R-R-Mv>R-R-Fa, but there were no significant differences ($P>0.05$) among other treatments. And there were no significant differences ($P>0.05$) in alkaline phosphatase activity under the R-R-Ry, R-R-Mv, R-R-Ra and R-R-Fa systems at the maturity stages. In the early and late rice seasons, the alkaline phosphatase activity of different soils collected at different growth stages was in the range of 108.34–186.37 and 92.02–169.12 μg *p*-nitrophenol g^{-1} soil h^{-1}, respectively. The highest activities were detected at the booting stage.

Dynamics of arylsulfatase activity during the rice growth period

In the early and late rice seasons, arylsulfatase activity in the soils was in the range of 20.61–29.08 and 15.86–23.25 μg *p*-nitrophenol g^{-1} soil h^{-1}, respectively (Fig. 4). At the tillering and booting stage, there was no significant difference in arylsulfatase activity among the R-R-Ry, R-R-Mv, R-R-Ra and R-R-Fa, but the activity of this enzyme in R-R-Ry, R-R-Mv, and R-R-Ra soils

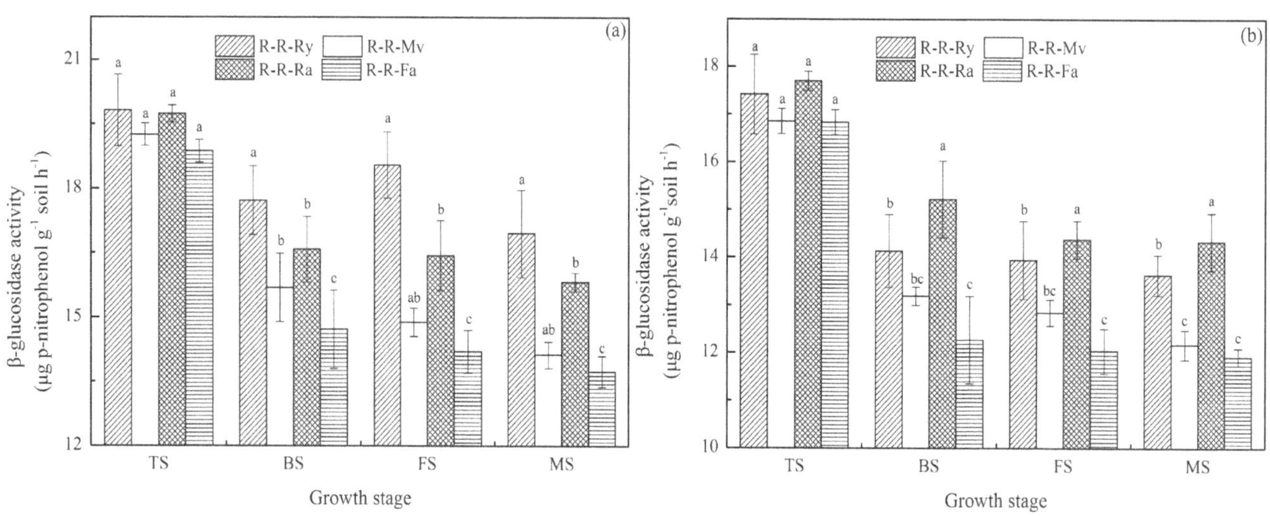

Figure 2. Dynamics of β-glucosidase activity in paddy fields during the rice growth period (a for the early rice season and b for the late rice season). R-R-Ry: rice–rice–ryegrass cropping system; R-R-Mv: rice–rice–Chinese milk vetch cropping system; R-R-Ra: rice–rice–rape cropping system; R-R-Fa: rice–rice cropping system with winter fallow. TS: tillering stage; BS: booting stage; FS: full heading stage; MS: maturity stage. Error bars represent the standard error of mean. Different letters indicate significance at $P<0.05$, according to the least significant difference test.

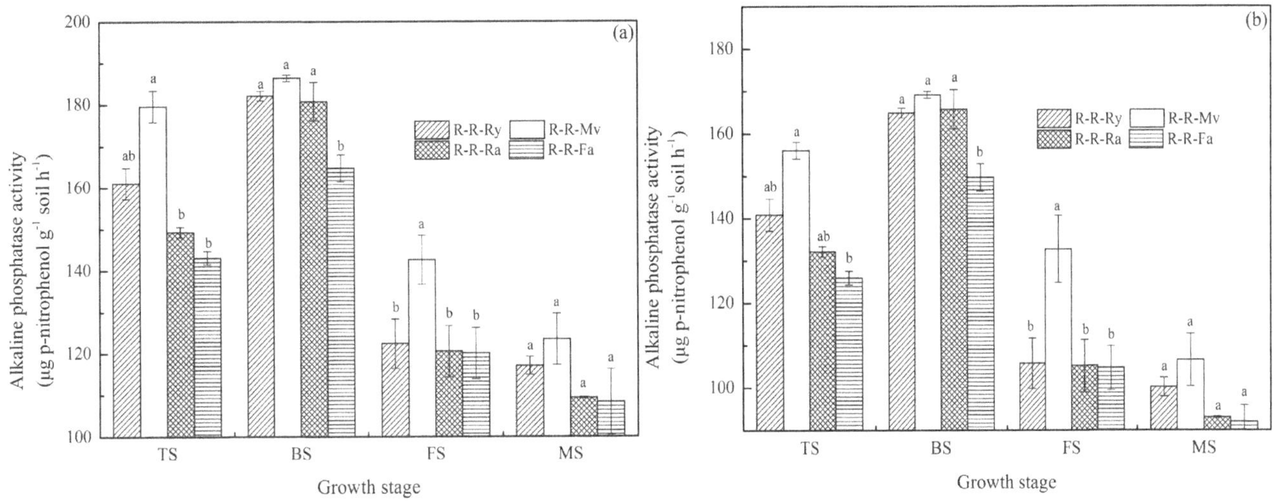

Figure 3. Dynamics of alkaline phosphatase activity in paddy fields during the rice growth period (a for the early rice season and b for the late rice season). R-R-Ry: rice–rice–ryegrass cropping system; R-R-Mv: rice–rice–Chinese milk vetch cropping system; R-R-Ra: rice–rice–rape cropping system; R-R-Fa: rice–rice cropping system with winter fallow. TS: tillering stage; BS: booting stage; FS: full heading stage; MS: maturity stage. Error bars represent the standard error of mean. Different letters indicate significance at $P<0.05$, according to the least significant difference test.

was significantly ($P<0.05$) higher than that in R-R-Fa at the full heading and maturity stage.

Dynamics of arylamidase activity during the rice growth period

Straw recycling of winter cover crops did not significantly affect arylamidase activity in the soil (Fig. 5). At 0–20 cm soil depth, there was no significant difference ($P>0.05$) in arylamidase activity under the R-R-Ry, R-R-Mv, R-R-Ra and R-R-Fa. In the early and late rice seasons, arylamidase activity changed in the range of 10.55–32.42 and 7.34–28.14 μg *p*-nitrophenol g^{-1} soil h^{-1}, respectively.

Enumeration of aerobic bacteria, anaerobic bacteria, actinomycetes and fungi in the paddy soil

Significant differences ($P<0.05$) were observed in the numbers of culturable aerobic bacteria, anaerobic bacteria, actinomycetes, and fungi between samplings throughout the rice growth season. The number of aerobic bacteria decreased in the order R-R-Ry> R-R-Mv>R-R-Ra>R-R-Fa in the early rice season and in the order R-R-Ra>R-R-Mv>R-R-Ry>R-R-Fa in the late rice season (Table 1). The number of anaerobic bacteria in the paddy soil under the R-R-Ry and R-R-Mv systems was significantly lower than that in the R-R-Ra system, while there was no significant difference between R-R-Ra and R-R-Fa. However, the numbers of actinomycetes and fungi in the soil with straw recycling of winter cover crops were significantly higher ($P<0.05$)

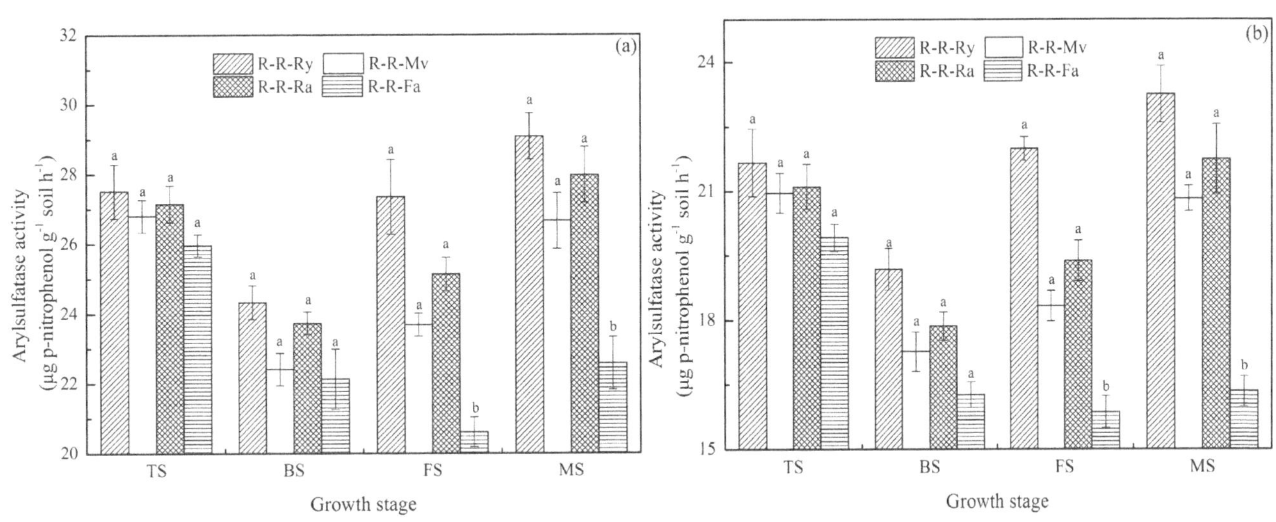

Figure 4. Dynamics of arylsulfatase activity during the rice growth period (a for the early rice season and b for the late rice season). R-R-Ry: rice–rice–ryegrass cropping system; R-R-Mv: rice–rice–Chinese milk vetch cropping system; R-R-Ra: rice–rice–rape cropping system; R-R-Fa: rice–rice cropping system with winter fallow. TS: tillering stage; BS: booting stage; FS: full heading stage; MS: maturity stage. Error bars represent the standard error of mean. Different letters indicate significance at $P<0.05$, according to the least significant difference test.

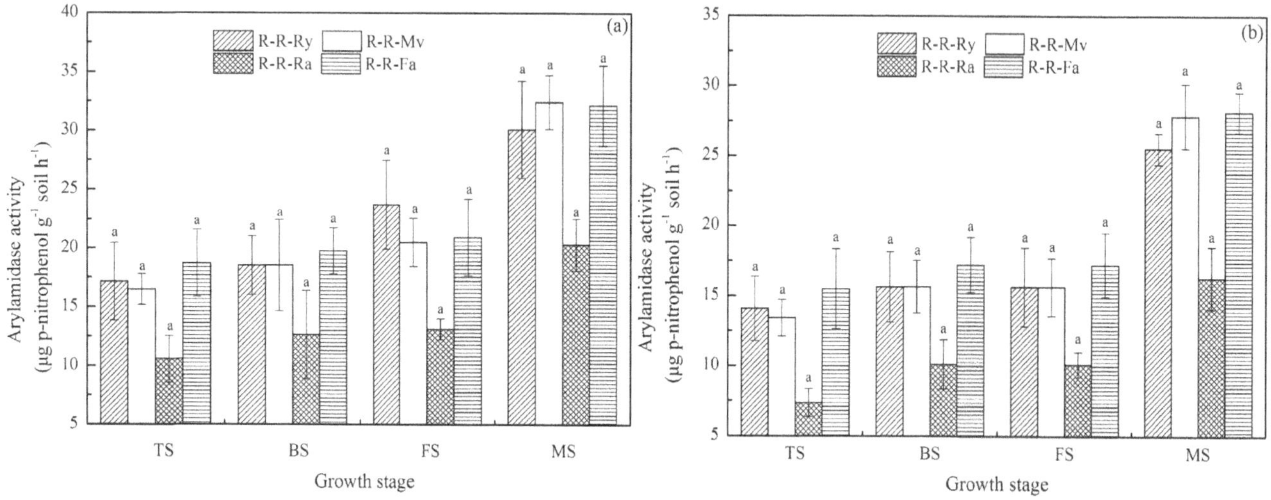

Figure 5. Dynamics of arylamidase activity during the rice growth period (a for the early rice season and b for the late rice season). R-R-Ry: rice–rice–ryegrass cropping system; R-R-Mv: rice–rice–Chinese milk vetch cropping system; R-R-Ra: rice–rice–rape cropping system; R-R-Fa: rice–rice cropping system with winter fallow. TS: tillering stage; BS: booting stage; FS: full heading stage; MS: maturity stage. Error bars represent the standard error of mean. Different letters indicate significance at $P<0.05$, according to the least significant difference test.

than that in the winter fallow soil throughout the main growth stages. And the number of fungi decreased in the order R-R-Ra> R-R-Ry>R-R-Mv>R-R-Fa in the early rice season and in the order R-R-Mv>R-R-Ry≈R-R-Ra>R-R-Fa in the late rice season. And the number of actinomycete decreased in the order R-R-Mv>R-R-Ry≈R-R-Ra>R-R-Fa in the early and late rice season.

Discussion

Soil quality and winter cover crops residue applied

Residue management in cropping systems is believed to improve the quality of soil [7]. In our study, adding winter cover crops and returning the straws to the field is benefit to increase the soil quality and rice yield in double-cropping rice. And it is beneficial in resources efficiency increase. Soil enzyme activities and soil microbial community play an important role in ecosystem function and cycling of soil nutrients. And the ability of maintain soil fertility or regulate nutrient cycling may be largely dependent on the composition of soil microbial communities and soil enzyme activities [7]. In our study, the activities of all four enzymes with winter cover crop residue were generally higher than the fallow fields because the increases of soil enzyme activity due to add organic matter to soils, this is same with other people's studies [4,24]. In addition, we were able to differentiate temporal variations in soil enzyme activities from changes in the enzyme activity in response to winter cover crop residue decomposition, different organic inputs (winter cover crop residue), straw incorporation or crop rotation.

Management practices (e.g. crop rotation, mulching, tillage and application of fertilizers and organic matter) may have diverse effects on the numbers of microorganisms in soil [14,15,17]. Thus, incorporation of winter cover crops into rotations may increase microbial community in soil compared to the R-R-Fa system. This might have been because there was more decomposable organic material in the adding winter cover crops soil which favored soil microbe. And the numbers of microorganisms were improved after returning straws of winter cover crops into the soil due to the supplements of carbon source. Soil microorganisms are mostly heterotrophic and use organic carbon as carbon and energy sources.

Soil enzyme activities, microbial community and winter cover crops residue applied

β-glucosidase is the rate-limiting enzyme in the microbial degradation of cellulose to glucose and plays a crucial role in the C cycle in soils. We found a significant difference in β-glucosidase activity between the R-R-Ry, R-R-Mv and R-R-Ra treatments at the rice growth stages, which is probably related to the different C inputs by the different winter cover crops [25]. In the early and late rice seasons, alkaline phosphatase activity decreased as follows: R-R-Mv>R-R-Fa, but there were no significant differences ($P>0.05$) between other treatments. There was no significant difference ($P>0.05$) in arylamidase activity under the R-R-Ry, R-R-Mv, R-R-Ra and R-R-Fa systems in the early and late rice seasons. Also, other investigators have reported that Chinese milk vetch residue contains enzymes (phosphatase) [26]. The different ranking of treatments in soil enzyme activities might be related to the kinds of winter cover crops straw type. This might have been because there was significant difference decomposable organic material in the winter cover crops straw-returned soil which favored soil enzyme activities [27,28]. Arylsulfatase is produced by both plants and microorganisms [15]. It is an extracellular enzyme that catalyses the hydrolysis of organic sulphate esters, releasing SO_4 that can be used by plants. In our study, the activity decreased in the following order: R-R-Ry>R-R-Ra>R-R-Mv in the early and late rice season. These observations indicate that ryegrass and rape made a higher contribution to the active pool of C in soil organic matter than Chinese milk vetch, which may explain the higher immobilization in rhizosphere soil containing ryegrass and rape residue than in rhizosphere soil containing Chinese milk vetch residue. Thus, there may be qualities unique to each cover crop that increases the activity of specific enzymes [28].

Yoshinari et al. [29] observed that rice straw affected soil microbial community structure. In the laboratory, we verified the increased numbers of bacteria, actinomycetes, and fungi in the soil with winter cover crop residue. In our study, the number of aerobic bacteria decreased in the order R-R-Ry>R-R-Mv>R-R-

Table 1. Variations in the number (10^4 CFUs g^{-1} dry soil) of aerobic bacteria, anaerobic bacteria, actinomycetes and fungi in a paddy soil under different cropping systems.

Items	Treatment	Early rice				Late rice			
		TS	BS	CS	MS	TS	BS	CS	MS
Aerobic bacteria	R-R-Ry	618.33±5.12a	453.47±7.00a	359.37±4.93a	173.07±5.96c	343.73±5.41c	219.13±5.32c	130.93±2.49c	58.33±1.06c
	R-R-Mv	566.03±4.76b	408.67±8.80b	345.03±3.39b	487.27±2.93b	461.83±4.76b	319.13±4.76b	225.63±3.18b	107.03±1.85b
	R-R-Ra	337.53±5.35c	212.60±3.74c	150.70±4.07c	751.87±5.41a	550.47±5.12a	394.37±5.42a	295.97±3.54a	127.37±2.05a
	R-R-Fa	258.63±6.09d	138.23±1.55d	98.60±1.79d	61.43±4.43d	211.93±6.09d	122.63±4.57d	62.43±1.67d	38.13±1.12d
Anaerobic bacteria	R-R-Ry	29.30±0.78b	23.80±0.71b	21.57±0.57c	27.33±0.61a	24.29±0.74b	19.42±0.72c	17.36±0.62c	17.19±0.62a
	R-R-Mv	31.93±0.55b	26.53±0.72b	23.07±0.58bc	18.07±0.43c	26.77±0.51b	22.33±0.88b	18.88±0.52bc	14.25±0.59c
	R-R-Ra	41.67±0.70a	35.83±0.76a	28.57±0.43a	24.93±0.82c	36.50±0.72a	31.50±0.71a	23.80±0.43a	18.11±0.59c
	R-R-Fa	39.80±1.96a	33.77±1.11a	24.23±0.58b	23.07±0.44b	34.79±1.96a	29.02±1.05a	19.60±0.56b	18.01±0.48b
Fungi	R-R-Ry	2.87±0.05a	2.03±0.03b	1.62±0.04b	0.65±0.03b	2.53±0.05b	1.75±0.04b	1.29±0.04b	0.40±0.03b
	R-R-Mv	2.74±0.03b	1.90±0.05c	1.42±0.03c	0.59±0.04b	2.42±0.03b	1.66±0.05b	1.06±0.02c	0.36±0.04b
	R-R-Ra	2.99±0.05a	2.44±0.02a	2.11±0.05a	1.09±0.03a	2.72±0.04a	2.11±0.02a	1.74±0.06a	0.81±0.03a
	R-R-Fa	2.20±0.02c	1.55±0.02d	1.34±0.03c	0.27±0.04c	1.94±0.02c	1.31±0.03c	0.99±0.04c	0.19±0.04c
Actinomycetes	R-R-Ry	9.07±0.29b	9.47±0.25b	9.93±0.43b	17.47±0.35a	6.93±0.28b	6.94±0.23b	7.30±0.46b	14.32±0.35a
	R-R-Mv	10.27±0.26a	11.50±0.32a	12.87±0.44a	18.47±0.35a	8.10±0.26a	8.92±0.32a	10.16±0.44a	15.23±0.35a
	R-R-Ra	8.10±0.21bc	8.73±0.38bc	9.17±0.23bc	12.77±0.43b	5.93±0.20c	6.53±0.41bc	6.40±0.23bc	9.73±0.43b
	R-R-Fa	7.13±0.41c	7.87±0.43c	8.50±0.36c	10.20±0.26c	4.99±0.41c	5.62±0.43c	5.87±0.36c	7.05±0.26c

R-R-Ry: rice–rice–ryegrass cropping system; R-R-Mv: rice–rice–Chinese milk vetch cropping system; R-R-Ra: rice–rice–rape cropping system; R-R-Fa: rice–rice cropping system with winter fallow.
TS: tillering stage; BS: booting stage; FS: full heading stage; MS: maturity stage.
Values are presented as mean ± SE (n = 3). Means in each column with different letters are significantly different at the $P<0.05$ level.

Ra>R-R-Fa in the early rice season and in the order R-R-Ra>R-R-Mv>R-R-Ry>R-R-Fa in the late rice season. This might have been because there was significant difference of decay rates in the winter cover crops straws-returned soil which favored aerobic bacteria. And the numbers of aerobic bacteria were increased under R-R-Ry and R-R-Ra systems in the early and late rice season, respectively. The numbers of anaerobic bacteria in the paddy soil under the R-R-Ry and R-R-Mv systems were significantly lower than that in the R-R-Ra system. The reason for this difference is being investigated. On the one hand, both of the soils in which rape straws had been incorporated showed considerable increases in the numbers of culturable anaerobic microorganisms compared to the ryegrass and Chinese milk vetch straws-added soil (Table 1). On the other hand, this might have been because there was more decomposable organic material in the rape straw-added soil resulting in more consumption of O_2 which favored anaerobes [17]. The limitations of cultivation based assessment of anaerobic microorganisms should be further studied.

Numerous studies indicate that fungi play an important role in both the formation and stabilization of soil aggregates but they are sensitive to disturbance, pollution and environmental change [30,31]. In this study, the number of fungi decreased in the order R-R-Ra>R-R-Ry>R-R-Mv in the early rice season and in the order R-R-Mv>R-R-Ry≈R-R-Ra in the late rice season. This may be that the changes of soil environmental on fungal diversity could influence ecosystem function via decomposition of different winter cover crops straws, fungi populations increased by returning rape and Chinese milk vetch straws in the early and late rice season, respectively. But the limitations of cultivation based assessment of fungi should be further studied. Furthermore, the number of actinomycete decreased in the order R-R-Mv>R-R-Ry≈R-R-Ra>R-R-Fa in the early and late rice season. Different ranking of treatments in the number of actinomycetes might be related to the decomposition rates of winter crop species [32]. In this study, it was believed that the use of Chinese milk vetch residue as organic manure resulted in an increase in the number of actinomycete. In general, the differences in the composition of the straws may have caused differences in the microbial and which resulted in the significant differences in the numbers of bacteria, actinomycetes, and fungi between the paddy soils added with ryegrass, Chinese milk vetch and rape straws. In our study, the approaches for enumeration of microbial groups are severely limited by the fact that only a fraction of microorganisms is amenable to cultivation using the traditional methods applied here. For example, it doesn't allow all aerobic microorganisms to grow by using the LB agar medium. Therefore, some advanced methods to study the aerobic microorganisms should be investigated. Further studies could be helpful to better understand on how these changes in the number of microbial might actually impact microbial functions and nutrient availability in paddy soils.

Conclusions

This study revealed that the soil enzyme activities and soil microbes were affected by residue management practices during the early and late rice growth stages, and that these two factors could be used as potential soil quality indicators. Application of winter cover crop residue can improve soil β-glucosidase, alkaline phosphatase and arylsulfatase activities. The activity of β-glucosidase in soils under R-R-Ry was increased 0.95, 3.01, 4.35, 3.22 μg *p*-nitrophenol g^{-1} soil h^{-1} in the early rice season, the activity of β-glucosidase in soils under R-R-Ra was increased 0.86, 2.95, 2.34, 2.42 μg *p*-nitrophenol g^{-1} soil h^{-1} compared with the R-R-Fa in the late rice season. Moreover, the activity of alkaline phosphatase under R-R-Mv was increased 36.50, 21.62, 22.53, 15.13 μg *p*-nitrophenol g^{-1} soil h^{-1} and 30.06, 19.43, 27.99, 14.58 μg *p*-nitrophenol g^{-1} soil h^{-1} in the early and late rice season, respectively. The activity of arylsulfatase in R-R-Ry was increased 1.55, 2.20, 6.74, 6.51 μg *p*-nitrophenol g^{-1} soil h^{-1} and 1.75, 2.93, 6.13, 6.91 μg *p*-nitrophenol g^{-1} soil h^{-1} in the early and late rice season, respectively. In contrast, no significant effect on arylamidase activity was found. The population of soil microbes also increased when winter cover crop residue was applied. The number of aerobic bacteria under the R-R-Ry systems was increased 359.70, 315.24, 260.77, 111.64×10^4 CFUs g^{-1} dry soil and 131.80, 96.50, 68.50, 20.20×10^4 CFUs g^{-1} dry soil compared with the R-R-Fa during early and late rice growth stage. The number of anaerobic bacteria under the R-R-Ra systems was increased 1.87, 2.06, 4.34, 1.86×10^4 CFUs g^{-1} dry soil and 1.71, 2.48, 4.20, 0.10×10^4 CFUs g^{-1} dry soil during early and late rice growth stage. The number of actinomycetes bacteria under the R-R-Mv systems was increased 3.14, 3.63, 4.37, 8.27×10^4 CFUs g^{-1} dry soil and 3.11, 3.30, 4.29, 8.18×10^4 CFUs g^{-1} dry soil during early and late rice growth stage. Thus, the results indicated that the management practices of winter cover crop residue returning with rotations can improve soil enzyme activities and the soil microbial community.

Author Contributions

Conceived and designed the experiments: XXP YGL. Performed the experiments: THM. Analyzed the data: THM TWG. Contributed reagents/materials/analysis tools: LYC WK. Wrote the paper: THM.

References

1. Benitez E, Melgar R, Sainz H, Gomez M, Nogales R (2000) Enzyme activities in the rhizosphere of pepper (CapsicumannuumL.) grown with olive cake mulches. Soil Biology & Biochemistry 32: 1829–1835.
2. Dick RP (1994) Soil Enzyme Activities as Indicators of Soil Quality. In: Doran JW, Coleman DC, Bezdicek DF, Stewart BA (Eds.). Defining soil quality for a sustainable environment. American Society of Agronomy, Madison, WI, 107–124.
3. Tabatabai MA (1994) Soil Enzymes. In: Page AL, Miller RH, Keeney DR. (Eds.). Methods of soil analysis. American Society of Agronomy, Madison, 775–833.
4. Nannipieri P, Kandeler E, Ruggiero P (2002) Enzyme Activities and Microbiological and Biochemical Processes in Soil. In: Burns RG, Dick RP. (Eds.). Enzymes in the environment. Activity, Ecology and Applications. Dekker, New York, 1–33.
5. Pastor J, Aber JD, McClaugherty CA, Melillo JM (1984) Above-ground production and N and P cycling along a nitrogen mineralization gradient on Blackhawk Island, Wisconsin. Ecology 65: 256–268.
6. Robertson GP, Paul EA, Harwood RR (2000) Greenhouse gases in intensive agriculture: contributions of individual gases to the radiative forcing of the atmosphere. Science 289: 1922–1925.
7. Singh BK, Bardgett RD, Smith P, Reay DS (2010) Microorganisms and climate change: terrestrial feedbacks and mitigation options. Nature Reviews Microbiology 8: 779–790.
8. Prosser JI (2002) Molecular and functional diversity in soil micro-organisms. Plant and Soil 244: 9–17.
9. Hermawan B, Bomke AA (1997) Effects of winter cover crops and successive spring tillage on soil aggregation. Soil & Tillage Research 44: 109–120.
10. McCracken DV, Smith MS, Grove JH, MacKown CT, Blevins RL (1994) Nitrate leaching as influenced by cover cropping and nitrogen source. Soil Science Society of America Journal 58:1476–1483.
11. Barnes JP, Putnam AR (1983) Rye residues contribute weed suppression in no-tillage cropping systems. Journal of Chemical Ecology 9: 1045–1057.
12. Powlson DS, Prookes PC, Christensen BT (1987) Measurement of soil microbial biomass provides an early indication of changes in total soil organic matter due to straw incorporation. Soil Biology & Biochemistry 19(2): 159–164.
13. Ekenler M, Tabatabai MA (2003). Tillage and residue management effects on β-glucosaminidase activity in soils. Soil Biology & Biochemistry 35: 871–874.
14. Carmine C, Magda C, Maria DRP, Patrizia R, Pacifico R (2004) Effects of municipal solid waste compost amendments on soil enzyme activities and bacterial genetic diversity. Soil Biology & Biochemistry 36: 1595–1605.

15. Tejada M, Garcia C, Gonzalez JL, Hernandez MT (2006) Use of organic amendment as a strategy for saline soil remediation: Influence on the physical, chemical and biological properties of soil. Soil Biology & Biochemistry 38: 1413–1421.
16. Hamido SA, Kpomblekou AK (2009) Cover crop and tillage effects on soil enzyme activities following tomato. Soil & Tillage Research 105: 269–274.
17. Wu WX, Ye QF, Min H, Duan XJ, Jin WM (2004) Bt-transgenic rice straw affects the culturable microbiota and dehydrogenase and phosphatase activities in a flooded paddy soil. Soil Biology & Biochemistry 36: 289–295.
18. Anna KB, Richard PD (1999) Field management effects on soil enzyme activities. Soil Biology and Biochemistry 31: 1471–1479.
19. Klose S, Koch J, Baucker E, Makeschin F (2001) Indicative properties of fly-ash affected forest soils in Northeastern Germany. Journal of Plant Nutrition and Soil Science 164: 561–568.
20. Acosta-Marti'nez V, Tabatabai MA (2000) Arylamidase activity of soils. Soil Science Society of America Journal 64: 215–221.
21. Oliveira AP, Pampulha ME, Bennett JP (2008) A two-year field study with transgenic *Bacillus thuringiensis* maize: Effects on soil microorganisms. Science of the Total Environment 405: 351–357.
22. Min H, Ye YF, Chen ZY, Wu WX, Du YF (2001) Effects of butachlor on microbial populations and enzyme activities in paddy soil. Journal of Environmental Science and Health B 36: 581–595.
23. SAS Institute (2003) SAS Version 9.1.2 2002–2003. SAS Institute Inc., Cary, NC.
24. Crecchio C, Curci M, Mininni R, Ricciuti P, Ruggiero P (2001) Short-term effects of municipal solid waste compost amendments on soil carbon nitrogen content, some enzyme activities and genetic diversity. Biology and Fertility of Soils 34: 311–318.
25. Huang C, Deng LJ, Gao XS, Zhang SR, Luo T, et al. (2010) Effects of fungal residues return on soil enzymatic activities and fertility dynamics in a paddy soil under a rice-wheat rotation in Chengdu Plain. Soil & Tillage Research, 108: 16–23.
26. Nannipieri P, Muccini L, Ciardi C (1983) Microbial biomass and enzyme activities: production and persistence. Soil Biology & Biochemistry 15: 679–685.
27. Luo YJ, Wang ZF, Gao M, Wei CF (2011) Effects of conservation tillage on organic carbon, nitrogen and enzyme activities in a hydragric anthrosol of Chongqing, China. Energy Procedia 5: 30–36.
28. Sharma RK, Arora DS (2011) Solid state degradation of paddy straw by *Phlebia floridensis* in the presence of different supplements for improving its nutritive status. International Biodeterioration & Biodegradation 65: 990–996.
29. Yoshinari E, Seigo O, Hiroshi Y, Yoshio N (2001) Influence of earthworm activity and rice straw application on the soil microbial community structure analyzed by PLFA pattern. European Journal of Soil Biology 37: 269–272.
30. David JB, Kurt AS, Juan CL, Jared LD (2012) Soil fungi influence the distribution of microbial functional groups that mediate forest greenhouse gas emissions. Soil Biology & Biochemistry 53: 112–119.
31. Li YJ, Chen X, Shamsi IH, Fang P, Lin XY (2012) Effects of irrigation patterns and nitrogen fertilization on rice yield and microbial community structure in paddy soil. Pedosphere 22(5): 661–672.
32. Zhang QC, Shamsi IH, Xua DT, Wang GH, Lin XY, et al. (2012) Chemical fertilizer and organic manure inputs in soil exhibit a vice versa pattern of microbial community structure. Applied Soil Ecology 57:1–8.

12

Green Manure Addition to Soil Increases Grain Zinc Concentration in Bread Wheat

Forough Aghili[1]*, Hannes A. Gamper[1], Jost Eikenberg[2], Amir H. Khoshgoftarmanesh[3], Majid Afyuni[3], Rainer Schulin[4], Jan Jansa[5], Emmanuel Frossard[1]

1 Institute of Agricultural Sciences, Department of Environmental Systems Science, Swiss Federal Institute of Technology (ETH) Zürich, Switzerland, **2** Paul Scherrer Institute (PSI), Radioanalytics Laboratory, Villigen, Switzerland, **3** College of Agriculture, Department of Soil Sciences, Isfahan University of Technology, Isfahan, Iran, **4** Institute of Terrestrial Ecosystems, Department of Environmental Systems Science, Swiss Federal Institute of Technology (ETH) Zürich, Switzerland, **5** Institute of Microbiology, Academy of Sciences of the Czech Republic, Prague, Czech Republic

Abstract

Zinc (Zn) deficiency is a major problem for many people living on wheat-based diets. Here, we explored whether addition of green manure of red clover and sunflower to a calcareous soil or inoculating a non-indigenous arbuscular mycorrhizal fungal (AMF) strain may increase grain Zn concentration in bread wheat. For this purpose we performed a multifactorial pot experiment, in which the effects of two green manures (red clover, sunflower), $ZnSO_4$ application, soil γ-irradiation (elimination of naturally occurring AMF), and AMF inoculation were tested. Both green manures were labeled with ^{65}Zn radiotracer to record the Zn recoveries in the aboveground plant biomass. Application of $ZnSO_4$ fertilizer increased grain Zn concentration from 20 to 39 mg Zn kg^{-1} and sole addition of green manure of sunflower to soil raised grain Zn concentration to 31 mg Zn kg^{-1}. Adding the two together to soil increased grain Zn concentration even further to 54 mg Zn kg^{-1}. Mixing green manure of sunflower to soil mobilized additional 48 μg Zn (kg soil)$^{-1}$ for transfer to the aboveground plant biomass, compared to the total of 132 μg Zn (kg soil)$^{-1}$ taken up from plain soil when neither green manure nor $ZnSO_4$ were applied. Green manure amendments to soil also raised the DTPA-extractable Zn in soil. Inoculating a non-indigenous AMF did not increase plant Zn uptake. The study thus showed that organic matter amendments to soil can contribute to a better utilization of naturally stocked soil micronutrients, and thereby reduce any need for major external inputs.

Editor: Raffaella Balestrini, Institute for Sustainable Plant Protection - C.N.R., Italy

Funding: This study was supported by the grant 123920 of the Swiss National Science Foundation (SNF) and the Swiss Agency for Development and Cooperation (SDC) in the framework of Research Partnerships with Developing Countries. The funding body and host institutions had no role in study-design, data collection, analysis, and manuscript preparation. JJ was further supported by the J.E. Purkyně Fellowship and the long-term development program RVO61388971.

Competing Interests: The authors have declared that no competing interests exist.

* Email: forough.aghili@usys.ethz.ch

Introduction

Wheat is the third most-grown cereal after maize and rice, draws on the largest cropping area, and ranks first in global cereal trade [1]. It is the crop contributing most calories to large segments of the global human population. Yet, wheat-based diets do not provide sufficient zinc (Zn), which is a leading cause of widespread Zn deficiency [2]. Low grain Zn concentration is cause of low plant-availability of Zn in many soils on which wheat is produced [3,4]. Efforts have been made to breed wheat cultivars efficient in acquiring Zn from soil and effective in (re-) translocating Zn to grains [5,6]. Zinc supply from soil has, however, to be increased for such elite crop cultivars to become fully effective, which can be achieved via appropriate soil fertility management, such as application of mineral Zn fertilizer [7], addition of green manure [8], and/or a strengthening of nutrient acquisition traits, including symbioses of roots with naturally occurring and newly introduced arbuscular mycorrhizal fungi (AMF) [9–12].

Plants and probably also AMF take up Zn from the soil solution in form of free Zn^{2+} and $ZnOH^{+}$ ions, or as Zn-organic ligand complexes [13,14]. The depleted soil solution is replenished with Zn via desorption or dissolution from the soil matrix, followed by dispersion by diffusion [15]. Desorption from the soil matrix is controlled by soil pH, salinity, and clay, carbonate, and organic matter contents [15]. To promote Zn dissolution plants and microorganisms exude soil-acidifying protons and Zn-chelating low-molecular-weight organic compounds [16,17]. These Zn mobilizing activities of roots and microorganisms are promoted under iron (Fe), Zn, and other nutrient deficiencies [18]. The most well known classes of exuded organic Zn ligands are carboxylates, amides, and non-proteinaceous amino acids (= siderophores) [19,20]. A recent study on Zn uptake by wheat in the field [8] suggests that residues from the previous crop in crop rotations and thus possibly also addition of green manure to soil can raise the plant-available Zn in soil and promote Zn uptake in crop plants, mainly by contributing Zn-chelating dissolved organic carbon and amino acids.

Arbuscular mycorrhizal fungi, phylum Glomeromycota), because of their ubiquity and crucial involvement in plant nutrient acquisition and health, are a plant production-relevant group of root symbionts [21,22]. Their hyphae form effective, large-distance conduits for the transport of temporarily and spatially heterogeneously available nutrients of low mobility in soil, such as phosphate (P) and Zn [23]. Crop residue or green manure addition to soil and sporadic rain and irrigation may create nutrient flushes that crop plants with root systems extended by extraradical AMF hyphae may much more effectively utilize [24]. Arbuscular mycorrhizal fungi, whether indigenous or deliberately introduced as inoculants, were repeatedly reported to benefit Zn uptake in crop plants [9,10]. While the phenomenon of increased plant Zn uptake of mycorrhized as opposed to non-mycorrhized plants has already often been demonstrated [10], we still lack information about controlling factors [8] and the possible reinforcing effect of combined enrichment of soils with organic matter and mineral Zn fertilizer.

Here we report findings of a cross-factorial pot experiment on Zn nutrition of bread wheat involving the enrichment of calcareous soil with green manure of red clover and sunflower and inoculating plants with a non-indigenous AMF strain. For comparison and to study its interactions with the other experimental factors also a water-soluble mineral Zn fertilizer ($ZnSO_4$) was applied. To compare the effect of all naturally occurring AMF and possibly associated other soil microorganisms, the soil of half of the experimental units was γ-irradiated and re-inoculated with an AMF-free microbial filtrate of a water suspension of non-sterilized soil. An AMF inoculation treatment was further included in an attempt to assess the potential of this rapidly growing technology and for controlling against non-AMF mediated biotic effects by the soil γ-irradiation and microbial re-inoculation procedure. We predicted that 1) green manure addition to soil improves Zn uptake by wheat, 2) green manure addition improves Zn uptake from mineral Zn fertilizer and 3) inoculation of wheat plants with a foreign AMF strain improves Zn uptake, particularly from soil whose indigenous AMF had been killed by γ-irradiation.

Materials and methods

Experimental design

A complete cross-factorial experiment with the factors green manure (3 levels: none, red clover, sunflower), mineral Zn fertilization (2 levels: none, $ZnSO_4$), γ-irradiation (2 levels: non-irradiated; γ-irradiated and re-inoculated with soil microbes, except AMF), and AMF inoculation (2 levels: none, inoculated) was set up with each treatment replicated five times. The 120 experimental pots were arranged randomly on a glasshouse bench.

Soil origin and physicochemical soil analyses

The soil was collected in September 2009 to a depth of 30 cm from an arable field at the Rudasht research station (32, 29/N, 52, 10/E, 1560 m above sea level), located 65 km south-east of Isfahan in central Iran [permission for soil sampling was given by the Isfahan Agricultural and Natural Resource Research Center and for its import to Switzerland by the Swiss Federal Department of Agriculture]. The soil was air-dried and passed through a 5 mm sieve, before being shipped to Switzerland, where 5 kg were further passed through a 2 mm sieve for physicochemical analyses.

This experimental soil was a Typic Haplocambid, according to the USDA soil taxonomy [25]. The field where the soil was sampled had been under agricultural use with periodical irrigation from a nearby river. The crop rotation prior to soil sampling comprised barley, wheat, sugar beet, and wheat, interrupted by dry fallow periods. The climate in central Iran is semi-arid with a long-term mean annual temperature of 16.8°C, a mean annual rainfall lower than 100 mm, and a potential annual evaporation higher than 2000 mm [26].

The soil texture was a silty clay loam based on the measurements by the hydrometer method [27] and according to the USDA soil classification scheme [6]. The pH was 7.9 and the electrical conductivity (EC) 23 dSm^{-1}, as determined by the procedure of Rhoades [28], using an ISE pH meter 720A (Mettler Toledo LLC, Columbus, OH, USA) and an E- 518 conductometer (Metrohm Herisau, Switzerland), respectively. The soil was low in organic carbon [5 g (kg soil) $^{-1}$] and nitrogen [0.6 g (kg soil) $^{-1}$], as estimated by the methods of Walkley and Black [29] and on a Carlo Erba flash combustion CN-analyzer (NA1500, Carlo Erba, Milano, Italy), respectively. The soil contained 300 mg $CaCO_3$ (kg soil)$^{-1}$ as determined after neutralization with hydrochloric acid and back titration with sodium hydroxide [30].

The total Zn concentration of the soil was 80.2 mg Zn (kg soil)$^{-1}$, according to energy dispersive X-ray fluorescence spectrometry (XRF) measurements on a Spectra X-Lab 2000 instrument (SPECTRO Analytical Instruments GmbH, Kleve, Germany). Potentially plant-available Zn was extracted from 25 g of soil, using 50 ml of 0.05 M diethylene triamine-penta-acetic acid (DTPA), adjusted to pH 7.2 [31], and measured on an Agilent 7500 C inductively coupled plasma mass spectrometer (ICP-MS, Agilent Technologies, Santa Clara, California, USA). The amount of DTPA-extractable Zn was 0.45 mg Zn (kg soil)$^{-1}$.

Partial soil desalination, γ-irradiation and fertilization

Preliminary tests showed that wheat growth on the untreated soil was severely impaired by salinity stress (Forough Aghili et al., unpubl. data). Therefore, we flushed the soil in five rounds, with a volume ratio of 1:1 soil:deionized water each prior to the experiment. This treatment can be considered comparable to repeated flood irrigation, used by farmers at the field site prior to sowing. Soil washing lowered the EC from 23 to 6.7 dS m^{-1}, which is close to the threshold value of 5.9 dS m^{-1} for salinity-induced growth depression in wheat [32]. Flushing with water also slightly decreased the soil pH from 7.9 to 7.7. Analyses of the leachate indicated that large amounts of sodium, chlorine and sulfur had been removed, but virtually none of the total soil Zn (data not shown). The DTPA-extractable concentration of Zn in the soil even increased slightly from 0.45 to 0.47 mg Zn (kg soil)$^{-1}$. No spores of AMF could be retrieved from the leachate. The partially desalinated soil was again air-dried and sieved to a particle size of ≤5 mm.

Half of the soil was γ-irradiated with a dose of 25–75 kGy using a ^{60}Co source (Studer-Hard, Däniken, Switzerland, http://www.leoni-irradiation-services.com) in order to kill the indigenous soil microbes. To reintroduce soil microbes, but not AMF, 50 ml of a filtrate of a 2.5% (w:v) soil water suspension was added per kilogram soil. This microbial filtrate was prepared by passing the soil suspension twice through Whatman No 1 filter papers, which size-excluded AMF propagules. This procedure was recommended by Thompson [33] for studying the effect of AMF on Zn nutrition and is also utilized in many comparable studies to analyze the relative effect of different soil microorganisms on plant nutrition [34,35]. Indeed, soil γ-irradiation is currently the only way to create a control treatment, free of naturally occurring AMF, since mycorrhiza-defective mutants of wheat are not available [36].

The flushed soils (non-irradiated and γ-irradiated) received 400 mg K, 200 mg P, and 80 mg Fe per kilogram soil in the form of finely ground K_2SO_4, $CaHPO_4$ and $FeSO_4$ $7H_2O$ salts.

Preparation and application of ^{65}Zn-labeled green manure

We chose red clover (*Trifolium pratense L.*) and sunflower (*Helianthus annuus* L.) for the production of experimental green manures, because they are also often grown before wheat in the region of Isfahan [8]. Furthermore, Soltani et al. [8] found that compared to other possible green manure plants, red clover and sunflower introduced significant amounts of dissolved organic carbon into the soil solution, which was probably important for Zn solubilization.

Seeds of clover and sunflower were surface sterilized in 15% H_2O_2 for 15 min, rinsed with distilled water and germinated on moistened filter papers for two days. The most vigorously growing seedlings were transferred to a combined sand bed-hydroponics system, which consisted of compartmented seedling trays filled with silica sand of a particle size of 0.7–1.2 mm, partially immersed in a full-strength Hoagland nutrient solution [37]. After 28 days, the nutrient solution was exchanged for one that contained radioactive ^{65}Zn for an additional 26 days of vegetative plant development. While 30 liters of Hoagland nutrient solution were labeled with 5.55 MBq of ^{65}Zn for about 400 red clover plantlets, 40 l of Hoagland nutrient solution were labeled with 9.25 MBq of ^{65}Zn for about 140 sunflower plantlets. The ^{65}Zn radiotracer was added as carrier-free $ZnCl_2$, dissolved in weak HCl (Amersham plc, GE Healthcare, Little Chalfont, UK). The photoperiod in the glasshouse was set to 14 h, the minimal light intensity to 12 kLux, the temperature to 22°C during the day and 16°C at night, and the relative air humidity was allowed to vary between 30 and 40%.

At harvest, the entire shoots of red clover and only the leaf blades of sunflower, without petioles and stems, were collected, air-dried and finely ground to powder. The total tissue C and N concentrations were determined in 4 mg tissue samples using the above-mentioned CN-analyzer. The total P and Zn concentrations were measured in 100 mg subsamples of each of the green manures, using the above-mentioned ICP-MS after incineration at 550°C for 6 h and extracted with 3 ml of hot 14.4 M nitric acid (HNO_3). The concentration of radioactive ^{65}Zn in the green manures was measured by γ-spectrometry in 300 mg subsamples using IGC2 high purity germanium detectors (ORTEC Advanced Measurements Technologies Inc, Oak Ridge, Tennessee, USA). Selected characteristics of both green manures are presented in Table 1.

The green manures (Table 1) were homogenously mixed into the soils at a dosage of 4 g dry matter per kilogram of dry soil. This dosage corresponds approximately to an addition of 18 tons of dry matter per hectare considering a soil bulk density of 1.5 t m^{-3} and incorporation to a soil depth of 30 cm.

Mineral Zn fertilization

Half of the soil in each green manure × irradiation treatment combination was fertilized with water soluble Zn in the form of $ZnSO_4$ at a dosage of 4.7 mg Zn (kg soil)$^{-1}$, corresponding approximately to 19 kg Zn ha^{-1}.

AMF inoculation

Finally, half of all pots in each combination of the above treatments were inoculated with approximately 250 spores of a pot culture of the Swiss isolate BEG155 of the AMF species *Claroideoglomus claroideum* [38]. The spores of the inoculum were extracted by wet-sieving and decanting [39] and directly pipetted onto the roots of the transplanted wheat seedlings. This careful spore isolation and inoculation procedure prevented co-inoculation of many other microbes and introduction of organic matter in the form of root pieces, which would have been the case with so-called whole inoculum {= growth substrate, containing fungal spores, hyphae and colonized root fragments [40]}.

Data on AMF root colonization were not systematically collected in this experiment, since microscopic and molecular genetic quantification in a previous experiment with the same wheat cultivar and soil clearly showed a high infection pressure by the naturally occurring AMF (Forough Aghili et al., unpublished data). However, successful colonization of the roots by the AMF isolate BEG155 and by naturally occurring AMF were confirmed in a subset of all experimental treatments by quantitative polymerase chain reaction as in the previous study and additionally by phylogenetic sequence analysis of a 1.6 kb-long ribosomal DNA amplicon of the AMF colonizing the roots of a subset of all experimental units (Anouk Guyer et al., unpublished data).

Planting, growth conditions, harvest and mineral nutrient analyses

Wheat kernels {*Triticum aestivum* cv. Kavir, a cultivar widely grown in Iran [41]} were surface-sterilized with 15% H_2O_2 for 15 min and pre-germinated on moistened filter paper for three days prior to transplantation. Four seedlings were planted into each pot with 550 g soil and later thinned to the most vigorously growing two individuals. Nitrogen in the form of NH_4NO_3 was applied weekly, six times, amounting to 600 mg N (kg soil)$^{-1}$ until tillering.

Table 1. Mineral nutrient concentrations of the two types of applied green manure prepared from red clover and sunflower.

	Red clover	**Sunflower**
	(Mean ± SE)	**(Mean ± SE)**
N mg (g DM*)$^{-1}$	43.3±0.6	58.4±0.4
C/N (mass ratio)	9.1±0.1	6.3±0.1
P mg (g DM)$^{-1}$	5.0±0.1	7.3±0.1
Zn µg (g DM)$^{-1}$	45.6±0.6	104.4±2.4
^{65}Zn kBq¶ (g DM)$^{-1}$	6.9±0.02	15.8±0.1

* DM: dry matter.
¶Radioactivity at the time of harvest of the experiment.
The means and associated standard errors (SE) of three replicate subsamples of the nitrogen (N) concentration, carbon (C) to N mass ratio, phosphorus (P) and zinc (Zn) concentrations, and ^{65}Zn concentration are listed.

The climatic conditions in the glasshouse were set to a photoperiod of 14 h with a minimum light intensity of 10 kLux, a 22/17°C day/night air temperature regime, and 40–45% relative air humidity. Soil moisture was kept at about 70% soil water holding capacity by daily watering to weight.

Harvest took place at grain maturity, 125 days after planting. The total weight of the grains and straw of the two plants in each pot was determined after drying to constant weight at 60°C for 72 h. Subsamples of 200 mg of finely ground wheat straw and grain were incinerated in porcelain crucibles at 550°C for 6 h and the ashes dissolved in 2 ml of boiling 13 M HNO_3. Zinc concentrations were measured using ICP-MS. Measurement accuracy and precision were ensured by regular inclusion of a certified plant reference standard in the analyses (hay powder IAEA-V-10, 24 mg Zn kg^{-1}) and by running blank samples with each sample batch. The concentration of radioactive ^{65}Zn was measured in the grains and straw by γ-spectrometry as described for the green manure above. Zinc uptake into roots could not be determined because it was not possible to recover the roots quantitatively from the clay-rich soil. Soil pH and the concentrations of total N and DTPA-extractable soil Zn were measured at the end of the experiment, using the above-described analytical procedures.

Calculations and data analyses

Plant Zn uptake from green manure, soil and mineral Zn fertilizer. Plant Zn uptake was calculated as the product of Zn concentration and biomass for both the grain and straw fractions and summed to obtain the amount of Zn contained in the entire aboveground biomass. All radioactivity measurements were related to the harvest time, taking the radioactive decay into account (^{65}Zn half-life time: 243.9 d). The percentage of Zn in the aboveground biomass derived from the green manure [Zn_{dgm} (%)] was calculated as:

$$Zn_{dgm}(\%) = 100 \times SA_p \times (SA_{gm})^{-1} \quad (1)$$

where SA_p [Bq (mg Zn^{-1})] is the specific activity (SA) of Zn in the aboveground biomass and SA_{gm} the specific activity of Zn in the added green manure. The specific activity of Zn in the plant was calculated as:

$$SA = [^{65}Zn] \times [Zn]^{-1} \quad (2)$$

where [^{65}Zn] stands for the concentration of radioactivity per kilogram of plant material [Bq (kg DM)$^{-1}$] and [Zn] stands for the concentration of total Zn in the plant [mg Zn (kg DM)$^{-1}$].

Using the equations 1 and 2, the amount of Zn taken up to the aboveground biomass from green manure [Zn_{dgm}, μg Zn (kg soil)$^{-1}$] was calculated as:

$$Zn_{dgm} = (Zn_{dgm} \times Zn_{upt}) \times 10^{-2} \quad (3)$$

where Zn_{upt} is the amount of Zn taken up to the aboveground plant biomass per kilogram of soil [μg Zn (kg soil)$^{-1}$].

Knowing Zn_{dgm}, it is possible to calculate the percentage of Zn recovered by the aboveground plant biomass from the added green manure [Zn_{rec_gm} (%)]:

$$Zn_{rec\_gm} = 100 \times Zn_{dgm} \times Zn_{added\_gm}^{-1} \quad (4)$$

where Zn_{added_gm} is the amount of Zn added with the green manure [μg Zn (kg soil)$^{-1}$].

Knowing Zn_{dgm}, it is also possible to calculate the amount of Zn in the plant that is derived from non-labeled Zn sources in soil, i.e. in the non-fertilized soil from the native soil Zn pool and in the mineral Zn-fertilized soil from the native soil Zn pool as well as the applied $ZnSO_4$ [Zn_{dsoil}, μg Zn (kg soil)$^{-1}$]:

$$Zn_{dsoil} = Zn_{upt} - Zn_{dgm} \quad (5)$$

Statistical analyses. The results were analyzed by fixed factor four-way or three-way analyses of variance (ANOVA) with two levels (γ-irradiated and non-irradiated) of 'soil γ-irradiation', three levels (none, red clover and sunflower) of 'green manure addition', two levels [0 and. 4.7 mg Zn (kg soil)$^{-1}$] of 'mineral Zn fertilization', and two levels (with and without inoculation with *C. claroideum*) of 'AMF inoculation'. The interaction terms were sequentially removed from the initially complete ANOVA model when not significant (backward elimination). In addition, a one-way ANOVA was carried out on the saved residuals of a three-way ANOVA with the factors 'soil γ-irradiation', 'mineral Zn fertilization' and 'AMF inoculation' in order to specifically analyze the effect of green manure addition. Compliance of data dispersion with ANOVA model assumptions was verified using Shapiro-Wilcoxon's normality and Levene's equal variance tests. If necessary, data were log_{10} or squareroot transformed. All statistical analyses were carried out using the software package SPSS version 17 (SPSS Inc., Chicago, Ill inois, USA). Means of factor levels were compared by least-significant-difference (LSD) tests at the significance level of $p<0.05$. The figures were prepared in SigmaPlot version 12 [Systat Software Inc. (SSI), San Jose, California, USA].

Results

Plant biomass and grain nitrogen concentration

The grain yield of the experimental wheat plants was not significantly affected by any of the treatments (Figure 1A, Table 2), but the large variance may have concealed a similar γ-irradiation effect as on the total aboveground biomass (Figure 1B). The total aboveground biomass (= straw + grains) was significantly larger for the wheat plants grown in γ-irradiated than non-irradiated soil ($F_{1,96} = 10.45$, $p = 0.002$). Inoculation with *C. claroideum* had no significant effect on aboveground biomass production (Table 2).

Both types of green manure significantly increased grain N concentration ($F_{2,70} = 31.41$, $p<0.001$, Table 2), which ranged between 30 and 45 g N (kg DM)$^{-1}$ in the plants grown on γ-irradiated soil and between 25 and 37 g N (kg DM)$^{-1}$ in the plants grown on non-irradiated soil (Figure 2). Soil γ-irradiation also led to a significant increase in grain N concentrations ($F_{1,70} = 10.04$, $p = 0.003$, Table 2). Inoculation with *C. claroideum* and $ZnSO_4$ addition had no significant effects on grain N concentration (Table 2).

Grain Zn concentration and Zn uptake to grains and aboveground biomass

Grain Zn concentration increased significantly with application of mineral Zn fertilizer ($F_{1,70} = 50.96$, $p<0.001$) and green manure ($F_{2,70} = 13.48$, $p<0.001$), but was not affected by inoculation of *C. claroideum* and soil γ-irradiation (Figure 3A, Table 2). The lowest grain Zn concentration [20±1.64 mg Zn (kg DM)$^{-1}$] was measured in the treatment with neither mineral Zn fertilization, nor addition of green manure, while the highest grain Zn

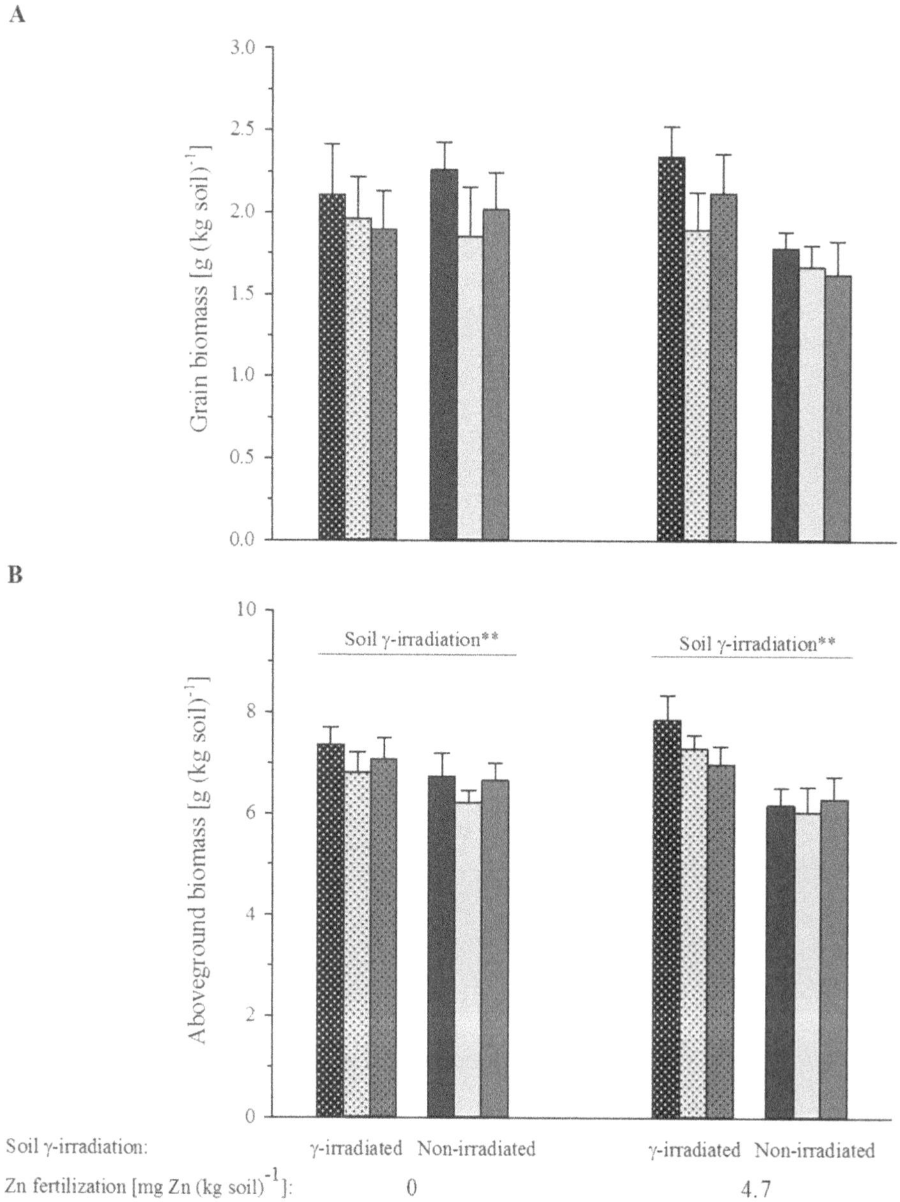

Figure 1. Grain yield (A) and aboveground biomass (B) of bread wheat grown in a calcareous soil. Black bar filling denotes the treatments without any added green manure, light grey bar filling the treatments in which the soil was amended with green manure of red clover, and dark grey bar filling those treatments in which green manure of sunflower was added to the soil. Bar hatching highlights the treatment missing the naturally occurring arbuscular mycorrhizal fungi after γ-irradiating the soil. Bars show mean values and associated standard errors of five experimental replicates. The significances (**, $p<0.01$) of soil γ-irradiation from a four-factorial analysis of variance are indicated. For full statistical details see Table 2.

concentration [54 ± 2.32 mg Zn (kg DM)$^{-1}$] was found in the treatment with combined application of $ZnSO_4$ fertilizer and green manure of sunflower (Figure 3A).

The highest Zn concentration in the aboveground biomass [46.1 ± 1.43 mg Zn (kg DM)$^{-1}$] was found in plants grown on non-irradiated soil amended with green manure of sunflower in combination with $ZnSO_4$ (Figure 3B). The lowest aboveground Zn concentration was observed in plants grown on γ-irradiated soil that had neither been fertilized with mineral Zn, nor amended with green manure (Figure 3B). Application of mineral Zn fertilizer reduced the differences between the aboveground biomass Zn concentrations of those plants that had received green manure and those that had not, resulting in a statistically significant interaction between green manure addition and mineral Zn fertilization ($F_{2,96}=6.24$, $p<0.01$; Table 2). Inoculation with *C. claroideum* had no significant effect on the concentration of Zn in the aboveground biomass (Table 2).

The amount of Zn taken up from the native soil Zn pool and mineral Zn fertilizer that was allocated to grains ranged between 46 and 88 μg Zn (kg soil)$^{-1}$, while the total amount of Zn accumulated in the aboveground biomass ranged between 117 and 367 μg Zn (kg soil)$^{-1}$ (Table 3). The lowest values were again observed in the absence of mineral Zn fertilization and green manure addition, while the highest values were observed after

Table 2. Results of fixed factor four-way analyses of variance (ANOVA) on plant and soil parameters.

ANOVA model	Grain biomass		Aboveground biomass		Grain N concentration		Grain Zn concentration		Aboveground Zn concentration		Zn uptake into grains	
Source of variance	**F value**	**df**	**F value**	**df**	**F value**	**df**	**F value**	**df**	**F value**	**df**	**F value**	**df**
Full model	0.29^{ns}	23	2.47*	23	4.45***	23	5.92***	23	8.91***	23	2.78**	23
Green manure (A)	0.33^{ns}	2	0.26^{ns}	2	31.41***	2	13.48***	2	13.71***	2	1.98^{ns}	2
Zn fertilization (B)	0.19^{ns}	1	0.12^{ns}	1	0.09	1	50.96***	1	125.23***	1	40.52***	1
Soil γ-irradiation (C)	0.20^{ns}	1	10.45**	1	10.04	1	2.99^{ns}	1	9.15**	1	0.17^{ns}	1
Inoculation (D)	0.76^{ns}	1	0.13^{ns}	1	2.97	1	1.95^{ns}	1	1.04^{ns}	1	0.95^{ns}	1
A×B	r	2	r	2	R	2	12.71**	2	6.24**	2	12.71***	2
A×C	r	2	0.61^{ns}	2	0.98^{ns}	2	3.36*	2	4.41*	2	1.22^{ns}	2
ANOVA model	**Zn uptake into aboveground biomass**		**Zn_{dgm}**		**Zn_{rec_gm}**		**Zn_{dsoil}**		**DTPA-extractable soil Zn**			
Source of variance	**F value**	**df**	**F value**	**df**	**F value**	**df**	**F value**	**df**	**F value**	**df**		
Full model	4.26***	23	5.95***	15	8.61***	15	6.15***	23	16.40***	23		
Green manure (A)	4.13*	2	39.92***	2	72.25***	2	3.50*	2	85.22***	2		
Zn fertilization (B)	53.49***	1	21.23***	1	22.31***	1	77.13***	1	168.84***	1		
Soil γ-irradiation (C)	0.01^{ns}	1	5.21*	1	8.85**	1	1.14^{ns}	1	0.64^{ns}	1		
Inoculation (D)	2.34^{ns}	1	8.25**	1	4.82*	1	1.26^{ns}	1	2.10^{ns}	1		
A×B	0.49^{ns}	2	2.12^{ns}	2	0.13^{ns}	2	2.70^{ns}	2	2.17^{ns}	2		
A×C	0.44^{ns}	2	0.63^{ns}	2	4.64*	2	2.95^{ns}	2	0.54^{ns}	2		

DTPA: diethylene triamine-penta-acetic acid.

na: not applied, r: removed from the statistical model, ns: not significant, *, $p<0.05$; **, $p<0.01$; ***, $p<0.001$.

The measurements of the plant parameters {grain biomass, total above ground biomass, grain N concentration, grain Zn concentration, aboveground Zn concentration, Zn uptake to grains, Zn uptake to the aboveground biomass, Zn derived from green manure in the aboveground biomass [Zn_{dgm}], proportion of the Zn recovered in the above ground biomass [Zn_{rec_gm}] of the Zn added with the green manure Zn derived from the soil and fertilizer in the aboveground biomass [Zn_{dsoil}]}and soil parameter (DTPA-extractable Zn from soil) were made at grain maturity. Effect sizes of the experimental factors and their significant interactions are indicated as F-values alongside with the statistical significance level. The interactions of the factors A×D, B×C, B×D, C×D, A×B×C, A×B×D, A×D×C, B×C×D, A×B×C×D were not significant in any of the 11 ANOVAs and thus these results are not listed.

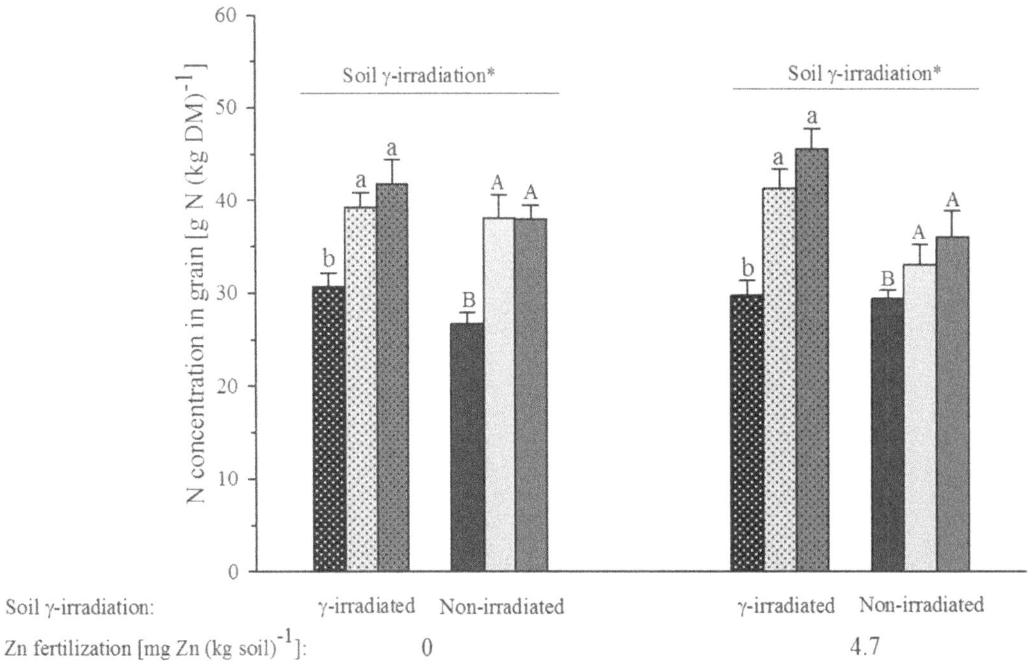

Figure 2. Grain nitrogen (N) concentration of wheat grown in a calcareous soil. Bar filling denoting the different green manure addition treatments is the same as in Figure 1. Bars show mean values and associated standard errors of five experimental replicates. The significances (*, $p < 0.05$) of the effect of soil γ-irradiation from a four-factorial analysis of variance are shown. Different letters indicate statistical differences of separate least significant difference tests at $p<0.05$ for the different green manure addition treatments within the combinations of soil γ-irradiation and mineral Zn fertilization. For full statistical details see Table 2.

addition of mineral Zn fertilizer and green manure of sunflower. Adding green manure to the soil tended to increase Zn accumulation in the grains, when no mineral Zn fertilizer had been applied, but this effect was not statistically significant (Table 2). Inoculation with *C. claroideum* had no significant effect on Zn uptake to the grains (Table 2).

Zn derived from green manure and both, soil and mineral Zn fertilizer, in the aboveground biomass

The Zn from the added green manure recovered in the aboveground biomass (Zn_{rec_gm}) ranged between 2.5% and 4.1% for red clover, and between 1.4% and 2.7% for sunflower (Figure 4). These Zn recoveries from the two types of green manures differed significantly across all treatments ($F_{1,79} = 72.25$, $p<0.001$). Mineral Zn fertilization lowered Zn_{rec_gm} for both forms of green manure ($F_{1,64} = 22.31$, $p<0.001$, Figure 4 & Table 2), as did soil γ-irradiation ($F_{1,64} = 8.85$, $p<0.01$). The recovered Zn fraction from sunflower residues was also reduced by the inoculation of *C. claroideum* when also mineral Zn fertilizer was applied ($F_{1,18} = 10.84$, $p<0.01$; Figure 4).

The amount of Zn in the aboveground biomass derived from green manure (Zn_{dgm}) ranged between 4.5 µg Zn (kg soil)$^{-1}$ in the treatment of combined soil γ-irradiation, mineral Zn fertilization, and addition of green manure of red clover, and 11 µg Zn (kg soil)$^{-1}$ in the treatment with green manure of sunflower but without mineral Zn fertilization, soil γ-irradiation, and AMF inoculation (Table 3). The amount of plant Zn derived from green manure of sunflower (Zn_{dgm}) was consistently higher than that derived from red clover ($F_{1,64} = 39.92$, $p<0.001$; Tables 2 & 3).

The amount of Zn derived from the native soil Zn pool (Zn_{dsoil}) was significantly increased with green manure addition, when no $ZnSO_4$ was applied ($F_{2,96} = 3.50$, $p = 0.034$, Table 2 & Figure 5). This apparent soil Zn mobilization was more pronounced for green manure produced from sunflower than from red clover (Figure 5). Under conditions of co-application of $ZnSO_4$, the effect of green manure addition was, however, variable and weak and statistically not significant. There was also no significant effect of γ-irradiation, nor any effect by the non-indigenous *C. claroideum* inoculant on the total amount of Zn derived from soil and mineral Zn fertilizer, either (Table 2).

Changes in DTPA-extractable Zn from soil, total N and pH in soil

The concentration of DTPA-extractable Zn from soil measured after plant harvest was significantly higher in the treatments with green manure than without green manure ($F_{1,96} = 169$, $p<0.001$). Application of mineral Zn fertilizer also significantly increased DTPA-extractable Zn ($F_{2,96} = 85.2$, $p<0.001$; Tables 2 & 3). There was a highly significant correlation between the total Zn taken up into the aboveground biomass (Y) and the DTPA-extractable Zn concentration in the soil (X) according to a double reciprocal regression model [$Y = 1/(0.004 + 2.05/X)$] with $r^2 = 0.56$ and $p<0.0001$.

Green manure addition slightly decreased soil pH, independently of the soil γ-irradiation treatment ($F_{2,96} = 139.4$, $p<0.001$; Table 4). The total soil N content was higher at harvest in the treatments that received green manure than in those that did not ($F_{2,96} = 321.7$, $p<0.001$, Table 2 & 4).

Discussion

The results confirmed our first prediction, because green manure addition to soil increased Zn uptake by bread wheat. The findings, however, did not unequivocally support our second

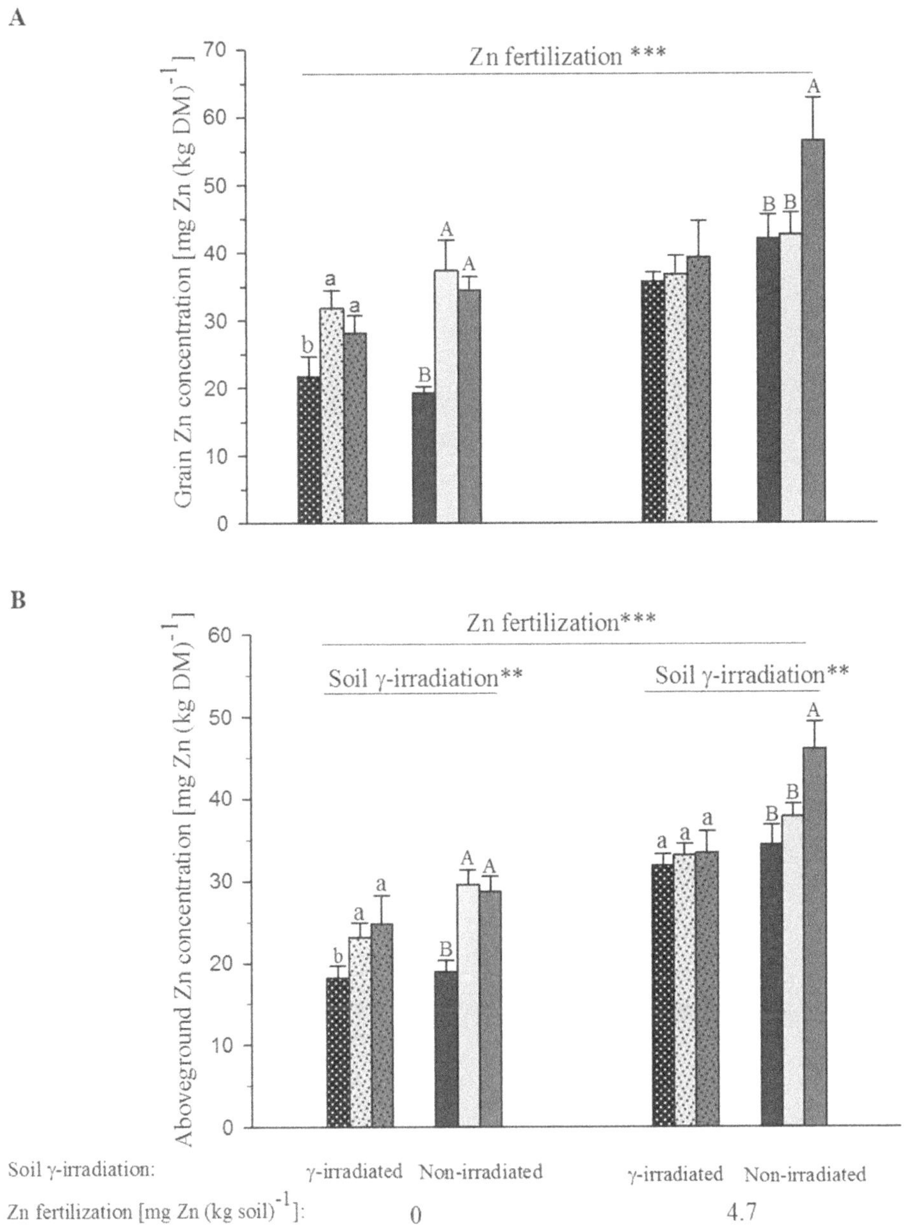

Figure 3. Grain (A) and aboveground (B) zinc (Zn) concentrations of wheat grown in a calcareous soil. Bar filling denoting the different green manure addition treatments is the same as in Figure 1. Bars show mean values and associated standard errors of five experimental replicates. The significances (**, $p<0.01$; ***, $p<0.001$) of the effect of mineral Zn fertilization, soil γ-irradiation of fixed factor four-way analyses of variance are shown. Different letters indicate statistical differences of separate least significant difference tests at $p<0.05$ for the different green manure addition treatments within the combinations of soil γ-irradiation and mineral Zn fertilization. For full statistical details see Table 2.

prediction, since utilization of the applied mineral Zn fertilizer by the plants did not improve in response to green manure addition. Also our last prediction that inoculation of a non-indigenous AMF strain would improve Zn transfer from soil to plants was not confirmed, not even in AMF- and thus competitor-free soil after removal of naturally occurring AMF by soil γ-irradiation. These two later findings appear to be related to the fact that Zn availability in the studied soil was neither limiting plant growth, nor yield.

In agreement with previous studies [43,44], we observed a clear increase in grain Zn concentration after application of mineral Zn fertilizer. However, application of mineral Zn fertilizer did neither translate into increased plant growth, nor higher grain yields, showing that Zn availability was not the primary plant growth-limiting factor in this soil. Sadeghzadeh [42] reports DTPA-extractable Zn levels below 0.6 mg Zn (kg soil)$^{-1}$ to limit wheat growth in calcareous soil. Since our soil contained 0.47 mg Zn (kg soil)$^{-1}$ and application of mineral Zn fertilizer did not promote wheat growth, it appears that the threshold for growth-limitation of soil Zn concentration may even be lower, although Zn uptake can obviously be raised by application of mineral Zn fertilizer.

Table 3. Effects of green manure addition and $ZnSO_4$ application to soil, soil γ-irradiation and inoculation of a non-indigenous arbuscular mycorrhizal fungus (AMF) on plant and soil parameters.

Green manure addition	Soil γ-irradiation	Zn added with $ZnSO_4$	Zn added with green manure	AMF inoculation	Zn uptake into aboveground biomass	Zn_{dgm}	Zn_{dsoil}	DTPA-extractable soil Zn
		µg Zn (kg soil)$^{-1}$			µg Zn (kg soil)$^{-1}$	µg Zn (kg soil)$^{-1}$		µg Zn (kg soil)$^{-1}$
None	**γ-irradiated**	0	0	no	159.3±12.4		159.3±12.4	450± 20
	γ-irradiated	0	0	yes	127.6± 6.9		127.6± 6.9	440± 32
	Non-irradiated	0	0	no	117.4± 13.2		117.4± 13.2	460± 35
	Non-irradiated	0	0	yes	127.2± 10.4		127.2± 10.4	480± 42
	γ-irradiated	4700	0	no	251.6± 16.6		251.6± 16.6	2150± 100
	γ-irradiated	4700	0	yes	227.3± 14.3		227.3± 14.3	1960± 93
	Non-irradiated	4700	0	no	206.6± 13.2		206.6± 13.2	2180± 110
	Non-irradiated	4700	0	yes	228.4± 27.5		228.4± 27.5	1990± 160
Red clover	**γ-irradiated**	0	183	no	152.9± 21.4	5.56± 0.59	147.4± 21.3	1220± 100
	γ-irradiated	0	183	yes	156.9± 25.3	5.86± 1.03	151.1± 19.6	1500± 120
	Non-irradiated	0	183	yes	205.0± 18.6	8.52± 0.35	196.5± 18.4	2010± 70
	Non-irradiated	0	183	yes	176.0± 23.8	6.10± 0.42	169.9± 20.2	2430± 110
	γ-irradiated	4700	183	no	224.0± 34.1	4.51± 0.39	219.5± 34.1	5050± 120
	γ-irradiated	4700	183	yes	243.9± 10.6	4.93± 0.38	239.0± 12.6	3550± 74
	Non-irradiated	4700	183	no	202.9± 25.2	5.32± 0.52	197.6± 23.8	4430± 32
	Non-irradiated	4700	183	yes	226.7± 22.4	6.11± 0.54	220.6± 21.5	3950± 92
Sunflower	**γ-irradiated**	0	418	no	227.9± 39.6	10.32± 1.39	217.6± 31.4	3310± 66
	γ-irradiated	0	418	yes	128.9± 11.4	8.38± 0.96	120.6± 22.6	1750± 120
	Non-irradiated	0	418	no	189.9± 14.3	11.05± 0.87	178.9± 16.4	3230± 86
	Non-irradiated	0	418	yes	174.9± 30.2	9.45± 1.61	165.5± 20.7	1950± 33
	γ-irradiated	4700	418	no	258.9± 25.8	8.44± 0.64	250.5± 41.3	3850± 82
	γ-irradiated	4700	418	yes	205.8± 20.4	5.94± 0.75	200.0± 19.5	4990± 36
	Non-irradiated	4700	418	no	367.0± 40.2	8.5± 0.64	358.5± 37.4	4540± 120
	Non-irradiated	4700	418	yes	227.9± 24.5	5.97± 0.80	222.0± 26.2	4680± 130

DTPA: diethylene-triamine-penta-acetic acid

Mean values and associated standard errors of five experimental replicates of Zn uptake to the aboveground biomass, Zn in the aboveground biomass derived from green manure (Zn_{dgm}), Zn derived from soil and fertilizer (Zn_{dsoil}) and DTPA-extractable Zn from soil at grain maturity are listed.

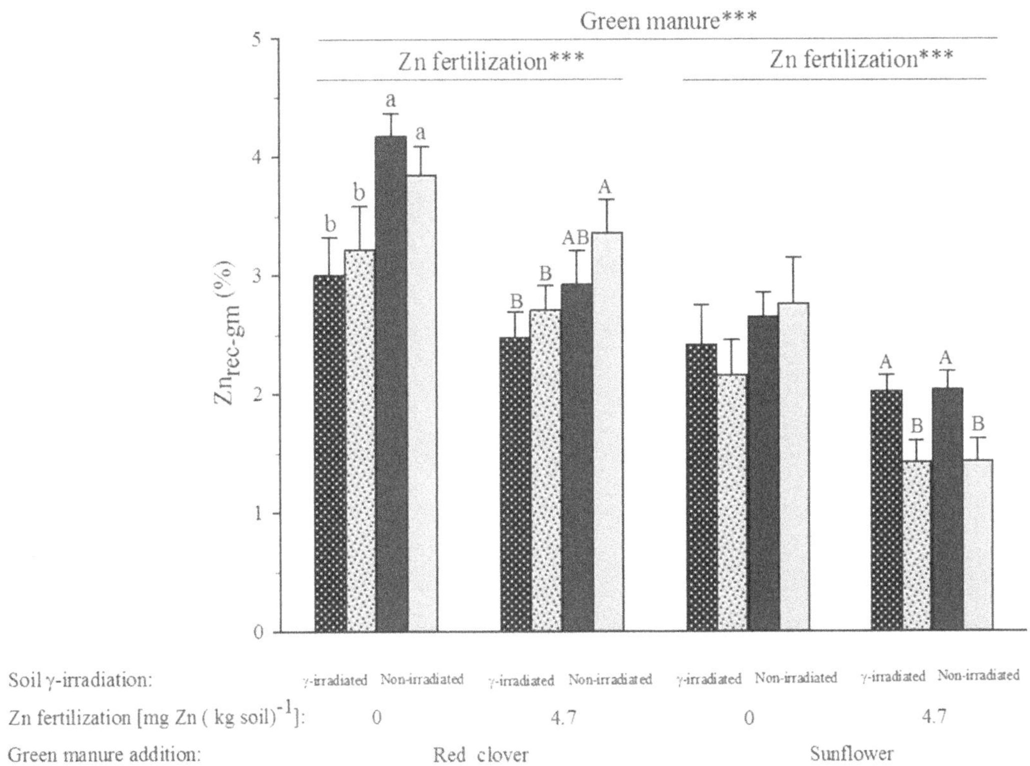

Figure 4. Percentage of zinc (Zn) recovered from green manure (Zn_{rec-gm}) in the aboveground wheat biomass. Black bar filling denotes treatments with arbuscular mycorrhizal fungus (AMF) inoculation and light grey filling treatments without AMF inoculation. Bar hatching highlights treatments in which the naturally occurring AMF were missing after γ-irradiating the soil. Bars show mean values and associated standard errors of five experimental replicates. The significances (***, $p<0.001$) of the effect of green manure addition and mineral Zn fertilization from a fixed factor four-way analysis of variance are indicated above the bars. For full statistical details see Table 2.

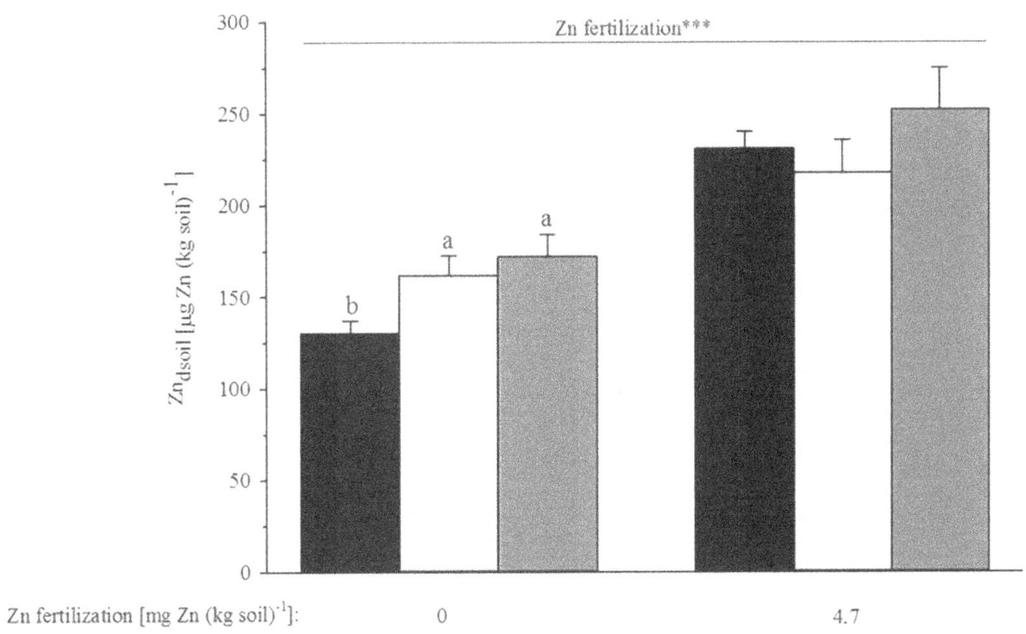

Figure 5. Zinc (Zn) uptake from soil to the aboveground biomass of bread wheat grown in a calcareous soil after addition of two types of green manure. The Zn taken up from soil either originated from the native soil Zn pool or from soil as well as applied $ZnSO_4$ fertilizer. Bar filling denoting the different green manure addition treatments is the same as in Figure 1. Bars show mean values and associated standard errors of five experimental replicates. The significance (***, $p<0.001$) of the effect of mineral Zn fertilization from a fixed factorial four-way analysis of variance is indicated above the bars. Different letters indicate statistical differences of separate least significant difference tests at $p<0.05$ for the different green manure addition treatments within the different Zn fertilization levels. For full statistical details see Table 2.

Table 4. Total soil nitrogen (N) concentration and pH in water at wheat harvest.

Soil	Green manure	Total N (g kg^{-1}) (Mean ± SE)	pH (H_2O) (Mean ± SE)
Non-irradiated	**None**	0.55 ± 0.02^{B}	7.75 ± 0.03^{A}
	+ Red clover	0.87 ± 0.02^{A}	7.50 ± 0.02^{B}
	+ Sunflower	0.88 ± 0.02^{A}	7.53 ± 0.01^{B}
γ-irradiated	**None**	0.63 ± 0.01^{b}	7.64 ± 0.02^{a}
	+ Red clover	0.89 ± 0.02^{a}	7.46 ± 0.01^{b}
	+ Sunflower	0.96 ± 0.02^{a}	7.48 ± 0.01^{b}

DTPA: diethylene-triamine-penta-acetic acid
The values represent the means and associated standard errors (SE) of 20 experimental units. Different superscript letters of the same type following the SE indicate statistical difference among the means of the three green manure treatments within each soil γ-irradiation treatment at $p<0.05$, according to least significance difference tests.

Mixing green manure to soil increased Zn uptake by wheat via Zn mobilizing effects

The observed increase in Zn uptake from the native soil Zn pool after green manure addition could originate from changes to the soil conditions, root growth, and/or plant physiology. Only addition of green manure to the soil increased the DTPA-extractable Zn from 0.47 to 0.5 or 2 mg Zn (kg soil)$^{-1}$, when green manure of red clover and sunflower were added, respectively. Combining $ZnSO_4$ application with green manure addition to soil increased the DTPA-extractable Zn concentration even further, namely up to 3.8 mg Zn (kg soil)$^{-1}$ when red clover was used and up to 5.0 mg Zn (kg soil)$^{-1}$ when sunflower was used, respectively (Table 3). We thus can conclude that the lack of a plant growth response to application of mineral Zn fertilizer (Figure 1A&B) must be explained by an alternative plant growth-limiting factor and not the >0.6 mg Zn (kg soil)$^{-1}$ of DTPA-extractable Zn [42].

The larger increase in DTPA-extractable soil Zn in response to addition of green manure of sunflower than such of red clover (Table 3) could have been due to the Zn-richer green manure of sunflower than the green manure of red clover. For production of the green manure of sunflower, but not that of red clover, only leaf laminae were used, which could explain the higher mineral nutrient concentrations of it (Table 1). Moreover, the relatively young age (54 day) and bigger sunflower than red clover seeds with own nutrient reserves may have further contributed to nutrient concentration differences of the two types of green manure (Table 1), translating into differences in plant Zn uptake.

There may have been further co-nutritional or root- and microbial growth-mediated Zn uptake stimulation, because the green manure produced of sunflower was also richer in N than that of red clover (Table 1). The green manure of the leaf laminae of the well N-supplied sunflower plants could have released N during its decomposition and stimulated root growth and possibly the synthesis of Zn facilitating N-rich plant and microbial compounds. Elevated root and microbial exudation and necromass in response to green manure of sunflower may also explain the higher DTPA-extractable Zn concentrations of the soil and increased plant Zn uptake. Root- and microbe-derived organic acids, amides and (phyto-) siderophores may have chelated and thereby mobilized Zn for uptake by the wheat plants. More N released from the decomposing green manure of more N-rich sunflower than red clover material (Table1) may have supported the N-demanding biosynthesis of Zn chelating compounds [8,14,45].

Elevated root and microbial activity in response to green manure addition may further have lowered soil pH (Table 4) and thereby mobilized soil Zn. Similarly, stimulation of nitrification upon higher N inputs with the green manure of sunflower than that of red clover and thus release of protons into the soil solution may have raised Zn bioavailability in soil for uptake by the mycorrhized wheat plants.

All in all, the effectiveness of green manure addition to soil to release Zn from the native soil Zn pool for plant uptake was remarkable (Figure 5). This stimulatory effect on plant Zn uptake diminished, however, when additionally $ZnSO_4$ fertilizer was applied. This is probably, because the applied $ZnSO_4$ fertilizer introduced already a high amount of Zn, compared to the amount solubilized by chelating compounds released upon addition of green manure, reducing Zn mobilization form soil.

Inoculation of a non-indigenous AMF strain did not increase Zn uptake in wheat

A previous experiment (Forough Aghili, unpublished) with the same soil and wheat cultivar and additional analyses on samples of this experiment (Anouk Guyer, unpublished data) showed that the roots of the wheat plants were colonized by naturally occurring AMF of this soil. 1.6 kb-long ribosomal DNA sequences confirmed root colonization by the inoculated Swiss strain of *C. claroideum* of the wheat plants raised in γ-irradiated soil. We suspect that maladaptation of the AMF inoculant to the living conditions in the experimental soil [46] must have compromised symbiotic Zn acquisition by wheat. The inoculated AMF was originally isolated from Swiss arable soil, not calcareous, and with a loamy texture, a lower, near neutral pH, and much richer in organic matter and bioavailable nutrients [47] than the study soil here. The improved Zn recovery from green manure of sunflower after application of mineral Zn fertilizer by the plants inoculated with the foreign AMF strain points at possibly complex soil-root-microbe interactions beneficial to mycorrhized plants. Lack, or just minor positive effects by the non-indigenous AMF inoculant should draw our attention to the fact that AMF inoculants can not be *a priori* considered effective in assisting crop plants to take up Zn from soil when introduced to soils of different physicchemical and biological properties.

γ-Irradiation has been shown to modify the physicochemical soil properties to some degree [48,49]. However, we believe that this did not invalidate our results. While the increased biomass (Figure 1) and N uptake (Figure 2) of the wheat plants grown in γ-irradiated soil can probably be related to released N from decomposing microbial necromass [49], the magnitude of this effect was certainly much smaller than that by the N added with the green manures (Figure 2, Table 2). Likewise, the levels of dissolved organic compounds [48] must have risen in response to γ-irradiation, but again this effect must have been considerably smaller than that caused by mixing 0. 4% (w/w) finely ground green manure to soil. Therefore, following Thompson [33], and until a mycorrhiza symbiosis-defective wheat mutant becomes available, we see no better approach than using γ-irradiation for the study of the influence of entire natural AMF assemblages on plant Zn uptake from real arable soil.

The finding of higher grain and aboveground Zn concentrations in plants raised on native than γ-irradiated soil (Figure 3) points at no Zn uptake stimulatory effect by elevated levels of dissolved organic compounds [48] and N after soil γ-irradiation [49]. The generally increased Zn tissue concentration after addition of green manure to soil (Figure 3) and higher Zn recovery from less N- and Zn-rich green manure of red clover than more N- and Zn-rich green manure of sunflower point at considerable influences by C- rather than N-limited saprotrophic soil microbes on Zn uptake of mycorrhized bread wheat plants. Being obligate biotrophs, AMF rely on saprotrophs for remobilization of mineral nutrients from decomposing green manure that they may as soon as available take up and then possibly deliver to their host plants as observed for the Zn from the green manure of sunflower (Figure 4). Much higher Zn recovery from the Zn- and N-poorer green manure of red clover than the Zn-richer and N-poorer green manure of sunflower (Figure 4) points at faster decomposition of the earlier than latter and thus also different speeds in nutrient recycling from different green manures. However, the two types of green manure appear not to have differed in Zn mobilization from soil (Figure 5).

Native as opposed to γ-irradiated soil after back-addition of an AMF propagule-free microbial filtrate of a water suspension of living soil does certainly not only differ in AMF occurrence. Besides the physicochemical changes discussed above [49], largely stochastic epidemic population growth of saprophytic organisms must occur [35]. Therefore, importance of fungi and microbes conferring similar nutritional benefits to plants as AMF [50] may increase in absence of otherwise naturally occurring AMF and much reduced population sizes of most of other just re-introduced soil microbes. Microbial nutrient competition with plants [34] may be particularly harsh when microbial population growth after inoculation coincides with release of nutrients from new microbial necromass and freshly applied mineral fertilizer, as it must have occurred in the γ-irradiated control soil of this experiment. The re-introduced soil microbes in the filtrate of the soil suspension and the newly inoculated foreign AMF inoculant may thus have mainly sequestered Zn and other nutrients in their newly formed biomass in γ-irradiated soil [34]. This would further explain why the inoculated foreign AMF strain did not assist much in plant Zn uptake.

In summary, this study advanced our knowledge on possible agronomic practices to biofortify cereal grains with Zn. i) It showed that addition of readily decomposable N-rich green manure to calcareous soil can raise grain Zn concentration of bread wheat to levels approaching those of only $ZnSO_4$ fertilizer-fed wheat plants; ii) The study showed that combined addition of N-rich green manure and $ZnSO_4$ fertilizer can considerably raise grain Zn concentration; iii) Mixing green manure to the soil showed that the native soil Zn pool can contribute considerably to total Zn accumulation in the aboveground plant biomass even when no additional mineral Zn fertilizer had been applied to the soil; iv) Lastly, the study emphasizes that care should be taken when trying to utilize AMF inoculants for supporting crop plants in their Zn uptake, when the fungi are not adapted to the prevailing soil conditions.

Acknowledgments

The authors would like to thank Dr. Laurie P. Mauclaire Schönholzer, Thomas Flura and Max Ruethi for their assistance in running the analytical equipment and Simon Ineichen for practical help in sample processing. We acknowledge critical comments by two anonymous reviewers that helped to clarify the main message of this study.

Author Contributions

Conceived and designed the experiments: JJ FA EF. Performed the experiments: FA. Analyzed the data: FA EF HAG. Wrote the paper: FA HAG EF RS. Provided the AM fungal inoculant: JJ. Provided the plant seeds and the soil: AHK MA. Enabled gamma-countingL JE. Approved the final manuscript version: JJ FA EF HAG AHK MA JE RS.

References

1. Curtis BC (2002) Wheat in the world. In B. C. . Curtis, S. . Rajaram, H. Gomez Macpherson (eds.), Bread wheat: Improvement and production. Food and Agriculture Organization of the United Nations, Rome, Italy.
2. Black RE, Victora CG, Walker SP, Bhutta ZA, Christian P, et al. (2013) Maternal and child undernutrition and overweight in low-income and middle-income countries. Lancet 382: 427–451.
3. Welch RM, Graham RD (2004) Breeding for micronutrients in staple food crops from a human nutrition perspective. J Exp Bot 55: 353–364.
4. Cakmak I, Kalayci M, Kaya Y, Torun AA, Aydin N, et al. (2010) Biofortification and localization of zinc in wheat grain. J Agr Food Chem 58: 9092–9102.
5. Cakmak I (2008) Enrichment of cereal grains with zinc: Agronomic or genetic biofortification? Plant Soil 302: 1–17.
6. Khoshgoftar AH, Shariatmadari H, Karimian N, Khajehpour MR (2006) Responses of wheat genotypes to zinc fertilization under saline soil conditions. J Plant Nutr 29: 1543–1556.
7. White PJ, Broadley MR (2011) Physiological limits to zinc biofortification of edible crops. Front Plant Sci 2: 80.
8. Soltani S, Khoshgoftarmanesh AH, Afyuni M, Shrivani M, Schulin R (2013) The effect of preceding crop on wheat grain zinc concentration and its relationship to total amino acids and dissolved organic carbon in rhizosphere soil solution. Biol Fert Soils 50: 239–247.
9. Cavagnaro TR (2008) The role of arbuscular mycorrhizas in improving plant zinc nutrition under low soil zinc concentrations: a review. Plant Soil 304: 315–325.
10. Lehmann A, Veresoglou SD, Leifheit EF, Rillig MC (2014) Arbuscular mycorrhizal influence on zinc nutrition in crop plants – A meta-analysis. Soil Biol Biochem 69: 123–131.
11. Ryan MH, Angus JF (2003) Arbuscular mycorrhizae in wheat and field pea crops on a low P soil: increased Zn-uptake but no increase in P-uptake or yield. Plant Soil 250: 225–239.
12. Ryan MH, Derrick JW, Dann PR (2004) Grain mineral concentrations and yield of wheat grown under organic and conventional management. J Sci Food Agr 84: 207–216.
13. Diesing WE, Sinaj S, Sarret G, Manceau A, Flura T, et al. (2008) Zinc speciation and isotopic exchangeability in soils polluted with heavy metals. Eur J Soil Sci 59: 716–729.
14. Gramlich A, Tandy S, Frossard E, Eikenberg J, Schulin R (2013) Availability of zinc and the ligands citrate and histidine to wheat: does uptake of entire complexes play a role? J Agric Food Chem 61: 10409–10417.
15. Alloway BJ (2009) Soil factors associated with zinc deficiency in crops and humans. Environ Geochem Hlth 31: 537–548.
16. Gao X, Zhang F, Hoffland E (2009) Malate exudation by six aerobic rice genotypes varying in zinc uptake efficiency. J Environ Qual 38: 2315–2321.
17. Daneshbakhsh B, Khoshgoftarmanesh AH, Shariatmadari H, Cakmak I (2013) Phytosiderophore release by wheat genotypes differing in zinc deficiency tolerance grown with Zn-free nutrient solution as affected by salinity. J Plant Physiol 170: 41–46.
18. Dakora FD, Phillips DA (2002) Root exudates as mediators of mineral acquisition in low-nutrient environments. Plant Soil 245: 35–47.
19. Hoffland E, Wei CZ, Wissuwa M (2006) Organic anion exudation by lowland rice (*Oryza sativa L.*) at zinc and phosphorus deficiency. Plant Soil 283: 155–162.
20. Ueno D, Rombola AD, Iwashita T, Nomoto K, Ma JF (2007) Identification of two novel phytosiderophores secreted by perennial grasses. New Phytol 174: 304–310.

21. Parniske M (2008) Arbuscular mycorrhiza: the mother of plant root endosymbioses. Nat Rev Microbiol 6: 763–775.
22. Bonfante P, Genre A (2010) Mechanisms underlying beneficial plant-fungus interactions in mycorrhizal symbiosis. Nat Commun 1: 48.
23. Fitter AH (1991) Costs and benefits of mycorrhizas - implications for functioning under natural conditions. Experientia 47: 350–355.
24. Joner EJ, Jakobsen I (1995) Growth and extracellular phosphatase-activity of arbuscular mycorrhizal hyphae as influenced by soil organic-matter. Soil Biol Biochem 27: 1153–1159.
25. Anonymous (2010). Keys to soil taxonomy. Natural Resources Conservation Service, United States Department of Agriculture, Washington, D.C., 338 p.
26. Mostafazadeh-Fard B, Mansouri H, Mousavi SF, Feizi M (2009) Effects of different levels of irrigation water salinity and leaching on yield and yield components of wheat in an arid region. J Irrig Drain E-Asce 135: 32–38.
27. Gee GW, Bauder JW (1986) Particle size analysis. In: A Klute, editor editors. Methods of soil analysis. Madison, Wisconsin, USA: American Society of Agronomy. pp. 383–411.
28. Rhoades JD (1996) Salinity: Electrical conductivity and total dissolved salts. Methods of soil analysis. Madison, Wisconsin, USA: Soil Science Society of America. pp. 417–435.
29. Walkey A, Black IA (1934) An examination of the Degtjareff method for determinig soil organic matter and proposed modification of the chromic acid titration method. Soil Sci 37: 29–38.
30. Burt R (2004) Soil survey laboratory methods manual: soil survey investigations report Nebraska: United States Department of Agriculture. Natural Resources Conservation Servic.
31. Lindsay WL, Norvell WA (1978) Development of a DTPA soil test for zinc, iron, manganese, and copper. Soil Sci Soc Am J 42: 421–428.
32. Maas EV (1986) Salt tolerance of plants. Appl Agric Res 1: 12–25.
33. Thompson JP (1990) Soil sterilization methods to show VA-mycorrhizae aid P and Zn nutrition of wheat in vertisols. Soil Biol Biochem 22: 229–240.
34. Nazeri N, Lambers H, Tibbett M, Ryan M (2013) Do arbuscular mycorrhizas or heterotrophic soil microbes contribute toward plant acquisition of a pulse of mineral phosphate? Plant Soil 373: 699–710.
35. van de Voorde TFJ, van der Putten WH, Bezemer TM (2012) Soil inoculation method determines the strength of plant–soil interactions. Soil Biol Biochem 55: 1–6.
36. Barker SJ, Stummer B, Gao L, Dispain I, O'Connor PJ, et al. (1998) A mutant in *Lycopersicon esculentum* Mill. with highly reduced VA mycorrhizal colonization, isolation and preliminary characterisation. Plant J 15: 791–797.
37. Hogland DR, Arnon OD (1938) The water-culture method of growing plants without soil. University of California college of agriculture, agricultural experiment station Berkeley, California. USA.
38. Schenck NC, Smith GS (1982) Additional new and unreported species of mycorrhizal fungi (Endogonaceae) from Florida. Mycologia 74: 77–92.
39. Gerdemann JW, Nicolson TH (1963) Spores of mycorrhizal *Endogone* species extracted from soil by wet sieving and decanting. T Brit Mycol Soc 46: 235–244.
40. Cavagnaro TR, Smith FA, Hay G, Carne-Cavagnaro VL, Smith SE (2004) Inoculum type does not affect overall resistance of an arbuscular mycorrhiza-defective tomato mutant to colonisation but inoculation does change competitive interactions with wild-type tomato. *New Phytol* 161: 485–494.
41. Karami M, Afyuni M, Khoshgoftarmanesh AH, Papritz A, Schulin R (2009) Grain zinc, iron, and copper concentrations of wheat grown in central Iran and their relationships with soil and climate variables. J Agr Food Chem 57: 10876–10882.
42. Sadeghzadeh B (2013) A review of zinc nutrition and plant breeding. J Soil Sci Plant Nutr 13: 905–927.
43. Kalayci M, Torun B, Eker S, Aydin M, Ozturk L, et al. (1999) Grain yield, zinc efficiency and zinc concentration of wheat cultivars grown in a zinc-deficient calcareous soil in field and greenhouse. Field Crop Res 63: 87–98.
44. Rengel Z (1999) Zinc deficiency in wheat genotypes grown in conventional and chelator-buffered nutrient solutions. Plant Sci 143: 221–230.
45. Neilands JB (1995) Siderophores: Structure and function of microbial iron transport compounds. J. Biol. Chem 270: 26723–26726.
46. Oliveira RS, Vosatka M, Dodd JC, Castro PML (2005) Studies on the diversity of arbuscular mycorrhizal fungi and the efficacy of two native isolates in a highly alkaline anthropogenic sediment. Mycorrhiza 16: 23–31.
47. Jansa J, Mozafar A, Anken T, Ruh R, Sanders IR, et al. (2002) Diversity and structure of AMF communities as affected by tillage in a temperate soil. Mycorrhiza 12: 225–234.
48. Marschner B, Bredow A (2002) Temperature effects on release and ecological properties of dissolved organic matter (DOM) in sterilized and biologically active soil samples. Soil Biol Biochem 34: 459–466.
49. McNamara NP, Black HIJ, Beresford NA, Parekh NR (2003) Effects of acute gamma irradiation on chemical, physical and biological properties of soils. Appl Soil Ecol 24: 117–132.
50. Kariman K, Barker SJ, Jost R, Finnegan PM, Tibbett M (2014) A novel plant–fungus symbiosis benefits the host without forming mycorrhizal structures. New Phytol 201: 1413–1422.

13

Ecosystem Carbon and Nitrogen Accumulation after Grazing Exclusion in Semiarid Grassland

Liping Qiu, Xiaorong Wei*, Xingchang Zhang, Jimin Cheng

State Key Laboratory of Soil Erosion and Dryland Farming in the Loess Plateau, Northwest A&F University, Yangling, China

Abstract

The grazing exclusion in degraded grassland has been extensively used to prevent the loss of grassland resources and to improve grassland services. The effects of grazing exclusion on C and N balance, however, have not been well addressed but are essential for assessing grassland C sinks, the sustainable use of grassland resources and the support of grassland services. To understand the response of ecosystem C and N to grazing exclusion in semiarid grassland, we determined the C and N in litter, aboveground biomass, roots and soils from ungrazed grassland fenced at different times in northwest China. Our results showed that the aboveground biomass, root biomass and plant litter were 70–92%, 56–151% and 59–141% higher, respectively, in grazer excluded grassland than in grazed grassland. Grazing exclusion significantly increased C and N stored in plant biomass and litter and increased the concentrations and stocks of C and N in soils. Grazing exclusion thus significantly increased the C and N stored in grassland ecosystems. The increase in C and N stored in soil contributed to more than 95% and 97% of the increases in ecosystem C and N storage. The highest C and N stocks in ecosystems were observed in 17-year grazer excluded grassland. The results from this study indicate that grazing exclusion has the potential to increase C and N storage in degraded semiarid grassland and that the recovery of ecosystem C and N was mainly due to the accumulation of C and N in soils.

Editor: Minna-Maarit Kytöviita, Jyväskylä University, Finland

Funding: This study was supported by the National Natural Science Foundation of China (40901145), Chinese Academy of Sciences (KZCX2-XB3-13-02), Science and Technology Development Research Program in Shaanxi Province (2011KJXX25), the Program for Youthful Talents in Northwest A & F University, and the Program from Northwest A & F University (QN2011147). The funders had no role in study design, data collection and analysis, decision to publish, or preparation of the manuscript.

Competing Interests: The authors have declared that no competing interests exist.

* E-mail: xrwei78@163.com

Introduction

Grasslands cover 20% of the terrestrial surface and account for 20% or more of total terrestrial production [1]. Owing to their importance in socio-economics, culture, ecology and environmental quality, grassland ecosystems have become one of the most active subjects of research by ecologists around the world [2]. On a global scale, grasslands store more than 10% of terrestrial biomass C and 10–30% of global soil organic carbon (OC), and they have been estimated to sequester C in soil at a rate of 0.5 Pg C yr^{-1}, accounting for about one-fourth of the potential carbon (C) sequestration in world soils [1,3]. Carbon and nitrogen (N) are key factors determining the properties of ecosystems, such as primary productivity and plant diversity [4–6] and rates of decomposition [7], and therefore could be important mediators of the richness of functional groups and ecosystem functioning. Understanding the relationships between grassland management and the status of C and N is thus not only of academic interest but also crucial for the sustainable use of grassland resources and support of grassland services.

Previous studies in semiarid regions have shown that any changes in land use, such as introducing grazing and conversion to shrubland and agricultural land, will decrease soil and ecosystem C and N [8–10]. Possible mechanisms of the loss of C and N include changes in composition of plant communities, changes in plant roots and their control of the soil microbial community [11,12], decreased input of organic material from aboveground biomass and roots due to biomass removal [13], decreased availability of soil resources in grasslands [14–17] and the disruption of soil structure resulting in losses of C and N through accelerated mineralization of soil organic matter and erosion by water and wind [9,18–23].

The grazing exclusion of degraded grassland has been extensively employed in China and other regions to prevent its degradation and to improve grassland services [2,24]. As an important conservation practice, grazing exclusion can have potential effects on the storage of C and N in grassland ecosystems due to the increased productivity of the biomass and the input of C and N to soils and ecosystems. The effects of grazing exclusion on C and N balance in grassland, however, have not been well addressed. We hypothesize that grazing exclusion in degraded grassland can increase C and N in soils and ecosystems because the removal of herbivores, particularly larger herbivores, can increase C and N stocks in ecosystems [25].

To test our hypothesis, we collected soil, plant, root and litter samples from fenced semiarid grasslands with fencing ages of 17-year, 22-year and 27-year (grazer excluded grassland) in a typical grassland of northwest China. We selected an adjacent unfenced grassland as a control (grazed grassland). The C and N concentrations in soils and plant samples were measured, and the stocks of C and N in different pooled samples and in the ecosystem were calculated to assess the response of C and N to

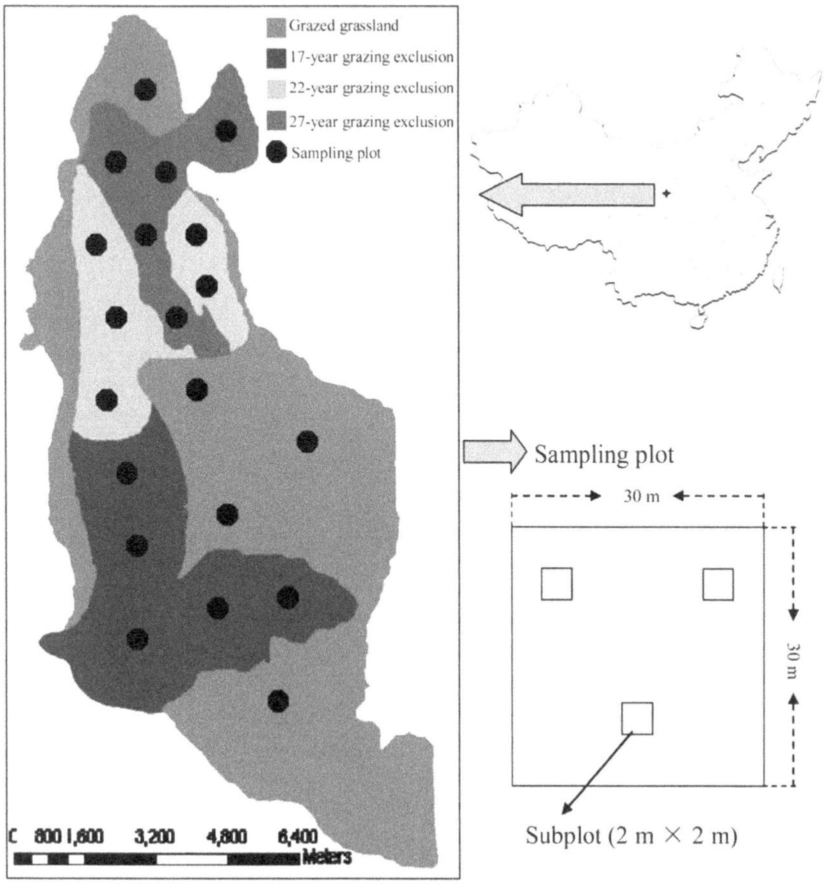

Figure 1. Location of the study site and sampling scheme.

grazing exclusion in semiarid grassland. The objective of this study was to understand how the C and N accumulate in semiarid grassland after grazing exclusion.

Materials and Methods

Ethics statement

The permit for soil and plant sampling was obtained from the Yunwushan Natural Grassland Management Bureau, which is responsible for the protection and management of the Yunwushan Natural Grassland Protection Zone.

Study sites

This study was conducted in the Yunwushan Natural Grassland Protection Zone (36°13′–36°19′N; 106°24′–106°28′E) at Guyuan City, Ningxia Hui Autonomous Region, China. The grassland protection zone has an area of 6700 ha and an elevation ranging from 1800 to 2148 m. The study area has a continental monsoon climate. The mean temperature is 6.9°C. The maximum and minimum temperatures occur in July (24°C) and January (−14°C), respectively. The frost-free period is 124 d, normally beginning in mid-April and ending in late September. The mean annual precipitation is 425 mm. The soil in the study area is a montane grey-cinnamon soil classified as a Calci-Orthic Aridisol according to the Chinese taxonomic system, which is equivalent to a Haplic Calcisol in the FAO/UNESCO system.

Field investigation and sampling

The study area was grazed by goats before protection. The intensity of grazing had varied over the years. In 1982, before protection, the physicochemical properties of the soils in the study area with the same soil type and similar natural conditions were not significantly different [26]. Three grazer excluded treatments established at different times were studied. Grassland sites were fenced with goat proof wire mesh year 1982, 1987 and 1992 and consequently goat grazing was excluded for 27 (GE27), 22 (GE22) and 17 years (GE17), respectively. The areas of the three treatments were 1200, 1000, and 1500 ha, respectively. A grazing area of 3000 ha was maintained as a control (CK). All four areas adjoined each other, and no fertilizer was applied. The dominant grass species in the grazer excluded and grazed grassland were bunge needlegrass (*Stipa bungeana*) and Russian wormwood (*Artemisia sacrorum Ledeb.*).

Most parts of the grazed grassland and grazer excluded grassland have similar topography and altitude. All the plots for both grazed and grazer excluded treatments have the same soil type and similar physiographic conditions (slope degree, slope direction, topography and altitude). Our previous study in the same control area showed that there were no changes in soil OC and N during past 27 years in the continuously grazed area (CK treatment) [9]. We therefore believe that the soils in each treatment had similar initial conditions and the existing grazed grassland can be used as the control to compare the effects of grazing exclusion on C and N accumulation in plants and soils.

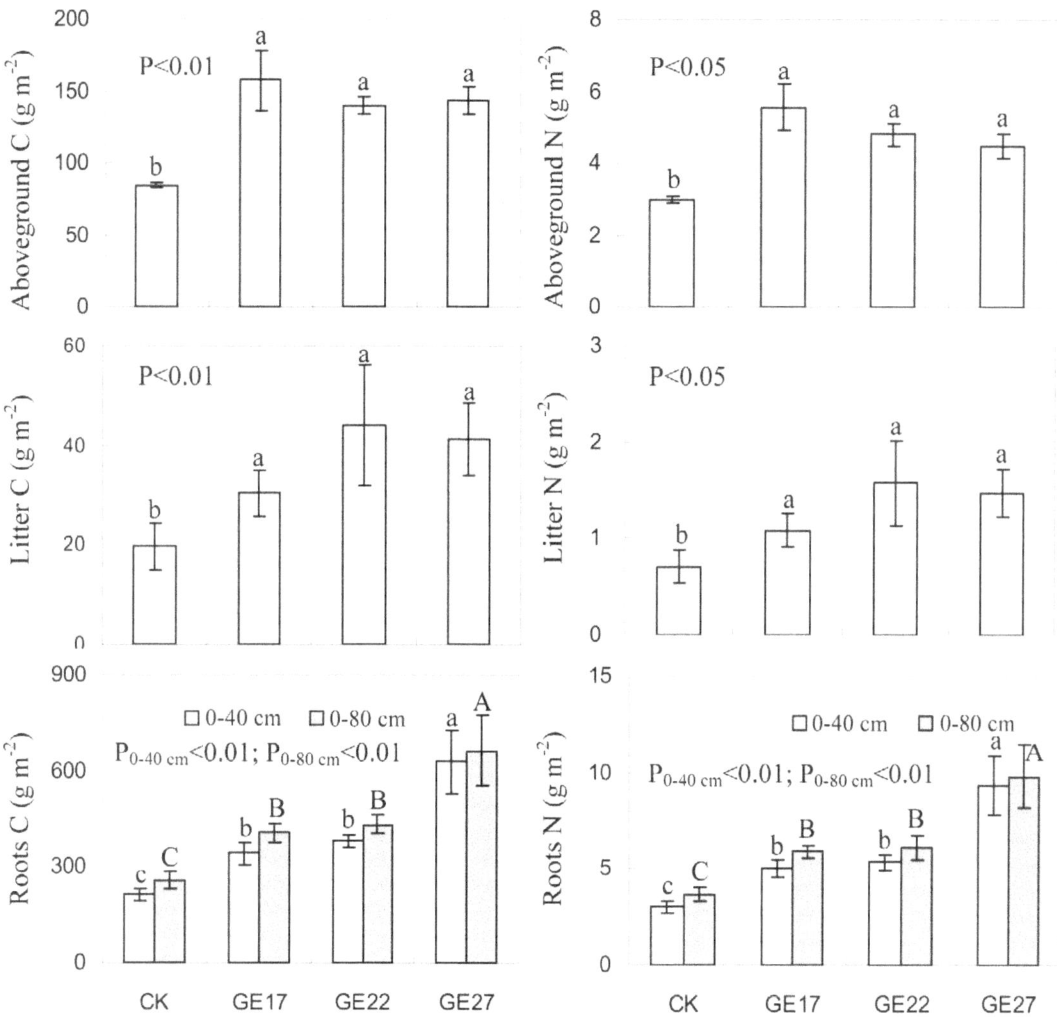

Figure 2. The effects of grazing exclusion on surface litter, aboveground biomass and roots biomass in grazed grassland. Error bars are the standard error of the mean. For aboveground biomass, surface litter and roots biomass in the 0–40 cm layer, means with different lowercase letters are different at P<0.05. For roots biomass in the 0–80 cm layer, means with different uppercase letters are different at P<0.05. (CK: grazed grassland; GE17: 17-year grazing exclusion; GE22: 22-year grazing exclusion; GE27: 27-year grazing exclusion).

Five pseudo-replicated plots (30 m×30 m) were randomly established within the grazed and grazer excluded treatments in August 2009 (Fig. 1). True replication was not possible in this study. The plots were at least 2 km apart. Considering that each treatment has the area larger than 1000 ha and the distance between each plot is larger than 2 km, our sampling design can reflect the effects of grazing exclusion on C and N in this ecosystem. Three 2 m×2 m subplots were established in each plot for field investigation and sampling. The subplots were at least 15 m apart and did not differ in physiographic conditions. In each subplot, the canopy cover and dominant species were determined. The canopy covers were 25%, 85%, 100% and 100% in grazed, 17-year grazing exclusion, 22-year grazing exclusion and 27-year grazing exclusion treatments, respectively. The aboveground biomass was harvested in August when the peak aboveground biomass of the two grass species occurred. The surface litter was also collected. Three profile samples were collected from each subplot for the measurement of root biomass. The samples (at depths of 0–10, 10–20, 20–40, 40–60, and 60–80 cm) were collected using a tube auger 9.0 cm in diameter. Roots were separated from the soil samples. The samples of litter, aboveground biomass and roots were washed with deionized water and dried at 65°C for 48 h. The dry samples were weighed and then ground to pass through a 0.25 mm sieve for measurement of C and N.

In each plot, a pit 1.0 m long×0.7 m wide×1.0 m deep was dug for measuring soil bulk density. The soil was sampled using a stainless-steel cutting cylinder (5.0 cm dia.×5.0 cm high) at depths of 0–10, 10–20, 20–40, 40–60 and 60–80 cm. The soil cores were dried at 105°C for 24 hours and then weighed for the calculation of soil bulk density. Three representative soil samples were randomly collected at the same depth increments as the bulk density samples in each subplot for the measurement of soil OC and N. All visible pieces of organic material were removed. The moist soil samples were brought to the laboratory, air-dried and ground to pass through a 0.25 mm sieve for the measurement of soil OC and N concentrations.

Laboratory and data analyses

The C and N concentrations of litter, aboveground biomass, root samples, and OC and N concentrations in soils were analyzed using a VARIO EL III CHON analyzer (Elementar, Germany) at

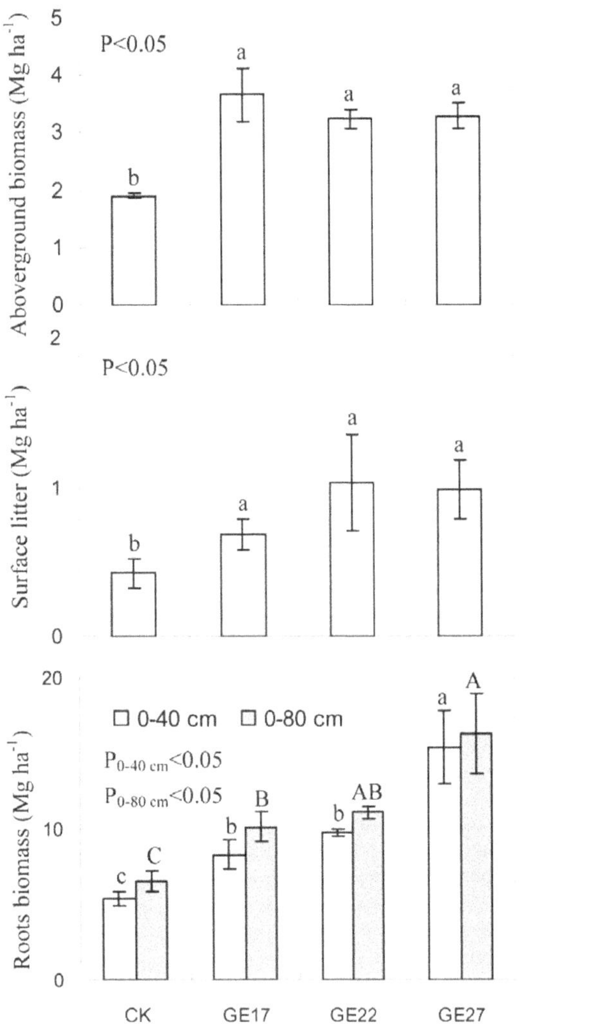

Figure 3. The effects of grazing exclusion on C and N stored in litter, aboveground biomass and roots in grazed grassland. Error bars are the standard error of the mean. For C and N stored in aboveground biomass, surface litter and roots biomass in the 0–40 cm layer, means with different lowercase letters are different at P<0.05. For C and N stored in roots biomass in the 0–80 cm layer, means with different uppercase letters are different at P<0.05. (CK: grazed grassland; GE17: 17-year grazing exclusion; GE22: 22-year grazing exclusion; GE27: 27-year grazing exclusion).

the Testing and Analysis Center of Northwest University, Xi'an, China.

The stocks of soil OC and N (Mg ha^{-1}) were calculated as follows:

Stocks of $OC_i = D_i \times BD_i \times OC_i/10$ (1)

Stocks of $N_i = Di \times BD_i \times N_i/10$ (2)

where D_i, BD_i, OC_i and N_i are the thickness (cm), bulk density (g cm^{-3}), organic C (g kg^{-1}) and N (g kg^{-1}), respectively, of the *i*th layer of the soil.

A one-way analysis of variance followed by a Turkey's post hoc test was conducted using SAS version 8.0 to examine the effects of grazing exclusion on the C and N stored in litter, aboveground biomass, roots, and soils.

Results

1. C and N in vegetation

Grazing exclusion significantly enhanced the accumulation of above- and belowground biomass and plant litter in the ecosystem (Fig. 2). The amount of aboveground biomass, belowground biomass and plant litter accumulated on the surface of the soils in the grazer excluded grassland ranged from 3.2 to 3.7, 10.2 to 16.3 and 0.7 to 1.0 Mg ha^{-1}, respectively, which were 70–92% ($P<0.05$), 56–151% ($P<0.05$) and 59–141% higher ($P<0.05$), respectively, than those in the grazed grassland. The root biomass was mainly distributed in the 0–40 cm soil layer, accounting for 75–96% of the root biomass in the top 80 cm of soil. The root biomass in the 0–40 cm soil layer was 53–183% higher in the grazer excluded grassland than in the grazed grassland ($P<0.05$).

Grazing exclusion significantly increased C and N stored in plant biomass and litter (Fig. 3). The C stocks in aboveground biomass, belowground biomass and litter were 64–86% ($P<0.01$), 58–157% ($P<0.01$) and 55–125% higher ($P<0.01$), respectively, in grazer excluded grassland than in grazed grassland, while the N stocks were 51–87% ($P<0.05$), 58–157% ($P<0.05$) and 55–125% higher ($P<0.05$), respectively. Similar to root biomass, the C and N stored in roots were mainly in the 0–40 cm layer, accounting for 83–95% of the root C and N in the top 80 cm of soil, and were 61–194% higher in grazer excluded grassland than in grazed grassland ($P<0.01$). The C and N stocks in the aboveground biomass were not affected by grazing exclusion age, but the C and N stocks in the root biomass increased with grazing exclusion age. These results suggest that grazing exclusion has the potential to accumulate C and N in grass biomass and litter in semiarid grassland.

2. C and N in soils

Grazing exclusion significantly increased OC and N concentrations (Fig. 4) and stocks (Fig. 5) in the 0–80 cm soil layer. The OC and N concentrations were 6–14% ($P<0.01$) and 15–37% higher ($P<0.01$), respectively, in grazer excluded grassland than in grazed grassland, and the OC and N stocks were 49–77% ($P<0.05$) and 125–199% higher ($P<0.05$), respectively, in grazer excluded grassland than in grazed grassland. Additionally, the OC and N stored in the 0–40 cm layer accounted for more than half of the OC and N stored in the top 0–80 cm of soil (53–65% for OC and 55–75% for N). These results suggest that grazing exclusion has the potential to sequester C and N in soils and that the potential was larger in the 0–40 cm soil layer than in deeper soils. The effects of grazing exclusion on soil C and N concentrations and stocks, however, varied with grazing exclusion age, with the highest C and N in 17-year grazing exclusion treatment, indicating a peak accumulation and maximum sequestration of C and N in soils around 17 years after grazing exclusion.

3. C and N in the ecosystem

Grazing exclusion significantly increased the storages of C and N in the ecosystem, which were 50–75% ($P<0.01$) and 55–106% higher ($P<0.01$), respectively, in grazer excluded grassland compared to grazing grassland (Fig. 6). For both grazer excluded and grazed systems, more than 95% of C and 97% of N were stored in the soil, while less than 1% of C and N were stored in aboveground biomass and litter. Additionally, the highest C and N stocks in the ecosystem were observed in 17-year grazing exclusion treatment, suggesting a peak accumulation of C and N in semiarid grassland around 17 years after grazing exclusion.

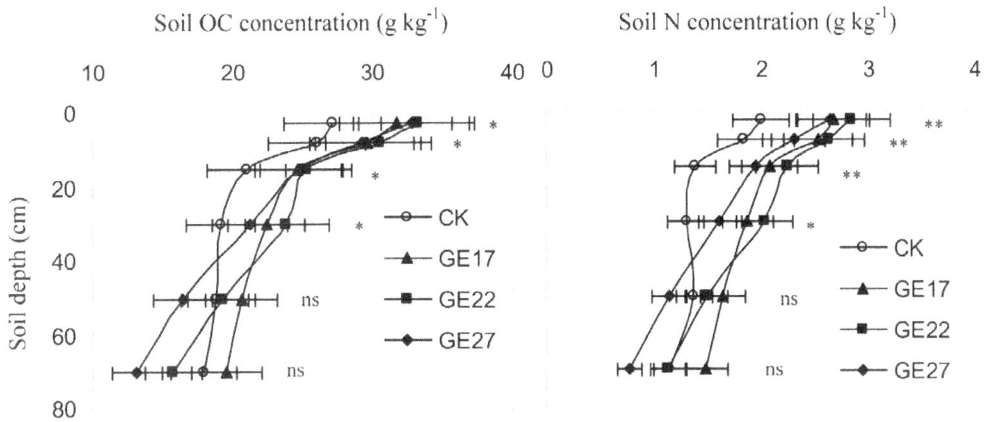

Figure 4. The effects of grazing exclusion on OC and N concentrations in soils of grazed grassland. Error bars are the standard error of the mean. (*: P<0.05; **: P<0.01; ns: P>0.05; CK: grazed grassland; GE17: 17-year grazing exclusion; GE22: 22-year grazing exclusion; GE27: 27-year grazing exclusion).

Discussion

Our results demonstrate a significant increase of C and N in soils and ecosystems after grazing exclusion in grassland, supporting our hypothesis that grazing exclusion has the potential to increase C and N storage in degraded semiarid grassland. Grazing exclusion resulted in an accumulation of 2.4–5.6 Mg C ha^{-1} yr^{-1}, greater than the estimated global sequestration rate of C for grassland and semiarid grassland, and an accumulation of 0.18–0.54 Mg N ha^{-1} yr^{-1}. The estimated global sequestration rate of C of grassland ranges from 0.003 to 0.057 Mg ha^{-1} yr^{-1} [27,28]. These estimates do not consider the effects of excluding grazing. Our higher values for the accumulation of C suggest that grazing exclusion may accelerate the sequestration of C, at least in semiarid regions. The accumulation of C and N in grassland ecosystems can be directly attributed to the increase of above- and belowground biomass, the accumulation of grass materials as litter

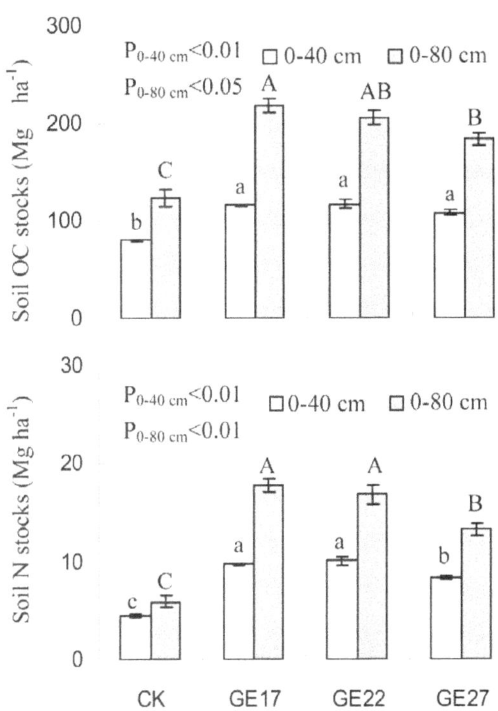

Figure 5. The effects of grazing exclusion on OC and N stocks in soils of grazed grassland. Error bars are the standard error of the mean. For soil OC and N stocks in 0–40 cm and 0–80 cm soil layers, means with different lowercase and uppercase letters are different at P<0.05, respectively. (CK: grazed grassland; GE17: 17-year grazing exclusion; GE22: 22-year grazing exclusion; GE27: 27-year grazing exclusion).

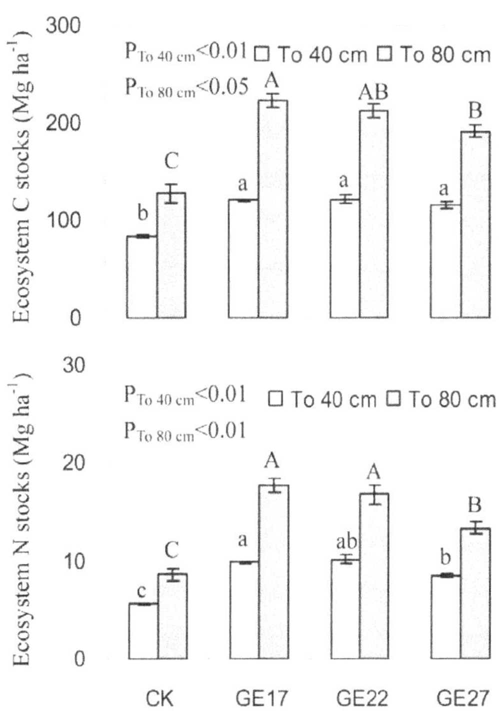

Figure 6. The effects of grazing exclusion on C and N in ecosystems (aboveground+belowground) in grazed grassland. Error bars are the standard error of the mean. For ecosystem C and N stocks down to 40 cm and 80 cm depths, means with different lowercase and uppercase letters are different at P<0.05, respectively. (CK: grazed grassland; GE17: 17-year grazing exclusion; GE22: 22-year grazing exclusion; GE27: 27-year grazing exclusion).

on the ground and the increased input of C and N into soils by litter and roots. This accumulation can also be indirectly attributed to the decreased loss of biomass C and N from the exclusion of disturbance, and the decreased loss of soil C and N by reducing soil respiration, and ammonium volatilization and N_2O emission resulting from nitrification [29–34].

The effect of grazing on changes in C and N in grassland has received much attention, but the accumulation of C and N in the ecosystem after grazing exclusion in grassland has rarely been reported. Grazing significantly reduces C and N stored in grassland ecosystems. Bagchi and Ritchie (2010) reported a 32–33% decrease in aboveground C input and a 21–63% decrease in belowground C input to ecosystems with grazing [12]. Schuman et al. (1999) observed 20–52% and 7–16% decreases of C and 15–30% and 18–52% decreases of N in aboveground biomass and roots (0–60 cm depth), respectively, after 12 years of grazing on a native mixed grassland in Wyoming [35]. In our grazer excluded grassland, the aboveground C and N input (including litter) to the ecosystem increased by 76–80% and 62–81%, respectively, while the belowground C and N input to the ecosystem increased by 50–75% and 55–106%, respectively. The relatively larger increase in C and N by grazing exclusion compared to the decrease in C and N by grazing suggest that the recovery of C and N in degraded grassland could be faster than expected.

Grazing exclusion significantly increased both concentrations and stocks of C and N in the soils of grassland. Concentrations increased by 0.05–0.18 g C kg^{-1} yr^{-1} and 0.01–0.03 g N kg^{-1} yr^{-1}, and stocks increased by 0.21–0.55 Mg C ha^{-1} 10 cm^{-1} yr^{-1} and 0.02–0.05 g N ha^{-1} 10 cm^{-1} yr^{-1}. These rates were relatively larger than the losses of soil C and N due to grassland grazing or conversion to other types of land use. For example, Wang et al. (2012) reported a 20–37% decrease in soil OC and a 5–37% decrease in soil N after light and heavy grazing on the northern Tibetan Plateau, corresponding to losses of 0.14–0.29 Mg C ha^{-1} 10 cm^{-1} yr^{-1} and 0.004–0.013 Mg N ha^{-1} 10 cm^{-1} yr^{-1} [36]. Ingram et al. (2008) reported that 21 years of heavy grazing resulted in losses of 0.10 Mg C ha^{-1} 10 cm^{-1} yr^{-1} and 0.009 Mg N ha^{-1} 10 cm^{-1} yr^{-1} in the 0–60 cm soil layer in a mixed-grass ecosystem [37]. Qiu et al. (2012) reported a 58% decrease in soil OC after conversion of grassland to cropland in the Loess Plateau, representing a loss of 0.30 Mg C ha^{-1} 10 cm^{-1} yr^{-1} [9]. The accumulation of soil OC and N after grazing exclusion in grassland should thus be greater than the loss of OC and N by grazing or conversion to agriculture, and rehabilitation of degraded grassland after stopping any disturbance should be rapid.

The accumulation of C and N in soils after grazing exclusion can be explained by increased input and decreased loss of C and N. We observed significant positive relationships between soil C and N concentrations and root biomass ($r = 0.653$, $P<0.001$ for C; $r = 0.669$, $P<0.001$ for N; $n = 100$), in agreement with the finding that grazing triggers the loss of soil C by altering plant roots and their control on soil microbes [38] and thus in support of our explanation that grazing exclusion enhances the input of soil C and N. Additionally, many studies have shown that disturbance of grassland increases the loss of C and N through respiration and N_2O emission and that these losses are reduced when the disturbance is stopped. For example, grazing can significantly increase soil C mineralization, nitrification and ammonification [29], enhancing soil respiration rates [30], ecosystem respiration rates [31,32] and N_2O emission [33,34], which result in the loss of soil OC and N. Additionally, the effects of grazing on mineralization of soil OC and N were seasonally dependent and were characterized by increased mineralization during the growing season and decreased mineralization during the non-growing season [39,40]. These findings indirectly support our explanation that grazing exclusion reduces losses of soil OC and N. The increase in OC and N stored in the soil contributed to 92–98% and 97–99% of the increases in ecosystem C and N storage, respectively, suggesting that the recovery of ecosystem C and N mainly occurred by the accumulation of C and N in soils after grazing exclusion in grassland.

The effects of grazing exclusion age on C and N stored in both soils and ecosystems indicated a maximum for C and N accumulation. Such a maximum amount of C and N accumulation represents the potential of C and N sequestration of degraded grassland at the study site. The existence of a maximum accumulation of C and N in soils and ecosystems also suggests that after certain periods of grazing exclusion (17-year in this study), the soils and ecosystems will no longer accumulate C and N, indicating that the grassland in the study area was stable or mature, in terms of their C and N storage, after 17-year of grazing exclusion. Other studies have found that grasslands in northern China subjected to 20-year grazing exclusion were stable in their productivity [41,42], soil respiration [43] and C and N storage [22,44]. The decrease of C and N after 17-year grazing exclusion may be due to increased productivity driving greater competition for resources such as nutrients, water, light and heat. Low availability of such resources can feedback to the increased microbial demand of resources, which in turn enhances the mineralization of soil organic material and thus the loss of C and N [22,45]. An increased accumulation of fresh litter and organic material may also prevent the infiltration of rainfall into the soil. The interception of rainfall by accumulated litter and organic material accelerates their decomposition and renders the rainfall unavailable for infiltration, plant growth and cycling of C and N in semiarid and arid regions [22,46,47]. This interception of rainfall results in increased losses of biomass C and N and a decreased return of C and N through litter and organic material to the soil. Grassland in arid and semiarid regions could thus be properly used after accumulating maximum C and N so as to maintain the function of C and N sequestration. Our results demonstrated the accumulation of C and N in semiarid grassland after relatively long-term (17 to 27 years) grazing exclusion, however, short-term effects of grazing exclusion on C and N in grassland should be further examined so as to establish C and N dynamics following grazing exclusion in semiarid grazed grassland.

Acknowledgments

We thank Chen J, Li P, Gao J, Liu Z, Li X, Jia X, and Yunwushan natural grassland management bureau for their help in field and laboratory experiments.

Author Contributions

Conceived and designed the experiments: LQ XW XZ JC. Performed the experiments: LQ XW. Analyzed the data: LQ XW XZ. Wrote the paper: LQ.

References

1. Scurlock JM, Hall DO (1998) The global carbon sink: A grassland perspective. Global Change Biol 4: 229–233.
2. Kang L, Han X, Zhang Z, Sun OJ (2007) Grassland ecosystems in China: review of current knowledge and research advancement. Phil Trans R Soc B 362: 997–1008.

3. Follett RF, Reed DA (2010) Soil carbon sequestration in grazing lands: societal benefits and policy implications. Rangeland Ecol Manage 63: 4–15.
4. Tilman D (1987) Secondary succession and the pattern of plant dominance along experimental nitrogen gradients. Ecol Monogr 57: 189–214.
5. De Deyn GB, Raaijmakers CE, van der Putten WH (2004) Plant community development is affected by nutrients and soil biota. J Ecol 92: 824–834.
6. Hooper DU, Chapin FS, Ewel JJ, Hector A, Inchausti P, et al. (2005) Effects of biodiversity on ecosystem functioning: a consensus of current knowledge. Ecol Monogr 75: 3–35.
7. Wardle DA (2002) Communities and Ecosystems: Linking the Aboveground and Belowground Components. Princeton: Princeton University Press. 400p.
8. Han GD, Hao X, Zhao MG, Wang MJ, Ellert BH, et al. (2008). Effect of grazing intensity on carbon and nitrogen in soil and vegetation in a meadow steppe in Inner Mongolia. Agr Ecosys Environ 125: 21–32.
9. Qiu L, Wei X, Zhang X, Cheng J, Gale W, et al. (2012) Soil organic carbon losses due to land use change in a semiarid grassland. Plant Soil355: 299–309.
10. Wei X, Huang M, Shao M, Li L, Zhang X, et al. (2013) Shrubs increase soil resources heterogeneity along semiarid grass slopes in the Loess Plateau. J Arid Environ 88: 175–183.
11. De Deyn GB, Quirk H, Yi Z, Oakley S, Ostle NJ, et al. (2009) Vegetation composition promotes carbon and nitrogen storage in model grassland communities of contrasting soil fertility. J Ecol 97: 864–875.
12. Bagchi S, Ritchie ME (2010) Introduced grazers can restrict potential soil carbon sequestration through impacts on plant community composition. Ecol Lett 13: 959–968.
13. Johnson LC, Matchett JR (2001) Fire and grazing regulate belowground processes in tallgrass prairie. Ecology 82: 3377–3389.
14. Wilson SD, Tilman D (1993) Plant competition and resource availability in response to disturbance and fertilization. Ecology 74: 599–611.
15. Bakker C, Blair JM, Knapp AK (2003) Does resource availability, resource heterogeneity or species turnover mediate changes in plant species richness in grazed grasslands? Oecologia 137: 385–391.
16. Renne IJ, Tracy BF, Colonna IA (2006) Shifts in grassland invisibility: effects of soil resources, disturbance, composition, and invader size. Ecology 87: 2264–2277.
17. Gendron F, Wilson SD (2007) Responses to fertility and disturbance in a low-diversity grassland. Plant Ecol 191: 199–207.
18. Hiernaux P, Bielders CL, Valentin C, Bationo A, Fernandez-Rivera S (1999) Effects of livestock grazing on physical and chemical properties of sandy soils in Sahelian rangelands. J Arid Environ 41: 231–245.
19. Belnap J (2003) The world at your feet: desert biological soil crusts. Front Ecol Environ 1: 181–189.
20. Liu YR, Yang C, Zhu ZM, Liu ML (2003) Soil C and N dynamics during desertification of grassland in Northern China. Chin J Appl Ecol 15: 1604–1606.
21. Neff JC, Reynolds RL, Belnap J, Lamothe P (2005) Multi-decadal impacts of grazing on soil physical and biogeochemical properties in southeast Utah. Ecol Appl 15: 87–95.
22. He N, Yu Q, Wu L, Wang Y, Han X (2008) Carbon and nitrogen store and storage potential as affected by land-use in a *Leymus chinensis* grassland of northern China. Soil Biol Biochem 40: 2952–2959.
23. Steffens M, Kölbl A, Totsche KU, Kögel-Knabner I (2008) Grazing effects on soil chemical and physical properties in a semiarid steppe of Inner Mongolia (P.R. China). Geoderma 143: 63–72.
24. Suttie JM, Reynolds SG, Batello C (2005) Grasslands of the World. FAO, Rome. 514p.
25. Tanentzap AJ, Coomes DA (2012) Carbon storage in terrestrial ecosystems: do browsing and grazing herbivores matter? Biol Rev Camb Philos Soc 87: 72–94.
26. Yunwushan Natural Grassland Protection Zone Bureau (YNGPZB) (2000) A Collected Works on Scientific Investigation and Management in Ningxia Yunwushan Reservation. Ningxia People's Press, Yinchuan. 283p.
27. Lal R (2004) Soil carbon sequestration impacts on global climate change and food security. Science 304:1623–1627.
28. Follett RF, Schuman GE (2005) Grazing land contributions to carbon sequestration. In McGillowayDA editor. Grazingland: A Global Resource. Wageningen: Dublin, Irelad. Wageningen Academic Publishers. 265–277.
29. Wu H, Wiesmeier M, Yu Q, Steffens M, Han X, et al. (2012) Labile organic C and N mineralization of soil aggregate size classes in semiarid grasslands as affected by grazing management. Biol Fertil Soils 48: 305–313.
30. Frank AB (2002) Carbon dioxide fluxes over a grazed prairie and seeded pasture in the Northern Great Plains. Environ Pollut 116: 397–403.
31. Owensby CE, Ham JM, Auen LM (2006) Fluxes of CO_2 from grazed and ungrazed tallgrass prairie. Rangeland Ecol Manag 59: 111–127.
32. Polley HW, Frank AB, Sanabria J, Phillips RL (2008) Interannual variability in carbon dioxide fluxes and flux-climate relationships on grazed and ungrazed northern mixed-grass prairie. Global Change Biol 14: 1620–1632.
33. Saggar S, Andrew RM, Tate KR, Hedley CB, Rodda NJ, et al. (2004) Modelling nitrous oxide emissions from dairy-grazed pastures. Nutr Cycl Agroecosys 68: 243–255.
34. Wang J, Cardenas LM, Misselbrook TH, Cuttle S, Thorman RE, et al. (2012) Modelling nitrous oxide emissions from grazed grassland systems. Environ Pollut 162: 223–233.
35. Schuman GE, Reeder JD, Manley JT, Hart RH, Manley WA (1999) Impact of grazing management on the carbon and nitrogen balance of a mixed-grass rangeland. Ecol Appl 9: 65–71.
36. Wang X, Yan Y, Cao Y. (2012) Impact of historic grazing on steppe soils on the northern Tibetan Plateau. Plant Soil 354:173–183.
37. Ingram LJ, Schuman GE, Buyer JS, Vance GF, Ganjegunte GK, et al. (2008) Grazing impacts on soil carbon and microbial communities in a mixed-grass ecosystem. Soil Sci Soc Am J 72: 939–948.
38. Klumpp K, Fontaine S, Attard E, Roux XL, Gleixner G, et al. (2009) Grazing triggers soil carbon loss by altering plant roots and their control on soil microbial community. J Ecol 97: 876–885.
39. Shan Y, Chen D, Guan X, Zheng S, Chen H, et al. (2011) Seasonally dependent impacts of grazing on soil nitrogen mineralization and linkages to ecosystem functioning in Inner Mongolia grassland. Soil Biol Biochem 43: 1943–1954.
40. Liu T, Nan Z, Hou F (2011) Grazing intensity effects on soil nitrogen mineralization in semi-arid grassland on the Loess Plateau of northern China. Nutr Cycl Agroecosys 91: 67–75.
41. Bai YF, Han XG, Wu JG, Chen ZZ, Li LH (2004) Ecosystem stability and compensatory effects in the Inner Mongolia grassland. Nature 431: 181–184.
42. Jia H, Jiang P, Zhao C, Hu Y, Li Y (2009) The influence of enclosing life on carbon distribution of grassland ecosystem. Agr Res Arid Areas 27: 33–36.
43. Li LH, Han XG, Wang QB, Bai WM, Bai YF, et al. (2002) Soil carbon balance in a native temperate grassland in the Xilin River basin of Inner Mongolia. Acta Bot Sin 44: 740–742.
44. Werth M, Brauckmann HJ, Broll G, Schreiber KF (2005) Analysis and simulation of soil organic carbon stocks in grassland ecosystems in SW Germany. J Plant Nutr Soil Sci 168: 472–482.
45. Billings SA, Ballantyne IVF (2012) How interactions between microbial resource demands, soil organic matter stoichiometry, and substrate reactivity determine the direction and magnitude of soil respiratory responses to warming. Global Change Biol doi:10.1111/gcb.12029.
46. Huxman TE, Snyder KA, Tissue D, Leffler AJ, Ogle K, et al. (2004) Precipitation pulses and carbon fluxes in semiarid and arid ecosystem. Oecologia 141: 254–268.
47. Schwinning S, Sala OE, Loik ME, Ehleringer JR (2004) Thresholds, memory, and seasonality: understanding pulse dynamics in arid/semi-arid ecosystems. Oecologia 141: 191–193.

14

Characterization and Molecular Interpretation of the Photosynthetic Traits of *Lonicera confusa* in Karst Environment

Geng Wu[1,2], Haibo Jia[1], Yongwei Huang[1], Lu Gan[1], Chunhua Fu[1], Libin Zhang[1], Longjiang Yu[1], Maoteng Li[1]*

1 School of Life Science and Technology, Huazhong University of Science and Technology, Wuhan, Hubei, China, **2** State Key Laboratory of Biogeology and Environmental Geology, China University of Geosciences, Wuhan, Hubei, China

Abstract

Lonicera confusa was a medical plant which could adapt to the Ca-rich environment in the karst area of China. The photosynthesis, relative chlorophyll content,differentially expressed genes (DEGs) and differentially expressed proteins (DEPs) of *L. confusa* that cultivated in calcareous and sandstone soils were investigated. The results showed that the relative chlorophyll content and net photosynthesis rate of *L. confusa* in calcareous soil are much higher than that planted in sandstone soil, the higher content of calcium might play a role in keeping the chloroplast from harm and showed higher photosynthesis rate. The transpiration and stomata conductance were decreased in calcareous soil, which might result from the closure of stomata. The GeneFishing and proteomic results showed that the expression of DEGs and DEPs were critical for photosynthesis and stomata closure, such as RuBisCO, photosynthetic electron transfer c and malate dehydrogenase varied in the leaves of *L. confusa* that cultivated in different soils. These DEGs or DEPs were further found to be directly or indirectly regulated by calcium sensor proteins. This study enriched our knowledge of the molecular mechanism of high net photosynthesis rate and lower transpiration of *L. confusa* that cultivated in the calcareous soil in some degree.

Editor: Ji-Hong Liu, Key Laboratory of Horticultural Plant Biology (MOE), China

Funding: The research was supported by the National Science Foundation of China (31270373), Natural Science Funds for Distinguished Young Scholars of Hubei Province of China (2010CDA097), New Century Talents Support Program by the Ministry of Education of China (NCET110172) and Hubei Key Laboratory of Economic Forest Germplasm Improvement and Resources Comprehensive Utilization (2011BLKF247). The funders had no role in study design, data collection and analysis, decision to publish, or preparation of the manuscript.

Competing Interests: The authors have declared that no competing interests exist.

* Email: limaoteng426@mail.hust.edu.cn

These authors contributed equally to this work.

Introduction

The highly ordered and complex process of plant leaf development is influenced by a variety of factors and constrained by several abiotic stresses. The research on the mechanisms of the adaptation to abiotic stress environment in plants is becoming a hotspot in recent years [1–3]. Carbonate rock is the bedrock of karst ecosystems, which occupies about 1/3 of mainland of China [4],the high calcium content and drought are the main characteristics of karst soils [5]. Accordingly, the plants in karst areas have obtained the ability to adapt to the high level of calcium and drought through long-term evolution [6–8]. Research on the mechanisms of plant adaptation to karst environments is necessary for it can provide new ways on the fragile karst ecosystem rehabilitation. *L. confusa* was one of Chinese medicinal plant with a typical ecological value and which was widely cultivated in calcium-rich karst areas of China [9]. Previous studies had showed that the cytosolic free Ca^{2+} could restrict to 10^{-7} M or less in *L. confusa* [10], because of it could excrete the excess of Ca^{2+} via stomata or stored in glands and trichomes under higher Ca^{2+} supplied in karst areas [8].

As the key process for neogenesis of biological material, photosynthesis plays a central role in plant performance under abiotic stresses, and the net photosynthesis rate (Pn) and transpiration rate (E) in most fruit crops could be reduced with a rapid closure of stomata, with the reduction of stomata conductance (Gs) as well [11–13]. As one of essential and major plant nutrients, Ca^{2+} is required to maintain cell wall structure and membrane function [14]. Soil Ca^{2+} depletion could affect some important physiological processes, such as carbohydrate storage, photosynthesis, chlorophyll content and antioxidant enzyme activity [15,16]. It was revealed that the decline of photosynthesis caused by the simulated acid rain treatment could be recovered with high concentration of Ca^{2+} treatment in *Lonicera formosana* [17]. Ca^{2+} are directly involved in several aspects of photosynthesis through modulating phosphatase enzymes activity and regulating chloroplast activity of NAD^+ kinase [18]. Tan *et al* (2011) revealed the photosynthesis is improved by exogenous calcium treatment in heat-stressed tobacco [19]. Research on photosynthesis of plants that adapt to karst areas was performed in recent years, Huang et al (2006) have studied the photosynthesis, transpiration and water use efficiency of pioneer specie *Cornus controversa*, *Zenia insigni* and *Lonicera maackii* in karst area, the results showed that these 3 karst species can process a physiecological drought and thermal adaptation and develop a strategy to escape environmental stress [20], but the molecular mechanisms of photosynthesis in *L. confusa* has not been mentioned [20,21].

In this paper, the variation in net photosynthetic rate (Pn), relative chlorophyll content (Chl), stomata conductance (Gs) and transpiration rate (E) were evaluated in *L. confusa* that cultivated on calcareous and sandstone soils, respectively. The differentially expressed genes (DEGs) and differentially expressed proteins (DEPs) were identified in the meanwhile by GeneFishing PCR and two-dimensional gel electrophoresis (2-DE) in order to explain the molecular mechanism of *L. confusa* that cultivated in different calcium cultivation conditions.

Results

Comparative analysis of Chl content, Pn, E and CI in leaves of *L. confusa* that cultivated in calcareous and sandstone soils

The diurnal variation of environment factors was exhibited in Fig. 1. The average photosynthetic active radiation (PAR) and air temperature were 762.2 μ mol m^{-2} s^{-1} and 35.4°C, respectively. The PAR value was increased continually from 461.2 μmol m^{-2} s^{-1} to 1119.0 μmol m^{-2} s^{-1} from 08:00 to 13:00 and decreased to 111.7 μmol m^{-2} s^{-1} at 18:00. The air temperature was up to highest point at 13:00 (37.12°C) and down at 15:00 (Fig. 1A).

Chlorophyll (Chl) is the molecule in photosynthesis that absorbs sunlight and uses its energy to synthesize carbohydrates from CO_2 and H_2O. Present results revealed that the relative content of Chl in *L. confusa* planted in calcareous soil was higher than that of planted in sandstone soil (Fig. 1A). Further research indicated that the relative Chl contents in two cultivated soils were both decreased to 11.1 and 13.8 at 12:00 noon, and after that time relative content of Chl was increased to 18.1 and 18.0, respectively. It also revealed that the relative Chl content in calcareous soil was relatively stable. Conversely, the relative content of Chl in sandstone soil varied a lot in diurnal, it indicated that the relatively higher calcium supply in calcareous soil is of importance for maintaining the stability of the chlorophyll. Further TEM observation revealed that more small bubble appeared in chloroplast grana of *L. confusa* leaves planted in calcareous soil when the TEM sections were treated with calcium chelator EGTA (Fig. 2A). Conversely, very few bubble appeared in the chloroplast of *L. confusa* that cultivated in sandstone soil (Fig. 2B). These results directly suggested the content of Ca^{2+} in chloroplast of leaves that cultivated in calcareous soil was much higher than that of *L. confusa* leaves planted in sandstone soil, which might keep the chloroplast unharmed under strong solar radiation. The Pn of *L. confusa* in calcareous soil was significantly higher than that cultured in sandstone soil (Fig. 1B), which indicated that *L. confusa* cultivated in calcareous soil might synthesize more carbohydrates daily. From 08:00 to 12:00, the Pn in two cultivated condition was maintained relatively stable and the tendency as a whole decreased after 12:00. The highest Pn of *L. confusa* in calcareous soil was found at 11:00. However, the highest Pn was observed at 9:30 in sandstone soil, and both represent at a lowest value at 14:00. The phenomena mentioned above suggested the relatively higher calcium supply in calcareous soil could improve the Pn of *L. confusa*, which was consistent with the higher relative Chl content in the leaves of *L. confusa* that planted in calcareous soil.

The E of *L. confusa* represented a typical double peak curve under the two culture conditions (Fig. 1C). When cultured in calcareous soil, the E of *L. confusa* arose rapidly from 8:00 AM and the double peak appeared at 9:30 AM and 1:00 PM with the value of 2.1 mmol m^{-2} s^{-1} and 2.6 mmol m^{-2} s^{-1}, respectively. However, the double peak appeared at 12:00 and 14:30 with the value of 3.5 mmol m^{-2} s^{-1} and 2.9 mmol m^{-2} s^{-1} respectively, when cultivated in sandstone soil. Further research revealed that the average value of E in *L. confusa* that cultivated in calcareous soil was 1.73 mmol $m^{-2}s^{-1}$, which was lower than that in sandstone soil (2.17 mmol $m^{-2}s^{-1}$). As we know, the E value were regulated by the degree of stomatal opening state, the stomatal closure would reduce the E value to decrease the water loss. The stomata conductance (Gs) analysis showed that the average Gs was 93.08 mmol m^{-2} s^{-1} and 117.45 mmol $m^{-2}s^{-1}$ under calcareous soil and sandstone soil, respectively. SEM analysis showed that most of stomata in the leaves of *L. confusa* were closed in the leaves planted in calcareous soil, otherwise, half of the stomata were in normal status in the leaves planted in sandstone soil (Fig. 2C, 2D), which was in accordance with E value of *L. confusa* planted in sandstone soil was higher than that of *L. confusa* planted in calcareous soil. The intercellular CO_2 concentration (CI) revealed that the average CI of *L. confusa* planted in calcareous soil was lower than that planted in sandstone soil (Fig. 1D), which was also consistent with the E value mentioned above.

The correlation between E, Gs, Pn were analyzed and showed in Table S3. On the conditions of sandstone, the Pn of *L. confusa* were significantly positive correlated with PAR ($p<0.05$) and significantly negative correlated with CI ($p<0.01$), other environment factors has little effect on Pn, no obvious correlation was found between the E, Gs and environment factors (Table S3). Otherwise, the E and Gs of calcareous soil planted *L. confusa* were significantly correlated with PAR, air temperature and leaf temperature, the Pn was also shown positive correlation with PAR (Table S3). All those results indicated that the *L. confusa* in relative higher level of calcium could enhance its sensitivity to environmental factors.

The identification of DEGs and DEPs through GeneFishing PCR and 2-DE analysis

GeneFishing PCR was performed on *L. confusa* that cultivated in calcareous and sandstone soils by using dT-ACP2 and 20 pairs of random primers, respectively. The PCR products were run on 2% agrose gel and these differentially expressed PCR products were cloned into a TOPO TA cloning vector and followed by sequence analysis (Fig. 3A). Totally, 23 DEGs were observed and which could be classified into 7 groups according to their functions, including DEGs involved in photosynthesis electron transfer chain, carbon fixation in photosynthesis, oxidation reduction reaction, plant stress resistance, chlorophyll synthesis, transposable element and some unknown genes (Fig. 3B, Table 1). Most of these DEGs were highly expressed in leaves of *L. confusa* that cultivated in calcareous soil. The main DEGs were described as fellows. DEG3 (RuBisCO activase, Rca) and DEG5 (glycerlde-3-phosphate dehydrogenase, GAPDH) are the genes involved in carbon fixation in photosynthesis [22]. DEG21 (Photosystem I reaction center subunit IV) and DEG23 (photosynthetic electron transfer c, PETC) are genes that are directly involved in the photosynthesis electron transfer chain. Those four DEGs were all up-regulated in the leaves of *L. confusa* that cultivated in calcareous soil, which could improve Pn and promote carbon fixation in photosynthesis or induce more production of ATP in *L. confusa* cultivated in calcareous soil. DEG17 (SAM synthase), DEG18 (NADH dehydrogenase subunit 2 homology), DEG24 (Hydroxy acid

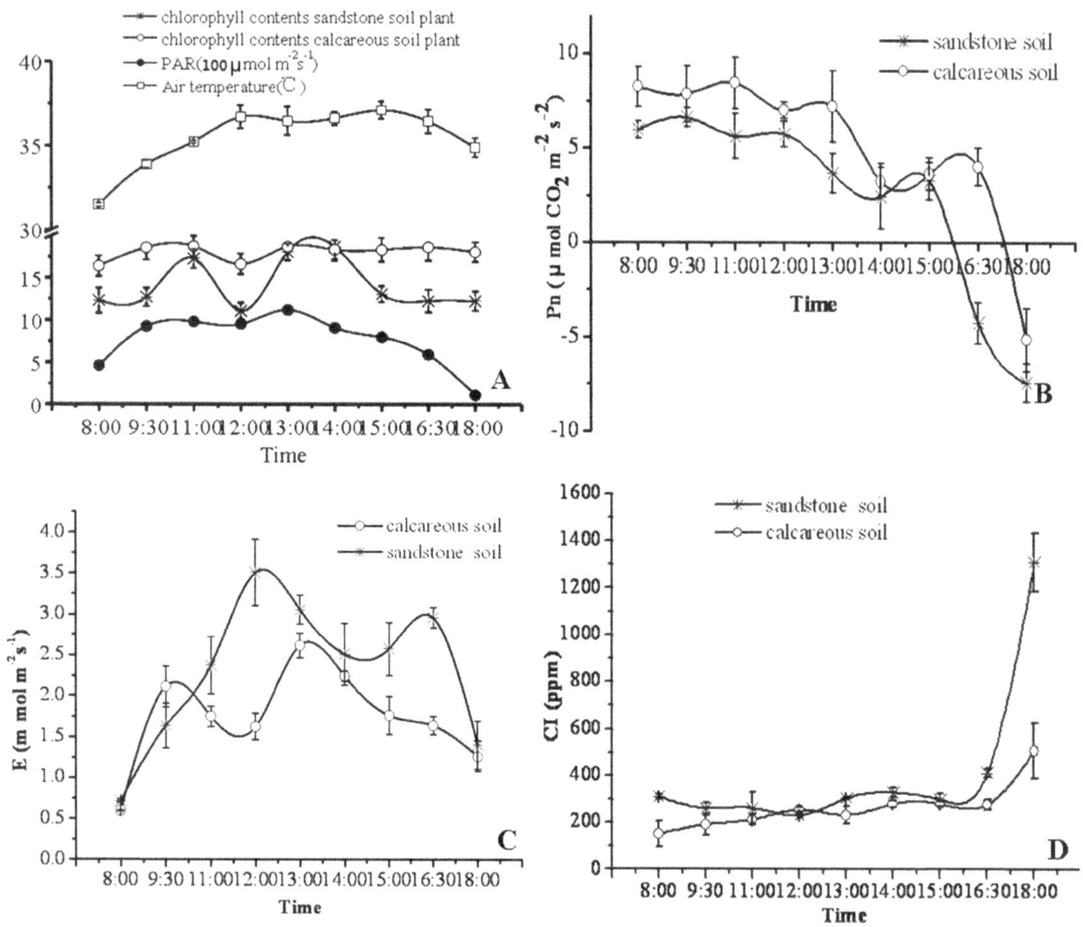

Figure 1. Diurnal variation of environment factors, chlorophyll content, Pn, E, CI. A. Diurnal variation of chlorophyll content, PAR, air temperature; B, C and D represent the Diurnal variation of CI, Pn and E, respectively.

oxidase, HAO) and DEG20 (SlUPTG1) are the DEGs involved in oxidation-reduction reaction and abiotic resistance. DEG17 plays an important role in the production of polyamine and the keep mobility of the membrane. It confirms that SAM syntheses might play a role for the adaptability of *L. confusa* to high calcium stresses. DEG18 located in the upper respiratory chain and transfers electrons to NADH ubiquinone and creates the transmembrane proton gradient to synthesize ATP. The results suggest that photorespiration is proportional to calcium concentration when *L. confusa* is faced with higher calcium content. DEG24 and DEG20 encode HAO and SlUPTG1, which are relate with abiotic resistance were also found.

2-DE was used to identify the DEPs in *L. confusa* that planted in calcareous soil and sandstone soils, respectively (Fig. 4). In all, 25 DEPs expressed more than 2 fold higher and 22 DEPs expressed less than 2 fold lower in calcareous soil were observed (Fig. 4A). The most significant 15 DEPs were performed for MALDI-TOF-MS-MS analysis and 10 DEPs were successfully identified (Table 2). Some up-regulated DEPs and up expressed gene were matched well base on the results that obtained by GeneFishing analysis. For example, the DEPs that identified as RuBisCO large subunit and RuBisCO large polypodiodes formosana subunit were increased about by 14 times (6601) and 27 times (8704), respectively (Fig. 4B). The expression of some DEPs that significantly decreased in calcareous soil environment was also observed (Fig. 4B). Among them, mRNA binding protein precursor decreased by 50%, cell cycle regulated protein 2 (CDC2) homolog decreased by 40%, Zinc finger domain-containing protein (ZBED) decreased by 90% and malate dehydrogenase (MDH) and protochlorophyllide reductase A (PORA) decreased to almost invisible. The highly expressed MDH indicates an increase the E of *L. confusa* that cultivated in sandstone soil. The cell division might be more active in the *L. confusa* that cultivated in sandstone soil for CDC2 were highly expressed.

Some DEGs or DEPs were selected out for RT-PCR analysis in the *L. confusa* that treated with different content of calcium (Fig. 5). The results revealed that the expression of DEG3 (Rca) and DEG5 (GAPDH) was improved with increasing concentration of Ca^{2+}, which is consistent with the Pn value of *L. confusa* that cultivated in calcareous soils. The expression of most DEGs were first increased with the increasing Ca^{2+} treatment no more than 75 mg/L, and then began to decrease when Ca^{2+} were more than 75 mg/L (for example, DEG17, DEG18, DEG21 and DEG23), this indicated that the expressions of most genes were inhibited in *L. confusa* when the Ca^{2+} content were higher than the calcareous soil. Few DEGs expression were decreased with the increased Ca^{2+} treatment, such as DEG24 that coded as thioredoxin H-type, which functions as a disulfide oxidoreductase and involved in lots of redox dependent cellular processes.

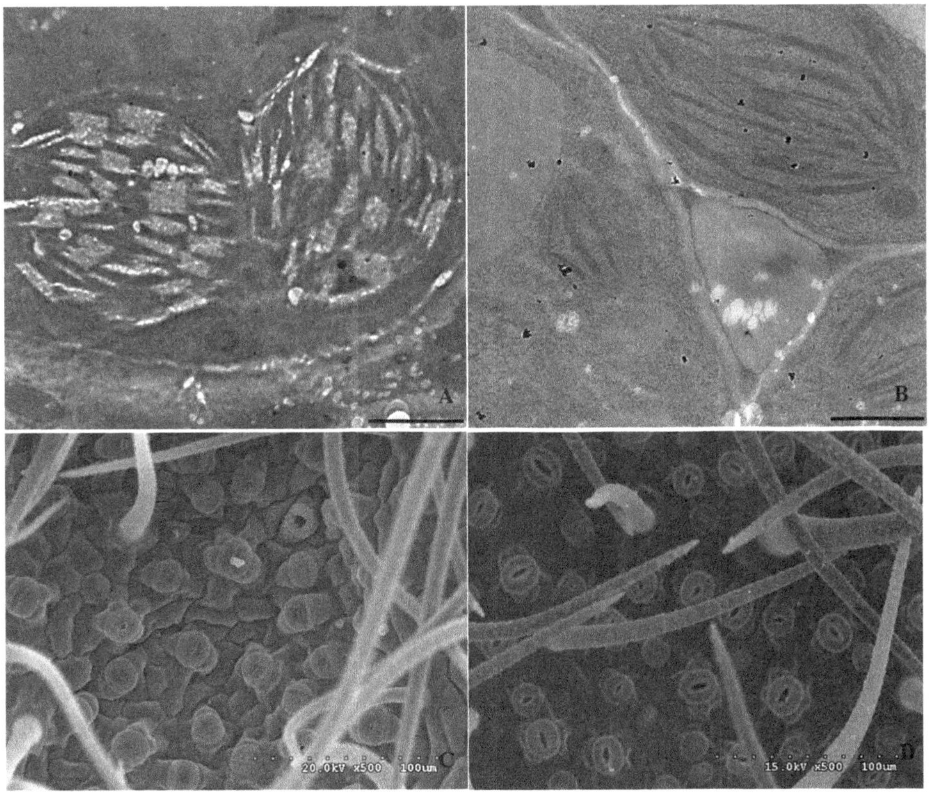

Figure 2. Observation of stomata status by using SEM and Ca^{2+} positioning analysis chelated by EGTA by using TEM. A and B represent the Ca^{2+} positioning analysis in the chloroplast of leaves that cultivated in calcareous soil and sandstone soil that chelated by EGTA, Bar in A and B represent 1000 nm; C and D represent the SEM observation of stomata status of leaves that cultivated in calcareous soil and sandstone soil.

The interaction between DEPs or DEGs with calcium sensor proteins and other genes

A better understanding of the calcium signaling network and the way for the genes that calcium regulated would be very important. Some genes have calcium binding site and could be regulated by calcium directly, and some genes were affected through calcium sensor proteins, such as calcium-dependent protein kinases (CDPKs), calmodulin (CaM) and calcineurin B-like (CBL) proteins [23,24]. Analyzing the interaction between DEGs with calcium sensor proteins would provide valuable information. Complicated network with 574 nodes and 983 edges was constructed using the Cytoscape software, and some of the DEGs or DEPs (for example, Rca, MDH, CDC_2, PORA and PETC) that directly or indirectly connected with calcium sensor proteins were observed (Fig. 6). Form the network, we deduced that the DEG3 (RuBisCO activase, Rca) was directly activated by the calcium sensor protein CDPK, in the meanwhile, the activation of CDPK were dependent on both Ca^{2+} and ATP. These results indicated that the activity of CDPK was improved by relatively higher calcium and then enhanced the expression of RuBisCO activase in the leaves, which would improve the carbon fixation and the Pn in *L. confusa* in calcareous soil. WNK2 was a cytoplasmic serine-threonine kinase, which would be active when its calcium-binding C2 domains was bound with Ca^{2+}, the CDC2 and PORA were both regulated by WNK2. Other calcium sensor proteins that indirectly interacted with DEGs and DEPs were shown in Fig. 6. The network results indicated that some of DEGs and DEPs were surely affected by calcium or calcium sensor proteins.

Discussion

The calcium content in calcareous soil was three to four times than that in non-karst soil [5,14]. A suitable concentration of calcium is required to maintain cell wall structure and membrane function and for photosynthesis [17,25]. Ca^{2+} depletion in soil may influence carbohydrate storage, photosynthesis, chlorophyll content and antioxidant enzyme activity [15,16]. The relative Chl content is one of most important factor in determining the Pn [26]. It demonstrated that low concentration of Ca^{2+} could slightly promote Chl accumulation [27], other research revealed that supplementary Ca^{2+} could ameliorate the negative effects of salinity on chlorophyll and dry mass production in strawberry [28]. The present results showed that the relative Chl content and Pn in *L. confusa* planted in calcareous soil were both higher than those planted in sandstone soil. The Chl was synthesized in chloroplast, when plant were exposed to light that higher than those required for photosynthesis, reactive oxygen species are generated in the chloroplasts and cause photodamage [29]. Plants have developed several protective mechanisms when facing the photodamage, one is chloroplast avoidance movement which actually has a role in reducing light absorption by photosystems under high light [29]. Exogenous Ca^{2+} inhibited the loss of chlorophyll under heat stress possibly by its reducing photo-oxidation or maintaining membrane integrity [30], Ca^{2+} treatment also could increase the synthesis of HS proteins, such as HSP26 and HSP70, which could protect the cells and tissues from damage after heat stress [31]. Our present research revealed the relative Chl content of *L. confusa* planted in calcareous soil changed very little compared with those planted in sandstone soil,

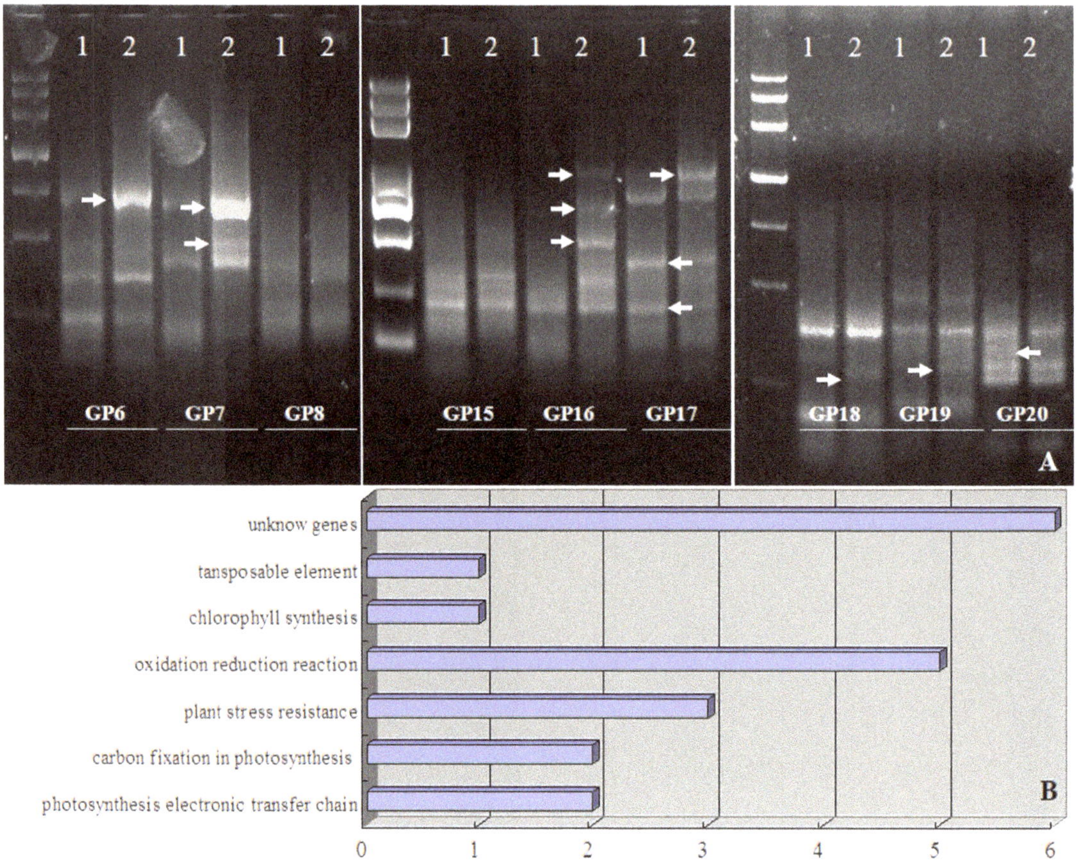

Figure 3. The Genefishing results of leaves and classification of DEGs. A represent the genefishing analysis of leaves that cultivated in sandstone soil (1) and calcareous soil (2), DEG6, DEG7, DEG8, DEG15, DEG16, DEG17, DEG18, DEG19 and DEG20 was the name of different primers; B represent the classification of all the DEGs.

indicating that the calcium has some effect on eliminating the reactive oxygen species and maintaining chloroplast stability [32], which was in accord with the higher expressed DEP of HAO (hydroxy-acid oxidase) in the *L. confusa* that planted in the calcareous soil.

Ca^{2+} plays a critical role in responding to environmental signals, activating or inactivating the expression of photosynthesis related genes [33], it was also suggested that Pn could be improved by $CaCl_2$ in heat stressed plants [19]. $CaCl_2$ pretreatment could improve RuBisCO activity under lower temperature, and its higher activity was associated with higher Pn [19,25], Previous study also revealed that Pn was inhibited by moderate heat stress due to the decrease of the activation of RuBisCO [34]. The present results showed that the Pn of *L. confusa* planted in calcareous soil was higher than that planted in sandstone soil at relatively higher temperature. Some up-regulated DEGs in *L.*

Table 1. Blastn analysis of DEGs that involved in photosynthesis in different cultivated conditions.

DEG	Accession	Description	Max score	E value	Max ident
3	HM773394.1	*Musa* AB Group RuBisCO activase (Rca) mRNA, partial cds	138	2e-29	80%
5	XM_002519612.1	*Ricinus communis* (S)-2-hydroxy-acid oxidase, putative	188	1e-44	82%
17	L36680.1	*Pisum sativum* S-adenosylmethionine synthase mRNA	322	1e-84	86%
18	AY059007.1	*Maticora bivirgata* NADH dehydrogenase subunit 2 gene, complete cds; mitochondrial gene for mitochondrial product	41.0	2.8	83%
20	AY622990.1	*Lycopersicon esculentum* UDP-glucose: protein transglucosylase-like protein SlUPTG1 mRNA, complete cds	66.2	1e-07	88%
21	XM_002521115.1	*Ricinus communis* photosystem I reaction center subunit IV A, chloroplast precursor, mRNA	134	2e-28	88%
23	NM_178964.2	*Arabidopsis thaliana* PETC (Photosynthetic Electron Transfer C)	266	2e-67	79%
24	XM_002534085.1	*Ricinus communis* Thioredoxin H-type, mRNA	141	1e-30	75%

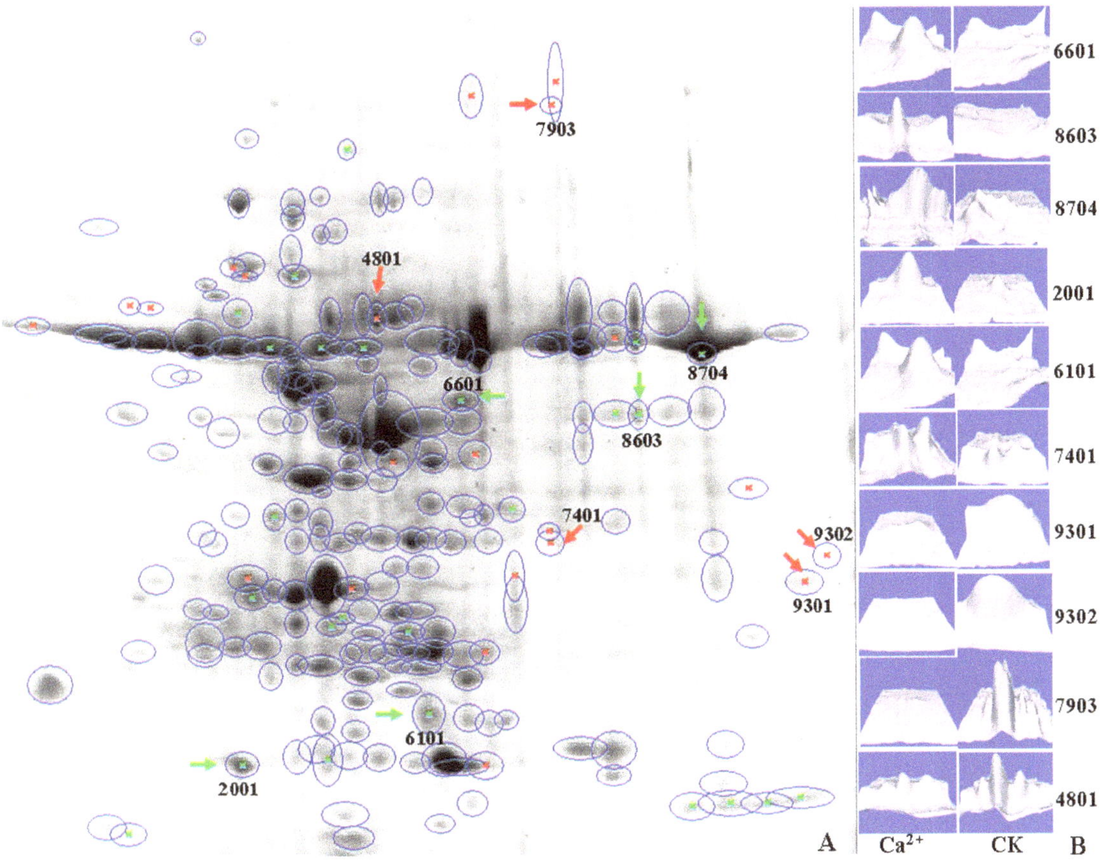

Figure 4. 2-DE analysis of *L. confusa* under different cultivated conditions. A represent the 2-DE map that showed the up and down expressed proteins in *L. confusa* that cultivated in calcareous soil (green and red arrow represent the protein that up-regulated and down-regulated in calcareous soil, respectively); B represent the three-dimensional profiles of the individual spots comparing control and Ca^{2+} treated profiles of each of the ten protein spots that showed significant changes.

confusa planted in calcareous soil were identified as RuBisCO activase (Rca) and glycerlde-3-phosphate dehydrogenase, which were considered to be the key limiting factors in recycling of mitochondrial CO_2 for carbon fixation in chloroplasts, this results indicated that relatively higher Ca^{2+} could improve photosynthesis and increase carbon fixation by improve the activity of Rca and

Table 2. The identity of DEPs of *L. confusa* that cultivated in different conditions.

Number	Accession number	The type of proteins	Molecular weight	PI	The relative contents of protein	
					sandstone soil	calcareous soil
6601	gi|8117180	RuBisCO large subunit	50625.5	6.37	234.4	3287.2
8603	gi|1707878	aminomethyltransferase	44248.7	8.77	0	1856.7
8704	gi|131971	RuBisCO large polypodiodes formosana submit	50910.8	6.23	332.4	9089.2
2001	gi|38344034	peroxiredoxin	28604.6	5.17	1902.9	5726.7
6101	gi|4490714	kinesin-related protein katB	84249.8	5.48	399.2	4515.4
7401	gi|26453355	mRNA binding protein precursor	43913.3	7.1	2331.3	1295
9301	gi|1170897	Malate dehydrogenase, glyoxysomal precursor	37714.9	8.82	4076.8	1686.3
9302	gi|15239574	Protochlorophyllide reductase A	43835.6	9.42	5517.6	0
7903	gi|42566188	Zinc finger domain-containing protein	73118.9	5.39	2769.2	25.3
4801	gi|1168812	cell cycle regulated protein 2 homolog (p34cdc2)	10664.7	8.82	8121.5	3033.3

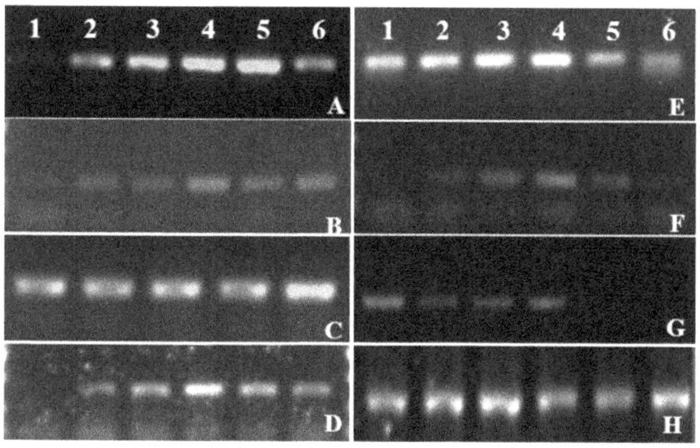

Figure 5. The RT-PCR analysis of DEGs in the leaves of *L. confusa* that treat with different concentration of Ca^{2+}. A to H represent the RT-PCR analysis of DEG1, DEG3, DEG5, DEG17, DEG21, DEG23, DEG24 and 18s rRNA, respectively. (1 to 6 represent the *L. confusa* was treated with 0 mg/L, 25 mg/L, 50 mg/L, 75 mg/L, 100 mg/L, 125 mg/L calcium chloride, respectively).

GAPDH. Some DEPs in *L. confusa* planted in calcareous soil were also identified as RuBisCO large subunit and RuBisCO large polypodiodes formosana subunit. Ca^{2+} addition could efficiently protect chrysanthemum leaves from the damage in photosynthetic apparatus under short term high temperature stress, and the Pn and PS II electron transport were increased by 31.11% and 21.88%, respectively [35]. Tan et al (2011) revealed that $CaCl_2$ pretreatment greatly increased PI_{ABS} under heat stress, they deduced that $CaCl_2$ pretreatment might play an important role in repairing the PS II complex and maintaining higher photosynthetic activity [19]. Some potential chloroplast targets of CaM could regulate include PSAN (a subunit of PSI) as well as the chloroplast homologue of CPN10 (Chaperonin 10), which was involved in the assembly of ribulose-1,5- bisphosphate carboxylase/oxygenase, thereby providing a further association of calcium regulation to CO_2 fixation [36,37]. Two up-regulated DEGs involved in photosynthesis electron transfer chain identified as PsaE and PETC were observed in *L. confusa* planted in calcareous soil, which further confirmed that higher Ca^{2+} in calcareous soil could be conducive to photosynthesis in higher temperature. Nunes-Nesi *et al* (2005) revealed that suppression of mitochondrial malate dehydrogenase in tomato plants leads to an unexpected increase in the rate of photosynthesis, which was consistent with low expression of DEPs that coded as malate dehydrogenase in *L. confusa* planted in calcareous soil [38].

Stomata form pores on leaf surfaces that regulate the uptake of CO_2 for photosynthesis and the loss of water vapor during transpiration [39]. Free cytosolic Ca^{2+} were increased in response to a high extracellular Ca^{2+} level through a CAS signalling pathway and finally leads to stomatal closure [40]. Present research revealed that E and Gs were decreased in *L. confusa* planted in calcareous soil, which indicated that long-term Ca^{2+} transients and lower malate more likely enhance stoma closure in the leaves of the *L. confusa*, which was also verified by SEM analysis on stomata status and highly expressed HAO and MDH in the *L. confusa* that cultivated in calcareous soil and sandstone soil, respectively. Cell-type specific changes in cytosolic calcium levels were observed in *Arabidopsis* root cells in response to drought [41], *SNAC1* play a important role in environmental stresses [32], the strong induction of *SNAC*1 in plants guard cells suggested that the increasing stomata closure is possible regulated by *SNAC*1, which could also reduce the water loss with increasing stomatal closure [42]. Increasing the expression of SAM synthase could improve the resistance to the chilling and salt stress [43], the plant photosynthesis is restrained when facing salt stresses, the expression of enzymes related with photorespiration is up-regulated to meet the energy requirements [44]. In this study, the highly expressed DEGs that coded as *SNAC* and SAM synthase were significantly induced when planted in calcareous soil, which indicated the higher abiotic resistance of *L. confusa* that planted in calcareous soil. The CI in leaves of *L. confusa* planted in calcareous soil was lower than that of planted in sandstone soil, it indicated more CO_2 was fixed, which was correlated with higher activity of RuBisCO in leaves of *L. confusa* planted in calcareous soil [19]. The stomata and non-stomata limitation during the diurnal course of photosynthesis was determined by the development trends of intercellular CO_2 concentration and Gs [45,46].

Calcium sensor proteins (such as, CDPK, CBL and CAM) were discovered to play a crucial role in abiotic stress signaling in plants, stimulus-specific Ca^{2+} signatures are decoded by Ca^{2+} binding proteins that function as Ca^{2+} sensors [47–49]. Qiu *et al* (2007) revealed that a higher level of CDPK activity occurred concurrently with the accumulation of photosynthetic enzymes [50], further study also showed that some transcripts linked to photosynthesis were observed when modulate the CDPK activity in sorghum [51]. The present network revealed that Rca was directly regulated by the downstream of Ca^{2+} target protein CDPK, it indicated that when Ca^{2+} was bound to Ca^{2+} sensors, the CDPKs would change their conformation and interact with RCA to regulate the Pn. ATP and NADPH synthesis via linear photosynthetic electron transfer or solely ATP production via cyclic electron flow was important in photosynthesis [52], the chloroplasts contribute to cellular Ca^{2+} signaling via the chloroplast-localized Ca^{2+} sensor protein CAS [48]. Petroutsos *et al* (2011) also demonstrated that CAS and Ca^{2+} are critically involved in the regulation of the high light response and particularly in the control of LHCSR3 expression [53]. The DEGs that confirmed as PETC and ND2 were observed in the present study, it indicated the high level of Ca^{2+} could bind to Ca^{2+} sensor protein and further improve the activity of PETC and ND2 in the leaves of *L. confusa*. The DEPs of ZBED and PORA were both highly expressed in *L. confusa* cultivated in the sandstone soil, and their functions in *L. confusa* cultivated in sandstone soil remained to be investigated in the future studies. The DEGs or DEPs that observed in this study had close relationship with photosynthesis regulation and calcium signaling could enrich the knowledge of plant adaptation to karst environments.

Materials and Methods

Plant materials and growth conditions

The *L. confusa* cultivars were taken from Nongla Karst Experimental Site, Institute of Karst Geology, Chinese Academy of Geological Science (108°19′ E,23°29′ N). The plants were cultivated in the greenhouse and were divided into two groups (15 plants per group): one group was transplanted into Ca^{2+}-rich calcareous soil that directly transported from Nongla Karst Experimental Site (Ca^{2+} content 3.1±0.05 g per 100 g soil, the pH value is 7.8±0.2), another group was transplanted into Ca^{2+}-poor sandstone soil (Ca^{2+} content 0.02±0.005 g per 100 g soil, the pH value is 6.7±0.3). Average cultivar height is almost 20 cm and one plant was cultivated per pot (The pot diameter is 25 cm). Thus leaves were becoming mature from the seventh leaf onward. During the experimental period from 10 March to 20 July, the average diurnal air temperatures between 19.84°C (day) and

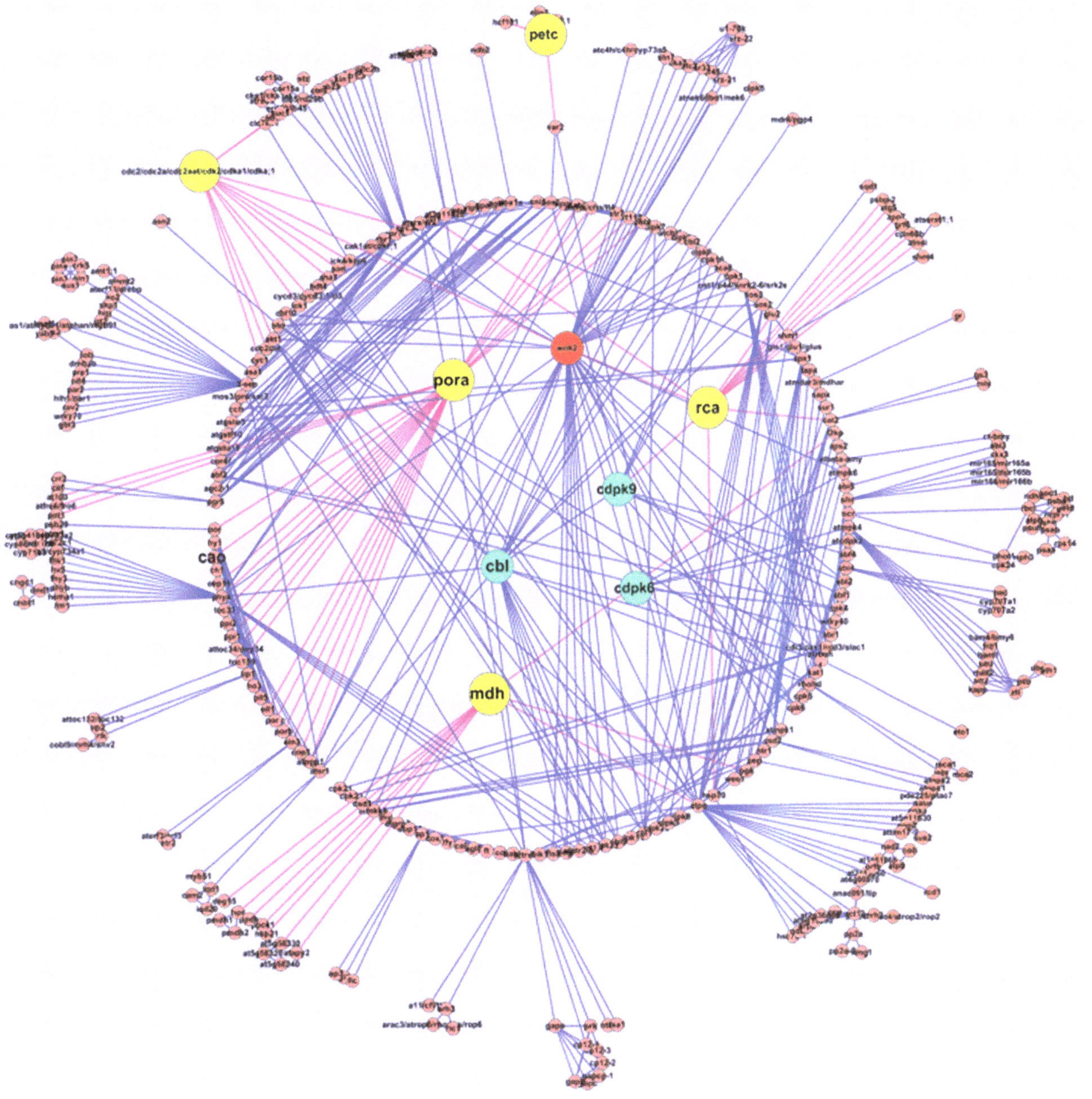

Figure 6. The network of DEGs or DEPs with calcium sensor proteins by using Cytoscape software. Yellow and green cycles represent the DEGs or DEPs and calcium sensor proteins, respectively; the pink lines represent the directly interacted genes with DEGs or DEPs.

9.5°C (night), respectively, with air humidity fluctuating between 45% and 85% average [8]. The mature leaves were used for photosynthesis, relative chlorophyll content in vivo. The same development stage mature leaves from each group were cut and mixed together, and transferred to liquid nitrogen immediately for Genefishing PCR and proteomics analysis.

Photosynthesis and relative chlorophyll measurement

Fully expanded mature leaves cultivated in Ca^{2+}-rich calcareous soil and Ca^{2+}-poor sandstone soil were sampled for measurement, respectively. The measurements were taken from 08:00 AM to 18:00 PM in July on plants under clear sky ambient sunlight, the values stated are mean values from three days. The fully expanded leaves of five different plants by using an open system photosynthetic gas analyzer (PP Systems Inc. model TPS-1, Amesbury, MA, USA) to measure the net photosynthesis rate (Pn), stomata conductance (Gs), transpiration rate (E) and intercellular CO_2 concentrations (CI) of plants in the greenhouse [54]. The chlorophyll content was measured (10 repeats per leaf) by using the CL-01 chlorophyll content meter, which determines relative chlorophyll content using dual wavelength optical absorbance (620 and 940 nm) measurements from leaf samples (Hansatech Instruments, Norfolk, UK) [55,56]. The measurement Data were analysed using SPSS version 12.0 statistical software. Probability (p) values of <0.05 were considered significant. The graph data were processed using origin 7.0 software (Microcal Software, Inc.,Northampton, MA, USA).

Transmission electron microscope (TEM) and scanning electron microscope (SEM) analysis of leaves

For TEM analysis, mmature leaf samples of *L. confusa* that cultivated with Ca^{2+}-rich calcareous soil and Ca^{2+}-poor sandstone soil were collected and washed with phosphate buffer solution, cut into 0.5 cm×0.5 cm slices and immediately immersed for fixation in 2.5% glutaraldehyde (v/v, pH = 7.8) with 0.2 M cacodylate and 2% (w/v) buffered osmium tetroxide. The samples were dehydrated through a graded ethanol series (30%, 50%, 70%, 85%, 95%, 100%) for 10 min each and in 100% isoamyl acetate twice for 20 min. Tissues were then vacuumed, incubated for 4 hours at 4°C and fixed with 2% potassium pyroantimonate (pH = 7.8) for 16 hours at 44°C. Tissues were then rinsed with PBS solution, dehydrated with graded acetone solution, embedded with Spurrs epoxy resin and sectioned with superfine section machine. For the calcium positioning, the slices were incubated in 100 mM EGTA (pH 8.0) solution at 60°C for 1 hour and stained with uranium acetate for observation followed by Tian *et al* [57].

For SEM analysis, mature leaves of *L. confusa* samples were first treated through 10%, 20%, 30%, 50% and 70% of ethanol and then immersed in 100% acetone twice. The dehydrated samples were treated with CO_2 critical point dryer. The dried samples were mounted on brass disks coated with Pt under vacuum. The ultrastructure of samples were observed by SEM [8].

Identification of differentially expressed genes (DEGs) by using GeneFishing PCR

Mature *L. confusa* leaves cultivated in Ca^{2+}-rich calcareous soil and Ca^{2+}-poor sandstone soil were chosen for RNA extraction using TRIzol (Invitrogen) according to the manufacturer's instructions. The RNA pellets were frozen and stored at −80°C until use. Differential display PCR was performed using the GeneFishing kit according to the manufacturer's instructions (Seegene, Inc.) and which was performed for three replications. The PCR products were separated in 1.2% agarose gel, only the differentially expressed bands that repeated existed in three replications were selected for further sequence analysis. The differentially expressed bands were extracted from the gel using a QIAquick Gel extraction kit (Qiagen) and directly cloned into a pGEM-T Easy vector for sequencing. Semi-quantitative RT-PCR was used for confirmation of the above Genefishing results. 18S rRNA was used as an internal control. The sequence of primers was showed in Table S1.

Identification of differentially expressed proteins (DEPs) by using two-dimensional gel electrophoresis (2-DE)

Mature leaves of *L. confusa* were collected and immediately freezed and stored at −80°C. For total protein extraction, samples were ground in liquid nitrogen to fine power, and the protein extraction method as outlined by Damerval *et al* [58]. Protein concentration was determined using the RC DC Protein Assay Kit I (Bio-Rad (500-0122), Hercules, CA, USA) with bovine serum albumin as standard according to the manufacturer's instructions. For 2-DE analysis, a volume of 300 μL rehydration buffer containing 1 mg of protein was loaded onto linear pH gradient (IPG) strips (ReadyStrip 170 mm, pH 3–10, Bio-Rad, Hercules, CA, USA). Isoelectric focusing (IEF) was performed by using a Protean IEF Cell (Bio-Rad, Hercules, CA, USA) system and followed their operation manual. The isoelectric focusing system were: 50 V for 1 h, 200 V for 1 h, 500 V for 1 h, 1000 V for 1 h, a linear increase of voltage to 10,000 V for 4 h, 10000 V for 8 h achieving approximately 80 000 Vhr. After IEF, the strips were then equilibrated twice for 15 min each in an equilibration buffer as described [59]. The second dimension separation of proteins was performed on 10% SDS-PAGE gels using a Bio-Rad PROTEIN II xi cell system. After electrophoresis, the gels were stained with colloidal Coomassie brilliant blue (CBB) G-250 according to Candiano et al [60]. The stained gels were scanned with a UMAX Powerlook 2100XL Imaging System with a resolution of 300 dpi and processed using the PDQuest 8.01 software (Bio-Rad Laboratories, Hercules, CA). The spots that changed in abundance more than two-fold and the least significant difference performed more than 95% ($p<0.05$) were selected for protein identification.

The MS/MS analysis (MALDI-TOF) for DEPs were performed as described by Li *et al* [61]. The DEPs were dug from the gels, placed in the 1.5 mL EP tube, washed twice with pure water, and decolorized with the same volume of decoloring liquid (30 mM, 100 mM potassium ferrocyanide and sodium thiosulfate, 100 mM ammonium bicarbonate, pH 8.0) for 20 min; Proteins were then washed with pure water again and dehydrated with acetonitrile, vacuum dried, and was added digested with 20 μL trypsin (20 ng/μL) at 37°C overnight; For dissolution, 50 μL serine extract (50% acetonitrile, 5% trifluoro ethyl) was added for 15 min, repeat twice, vacuum dry and 10 μL 0.1% trifluoro ethyl was added for dissolution. Then mix 2 μl of the above extraction liquid with 2 μl Mass spectrum loading buffer (Alpha-cyanogen-4-hydroxyl, 50% acetonitrile, 0.1% three fluorine acetic acid) and analyzed with MALDI-TOF-MS Autoflex II. The settings were as follows: 20 kV accelerating voltage, 60–65% grid voltage, 160 ns delay, 200/spectrum of the laser make protein point into peptides about 1000–4000 Da and obtain mass spectrometry results. The protein was identified according to peptide fingerprint spectra combined with the MASCOT search engine (http://www.matrixscience.com). Search parameters were set as: taxonomy, Virdiplantae; enzyme, trypsin; max missed cleavages, 1; fixed modification, carbamidometyl (C); variable modifications, oxidation (M); fragment mass tolerance, ±0.2 Da; and mass accuracy, 50 ppm.

RT-PCR analysis of DEGs and DEPs of *L. confusa* with different concentration of $CaCl_2$ treatment

2Hydrochemistry analysis showed that the average concentration of Ca^{2+} in Landiantang spring of karst area of Nongla was 75.20 mg/L in the year of 2003 and 2004 [62]. In this experiment, the successive concentration of $CaCl_2$ solution was designed to treat the *L. confusa* plants, and use 75 mg/L as tipping point. Twenty four pots of materials were separated into six groups, washed with pure water for a few times. For RT-PCR analysis of the DEGs that obtained by Genefishing in different $CaCl_2$ solutions (treated with 0 mg/L, 25 mg/L, 50 mg/L, 75 mg/L, 100 mg/L, 125 mg/L $CaCl_2$ solutions), fresh mature leaves mixed samples for treatment with 36 h were taken and preserved in −80°C fridge immediately. The primers that used for RT-PCR was listed in Table S2. The RT-PCR conditions was as follows, the samples were first treated with 95°C and followed with 30 cycles of 60 s of 94°C denaturation step, 40 s of 68°C annealing step and 40 s of 72°C extension. 18S rRNA was used as an internal control.

Network analysis of DEGs or DEPs with calcium sensor proteins

All sequences of DEGs were used to perform BLAST analysis. According to the conservatism of the gene, we deemed that DEGs should be the highest homology gene. The functions of DEGs or DEPs were analyzed using Kyoto Encyclopedia of Genes and Genomes (KEGG) database (http://www.genome.jp/tools/blast). The gene interaction between these Ca^{2+} induced DEGs or DEPs with calcium sensor proteins was as follows. The plugin of Agilent Literature Search in Cytoscape software (http://www.cytoscape.org/plugins/index.php) was used.

Author Contributions

Conceived and designed the experiments: ML. Performed the experiments: GW HJ YH. Analyzed the data: LG LZ ML. Contributed reagents/materials/analysis tools: LG CF LY. Contributed to the writing of the manuscript: GW HJ ML.

References

1. Miller A, Schlagnhaufer C, Spalding M, Rodermel S (2000) Carbohydrate regulation of leaf development: Prolongation of leaf senescence in Rubisco antisense mutants of tobacco. Photosynth Res 63: 1–8.
2. Ouyang B, Yang T, Li H, Zhang L, Zhang Y, et al. (2007) Identification of early salt stress response genes in tomato root by suppression subtractive hybridization and microarray analysis. J Exp Bot 58: 507–20.
3. Davis PA, Caylor S, Whippo CW, Hangarter RP (2011) Changes in leaf optical properties associated with light-dependent chloroplast movements. Plant Cell Environ 34: 2047–59.
4. Yue Y, Wang K, Zhang B, Chen Z, Jiao Q, et al. (2010) Exploring the relationship between vegetation spectra and eco-geo-environmental conditions in karst region, Southwest China. Environ Monit Assess 160: 157–68.
5. Cao J, Yuan D, Pan G (2003) Some soil features in karst ecosystem. Adv in Earth Sci 18: 37–44.
6. Yuan DX (2001) On the karst ecosystem. Acta Geol Sin-Engl 75: 336–338.
7. Wu G, Fu C, Huang Y, Li W, Yu L, et al. (2011) Calcium Salt Excreted by Stoma and Its Biomineralization in *Lonicera confusa* Under a Calcium-rich Environment. Chin Bull Bot 46: 658–664.
8. Wu G, Li M, Zhong F, Fu C, Sun J, et al. (2011) *Lonicera confusa* has an anatomical mechanism to respond to calcium-rich environment. Plant Soil 338: 343–353.
9. Yao X, Gongyu G, Chen G (2006) Determination of active constituents in *Lonicera confusa DC* by capillary electrophoresis with amperometric detection. Biomed Chromatogr 20: 1192–9.
10. Kretsinger RH (1976) Evolution and function of calcium-binding proteins. Int Rev Cytol 46: 323–93.
11. Yang Y, Zheng Q, Liu M, Long X, Liu Z, et al. (2012) Difference in sodium spatial distribution in the shoot of two canola cultivars under saline stress. Plant Cell Physiol 53: 1083–92.
12. Deng Y, Li C, Shao Q, Ye X, She J (2012) Differential responses of double petal and multi petal jasmine to shading: I. Photosynthetic characteristics and chloroplast ultrastructure. Plant Physiol Biochem 55: 93–102.
13. Roy A, Sahoo D, Tripathy BC (2013) Involvement of phytochrome A in suppression of photomorphogenesis in rice seedling grown in red light. Plant Cell Environ 36: 2120–34.
14. Liu Z, Groves C, Yuan D, Meiman J (2004) South china karst aquifer storm-scale hydrogeochemistry. Ground water 42: 491–499.
15. Gabara B, Sklodowska MW, Glinska S, Gapinska M (2003) Changes in the ultrastructure of chloroplasts and mitochondria and antioxidant enzyme activity in *Lycopersicon esculentum Mill.* leaves sprayed with acid rain. Plant Sci 164: 507–516.
16. Kobe RK, Likens GE, Eagar C (2002) Tree seedling growth and mortality responses to manipulations of calcium and aluminum in a northern hardwood forest. Can J For Res 32: 954–966.
17. Liu TW, Wu FH, Wang WH, Chen J, Li ZJ, et al. (2011) Effects of calcium on seed germination, seedling growth and photosynthesis of six forest tree species under simulated acid rain. Tree Physiol 31: 402–13.
18. Brand JJ, Becker DW (1984) Evidence for direct roles of calcium in photosynthesis. J Bioenerg Biomembr 16: 239–49.
19. Tan W, Meng Q, Brestic M, Olsovska K, Yang X (2011) Photosynthesis is improved by exogenous calcium in heat-stressed tobacco plants. J Plant Physiol 168: 2063–71.
20. Huang Y, Wang X, Lu S, Wang Q, P Z (2006) Studies of photosynthesis, transpiration and water use efficiency of some dominant species in rocky desert area, Guangxi, China. Guihaia 26: 171–177.
21. He C, Huang Y, Li X, Wang X, Wang Q (2007) The ecophysiological traits of three karst rockey desert restoration species. Guihaia 27: 53–61.
22. Taiz L, Zeiger E (2010) Photosynthesis: The Carbon Reactions. Plant Physiology, Fifth Edition.
23. Luan S, Kudla J, Rodriguez-Concepcion M, Yalovsky S, Gruissem W (2002) Calmodulins and calcineurin B-like proteins: calcium sensors for specific signal response coupling in plants. Plant Cell 14 Suppl: S389–400.
24. Hrabak EM, Chan CW, Gribskov M, Harper JF, Choi JH, et al. (2003) The *Arabidopsis* CDPK-SnRK superfamily of protein kinases. Plant Physiol 132: 666–80.
25. You J LJ, Yang W (2002) Effects of Ca^{2+} on photosynthesis and related physiological indexes of wheat seedlings under low temperature stress. Acta Agro Sin 28: 693–696.
26. Garty J, Tamir O, Hassid I, Eshel A, Cohen Y, et al. (2001) Photosynthesis, chlorophyll integrity, and spectral reflectance in lichens exposed to air pollution. J Environ Qual 30: 884–93.
27. Tanaka A, Tsuji H (1980) Effects of calcium on chlorophyll synthesis and stability in the early phase of greening in cucumber cotyledons. Plant Physiol 65: 1211–5.
28. Khayyat M, Rajaee S, Sajjadinia A, Eshghi S, Tafazoli E (2009) Calcium Effects on Changes in Chlorophyll Contents, Dry Weight and Micronutrients of Strawberry (*Fragaria × ananassa Duch.*) Plants Under Salt-stress Conditions. Fruits 64: 53–59.
29. Kasahara M, Kagawa T, Oikawa K, Suetsugu N, Miyao M, et al. (2002) Chloroplast avoidance movement reduces photodamage in plants. Nature 420: 829–32.
30. Coria N, Sarquis J, Penalosa I, Urzua M (1998) Heat-induced damage in potato (*Solanum tuberosum*) tubes: membrane stability, tissue viability, and accumulation of glycoalkaloids. J Agric Food Chem 46: 4524–4528.
31. Liu HT, Li B, Shang ZL, Li XZ, Mu RL, et al. (2003) Calmodulin is involved in heat shock signal transduction in wheat. Plant Physiol 132: 1186–95.
32. Selth LA, Dogra SC, Rasheed MS, Healy H, Randles JW, et al. (2005) A NAC domain protein interacts with tomato leaf curl virus replication accessory protein and enhances viral replication. Plant Cell 17: 311–25.
33. Zhu S, Zhou X, Wu X, Jiang Z (2013) Structure and function of the CBL-CIPK Ca^{2+} -deconding system in plant calcium signaling. Plant Mol Biol Rep 31: 1193–1202.
34. Crafts-Brandner SJ, Salvucci ME (2002) Sensitivity of photosynthesis in a C4 plant, maize, to heat stress. Plant Physiol 129: 1773–80.
35. Sun XZ, Guo XF, Zheng CS, Wang WL, Liang F (2008) Effects of exogenous Ca^{2+} on leaf photosynthetic apparatus and active oxygen scavenging enzyme system of chrysanthemum under high temperature stress. Chin J Appl Ecol 19: 1983–8.
36. Rocha AG, Vothknecht UC (2012) The role of calcium in chloroplasts–an intriguing and unresolved puzzle. Protoplasma 249: 957–66.
37. Bertsch U, Soll J (1995) Functional analysis of isolated cpn10 domains and conserved amino acid residues in spinach chloroplast co-chaperonin by site-directed mutagenesis. Plant Mol Biol 29: 1039–55.
38. Nunes-Nesi A, Carrari F, Lytovchenko A, Smith AM, Loureiro ME, et al. (2005) Enhanced photosynthetic performance and growth as a consequence of decreasing mitochondrial malate dehydrogenase activity in transgenic tomato plants. Plant Physiol 137: 611–22.
39. Ng CK, Carr K, McAinsh MR, Powell B, Hetherington AM (2001) Drought-induced guard cell signal transduction involves sphingosine-1-phosphate. Nature 410: 596–9.
40. Wang WH, Yi XQ, Han AD, Liu TW, Chen J, et al. (2012) Calcium-sensing receptor regulates stomatal closure through hydrogen peroxide and nitric oxide in response to extracellular calcium in *Arabidopsis*. J Exp Bot 63: 177–90.
41. Kiegle E, Moore CA, Haseloff J, Tester MA, Knight MR (2000) Cell-type-specific calcium responses to drought, salt and cold in the *Arabidopsis* root. Plant J 23: 267–78.
42. Hu H, Dai M, Yao J, Xiao B, Li X, et al. (2006) Overexpressing a NAM, ATAF, and CUC (NAC) transcription factor enhances drought resistance and salt tolerance in rice. Proc Natl Acad Sci U S A 103: 12987–92.

43. Roxas VP, Smith RK Jr, Allen ER, Allen RD (1997) Overexpression of glutathione S-transferase/glutathione peroxidase enhances the growth of transgenic tobacco seedlings during stress. Nat Biotechnol 15: 988–91.
44. Ludlow M, Sommer K, Flower D, Ferraris R, So H (1989) Influence of root signals resulting from soil dehydration and high soil strength on the growth of crop plants. Current Topics Biochem. Physiol. 8: 81–99.
45. Xu D (1997) Some problems in stomatal limitation analysis of photosynthesis. Plant Physiol Commun. 33: 241–244.
46. Farquhar G, Sharkey T (1982) Stomatal conductance and photosynthesis. Annu. Rev. Plant Physiol. 33: 317–345.
47. Cheng SH, Willmann MR, Chen HC, Sheen J (2002) Calcium signaling through protein kinases. The *Arabidopsis* calcium-dependent protein kinase gene family. Plant Physiol 129: 469–85.
48. Kudla J, Batistic O, Hashimoto K (2010) Calcium signals: the lead currency of plant information processing. Plant Cell 22: 541–63.
49. Dodd AN, Kudla J, Sanders D (2010) The language of calcium signaling. Annu Rev Plant Biol 61: 593–620.
50. Qiu QS, Hardin SC, Mace J, Brutnell TP, Huber SC (2007) Light and metabolic signals control the selective degradation of sucrose synthase in maize leaves during deetiolation. Plant Physiol 144: 468–78.
51. Mall TK, Dweikat I, Sato SJ, Neresian N, Xu K, et al. (2011) Expression of the rice CDPK-7 in sorghum: molecular and phenotypic analyses. Plant Mol Biol 75: 467–79.
52. Terashima M, Petroutsos D, Hudig M, Tolstygina I, Trompelt K, et al. (2012) Calcium-dependent regulation of cyclic photosynthetic electron transfer by a CAS, ANR1, and PGRL1 complex. Proc Natl Acad Sci U S A 109: 17717–22.
53. Petroutsos D, Busch A, Janssen I, Trompelt K, Bergner SV, et al. (2011) The chloroplast calcium sensor CAS is required for photoacclimation in *Chlamydomonas reinhardtii*. Plant Cell 23: 2950–63.
54. Hughes NM, Morley CB, Smith WK (2007) Coordination of anthocyanin decline and photosynthetic maturation in juvenile leaves of three deciduous tree species. New Phytol 175: 675–85.
55. Chaerle L, Pineda M, Romero-Aranda R, Van Der Straeten D, Baron M (2006) Robotized thermal and chlorophyll fluorescence imaging of pepper mild mottle virus infection in *Nicotiana benthamiana*. Plant Cell Physiol 47: 1323–36.
56. Peng T, Li P, Jia Y, Gao H (2006) Two noninvasive methods for determining pigment content in plant leaves. Plant Physiol Commun 42: 83–86.
57. Tian H, Russell S (1997) Calcium distribution in fertilized and unfertilized ovules and embryo sacs of *Nicotiana tabacum L.* Planta 202: 93–105.
58. Damerval C, Vartanian N, de Vienne D (1988) Differential Two-Dimensional Protein Patterns as Related to Tissue Specificity and Water Conditions in *Brassica napus* var oleifera Root System. Plant Physiol 86: 1304–9.
59. Lasserre JP, Menard A (2012) Two-dimensional blue native/SDS gel electrophoresis of multiprotein complexes. Methods Mol Biol 869: 317–37.
60. Candiano G, Bruschi M, Musante L, Santucci L, Ghiggeri GM, et al. (2004) Blue silver: a very sensitive colloidal Coomassie G-250 staining for proteome analysis. Electrophoresis 25: 1327–33.
61. Li XH, Wu XF, Yue WF, Liu JM, Li GL, et al. (2006) Proteomic analysis of the silkworm (*Bombyx mori L.*) hemolymph during developmental stage. J Proteome Res 5: 2809–14.
62. Li Q, Sun H, Han J, Liu Z, Yu L (2006) Effect of precipitation on the hydrochemical variations in Landiantang spring at Nongla, Mashan, China. Adv in Water Sci 17: 733–737.

15

Responses of Bacterial Communities in Arable Soils in a Rice-Wheat Cropping System to Different Fertilizer Regimes and Sampling Times

Jun Zhao[1,2], Tian Ni[1,2], Yong Li[3], Wu Xiong[1,2], Wei Ran[1,2], Biao Shen[1,2], Qirong Shen[1,2,4], Ruifu Zhang[1,2,4*]

1 Key Laboratory of Plant Nutrition and Fertilization in Low-Middle Reaches of the Yangtze River, Ministry of Agriculture, Nanjing, Jiangsu, China, **2** Jiangsu Key Lab and Engineering Center for Solid Organic Waste Utilization, Nanjing Agricultural University, Nanjing, Jiangsu, China, **3** Soil and Fertilizer Technical Guidance Station of Jintan City, Jintan Agricultural and Forestry Bureau, Jintan, Jiangsu, China, **4** Jiangsu Collaborative Innovation Center for Solid Organic Waste Resource Utilization, Nanjing Agricultural University, Nanjing, Jiangsu, China

Abstract

Soil physicochemical properties, soil microbial biomass and bacterial community structures in a rice-wheat cropping system subjected to different fertilizer regimes were investigated in two seasons (June and October). All fertilizer regimes increased the soil microbial biomass carbon and nitrogen. Both fertilizer regime and time had a significant effect on soil physicochemical properties and bacterial community structure. The combined application of inorganic fertilizer and manure organic-inorganic fertilizer significantly enhanced the bacterial diversity in both seasons. The bacterial communities across all samples were dominated by *Proteobacteria*, *Acidobacteria* and *Chloroflexi* at the phylum level. Permutational multivariate analysis confirmed that both fertilizer treatment and season were significant factors in the variation of the composition of the bacterial community. Hierarchical cluster analysis based on Bray-Curtis distances further revealed that bacterial communities were separated primarily by season. The effect of fertilizer treatment is significant ($P = 0.005$) and accounts for 7.43% of the total variation in bacterial community. Soil nutrients (e.g., available K, total N, total P and organic matter) rather than pH showed significant correlation with the majority of abundant taxa. In conclusion, both fertilizer treatment and seasonal changes affect soil properties, microbial biomass and bacterial community structure. The application of NPK plus manure organic-inorganic fertilizer may be a sound fertilizer practice for sustainable food production.

Editor: Hauke Smidt, Wageningen University, Netherlands

Funding: This research was financially supported by the Agricultural Ministry of China (201103004) and the Chinese Ministry of Science and Technology (2011BAD11B03), RZ and QS were also supported by the 111 Project (B12009) and the Priority Academic Program Development (PAPD) of Jiangsu Higher Education Institutions. The funders had no role in study design, data collection and analysis, decision to publish, or preparation of the manuscript.

Competing Interests: The authors have declared that no competing interests exist.

* E-mail: rfzhang@njau.edu.cn

Introduction

Soils are considered to be the most diverse microbial habitats on Earth with respect to species diversity, community size and microbial biomass [1,2]. Bacteria are the most abundant and diverse group of soil microorganisms [3], playing a vital role in agroecosystems through participation in recycling soil nutrients, maintaining soil structure and promoting plant growth [4,5]. It has long been recognized that the appropriate community structure, abundant diversity and a high activity of microorganisms are significant factors in maintaining the sustainability and productivity of ecosystems [6–9].

Fertilization is an important agricultural practice for improving plant nutrition, increasing soil organic matter and achieving high crop yields, but it can also affect soil microbial abundance, activity and community structure [10–13]. Significant differences in soil microbial biomass and microbial diversity have been observed following fertilization [14–16]. Organic fertilizers usually increase soil microbial biomass [17,18], enzyme activities [13,19] and functional diversity [16], while microbial biomass and enzyme activities are decreased in response to inorganic fertilizers [20–22].

Previous studies have indicated that seasonal changes impact soil microbial communities in agroecosystems [23–27]. Moreover, temporal variation of conditions is a very common feature of agroecosystems, including temperature, rainfall, plant type, nutrient levels, etc. Hence, we reason that study using single time point sampling cannot tease apart the effect of temporal variation on microbial community structure.

Soil microorganisms respond quickly to environmental changes (e.g., application of fertilizer or herbicide, tillage, crop rotation and seasonal variation), resulting in dynamic changes in microbial biomass, activity, diversity, abundance and composition. Microbial biomass, activity and diversity are effective indicators of soil quality and health [28,29]. Therefore, understanding the shifts of microbial biomass, community structure and diversity following different agricultural management practices is important for selecting suitable management strategies to improve ecosystem service [30,31].

The rice-wheat cropping system is one of the main cropping systems for cereal food production in China. Understanding the bacterial community in response to the different fertilizer treatments and seasonal variations will help us disclose the "real" effect of fertilizer regimes and guide us in selecting an appropriate

Table 1. Effects of fertilizer regime, sample time and the interaction between them on soil physicochemical characteristics.

	pH	Organic matter (g/kg)	Available N (mg/kg)	Available P (mg/kg)	Available K (mg/kg)	Total N (g/kg)	Total P (g/kg)	Total K (g/kg)
Fertilizer Regime[§](FR)								
CK	7.01±0.15 ab	24.9±3.5 a	117.0±7.6 b	8.1±2.9 b	57.0±15.9 c	1.43±0.11 b	0.83±0.24 a	15.3±1.4 a
NPK	6.97±0.14 b	26.4±4.2 a	134.2±9.7 a	13.8±4.3 ab	64.5±21.2 bc	1.52±0.10 ab	0.87±0.25 a	15.3±1.2 a
NPKM	7.11±0.19 ab	26.6±3.4 a	136.6±7.5 a	14.5±1.8 ab	66.1±21.7 bc	1.59±0.09 a	0.90±0.24 a	16.1±1.5 a
NPKS	7.13±0.07 ab	27.3±4.5 a	137.5±7.0 a	15.2±3.5 a	76.3±28.3 ab	1.62±0.11 a	0.89±0.21 a	15.4±2.2 a
NPKMS	7.27±0.19 a	27.7±4.2 a	135.0±6.5 a	16.6±4.4 a	70.7±23.9 b	1.64±0.17 a	0.92±0.22 a	16.0±1.7 a
NPKMOI	7.07±0.16 ab	27.3±3.6 a	138.5±5.9 a	15.1±3.1 a	86.5±36.0 a	1.59±0.07 a	0.89±0.22 a	15.6±1.7 a
Sample Time (ST)								
June	7.04±0.17 a	24.0±2.2 b	131.2±10.2 a	12.4±3.1 a	91.9±17.1 a	1.50±0.10 b	1.09±0.07 a	15.2±1.0 a
October	7.12±0.18 a	29.3±3.0 a	137.7±8.0 a	13.8±4.1 a	48.5±5.4 b	1.63±0.12 a	0.68±0.05 b	16.0±1.9 a
ANOVA *P*-values								
FR	0.021	NS	< 0.001	0.009	< 0.001	0.003	NS	NS
ST	NS	< 0.001	NS	NS	< 0.001	< 0.001	< 0.001	NS
FR × ST	NS	NS	NS	NS	0.008	NS	NS	NS

Values are means ± standard deviation (n = 6 or n = 18).
NS: not significant (P>0.05).
[§]Fertilizer regimes: CK: control without fertilizers; NPK: chemical fertilizers; NPKM: 50% NPK fertilizer plus 400 kg/ha manure; NPKS: 100% NPK fertilizer plus crop straw; NPKMS: 50% NPK fertilizer plus 400 kg/ha manure and crop straw; NPKMOI: 30% NPK fertilizer plus 240 kg/ha manure organic-inorganic compound fertilizer.
Means followed by the same letter for a given factor are not significantly different (P<0.05; Turkey's HSD test where there are more than two treatment levels).

management practice for more stable and sustainable agroecosystem for food production. In the present study, we aimed 1) to study the effect of different fertilizer regimes and seasonal variations on the bacterial community and microbial biomass; 2) to determine whether the fertilizer regime or seasonal change is the major determinant of bacterial community structure; and 3) to determine the contribution of environmental factors on changes in the bacterial community. To address this aim, measurement of biological properties (microbial biomass) and molecular analysis (pyrosequencing) were used to assess the variation in bacterial community.

Materials and Methods

Ethic statement

No specific permits were required for the described field studies. The locations are not protected. The field studied did not involve endangered or protected species.

Field description and experimental design

The field experiment is located in Jintan city, Jiangsu Province, China (31°39′N, 119°28′E, 3 m a.s.l). This region has a northern subtropical monsoon climate, with an average annual temperature of 15.3°C and a mean annual precipitation of 1063.6 mm. The soil type is classified as typical Clay loamy Fe-leachic-gleyic-stagnic anthrosol, with a long-term annual rotation of winter wheat (*Triticum aestivum* L.) and summer rice (*Oryza sativa* L.), which is the typical cropping system in this region. The fertilization experiment has been in operation since 2010, including four replicates of six treatments in a random block design. The treatments were control without fertilizers (CK), chemical fertilizers (NPK), 50% NPK fertilizer plus 400 kg/ha manure (NPKM), 100% NPK fertilizer plus crop straw (NPKS), 50% NPK fertilizer plus 400 kg/ha manure and crop straw (NPKMS) and 30% NPK fertilizer plus 240 kg/ha manure organic-inorganic compound fertilizer (NPKMOI). It is noted that the manure organic-inorganic compound fertilizer were comprised of pig manure compost and suitable amount of chemical fertilizer. Each plot was 5 m × 8 m and received the same levels (except for the control plot) of nitrogen, phosphorus and potassium (N: 240 kg/ha, P_2O_5: 120 kg/ha and K_2O: 100 kg/ha) from fertilizers in each cropping season. All P, K and manure fertilizers were applied as basal fertilizers before planting, whereas N fertilizer (urea) was used both as basal fertilizer before planting and supplementary fertilizer at tillering and panicle stage. The wheat was planted in October and harvested in June, followed by rice, which was planted in June and harvested in October.

Soil sampling and analysis

Soil samples were taken in summer (June 2012) after the wheat harvest and in autumn (October 2012) after the rice harvest. Ten cores (2.5 cm in diameter) were randomly collected from the plough layer of soil (0–20 cm) in each replicate plot. The cores from each replicate plot were mixed together, pooled in a sterile plastic bag and transported to the laboratory on ice. All the samples were sieved (2 mm), thoroughly homogenized, and divided into three subsamples: one was air-dried for soil characteristic analysis; another was stored at −4°C for the measurement of biological properties; the rest was stored at −80°C for DNA extraction and subsequent molecular analysis. The soil analysis of each sample was performed by the soil testing lab in Qiyang at the red soil experimental station of the Chinese Academy of Agricultural Sciences.

Microbial biomass

Microbial biomass C (MBC) and biomass N (MBN) were measured using the chloroform fumigation-extraction method [32,33]. After 24 h fumigation, soils were extracted using 0.5 M K_2SO_4 with a 1:5 ratio for 60 min on a rotary shaker, and C and N were determined on a LiquiTOC element analyzer II

Table 2. Effects of fertilizer regime, sample time and the interaction between them on soil microbial biomass carbon and nitrogen.

	Microbial biomass C (mg/kg)	Microbial biomass N(mg/kg)
Fertilizer Regime§ (FR)		
CK	310.76±89.51 b	20.29±6.45 d
NPK	387.75±88.13 ab	28.36±2.49 c
NPKM	428.73±100.23 a	33.71±7.24 abc
NPKS	398.00±117.64 ab	30.22±3.07 bc
NPKMS	447.68±103.02 a	36.00±5.48 ab
NPKMOI	471.20±87.32 a	38.85±4.93 a
Sample Time (ST)		
June	479.04±96.25 a	32.77±8.18 a
October	325.04±93.96 b	29.71±7.29 b
ANOVA *P*-values		
FR	< 0.001	< 0.001
ST	< 0.001	0.043
FR × ST	NS	NS

Values are means ± standard deviation (n = 8 or n = 24).
NS: not significant ($P>0.05$).
Means followed by the same letter for a given factor are not significantly different ($P<0.05$; Turkey's HSD test where there are more than two treatment levels).
§Fertilizer regimes as described in Table 1.

(Elementar Analysen-systeme GmbH,Hanau, Germany). Finally, the result was calculated using 0.45 (k_{EC}) and 0.54 (k_{EN}) correction factors [34,35].

DNA extraction

Three samples (biological replicates) of each fertilizer treatment for each season were used for DNA extraction. Total soil DNA was extracted from 0.5 g of fresh soil using a PowerSoil DNA Isolation Kit (Mo Bio Laboratories Inc., Carlsbad, CA, USA) according to the manufacturer's instructions. The extracted DNA was evaluated on a 1% agarose gel, and the concentration and quality (A_{260}/A_{280}) of the extracts were determined using a NanoDrop ND-2000 spectrophotometer (NanoDrop, Wilmington, DE, USA). To minimize the DNA extraction bias, three successive DNA extracts of each sample were pooled. The DNA of each sample was diluted 50-fold for further use.

16S rRNA gene amplification and 454 pyrosequencing

An aliquot (10 ng) of purified DNA from each sample (one biological replicate) was used as template for amplification. The V1-V3 hypervariable region of the bacterial 16S rRNA was amplified using the primer set 27F: AGAGTTTGATCCTGGCT-CAG, with the Roche-454 B sequencing adapter, and 533R: TTACCGCGGCTGCTGGCAC, which contained the Roche-454 A sequencing adapter and a unique, error-correcting 10-bp barcode sequence at the 5′-end of each primer. Each sample was amplified in triplicate in a 20 µl reaction using the following program: 2 min of initial denaturation at 95°C followed by 25 cycles of denaturation (95°C for 30 s), annealing (55°C for 30 s), extension (72°C for 30 s), and a final extension at 72°C for 5 min. The PCR products of each sample were pooled together and purified using a PCR Purification Kit (Axygen Bio, USA). The amplicons of each sample were then combined in equimolar concentrations into a single tube prior to 454 pyrosequencing. Pyrosequencing was performed on a 454 GS-FLX Titanium System (Roche, Switzerland) by Majorbio Bio-pharm Technology Co., Ltd (Shanghai, China).

Processing of pyrosequencing data

Raw pyrosequencing data were processed using Mothur (version 1.29.2) [36] following the Schloss standard operating procedure [37]. Specially, sequences having a minimum flow length of 360 flows and a maximum flow length of 450 flows were de-noised using the Mothur based re-implementation of the PyroNoise algorithm [38] with the default parameters. De-noised sequences with more than two mismatches to the forward primer sequence, one mismatch to the barcode sequence, containing a homopolymer longer than eight nucleotides, any ambiguous base calls (Ns) or a sequence length shorter than 200 bases were eliminated. Then, the filtered sequences were assigned to soil samples based on the unique 10-bp barcodes. After removing the barcode and primer sequences, the unique sequences were aligned against the Silva 106 database [39]. Through screening, filtering, pre-clustering processes and chimera removal using UCHIME [40], the retained sequences were used to build a distance matrix with a distance threshold of 0.2. Using the average neighbor algorithm with a cutoff of 97% similarity, bacterial sequences were clustered to operational taxonomic units (OTUs) (hereby defined as identified OTUs) and the most abundant sequence in each OTU was selected as the representative sequence. Representative sequences were taxonomically classified using an RDP naïve Bayesian rRNA Classifier [41] with a confidence threshold of 60%. Relative abundance of a given phylogenetic group was set as the number of sequences affiliated with that group divided by the total number of sequences per sample. To correct for sampling effort, we used a randomly selected subset of 4,000 sequences per sample for α-diversity analysis. To increase the sampling intensity, we pooled the sequences belonging to each fertilizer treatment for each season. In addition, to downweight the effects of rare species on ß-diversity analysis, only OTUs with ≥20 counts summed across all samples were retained (hereby defined as retained OTUs).

Table 3. Effect of fertilizer regime, sample time and the interaction between them on the relative abundance (%) of abundant bacterial classes (subgroups) £ (relative abundance >1%).

Phylum	*Proteobaceria*			*Acidobacteria*				*Chloroflexi*	*Bacteroidetes*
Class (subgroups)	***α-proteobacteria***	***ß-proteobacteria***	***γ-proteobacteria***	**6**	**4**	**7**	**3**	***Anaerolineae***	***Sphingobacteria***
Fertilizer Regime § (FR)									
CK	4.8±0.9 ab	13.4±0.9 ab	4.3±0.6 c	6.1±0.4 a	3.4±1.2 a	2.2±0.5 a	1.3±0.2 a	10.3±2.6 a	4.6±1.6 a
NPK	4.5±1.1 b	13.6±1.7 a	5.3±0.9 bc	5.1±0.9 ab	3.0±1.2 a	1.8±0.4 ab	1.5±0.5 a	11.3±3.0 a	5.2±1.4 a
NPKM	4.2±1.2 b	14.0±1.2 a	5.8±1.0 b	5.9±0.6 a	3.3±1.0 a	2.1±0.2 a	1.5±0.3 a	9.8±1.8 a	5.4±1.4 a
NPKS	4.0±1.4 b	14.8±0.7 a	5.8±0.9 b	5.9±0.6 a	3.3±0.9 a	2.1±0.2 a	1.2±0.4 a	9.0±1.9 a	5.2±1.6 a
NPKMS	5.9±2.1 ab	13.2±1.0 ab	5.2±0.9 bc	5.4±0.6 a	2.8±1.1 ab	1.8±0.3 ab	1.7±0.4 a	10.8±3.6 a	5.0±1.7 a
NPKMOI	7.1±1.6 a	11.7±1.0 b	7.6±1.4 a	4.0±1.1 b	2.1±0.7 b	1.5±0.2 b	1.8±0.4 a	9.7±3.1 a	5.1±2.1 a
Sample Time (ST)									
June	4.5±1.9 b	13.9±1.2 a	6.3±1.3 a	5.5±0.8 a	3.8±0.7 a	2.0±0.4 a	1.4±0.3 b	8.3±1.3 b	6.6±0.6 a
October	5.7±1.4 a	13.0±1.4 b	5.0±1.1 b	5.2±1.1 a	2.1±0.5 b	1.9±0.3 a	1.7±0.5 a	12.0±2.3 a	3.7±0.5 b
ANOVA *P*-values									
FR	0.005	0.001	< 0.001	< 0.001	< 0.001	0.005	NS	NS	NS
ST	0.019	0.013	< 0.001	NS	< 0.001	NS	0.026	< 0.001	<0.001
FR × ST	NS	NS	NS	NS	NS	NS	NS	NS	NS

Values are means ± standard deviation (n = 6 or n = 18).
NS: no significant (*P*>0.05).
Means followed by the same letter for a given factor are not significantly different (*P*<0.05; Turkey's HSD test where there are more than two treatment levels).
£*Actinobaceria, Gemmatimonadetes, Nitrospira* were not included in this analysis as they had only one class.
§Fertilizer regimes as described in Table 1.

Statistical analysis

An OTU-based analysis was performed to calculate the richness, diversity and coverage in our samples with a cutoff of 3% dissimilarity per 4,000 sequences. Richness indices, the Chao1 estimator [42] and the abundance-based coverage estimator (ACE) [43] were calculated to estimate the number of observed OTU that were present in the sampling assemblage. The diversity within each individual sample was estimated using the non-parametric Shannon diversity index [44] and the Simpson diversity index [45]. Good's non-parametric coverage estimator [46] was used to estimate the percentage of the total species that were sequenced in each sample. Rarefaction curves generated using Mothur [36] were also used to compare relative levels of bacterial OTU diversity across all fertilizer treatments in each season. Significant differences were considered when there was no overlap between the 95% confidence intervals.

In all tests, a *P*-value <0.05 was considered statistically significant. All data were tested for normality and homogeneity and the data were $\log_2$ (x+1)-transformed when necessary to meet the criteria for a normal distribution. A multiple analysis of variance (MANOVA) using PASW Statistics 18 (SPSS Inc.) was used to determine the effects of fertilizer treatment and sampling time on the dependent variables, soil characteristics, bacterial abundance, microbial biomass, relative abundance of abundant taxa and the α-diversity indices. If the multivariate *F* was significant, we then proceeded with the individual univariate analysis.

To describe the changes in soil bacterial assemblages among treatments and seasons without biases resulting from differences in sequencing depth, all samples were first rarefied to 3,000 counts using the rrarefy function in the vegan [47] package of R (version 2.15.0) [48] based on the retained OTUs tables. A permutational multivariate analysis of variance [49] was performed to assess the effect of fertilizer regime, sampling time and its interaction on bacterial community structure (abundance of OTUs and phyla) using the adonis function of the R vegan [47] package with 999 permutations. To compare bacterial community structures across all samples based on the OTU composition, hierarchical cluster analysis was carried out using the hclust function with the average linkage algorithm in the stats package of R [48]. Then, the Bray-Curtis and Euclidean distances were used to construct a community dissimilarity matrix and an environmental dissimilarity matrix, respectively. The effects of rare species were down-weighted by applying the Hellinger transformation for community data [50]. A Mantel test, using the mantel function in the vegan [47] package of R [48], was used to calculate the correlation between the Bray-Curtis distances of bacterial community and soil characteristics. To visualize the relationship between bacterial communities and environmental factors, a redundancy analysis (RDA) was carried out using the rda function, and the environmental factors were fitted to the ordination plots using the envfit function of the vegan [47] package in R with 999 permutations. Pearson correlations between soil characteristics and abundant phyla (classes) were calculated using PASW Statistics 18 (SPSS Inc.). The BioEnv procedure was used to select the subset of environmental properties best correlated (highest Pearson correlation) with bacterial assemblage dissimilarity, and they were used to construct the soil property matrix for variation partitioning analysis (VPA) in the vegan [47] package of R [48].

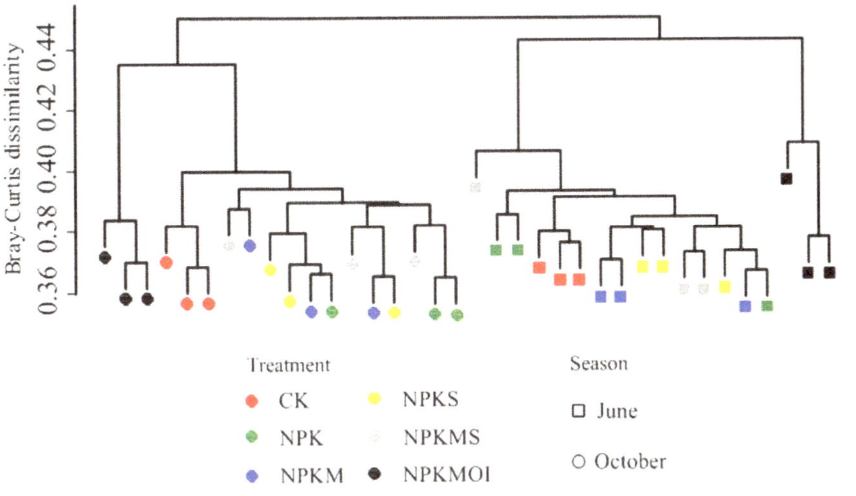

Figure 1. Hierarchical cluster dendrogram of bacterial communities. Pairwise Bray-Curtis dissimilarity of samples collected from six fertilizer treatments in June (square) and October (circle). OTU counts were rarefied to 3,000 counts per sample, and only OTUs ≥20 counts summed across all samples were included in the analysis with the Hellinger-transformation. Symbols of fertilizer regimes were as described in Table 1.

Sequence accession numbers

Sequence data have been deposited in the NCBI Sequence Read Archive (SRA) database with accession number SRA073640.

Results

Soil physicochemical characteristics

The soil pH, available N (AN) and available P (AP) were significantly ($P<0.05$) affected by different fertilizer treatments, while organic matter (OM) and total P (TP) were significantly ($P<0.001$) affected by sample time (Table 1). The soil OM content was higher in October (29.3 g/kg) than in June (24.0 g/kg), while the soil TP content was higher in June (1.09 g/kg) than in October (0.68 g/kg). Both treatment and season showed a significant ($P<0.01$) effect on soil total N (TN). In addition, the treatment, sample time and treatment × sample time interaction terms were all significant affected soil available K (AK) content and separate analyses for each season showed that soil AK content was highest in treatment NPKMOI (118.4 mg/kg, 54.6 mg/kg) and lowest in treatment CK (71.4 mg/kg, 42.7 mg/kg) both in June and October (Table S1). No significant differences of soil total K (TK) were observed between different treatments, seasons and the interaction terms.

Table 4. Permutational multivariate analyses of variance of the Bray dissimilarity between bacterial communities.

Source	df	Abundance of OTUs		Abundance of phylum	
		Sums of sqs	Pseudo-F	Sums of sqs	Pseudo-F
Fertilizer regime (FR)	5	0.6079	1.7276***	0.014495	2.6929**
Sample time (ST)	1	0.40962	5.8206***	0.033321	30.9527***
FR × ST	5	0.38064	1.0818	0.006292	1.1690
Residuals	24	1.68897		0.025836	

*indicate significant correlations ($P<0.05$); ** indicate significant correlations ($P<0.01$); *** indicate significant correlations ($P<0.001$).

Microbial biomass

No significant differences ($P<0.05$) of soil MBC and MBN were observed between the fertilizer treatment × sample time interactions (Table 2). The soil MBC changed significantly ($P<0.001$) between different treatments, being highest in treatment NPKMOI (541 mg/kg, 402 mg/kg) and lowest in treatment CK (359 mg/kg, 234 mg/kg) both in June and October (Table S2). Soil MBC also varied significantly ($P<0.001$) over time, being higher in June (479 mg/kg dry soil) compared with October (325 mg/kg dry soil). The greatest differences ($P<0.001$) in MBN were between different treatments, with season having a smaller but significant ($P=0.043$) effect. The soil MBN was higher in June (32.8 mg/kg dry soil) than in October (29.7 mg/kg dry soil). The highest and lowest MBN were also observed in treatment NPKMOI (42 mg/kg, 36 mg/kg) and CK (21 mg/kg, 19 mg/kg) both in June and October, respectively (Table S2).

Bacterial community composition

Across all soil samples, we obtained 262,299 quality sequences in total, with 4,366–12,538 sequences per sample (mean = 7,284) and were able to classify 86.5% of those sequences at the phylum level. The dominant phyla across all samples were *Proteobacteria*, *Acidobacteria*, *Chloroflexi*, *Bacteroidetes*, *Actinobacteria*, *Gemmatimonadetes*, *Verrucomicrobia* and *Nitrospira* (>1%), accounting for more than 74% of the bacterial sequences from each of the soils (Table S3; Table S4). In addition, *WS3*, *Firmicutes*, *Armatimonadetes*, *TM7*, *Planctomycetes* and *Chlorobi* were in low abundance, and 10 other rare phyla were identified (Table S3).

No significant ($P<0.05$) interaction between treatment and time was observed for abundant phyla (>1%) (Table S4), while the phylum distribution fluctuated under different fertilizer treatments and seasons. *Chloroflexi*, *Verrucomicrobia* and *Nitrospira* were significantly ($P<0.001$) affected by sample time, while only phylum *Gemmatimonadetes* was significantly ($P<0.05$) affected by fertilizer treatments. In addition, phyla *Acidobacteria*, *Bacteroidetes* and were affected by both treatment and time. No significant differences of

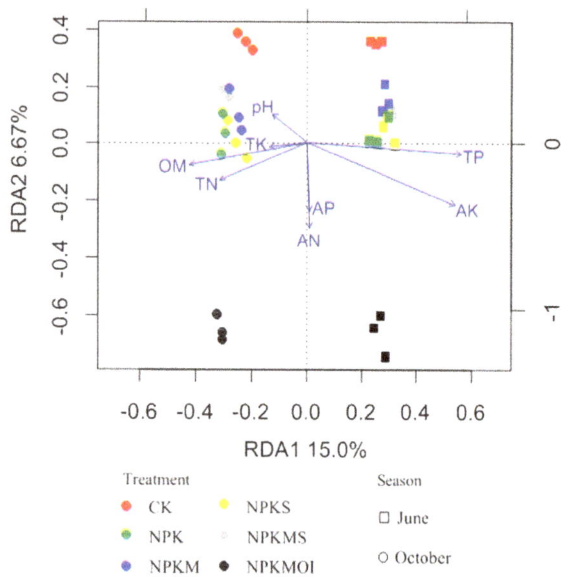

Figure 2. Redundancy analysis of soil bacterial communities and soil characteristics for individual samples. Samples from different fertilizer regimes both in June (square) and October (circle) were marked by different colors. Symbols of fertilizer regimes were as described in Table 1.

Proteobacteria and *Actinobacteria* were observed between different treatments, seasons and the interaction terms.

The relative abundance of abundant bacterial classes (subgroups) demonstrating a significant difference under different fertilizer treatments and times is outlined in Table 3. Likewise, no significant ($P<0.05$) interactions between treatment and time were observed for abundant classes (subgroups). Specifically, we observed that both treatment and time had a significant ($P<0.05$) effect on the *α-Proteobacteria*, *β-Proteobacteria*, and *γ-Proteobacteria*, while the phylum *Proteobacteria* was not affected by treatment, time or the interaction term. Moreover, the relative abundances of different proteobacterial classes were different in response to different fertilizer regimes or seasons. Classes *β-Proteobacteria* and *γ-Proteobacteria* were higher in June, while the *α-Proteobacteria* was higher in October (5.7%) than in June (4.5%). The treatment NPKMOI cultures higher relative abundances of *α-Proteobacteria* and *γ-Proteobacteria*, while the *β-Proteobacteria* was higher in treatment NPK, NPKM, NPKS. Within the *Acidobacteria*, subgroup 3 was only affected by time, while subgroups 4 and 6 were affected by treatment. This is not consistent with the phylum *Acidobacteria*, which is affected by both treatment and time. We also observed that the relative abundance of *Acidobacteria* subgroup 3 was higher in October than in June, which was opposite in the phylum level. In addition, *Anaerolineae* and *Sphingobacteria* were the most abundant subgroups in phyla *Chloroflexi* and *Bacteroidetes*, occupying the majority relative abundance in each phylum, respectively. So the effects of the treatment, time and interaction term were consistent at the phylum and class level. However, we also found that the phylum *Bacteroidetes* was significant ($P<0.05$) affected by both treatment and time, while the subgroups *Sphingobacteria* was only affected by time.

Bacterial α-diversity

Bacterial diversity and richness in individual samples under different fertilizer regimes of both seasons were calculated based on 4,000 sequences (Table S5). Statistically significant differences in richness and diversity ($P\leq0.01$) were observed only at the season level for observed OTUs, coverage, ACE, Chao1 and Simpson but not for Shannon. The number of OTUs, ACE, Chao1 and Simpson were higher in October in comparison to June, while the percentage showed the opposition pattern, being higher in June (80%) than October (76%). Although the fertilizer regimes had no significant effect on the bacterial diversity and richness, a higher number of observed OTUs, ACE, Chao1 and Shannon were observed with treatment NPKMS and NPKMOI in both seasons, indicating that rational fertilization can help increase the biodiversity in agricultural soils. Further analysis of rarefaction curves demonstrated that the number of observed OTUs for treatment NPKMS and NPKMOI were significantly higher than for other fertilizer treatments in June and treatment NPKMOI had a highest observed OTU numbers in October (Fig. S1), calculated based on 10,000 sequences randomly selected from the pooled sequences of each treatment in each season.

Bacterial community structure

A permutational multivariate analysis of variance confirmed that treatment and time were significant factors of variation for the composition of bacterial community in terms of both the relative abundance of OTUs and relative abundance of phylum (Table 4). No significant interaction between treatment and time were observed. Hierarchical cluster analysis of the similarity of bacterial community further confirmed that sampling time is the major determinant of bacterial community structure (Fig. 1). We also found that the treatment NPKMOI was separated from other treatments in both seasons, indicating this fertilizer treatment may have the greatest effect on the bacterial community.

The Mantel test showed significant ($r=0.44$, $P<0.001$) correlation between bacterial community and soil properties. The RDA (redundancy analysis) biplot clearly showed the relationship between soil characteristics and bacterial community structure (Fig. 2). The first two axes of RDA can explain 15.0% and 6.67% of the total variation in the data. The June bacterial communities were grouped separately from the October bacterial communities along the first axis, with the June bacterial communities more relating to higher contents of soil AK and TP, while the October bacterial communities were associated with higher contents of soil pH, OM, TK and TN. Along the second RDA axis, the bacterial communities of treatment NPKMOI and CK were separated from the other fertilizer treatments in both seasons. The treatments NPK, NPKM, NPKS and NPKMS were grouped together in both seasons, indicating that they may have approximately the same effect on the bacterial assembles. The effect of soil properties on bacterial community is shown by the direction and the length of the vectors. The results showed that the soil OM, TP, AK and TN had a significant ($P<0.01$) correlation between each variable and the ordination scores.

In addition, we used Pearson's correlation coefficient to evaluate relationships between abundant phyla and classes (subgroups) (Table 5). Most of the abundant phyla or classes were positively or negatively correlated with soil OM, AK, TN and TP, while only phylum *Verrucomicrobia* showed a significant positive correlation with soil pH and TN. We also found that soil AP content had no significant correlation with any of the abundant taxa. The relative abundances of phylum *Proteobacteria* and class *γ-Proteobacteria* were positively correlated with soil AN and AK, while the *γ-Proteobacteria* was positively correlated with soil TP. The abundance of *Acidobacteria* was negatively correlated with soil OM, AN, TN and TP while positively correlated with soil AK content.

Table 5. Pearson's correlation coefficient between soil characteristics and abundant phyla (classes).

Phylum (class)	Correlation							
	pH	Organic matter	Available N	Available P	Available K	Total N	Total P	Total K
Proteobacteria	0.203	0.091	**0.393***	0.221	**0.479****	0.178	0.327	0.100
Alphaproteobacteria	0.159	0.220	0.042	0.214	−0.156	0.116	−0.282	−0.003
Betaproteobacteria	0.096	−0.103	0.066	−0.083	0.197	−0.041	0.282	0.058
Deltaproteobacteria	−0.046	−0.023	0.056	−0.160	0.101	−0.032	0.187	−0.018
Gammaproteobacteria	−0.079	−0.152	**0.407***	0.269	**0.680****	0.064	**0.472****	0.030
Acidobacteria	−0.304	**−0.451****	**−0.455****	−0.209	**0.420***	**−0.418***	**−0.541****	−0.237
Acidobacteria_subgroups6	0.060	−0.146	−0.127	−0.182	0.025	−0.172	0.151	−0.060
Acidobacteria_subgroups4	−0.212	**−0.673****	−0.107	−0.105	**0.644****	**−0.569****	**0.787****	−0.262
Acidobacteria_subgroups7	−0.155	−0.130	−0.277	−0.193	−0.070	−0.159	0.079	0.066
Acidobacteria_subgroups3	−0.042	0.261	−0.093	0.036	−0.292	0.288	**−0.343***	−0.063
Chloroflexi	0.035	**0.452****	−0.145	−0.153	**−0.753****	**0.342***	**−0.720****	0.019
Bacteroidetes	−0.138	**−0.639****	0.160	0.175	**0.836****	**−0.461****	**0.898****	−0.220
Actinobacteria	0.093	−0.179	0.118	0.197	0.170	−0.201	0.046	−0.037
Gemmatimonadetes	−0.272	0.120	0.102	0.108	−0.240	0.041	−0.248	0.021
Verrucomicrobia	**0.354***	**0.368***	−0.095	0.120	**−0.580****	0.261	**−0.591****	**0.550****
Nitrospira	−0.152	**−0.568****	−0.011	0.073	**0.782****	**−0.447****	**0.838****	−0.202

Values in boldface type indicate significant correlations with *P* values indicated in superscript.
*indicate significant correlations ($P<0.05$); ** indicate significant correlations ($P<0.01$).

Linking the bacterial communities to soil characteristics, sample time and fertilizer regime

Variance partitioning analysis (VPA) was used to determine the relative contributions of sample time, fertilizer regime and soil characteristics to the bacterial communities. A subset of environmental parameters (OM, TP, AK and TK) was selected by the BioEnv procedure, which provides the highest Pearson correlation with bacterial communities. The variation in bacterial community structure was partitioned among soil characteristics, sample time and fertilizer regime, as well as interactions between them. These variables explained 23.58% of the observed variation, leaving 76.42% of the variation unexplained (Fig. 3). Soil characteristics and sample time explained small portions of the observed variation, which accounted for 0.67% ($P=0.18$) and 0.33% ($P=0.255$), respectively, while the fertilizer regime could explain 7.43% ($P=0.005$) of the total variation. The variation was mostly explained by interactions between soil characteristics and sample time, which accounted for 14.79%. The interactions between soil characteristics and fertilizer regime and sample time and fertilizer regime accounted for 2.81% and 0.40% of the variation, respectively. Thus, the fertilizer regimes are important factors in explaining the shifts in bacterial community structures.

Discussion

The present study attempts to assess the effect of fertilizer regime and sampling time on bacterial communities in a rice-wheat cropping system. The results demonstrate that both fertilizer treatment and season were found to impact soil chemical parameters, microbial biomass, bacterial abundance and bacterial community structure. Soil pH was lower in treatment NPK when compared to other fertilizer treatments and CK (Table 1). This might be due to the application of chemical fertilizer alone, which could acidify the soil [51], while the application of crop straw or pig manure in combination with chemical fertilizer can stabilize the soil pH [52]. We observed that soil available nutrients (AN, AP and AK) were significantly altered by fertilizer treatment in comparison to CK, with the treatment NPKMOI having the highest concentration of soil available nutrients. Application of organic-inorganic compound fertilizer can slowly release of nutrients into the soil, promote plant growth and increase crop yields [53]. However, the treatment NPKMOI did not show any significant differences in soil AN and AP contents compared to other fertilizer regimes. We also did not observe any significant difference in soil organic matter content not only between different fertilizer treatments but also between treatments and CK. The results of soil characteristics indicated that different fertilizer regimes gradually but not dramatically changed soil attributes.

Soil MBC and MBN were found to be higher in June than that in October, suggesting that the soil MBC and MBN contents may be affected by the atmospheric temperature. This result is consistent with previous studies [10,54]. Kikuchi et al. [55] and Noll et al. [56] found that bacterial community structure and diversity in a paddy soil can be affected by flooded/upland conditions and oxygen gradient. So we consider that the flooding period during rice cultivation in our study is also an important factor in soil microbial biomass shifting. We also found that the application of fertilizer can markedly affect the soil MBC and MBN both in June and October when compared with CK. This suggests the different fertilizers can also activate the microorganisms in soil.

In this study, we applied 454-pyrosequencing of the 16S rRNA gene to characterize the bacterial community under different fertilizer treatments and sampling times. The sequence analysis reveals that phyla *Proteobacteria*, *Acidobacteria*, *Chloroflexi* and *Bacteroidetes* were the most abundant phyla in all of the samples. This result roughly corresponded with previous studies demonstrating that *Proteobacteria*, *Acidobacteria* and *Bacteroidetes* are domi-

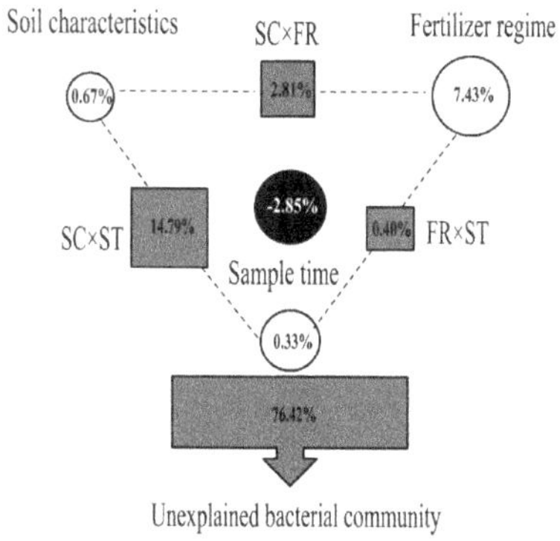

Figure 3. Variation partitioning analysis of the bacterial communities. Effects of soil characteristics (SC), fertilizer regime (FR), sample time (ST) and the interactions between them on the bacterial community structure. Circles on the edges of the triangle show the percentage of variation explained by each factor alone. The percentage of variation explained by interactions between two or three of the factors is shown as squares on the sides and as circle in the center of the triangle. The unexplained variation is depicted in square on the bottom.

nant soil bacterial taxa using sequencing of 16S rRNA gene clone libraries or pyrosequencing [57,58], while the phylum *Chloroflexi* showed an unexpectedly high relative abundance in our study. Specifically, a higher average abundance of *Chloroflexi* was observed in October (12.37%) than in June (8.66%) (Table S3), which suggests that the practice of periodic flooding (resulting in an anaerobic environment) in the rice season can significantly impact this facultative anaerobic phylum. Significant differences in soil bacterial community composition were observed both among fertilizer treatments and between sampling times. In particular, the highest relative abundance of *α-Proteobacteria* and *γ-Proteobacteria*, but the lowest of *β-Proteobacteria*, was observed in treatment NPKMOI in both seasons. *β-Proteobacteria* are considered as copiotrophic bacteria and always flourish in soils with large amounts of available nutrients [59], while the available nutrients in the treatment NPKMOI are higher than other fertilizer treatments and CK. Therefore, lower relative abundance of *β-Proteobacteria* in the treatment NPKMOI could be due to other factors. Interestingly, the phylum and its classes or subgroups had different patterns in response to the different treatments and seasons. For instance, classes *α-Proteobacteria*, *β-Proteobacteria* and *γ-Proteobacteria* were all affected by both treatment and season, while the phylum *Proteobacteria* was not affected by treatment, season or the interaction between fertilizer and season. In contrast, the phylum *Acidobacteria* was affected by both treatment and season, subgroups 6, 7 and 3 were only affected by treatment or season. This indicates that finer taxa may have more sensitive responses to different treatments and seasons due to its own characteristics.

Many studies have shown that environmental factors determine the soil microbial community [5,60,61], especially the soil pH, which has been demonstrated in several studies to be the strongest factor shaping microbial community structures [60,62–64]. Chu et al. [65] and Rousk et al. [60] have reported that the relative abundances of *α-Proteobacteria* and *γ-Proteobacteria* increase with higher soil pH, but we did not find the same trend, which may be due to the mild fluctuations (range from 6.86 to 7.37 in all samples) in the soil pH in our study. Jones et al. [66] reported that relative abundances of Acidobacterial subgroups 4, 6 and 7 were positively correlated with soil pH, which appears inconsistent with our results, being higher in treatment CK than that in treatment NPKMOI. The soil pH values varied significantly from 3 to 9 in those studies allowing insight into the relationship between pH and soil bacterial communities besides that the soil factors assessed in this study were not measured in the studies of Chu et al., [65], Roush et al., [60] and Jones et al., [66]. Thus, we hypothesize that there are important factors other than pH in shaping soil bacterial communities, which was similar with Navarrete et al., [67], demonstrating that abiotic soil factors (not only related to soil acidity) such as Al, Ca, Mg, Mn and B can also drive the acidobacterial populations. Further Pearson correlation analysis showed that only phylum *Verrucomicrobia* had a small correlation with soil pH. In contrast, we found that most of the proportions of abundant phyla and classes (subgroups) were highly correlated with soil OM, AK, TN and TP contents (Table 5). Lauber et al. [62] considered that the soil pH may not directly alter bacterial community structure but may instead function as a variable that provides an integrated index of soil conditions. In fact, there are a number of soil characteristics (e.g., nutrient availability, organic C characteristics, soil moisture regimen and salinity) that are often directly or indirectly related to soil pH [68], and these factors may drive the observed changes in community composition as the hydrogen ion concentration varies by many orders of magnitude across the soils [60,62–64]. However, the soil pH may not represent an integrating variable in our study due to the mild variation of soil pH, which suggests that soil characteristics such as soil nutrients, organic matter, soil moisture, etc. rather than soil pH can shape the bacterial communities by themselves. In addition, Nacke et al. [5] found that the phylum *Actinobacteria* was positively correlated with TN and class *δ-Proteobacteria* was negatively correlated with organic C. However, these two bacterial taxa had no significant correlation with any soil properties in our study (Table 5). Hence, we hypothesize that soil nutrients can also drive the bacterial community composition and that the correlation patterns are varied in different soils.

Soil microbial diversity is considered to be critical to the integrity, function and long-term sustainability of soil ecosystems [69]. Moreover, greater biodiversity in soil can lead to a more stable system and enhance the combination of vital microbial functions and processes [9]. In the present study, higher biodiversity was observed in treatment NPKMOI at both seasons, indicating that combined application of chemical fertilizer and organic-inorganic compound fertilizer can maintain a higher biodiversity. It may result in a more stable agroecosystem and may contribute to sustainable crop production.

Considerable temporal variation in soil bacterial communities has been previously been reported for agricultural soils [24,26,27]. In the present study, the greatest community variation was also found to be temporal (Fig. 1). All the soil samples showed temporal clustering and higher similarity was observed within each season. Bremer et al. [70] considered that plant presence may result in seasonal dynamics of bacterial communities. In our study, we collected soil samples after crop harvest to avoid this bias. However, we also found that within each season, the bacterial community structure of treatment NPKMOI differed from other treatments or CK (Fig. 1 and Fig. 2), indicating that fertilization is also an important factor in shaping bacterial community. This was verified by the permutational multivariate analysis of variance of the bacterial community (Table 4). In addition, soil properties were also significantly correlated with bacterial communities in the present study. Further analysis revealed that sampling time could only explain a smaller variation (0.33%) in the total bacterial community. In contrast, fertilizer regime can explain a greater variation (7.43%) than sampling time and soil properties.

Moreover, only fertilizer regime can significantly ($P = 0.005$) explain the variation in bacterial community structure.

In conclusion, fertilizer practice and seasonal changes can affect soil properties, microbial biomass, and bacterial community structure. We observed a significant and distinct seasonal shift for bacterial community while the fertilizer type also lead to a significant but much smaller variation in bacterial community structure. The treatment NPKMOI significantly increased the microbial biomass and biodiversity and may be more suitable for sustainable crop production.

Further study should pay more attention to the impact of the changes of bacterial community induced by fertilization and seasonal fluctuations on soil functionality.

Supporting Information

Figure S1 Rarefaction curves of bacterial communities for June (A) and October (B) based on the number of observed OTUs at 3% distance calculated from the randomly selected 10,000 pooled sequences of each treatment. Error bars indicate 95% confidence intervals.

Table S1 Soil available K of all samples from different fertilizer regimes both in June and October.

Table S2 Soil microbial biomass C (MBC) and microbial biomass N (MBN) of all samples from different fertilizer regimes both in June and October.

Table S3 Relative average abundances of phyla across all soils and soils grouped into fertilizer regime and sampling time categories, respectively (values represent % of total non-redundant sequences). Asterisks mean values less than 0.01.

Table S4 Effects of fertilizer regime, sample time and the interaction between them on the abundant phyla (relative abundance >1%).

Table S5 Effects of fertilizer regime, sample time and the interaction between them on the OTUs, coverage, richness and diversity calculated using a random 4,000 sequences per sample.

Acknowledgments

We would like to thank the Soil and Fertilizer Technical Guidance Station of Jintan City for managing the field experiment. We also gratefully acknowledge the anonymous reviewers for their helpful and constructive comments.

Author Contributions

Conceived and designed the experiments: WR BS RZ QS. Performed the experiments: JZ TN WX YL. Analyzed the data: JZ. Contributed reagents/materials/analysis tools: RZ QS. Wrote the paper: JZ RZ QS.

References

1. Brodie EL, DeSantis TZ, Parker JPM, Zubietta IX, Piceno YM, et al. (2007) Urban aerosols harbor diverse and dynamic bacterial populations. PNAS 104: 299–304.
2. Nannipieri P, Ascher J, Ceccherini M, Landi L, Pietramellara G, et al. (2003) Microbial diversity and soil functions. Eur J Soil Sci 54: 655–670.
3. Gans J, Wolinsky M, Dunbar J (2005) Computational Improvements Reveal Great Bacterial Diversity and High Metal Toxicity in Soil. Science 309: 1387–1390.
4. Glick BR (1995) The enhancement of plant growth by free-living bacteria. Can J Microbiol 41: 109–117.
5. Nacke H, Thürmer A, Wollherr A, Will C, Hodac L, et al. (2011) Pyrosequencing-based assessment of bacterial community structure along different management types in German forest and grassland soils. PLoS One 6: e17000.
6. Bell T, Newman JA, Silverman BW, Turner SL, Lilley AK (2005) The contribution of species richness and composition to bacterial services. Nature 436: 1157–1160.
7. Naeem S, Li S (1997) Biodiversity enhances ecosystem reliability. Nature 390: 507–509.
8. Cardinale BJ, Srivastava DS, Duffy JE, Wright JP, Downing AL, et al. (2006) Effects of biodiversity on the functioning of trophic groups and ecosystems. Nature 443: 989–992.
9. Chaer G, Fernandes M, Myrold D, Bottomley P (2009) Comparative resistance and resilience of soil microbial communities and enzyme activities in adjacent native forest and agricultural soils. Microb Ecol 58: 414–424.
10. Gu Y, Zhang X, Tu S, Lindström K (2009) Soil microbial biomass, crop yields, and bacterial community structure as affected by long-term fertilizer treatments under wheat-rice cropping. Eur J Soil Bio 45: 239–246.
11. Saha S, Prakash V, Kundu S, Kumar N, Mina BL (2008) Soil enzymatic activity as affected by long term application of farm yard manure and mineral fertilizer under a rainfed soybean–wheat system in NW Himalaya. Eur J Soil Bio 44: 309–315.
12. Islam MR, Chauhan PS, Kim Y, Kim M, Sa T (2011) Community level functional diversity and enzyme activities in paddy soils under different long-term fertilizer management practices. Biol Fert Soils 47: 599–604.
13. Chu H, Lin X, Fujii T, Morimoto S, Yagi K, et al. (2007) Soil microbial biomass, dehydrogenase activity, bacterial community structure in response to long-term fertilizer management. Soil Biol Biochem 39: 2971–2976.
14. Yu WT, Bi ML, Xu YG, Zhou H, Ma Q, et al. (2013) Microbial biomass and community composition in a Luvisol soil as influenced by long-term land use and fertilization. CATENA.
15. Esperschütz J, Gattinger A, Mäder P, Schloter M, Fließbach A (2007) Response of soil microbial biomass and community structures to conventional and organic farming systems under identical crop rotations. FEMS Microbiol Ecol 61: 26–37.
16. Orr CH, Leifert C, Cummings SP, Cooper JM (2012) Impacts of Organic and Conventional Crop Management on Diversity and Activity of Free-Living Nitrogen Fixing Bacteria and Total Bacteria Are Subsidiary to Temporal Effects. PLoS One 7: e52891.
17. Ebhin Masto R, Chhonkar P, Singh D, Patra A (2006) Changes in soil biological and biochemical characteristics in a long-term field trial on a sub-tropical inceptisol. Soil Biol Biochem 38: 1577–1582.
18. Lv M, Li Z, Che Y, Han F, Liu M (2011) Soil organic C, nutrients, microbial biomass, and grain yield of rice (Oryza sativa L.) after 18 years of fertilizer application to an infertile paddy soil. Biol Fert Soils 47: 777–783.
19. Ge G, Li Z, Fan F, Chu G, Hou Z, et al. (2010) Soil biological activity and their seasonal variations in response to long-term application of organic and inorganic fertilizers. Plant Soil 326: 31–44.
20. Plaza C, Hernandez D, Garcia-Gil J, Polo A (2004) Microbial activity in pig slurry-amended soils under semiarid conditions. Soil Biol Biochem 36: 1577–1585.
21. Ramirez KS, Craine JM, Fierer N (2012) Consistent effects of nitrogen amendments on soil microbial communities and processes across biomes. Global Change Biol 18: 1918–1927.
22. Zhong W, Cai Z (2007) Long-term effects of inorganic fertilizers on microbial biomass and community functional diversity in a paddy soil derived from quaternary red clay. Applied Soil Ecol 36: 84–91.
23. Griffiths RI, Whiteley AS, O'Donnell AG, Bailey MJ (2003) Influence of depth and sampling time on bacterial community structure in an upland grassland soil. FEMS Microbiol Ecol 43: 35–43.
24. Schutter M, Sandeno J, Dick R (2001) Seasonal, soil type, and alternative management influences on microbial communities of vegetable cropping systems. Biol Fert Soils 34: 397–410.
25. Kennedy NM, Gleeson DE, Connolly J, Clipson NJ (2005) Seasonal and management influences on bacterial community structure in an upland grassland soil. FEMS Microbiol Ecol 53: 329–337.
26. Hannula SE, de Boer W, van Veen J (2012) A 3-year study reveals that plant growth stage, season and field site affect soil fungal communities while cultivar and GM-trait have minor effects. PLoS One 7: e33819.
27. Kennedy N, Brodie E, Connolly J, Clipson N (2006) Seasonal influences on fungal community structure in unimproved and improved upland grassland soils. Canadian J Microbiol 52: 689–694.

28. Bending GD, Turner MK, Rayns F, Marx MC, Wood M (2004) Microbial and biochemical soil quality indicators and their potential for differentiating areas under contrasting agricultural management regimes. Soil Biol Biochem 36: 1785–1792.
29. Sparling G, Pankhurst C, Doube B, Gupta V (1997) Soil microbial biomass, activity and nutrient cycling as indicators of soil health. Biological indicators of soil health: 97–119.
30. Acosta-Martinez V, Dowd S, Sun Y, Allen V (2008) Tag-encoded pyrosequencing analysis of bacterial diversity in a single soil type as affected by management and land use. Soil Biol Biochem 40: 2762–2770.
31. Singh BK, Campbell CD, Sorenson SJ, Zhou J (2009) Soil genomics. Nat Rev Microbiol 7: 756–756.
32. Brookes P, Landman A, Pruden G, Jenkinson D (1985) Chloroform fumigation and the release of soil nitrogen: a rapid direct extraction method to measure microbial biomass nitrogen in soil. Soil Biol Biochem 17: 837–842.
33. Vance E, Brookes P, Jenkinson D (1987) An extraction method for measuring soil microbial biomass C. Soil Biol Biochem 19: 703–707.
34. Joergensen RG, Mueller T (1996) The fumigation-extraction method to estimate soil microbial biomass: Calibration of the k_{EN} value. Soil Biol Biochem 28: 33–37.
35. Joergensen RG (1996) The fumigation-extraction method to estimate soil microbial biomass: Calibration of the k_{EC} value. Soil Biol Biochem 28: 25–31.
36. Schloss PD, Westcott SL, Ryabin T, Hall JR, Hartmann M, et al. (2009) Introducing mothur: Open-Source, Platform-Independent, Community-Supported Software for Describing and Comparing Microbial Communities. Appl Environ Microbiol 75: 7537–7541.
37. Schloss PD, Gevers D, Westcott SL (2011) Reducing the Effects of PCR Amplification and Sequencing Artifacts on 16S rRNA-Based Studies. PloS one 6: e27310.
38. Quince C, Lanzen A, Davenport RJ, Turnbaugh PJ (2011) Removing noise from pyrosequenced amplicons. BMC Bioinformatics 12: 38.
39. Pruesse E, Quast C, Knittel K, Fuchs BM, Ludwig W, et al. (2007) SILVA: a comprehensive online resource for quality checked and aligned ribosomal RNA sequence data compatible with ARB. Nucleic Acids Res 35: 7188–7196.
40. Edgar RC, Haas BJ, Clemente JC, Quince C, Knight R (2011) UCHIME improves sensitivity and speed of chimera detection. Bioinformatics 27: 2194–2200.
41. Wang Q, Garrity GM, Tiedje JM, Cole JR (2007) Naïve Bayesian Classifier for Rapid Assignment of rRNA Sequences into the New Bacterial Taxonomy. Appl Environ Microbiol 73: 5261–5267.
42. Chao A (1984) Nonparametric estimation of the number of classes in a population. Scand J Stat: 265–270.
43. Eckburg PB, Bik EM, Bernstein CN, Purdom E, Dethlefsen L, et al. (2005) Diversity of the human intestinal microbial flora. Science 308: 1635–1638.
44. Washington H (1984) Diversity, biotic and similarity indices: a review with special relevance to aquatic ecosystems. Water Res 18: 653–694.
45. Simpson EH (1949) Measurement of diversity. Nature 163: 688.
46. Bunge J, Fitzpatrick M (1993) Estimating the number of species: a review. J Am Stat Assoc: 364–373.
47. Oksanen J, Kindt R, Legendre P, O'Hara B, Simpson GL, et al. (2009) Vegan: Community ecology package. R package version 1.15–2.
48. R Development Core Team (2012) R: A Language and Environment for Statistical Computing. Vienna, Austria: R Foundation for Statistical Computing.
49. Anderson MJ (2001) A new method for non-parametric multivariate analysis of variance. Austral Ecol 26: 32–46.
50. Legendre P, Gallagher ED (2001) Ecologically meaningful transformations for ordination of species data. Oecologia 129: 271–280.
51. van Diepeningen AD, de Vos OJ, Korthals GW, van Bruggen AH (2006) Effects of organic versus conventional management on chemical and biological parameters in agricultural soils. Appl Soil Ecol 31: 120–135.
52. Li R, Khafipour E, Krause DO, Entz MH, de Kievit TR, et al. (2012) Pyrosequencing reveals the influence of organic and conventional farming systems on bacterial communities. PloS one 7: e51897.
53. Chunyu S, Fudao Z, Shuqing Z, Hui L, Chenggao F (2004) Effects of organic-inorganic slow release fertilizers on yield and nitrogen recovery in tomato. Plant Nutrition and Fertilizer Science 10: 584.
54. Diaz-Ravina M, Acea M, Carballas T (1995) Seasonal changes in microbial biomass and nutrient flush in forest soils. Biol Fert Soils 19: 220–226.
55. Kikuchi H, Watanabe T, Jia Z, Kimura M, Asakawa S (2007) Molecular analyses reveal stability of bacterial communities in bulk soil of a Japanese paddy field: estimation by denaturing gradient gel electrophoresis of 16S rRNA genes amplified from DNA accompanied with RNA. Soil Sci Plant Nutr 53: 448–458.
56. Noll M, Matthies D, Frenzel P, Derakshani M, Liesack W (2005) Succession of bacterial community structure and diversity in a paddy soil oxygen gradient. Environ Microbiol 7: 382–395.
57. Roesch LF, Fulthorpe RR, Riva A, Casella G, Hadwin AK, et al. (2007) Pyrosequencing enumerates and contrasts soil microbial diversity. The ISME J 1: 283–290.
58. Janssen PH (2006) Identifying the dominant soil bacterial taxa in libraries of 16S rRNA and 16S rRNA genes. Appl Environ Microbiol 72: 1719–1728.
59. Fierer N, Bradford MA, Jackson RB (2007) Toward an ecological classification of soil bacteria. Ecology 88: 1354–1364.
60. Rousk J, Bååth E, Brookes PC, Lauber CL, Lozupone C, et al. (2010) Soil bacterial and fungal communities across a pH gradient in an arable soil. The ISME J 4: 1340–1351.
61. Lauber CL, Strickland MS, Bradford MA, Fierer N (2008) The influence of soil properties on the structure of bacterial and fungal communities across land-use types. Soil Biol Biochem 40: 2407–2415.
62. Lauber CL, Hamady M, Knight R, Fierer N (2009) Pyrosequencing-based assessment of soil pH as a predictor of soil bacterial community structure at the continental scale. Appl Environ Microbiol 75: 5111–5120.
63. Xiong J, Liu Y, Lin X, Zhang H, Zeng J, et al. (2012) Geographic distance and pH drive bacterial distribution in alkaline lake sediments across Tibetan Plateau. Environ Microbiol 14: 2457–2466.
64. Shen C, Xiong J, Zhang H, Feng Y, Lin X, et al. (2012) Soil pH drives the spatial distribution of bacterial communities along elevation on Changbai Mountain. Soil Biol Biochem 57:204–211.
65. Chu H, Fierer N, Lauber CL, Caporaso J, Knight R, et al. (2010) Soil bacterial diversity in the Arctic is not fundamentally different from that found in other biomes. Environ Microbiol 12: 2998–3006.
66. Jones RT, Robeson MS, Lauber CL, Hamady M, Knight R, et al. (2009) A comprehensive survey of soil acidobacterial diversity using pyrosequencing and clone library analyses. The ISME J 3: 442–453.
67. Navarrete AA, Kuramae EE, Hollander M, Pijl AS, Veen JA, et al. (2012) Acidobacterial community responses to agricultural management of soybean in Amazon forest soils. FEMS Microbiol Ecol 83: 607–621.
68. Brady NC, Weil RR (1996) The nature and properties of soils: Prentice-Hall Inc. 740p.
69. Kennedy A, Smith K (1995) Soil microbial diversity and the sustainability of agricultural soils. Plant Soil 170: 75–86.
70. Bremer C, Braker G, Matthies D, Beierkuhnlein C, Conrad R (2009) Plant presence and species combination, but not diversity, influence denitrifier activity and the composition of nirK-type denitrifier communities in grassland soil. FEMS Microbiol Ecol 70: 377–387.

16

Genome Sequence of *Candidatus* Nitrososphaera evergladensis from Group I.1b Enriched from Everglades Soil Reveals Novel Genomic Features of the Ammonia-Oxidizing Archaea

Kateryna V. Zhalnina[1], Raquel Dias[1], Michael T. Leonard[1], Patricia Dorr de Quadros[2], Flavio A. O. Camargo[2], Jennifer C. Drew[1], William G. Farmerie[3], Samira H. Daroub[4], Eric W. Triplett[1]*

1 Microbiology and Cell Science Department, Institute of Food and Agricultural Sciences, University of Florida, Gainesville, Florida, United States of America, **2** Soil Science Department, Federal Unviersity of Rio Grande do Sul, Porto Alegre, RS, Brazil, **3** Genome Sequencing Services Laboratory, Interdisciplinary Center for Biotechnology Research, University of Florida, Gainesville, Florida, United States of America, **4** Everglades Research and Education Center, University of Florida, Belle Glade, Florida, United States of America

Abstract

The activity of ammonia-oxidizing archaea (AOA) leads to the loss of nitrogen from soil, pollution of water sources and elevated emissions of greenhouse gas. To date, eight AOA genomes are available in the public databases, seven are from the group I.1a of the Thaumarchaeota and only one is from the group I.1b, isolated from hot springs. Many soils are dominated by AOA from the group I.1b, but the genomes of soil representatives of this group have not been sequenced and functionally characterized. The lack of knowledge of metabolic pathways of soil AOA presents a critical gap in understanding their role in biogeochemical cycles. Here, we describe the first complete genome of soil archaeon *Candidatus* Nitrososphaera evergladensis, which has been reconstructed from metagenomic sequencing of a highly enriched culture obtained from an agricultural soil. The AOA enrichment was sequenced with the high throughput next generation sequencing platforms from Pacific Biosciences and Ion Torrent. The *de novo* assembly of sequences resulted in one 2.95 Mb contig. Annotation of the reconstructed genome revealed many similarities of the basic metabolism with the rest of sequenced AOA. *Ca.* N. evergladensis belongs to the group I.1b and shares only 40% of whole-genome homology with the closest sequenced relative *Ca.* N. gargensis. Detailed analysis of the genome revealed coding sequences that were completely absent from the group I.1a. These unique sequences code for proteins involved in control of DNA integrity, transporters, two-component systems and versatile CRISPR defense system. Notably, genomes from the group I.1b have more gene duplications compared to the genomes from the group I.1a. We suggest that the presence of these unique genes and gene duplications may be associated with the environmental versatility of this group.

Editor: Mark R. Liles, Auburn University, United States of America

Funding: This work was supported by the National Science Foundation (grant number MCB-0454030); and the United States Department of Agriculture (grant numbers 2005-35319-16300, 00067345). University of Florida, Interdisciplinary Center for Biotechnology Research, Electron Microscopy and Bio-Imaging lab for assistance with SEM and TEM. The funders had no role in study design, data collection and analysis, decision to publish, or preparation of the manuscript.

Competing Interests: The authors have declared that no competing interests exist.

* Email: ewt@ufl.edu

Introduction

The ammonia-oxidizing archaea (AOA) are an abundant group of nitrifiers that plays important environmental roles in the open oceans, soils, the arctic, hot springs and marine sponges [1–8]. AOA oxidize ammonia (NH_3) to nitrite (NO_2^-) with further oxidation to nitrate (NO_3^-) by nitrite-oxidizing bacteria [9,10]. In soils, nitrification can increase mobility of inorganic N, hence it may cause NO_3^- leaching from soils, pollution of ground and surface waters, and an increased cost of applied N fertilizers in agricultural areas [11–13].

Another possible negative consequence of AOA activity in marine environments and soil, particularly in agricultural areas, is the increased pollution of the atmosphere by nitrous oxide (N_2O). Nitrous oxide is one of the most stable greenhouse gases, and agricultural soil management is the largest source of N_2O emissions in the United States (69% of total U.S. N_2O emissions) [14]. Several studies demonstrate that AOA produce N_2O [15–18]. However, the underlying pathways for biogeoproduction of N_2O remain unknown.

AOA are difficult to culture. Only a few AOA have been cultured and sequenced from either pure or enrichment cultures [2–6,19,20]. When AOA were first discovered, major AOA groups (I.1a, I.1b and Hot spring cluster) were proposed by 16S rRNA gene and *amoA* gene identities [10]. Group I.1b (or *Nitrososphaera* cluster) is mostly represented by AOA from soil and some other habitats, including hot springs, freshwater, and freshwater sediments [8,21–23]. Group I.1a (or *Nitrosopumilus* cluster) is mainly represented by marine archaea. However it has also been found in other environments including soil, hot springs, and freshwater [1–6,19,22,24]. Seven genome sequences from group I.1a (*Nitrosopumilus mritimus*, *Candidatus* Nitrosopumilus sediminis, *Candidatus* Nitrosopumilus salaria, *Candidatus* Nitrosoarchaeum limnia, *Candidatus* Nitrosopumilus koreensis, *Cenarchaeum symbiosum*, and *Candidatus* Nitrosotenuis uzonensis) and only one from group I.1b (*Candidatus* Nitrososphaera gargensis) are available in public databases.

The lack of genomic information limits our understanding of the physiology and biochemistry of AOA, particularly Thaumarchaeota from group I.1b. Furthermore, the recently sequenced genome of moderate thermophile *Ca.* Nitrososphaera gargensis from the group I.1b revealed some differences between I.1b and I.1a groups of Thaumarchaeota [20]. For example, the genome size of *Ca.* Nitrososphaera gargensis is much bigger than other sequenced genomes from group I.1a. Additionally, the sequence analysis indicated higher G+C content, more thermosome genes, and a different chemical structure of membrane lipids [20,25]. However, *Ca.* Nitrososphaera gargensis was isolated from hot springs, and it is unknown whether these features are specific only to thermophilic archaeon, or if mesophilic AOA, widely distributed in soils, from group I.1b also share these features.

From a previous study we found that AOA closely related to *Nitrososphaera* genus are highly abundant in the Everglades Agricultural Area, and their abundance significantly increases with agricultural management [26]. In this paper, we present (a) the preparation of an enriched culture of AOA from an Everglades histosol soil; (b) the sequencing and genome reconstruction of the first mesophilic AOA from the group I.1b enriched from the soil; (c) the genome annotation and analysis of main physiological features; and (d) the major metabolic differences between group I.1b and group I.1a.

For the first time, we report genome analysis of AOA group I.1b isolated from soil. This genome provides insight into genomic features present in *Ca.* N. evergladensis but not in other sequenced AOA. The genome analysis reveals features that distinguish AOA from I.1b and I.1a groups. This study provides important insight to guide our understanding of the role of AOA in terrestrial and marine environments.

Results and Discussion

Preparation of ammonia-oxidizing enrichment culture

An AOA enrichment culture was prepared from soil collected from the Everglades Agricultural Area using AOA medium and culture conditions described previously [21]. However, the addition of antibiotics to the enrichment culture did not result in pure culture as some other microorganisms remained. Preliminary genetic analysis of the AOA enrichment was performed by 16S rRNA amplification and Sanger sequencing of the clone library. Approximately 50% of all 16S rRNA clones were assigned to *Nitrososphaera* genus.

Enrichment was tested for the presence of ammonia-oxidizing bacteria (AOB) by PCR-amplification of the bacterial *amoA* genes. No amplification of the bacterial *amoA* was observed. In addition, sequence search of the bacterial *amoA* and 16S rRNA of AOB in the metagenomic sequences of the enrichment was performed against a customized database of bacterial *amoA* sequences and the reference Ribosomal Database Project (RDP) 16S SSU rRNA database [27]. This search did not reveal either bacterial *amoA* or known 16S rRNA genes affiliated with AOB. Further metagenomic analysis of the AOA enrichment showed that all present archaeal *amoA* (12 gene copies) corresponded to *amoA* from the *Ca.* N. evergladensis genome (NTE_00961) at the level of amino-acid identity 99.1–100%. Consistent with these results, all identified archaeal 16S rRNA (9 gene copies) in the enrichment displayed 99.1–100% of nucleotide identity with Ca. N. evergladensis 16S rRNA (NTE_02406).

Ammonia consumption and nitrite (NO_2^-) production, as well as archaeal *amoA* gene copy number, were measured every three days after inoculation (Figure 1A and 1B). Ammonia was converted to NO_2^- over a period of about 21 days (Figure 1A). Simultaneous oxidation of NH_3 and production of NO_2^- was accompanied by the increase of archaeal *amoA* gene copy number (Figure 1B).

Gene prediction and annotation

The genome of the enriched AOA was reconstructed from assembled reads generated using data from the Pacific Biosciences (PacBio) platform (Figure S1A in File S1). Genome assembly was verified by PCR of selected regions in the assembled genome and by alignment of the genome with the contigs obtained from sequencing results from an Ion Torrent platform (Figure S1B in File S1). The 2.95 Mb genome sequence included 3555 genes, 50% G+C content, and 43 RNA genes. Over eighty percent (83.6%) of the assembled bases were predicted to code for proteins. Only 52% of protein coding sequences had functional assignment. Moreover, 60.6% of identified genes were in paralog clusters.

Phylogeny and general genome features of *Ca.* N. evergladensis

Based on 16S rRNA and *amoA* classification, the mesophilic AOA *Ca.* N. evergladensis is phylogenetically affiliated with Thaumarchaeota from group I.1b (Figure 2, Figure S2 in File S1). The closest cultured relatives from the group I.1b are *Ca.* N. gargensis, *N. viennensis*, and *Nitrososphaera sp.* JG1 (Figure 2). *Ca.* N. evergladensis shares 97% and 85% 16S rRNA identity with *Ca.* N. gargensis and the AOA from group I.1a, respectively (Table A in File S1). Nucleotide identity of *amoA* genes were less conserved. *Ca.* N. evergladensis *amoA* was 87% and 71–74% identical to *Ca.* N. gargensis and group I.1a, respectively. AOA from the group I.1b have larger genomes and almost twice the number of protein coding sequences (CDS) compared to the group I.1a (Table A in File S1). Sixty-four percent of CDS from the *Ca.* N. evergladensis genome share 35% identity with *Ca.* N. gargensis and less than 34% of CDS were found in common with *N. maritimus* (Figure 3A). Overall, I.1a and I.1b groups shared about 30% CDS (Figure 3B). Whole-genome alignment of *Ca.* N. evergladensis to *Ca.* N. gargensis revealed 40% of conserved sites between two genomes. *Ca.* N. evergladensis shared a much smaller degree of genome synteny with *N. maritimus* than *Ca.* N gargensis (Figures 4A, 4B). An average nucleotide identity of 82.9% between both *Nitrososphaera* genomes confirmed that both genomes represent different species.

Carbon metabolism

The *Ca.* N. evergladensis genome codes for the key enzymes of a 3-hydroxypropionate/4-hydroxybutyrate pathway of CO_2 fixation, enzymes for forward and reverse tricarboxylic acid cycle (TCA) cycle, gluconeogenesis, a modified glycolytic pathway, and hexose monophosphate pathway (Table S1 in File S2, Figures S3, S4, S5, S6 in File S1).

3-hydroxypropionate/4-hydroxybutyrate carbon fixation pathway. *Ca.* N. evergladensis, much like all known chemolithotrophic AOA, is predicted to fix inorganic carbon via a modified 3-hydroxypropionate/4-hydroxybutyrate cycle [3]. Despite the fact that key enzymes for this pathway were found in other sequenced AOA genomes, some steps of the AOA 3-hydroxypropionate/4-hydroxybutyrate cycle remain undescribed

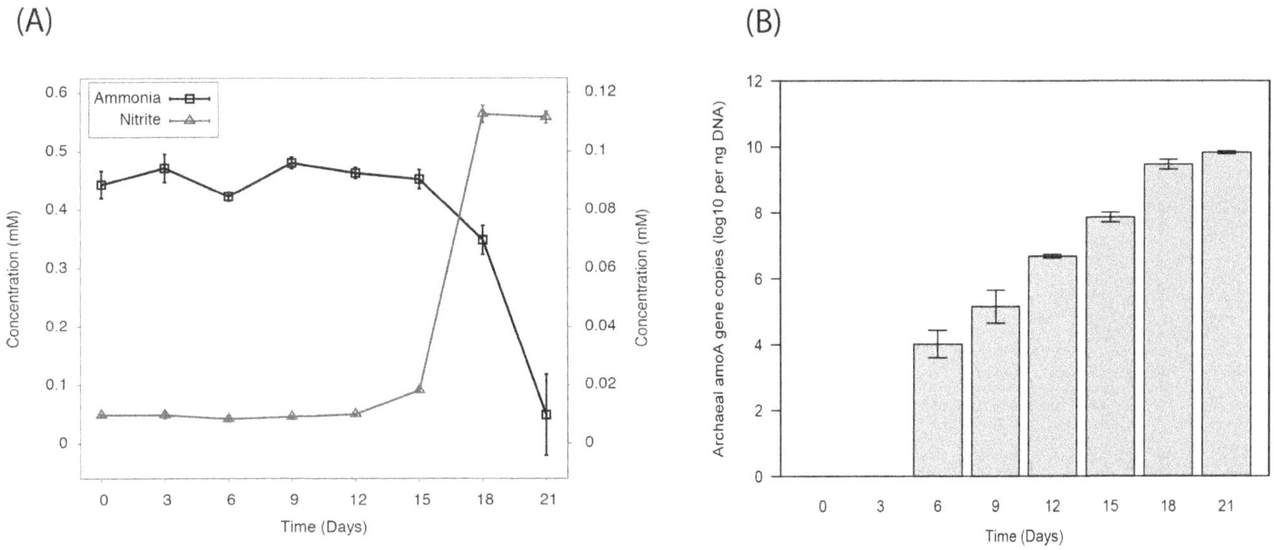

Figure 1. Correlation of ammonia oxidation with the growth of *Ca.* Nitrososphaera evergladensis enrichment culture incubated for 21 days. (A) Concentrations of ammonia and nitrite were determined spectrophotometrically. (B) Archaeal *amoA* gene copies were measured by quantitative PCR.

[1,2,20,28]. Genes for the key enzymes for this pathway are found in the *Ca.* N. evergladensis genome (Figure S3 in File S1, Table S1 in File S2). These genes include alpha and beta subunits of acetyl-CoA carboxylase, acetyl/propionyl-CoA carboxylase, methylmalonyl-CoA epimerase and two domains of mutase. Also, biotin-(acetyl-CoA-carboxylase) ligase, which is responsible for assembly of carboxylase subunits, was identified in *Ca.* N. evergladensis. Candidates for missing enzymes that catalyze reactions of malonyl-CoA to propionyl-CoA were suggested by functional similarity and gene clustering (Table S1 in File S2).

The Thaumarchaeota were recently shown to possess a very efficient, aerobic pathway for CO_2 fixation that differs from that

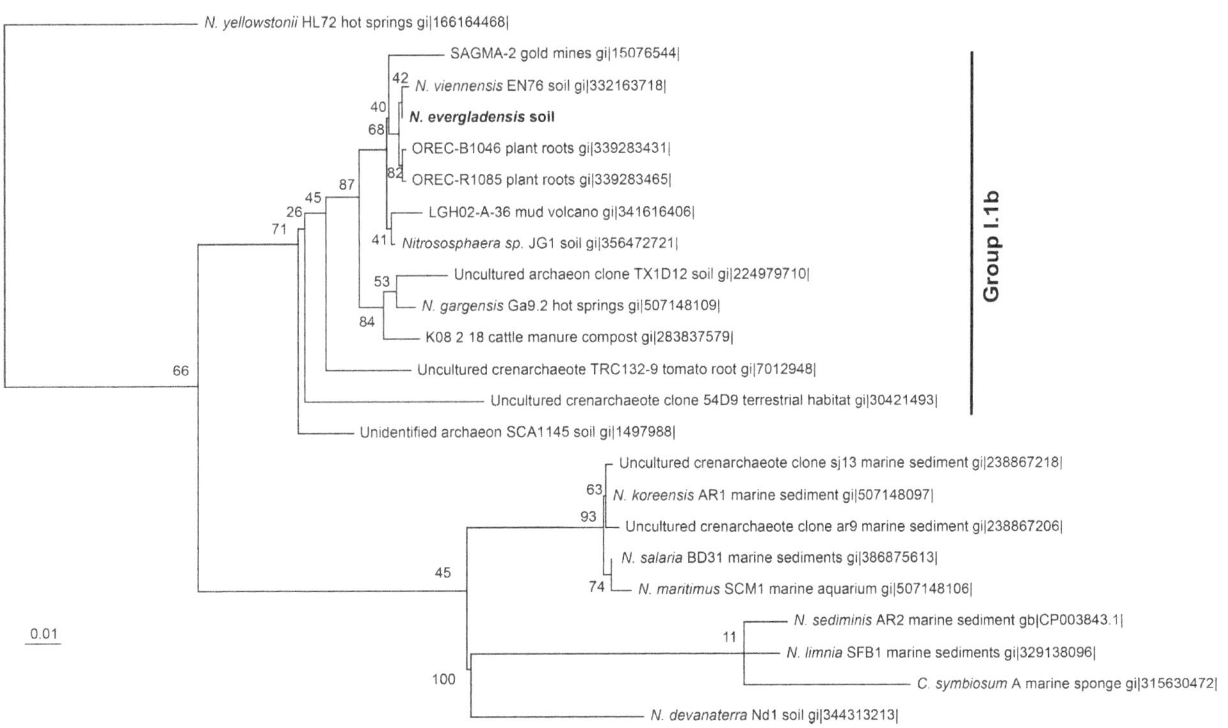

Figure 2. A phylogenetic tree of ammonia-oxidizing archaea 16S rRNA gene sequences (about 1.4 kb). 23 16S rRNA sequences of AOA were randomly selected from the National Center for Biotechnology Information databases. Conservative sites (1.08 kb) were selected using Gblocks. The branching patterns in the maximum-likelihood tree are denoted by their respective bootstrap values (1000 iterations).

Table 1. Two component systems (TS) identified in the genomes of AOA representatives from the group I.1a, I.1b and AOB using Conserved Domain Search and TIGRfam database.

Stimulus or related function	TS gene	Group I.1b		Group I.1a			AOB
		N. everg.	*N. garg.*	*N. kor.*	*N. lim.*	*N. marit.*	*N. europaea*
Cell-envelope stress	*BaeS*	17	20	8	11	4	6
	BaeR	0	0	0	0	0	0
Levels of phosphate	*PhoR*	0(12)*	0(14)	0(6)	0(6)	0(1)	0(5)
	PhoP(PhoB)	18(3)	21(7)	8(5)	9(9)	11(4)	4(8)
	PhoU	18(4)	13(5)	9(3)	10(3)	7(2)	4(1)
Chemotaxis	*CheA*	2	2	0	0	0	3
	CheB	2	1	0	1	0	2
	CheC	2	1	0	1	0	0
	CheY	5	3	0	1	1	3
	CheR	0	2	0	3	0	0
	CheX	1	0	0	1	0	0
Nitrogen levels	*NtrC*	1(14)	0(24)	0(8)	0(17)	0(6)	1(7)
	GlnK	4	4	2	2	2	0
	NtrB	1(2)	0(2)	0(1)	0	0(1)	1(3)
Osmotic stress	*EnvZ*	0	0	0	0	0	0
	OmpR	1	1	1	0	0	5
Osmotic stress	*MtrB*	0	0	0	0	0	0
	MtrA	0(1)	0(1)	0(1)	0(1)	0(1)	0
Trimethylamine N-oxide as an alternate electron acceptor	*TorS*	0(6)	0(7)	0	0(4)	0(4)	0(2)
	TorR	0	0	0	0	0	0
Oxidative, weak acid, and osmotic stress	*BarA*	0	0	0	0	0	0
	SirA	1	1	1	1	1	2
Copper resistance (CopRS)	*CopS*	0	0(1)	0(1)	0(5)	0(6)	0(5)
	CopR	0(2)	0(3)	0(2)	0(2)	0	0(4)
Quorum sensing	*ComP*	0	0	0	0	0	0
	ComA	0(2)	0(2)	0	0	0	0
Pyruvate response	*YpdA*	0(4)	0(6)	0(3)	0(3)	0(3)	0(10)
	YpdB	0	0	0	0	0	0
	Total	69(52)	70(72)	27(30)	26(50)	26(28)	24(45)

* components of TS found by TIGRfam database.

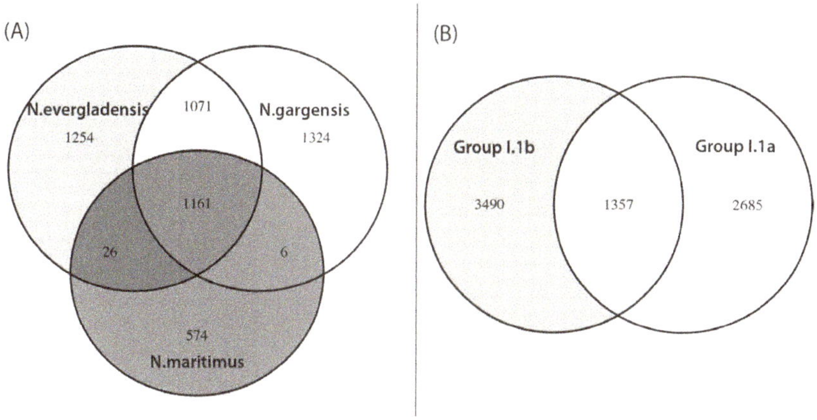

Figure 3. Comparison of protein coding sequences (CDS) of *Ca.* Nitrososphaera evergladensis with CDS of other ammonia-oxidizing archaea. (A) CDS of *Ca.* Nitrososphaera evergladensis were compared to CDS of *Ca.* N. gargensis. (B) CDS of the group I.1a (*N. maritimus*, *Ca.* N. sediminis, *C. symbiosum*, *Ca.* N. limnia, *Ca.* N. koreensis) were compared to CDS of the group I.1b (*Ca.* N. evergladensis and *Ca.* N. gargensis). Overlapping regions represent CDS with amino acid sequence identity 35% and higher.

found in the Crenarchaeota [29]. The *Ca.* N. evergladensis genome has all eleven of the genes identified to date in this pathway [29].

Tricarboxylic acid cycle. The reductive tricarboxylic acid cycle (TCA) is another potential pathway through which AOA may fix CO_2 autotrophically [30]. Enzymes for both oxidative and reductive TCA cycles are predicted for *Ca.* N. evergladensis (Figure S4 in File S1, Table S1 in File S2). Two genes coding for the two subunits of 2-oxoglutarate oxidoreductase, proposed to catalyze interconversion of a-ketoglutarate and succinyl-CoA in the TCA cycle of the hyperthermophilic crenarchaeote *Thermoproteus tenax* [31,32], were found adjacent to the gene coding for aconitase. Gene homologs for four subunits of the reversible succinate dehydrogenase/fumarate reductase were also detected in *Ca.* N. evergladensis. By contrast, the marine AOA *N. maritimus* lacks genes coding for citrate lyase and is thus thought to have an incomplete reductive TCA, or only the oxidative TCA cycle [2]. Spang *et al.* (2012) demonstrated the presence of all candidate enzymes for oxidative TCA in the hot spring AOA *Ca.* N. gargensis. Finding of a gene homolog of isocitrate lyase in the *Ca.* N. gargensis genome is evidence of possible usage of the glyoxylate bypass. Replenishment of the TCA intermediates in the *Ca.* N. evergladensis is mediated by either a pyruvate carboxylase or possibly via glyoxylate bypass. Isocitrate lyase as key enzyme of the glyoxylate bypass was identified in the sequenced genome, but malate synthase was not identified.

AOA have a genomic potential to uptake small organic molecules, and the addition of pyruvate stimulates growth of the soil archaeon *N. viennensis* [21]. However, the question of whether AOA are autotrophs or mixotrophs remains unanswered. Green

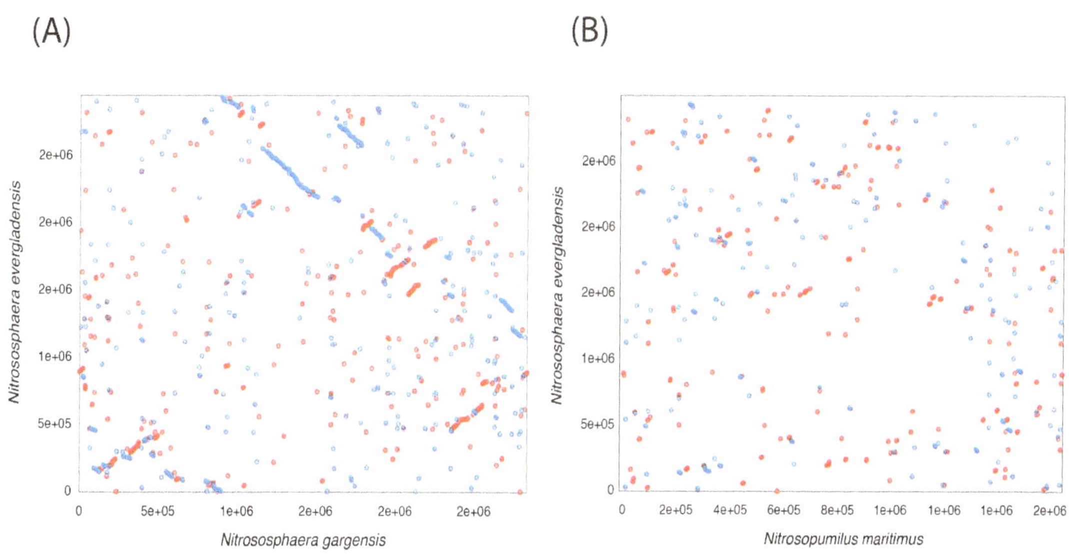

Figure 4. Genome synteny alignments of *Ca.* N. evergladensis, (A) *Ca.* N. gargensis and (B) *N. maritimus.* Axes X and Y represent topology of coding sequences in the comparing genomes. Entire genomes were compared by MUMmer 3.0 package using Promer tool [114]. Each dot represents a match of at least six amino acids from compared genomes. Forward matching amino acid sequences are plotted as red lines/dots while reverse are plotted as blue lines/dots. A line of dots with slope = 1 represents an undisturbed segment of conservation between the two sequences, while a line of slope = −1 represents an inverted segment of conservation between the two sequences.

sulfur bacteria can operate the TCA cycle in both directions [33], but compared to autotrophic growth (reductive TCA), green sulfur bacteria prefer mixotrophic growth (oxidative TCA) enhanced with pyruvate. The presence of the gene homologs of the complete oxidative TCA cycle, and encoded amino acid and di- and tricarboxylate transporters in the genome of *Ca.* N. evergladensis suggest the capacity of this AOA to metabolize small organic compounds via this pathway. The presence of 3-hydroxypropionate/4-hydroxybutyrate and reductive TCA cycles in the genome of *Ca.* N. evergladensis may give an advantage to the AOA to survive under different oxygen concentrations. Under the limited oxygen reductive TCA may be used to fix CO_2 and generate NADH. Reductive TCA is an efficient means to fix CO_2 (four ATPs per one molecule of pyruvate) but it is an oxygen sensitive pathway. It may operate under restricted oxygen, where NO_2^- produced during aerobic ammonia oxidation may be used as a terminal electron acceptor [34,35]. Conversely, under high oxygen availability AOA may shift to the less efficient but oxygen insensitive 3-hydroxypropionate pathway (five-nine ATPs per one molecule of pyruvate) [34]. Another evidence that suggests potential of AOA to live in the low-oxygen conditions, where they may operate reductive TCA cycle, is high AOA affinities to oxygen determined in the AOA cultures [9,16,36]. If TCA cycle present in *Ca.* N. evergladensis is solely utilized for the biosynthetic purposes, than 3-hydroxypropionate cycle will be the only pathway used for CO_2 fixation in this archaeon. Herein lies the significance of the results of that Thaumarchaeota can fix CO2 very efficiently under aerobic conditions [29].

Gluconeogenesis and glycolysis. *Ca.* N. evergladensis has a complete gluconeogenic pathway (Figure S5 in File S1, Table S1 in File S2). Archaea operate a variety of modified Embden-Meyerhof-Parnas (EMP) pathways, which differ from the classic glycolytic pathway [37]. Unusual enzymes for glycolysis were found in the genome of *Ca.* N. evergladensis such as multiple kinases (NTE_03124, NTE_00636, NTE_01922) from the ribokinase superfamily that have broad substrate specificity (e. g. glucose, fructose and mannose) and can be candidates for hexokinase and phosphofructokinase enzymes for glycolytic pathway in AOA. The isomerization of glucose-6P/fructose-6P in *Nitrososphaera* may be catalyzed by either metal-dependent phosphoglucose isomerase (NTE_01540), which belongs to the cupin superfamily and found in the Euryarchaeota or with bifunctional phosphoglucose/phosphomannose isomerase from the sugar isomerases family (NTE_02296). A homolog of pyruvate dikinase was detected in the genome (NTE_02861). Pyruvate dikinase catalyzes reversible interconversion of PEP and pyruvate in Thermopreotei [33,37]. Genes encoding glucose-6-phosphate isomerase, sugar kinases and phosphoglycerate kinase were found only in the *Nitrososphaera* species and did not show any close similarity with other Thaumarchaeota.

Hexose monophosphate pathway. All enzyme homologs of the non-oxidative phase of the hexose monophosphate pathway were identified in *Ca.* N. evergladensis except 6-phosphogluconate dehydrogenase is missing. This enzyme is one of the key enzymes of the oxidative phase (Figure S6 in File S1, Table S1 in File S2). This enzyme was found in the genome of *Ca.* N. gargensis but not in other Thaumarchaeota. Two other enzymes of the oxidative branch of HMP, F420-dependent oxidoreductase and glucono-lactonase, were identified in the genome. F420-dependent oxidoreductase, G6PDH family has been found in archaeal methanogens, *Streptomyces*, and *Mycobacteria* [38,39]. Similar to other AOA, *Ca.* N. evergladensis does not use the Entner-Doudoroff pathway [3].

Energy metabolism

An autotrophic lifestyle of AOA, in which NH_3 and O_2 are used to generate energy, was demonstrated in multiple studies [1,30]. Recently it was indicated that a group of polar Thaumarchaeota had the genomic potential to use urea to fuel a key step of nitrification [40,41]. *Ca.* N. gargensis [20] and *N. viennensis* [21] showed potential to utilize urea as source of NH_3. Coding sequences for multiple subunits of urease (*ureA, ureB, ureC, ureG, ureH, ureF, ureE*) were found clustered together in the genome with passive and electrochemically-driven urea transporters (Figure 5A). These gene homologs provide evidence that *Ca.* N. evergladensis can use urea as NH_3 source. Moreover, all subunits of urease have two copies in the genome and were identified only in the *Ca.* Nitrososphaera genomes. Some of AOA from the group I.1a (*Ca.* Nitrosopumilus sp. AR2 and *Cenarchaeum symbiosum*) showed similarity with *Nitrososphaera* ureases [1,5,6]. However, the majority of sequenced AOA from the group I.1a (*N. maritimus, Ca.* N. limnia, *Ca.* N. salaria, *Ca.* N. koreensis, *Ca.* N. uzonensis) do not have any signatures of urea degradation [1,4,16,19].

Ammonia is oxidized to NO_2^- in a two-step reaction. The first reaction is likely catalyzed by archaeal ammonia momooxygenase (AMO). Several genes (*amoA, amoB, amoC* and *amoX*-like) are predicted to encode subunits of AMO in the *Ca.* N. evergladensis genome (Figure 5B, Table S1 in File S2). The *amoA* gene is the most conserved of the *amo* genes and shares 99% amino acid identity with *amo* genes of *N. viennensis* and 95% with *Ca.* N. gargensis (Table S1 in File S2). Similar to ammonia-oxidizing bacteria (AOB), archaea of group I.1b encode several *amoC* subunits of AMO [42]. However, majority representatives of group I.1.a have only one copy of *amoC* (Figure S7 in File S1). Little is known regarding the function of *amoC*. Previous studies have revealed that *amoC* may stabilize the AMO under stress conditions such as starvation and heat shock [43]. It is noteworthy that multiple copies of *amoC* appear more often in AOA and AOB associated with soil environments, which harbor more diverse stressors than marine environments and require more adaptations for organisms to survive and successfully compete [42]. For example, ammonia oxidizers from soil have up to seven *amoC* copies (Figure S7 in File S1), while marine AOA and AOB usually encode only one *amoC* [20,42]. The amino acid alignment identity between seven *amoC* copies ranges between 72% and 97% in the *Ca.* N. evergladensis genome.

Bacterial AMO oxidizes NH_3 to hydroxylamine (NH_2OH) and it is further oxidized to NO_2^- by hydroxylamine oxidoreductase (HAO) [44,45]. However, no homologs of the bacterial HAO are found in AOA genomes [1,2,20,28,46]. Two hypothetical pathways of NH_3 oxidation to NO_2^- were proposed [2]. The first suggests nitroxyl (HNO) is produced a reactive intermediate that is further is oxidized to NO_2^- by nitroxyl oxidoreductase. In latter pathway, NH_2OH is a possible intermediate in the reaction and is oxidized to NO_2^- by periplasmic multicopper oxidases. Recently, Vajrala et al. [47] provided direct evidence that *N. maritimus* oxidizes NH_2OH to NO_2^-. Therefore, the alternative pathway with NH_2OH as the intermediate is possible. Similar to other AOA, the *Ca.* N. evergladensis genome encodes genes for six periplasmic multicopper oxidase proteins that may be candidates for HAO (Figure S8 in File S1). Moreover, two of these oxidases are dissimilatory copper-containing nitrite reductases (NirK).

AOB channel two electrons from HAO through cytochromes c_{554} to c_{m552} [44]. Similarly to other AOA no homologs for cytochromes c_{554} and c_{m552} were predicted for *Ca.* N. evergladensis. Instead multiple copper-containing plastocyanin-like electron carriers are candidates for transferring electrons to O_2 (Table S1 in File S2, Figure S8 in File S1). NAD-quinone

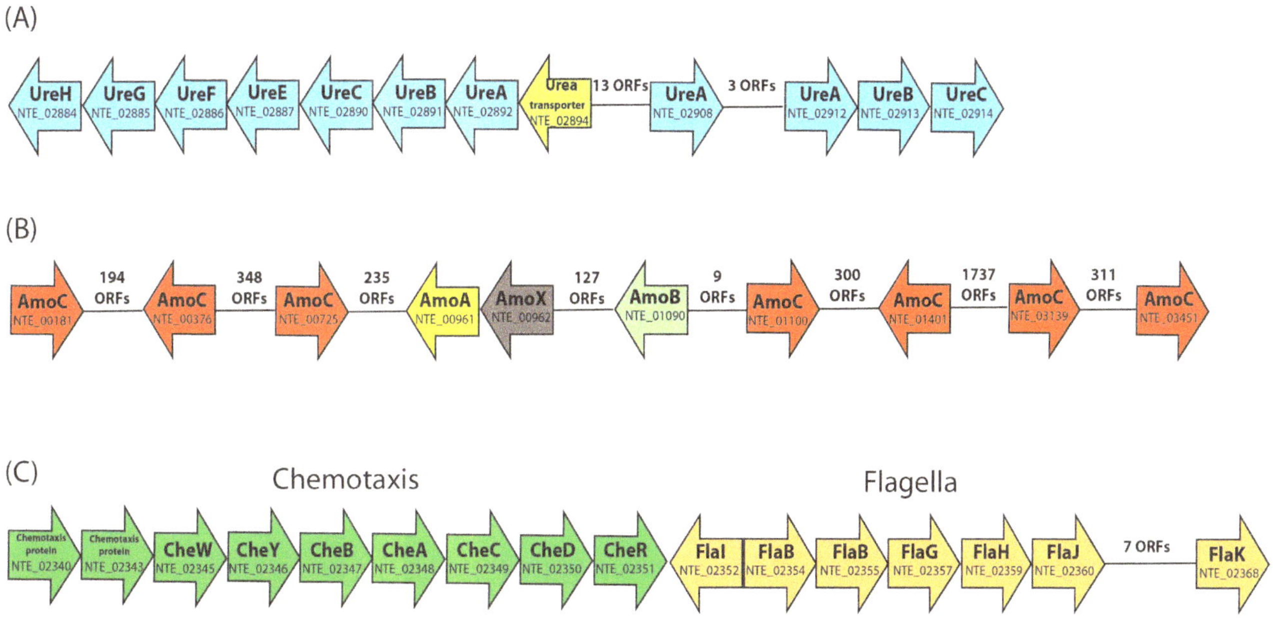

Figure 5. Organization of (A) urease and urea transporter genes; (B) ammonia monooxygenase gene order of *Ca.* N. evergladensis; (C) flagella and chemotaxis genes in the present genome.

oxidoreductase (Complex I) catalyzes the transfer of electrons from NADH to ubiquinone. *Ca.* N. evergladensis has 11 genes encoding subunits of NAD-quinone oxidoreductase, but it is missing genes that encode the E, F and G subunits. Further, a proton motive force (PMF) may be generated through complexes III (Rieske Fe-S proteins, plastocyanines), complex IV (proton-pumping oxygen-reducing plastocyanin-copper oxidases), and by complex V (Archaeal/vacuolar-type H^+-ATPase). Copper-containing nitrite reductases may reduce NO_2^- to nitric oxide (NO). NO was shown to have a stimulating effect on ammonia oxidation in the AOB [48]. Hence, NO may be involved in the regulation of the AMO activity in AOA.

Triacylglycerols and Polyhydroxyalkanoates as lipid reserve materials

Although many archaea store carbon in the form of polyhydroxyalkanoates (PHAs), some archaea and other organisms preserve carbon in the form of triacylglycerols (TAGs) [49]. *Ca.* N. evergladensis and *Ca.* N. gargensis possess lipases (lysophospholipase, monoglyceride lipase) that may hydrolyze ester bonds in triacylglycerides of long chain fatty acids. Extracellular lipases may also be involved in utilization of monoglycerides from the soil. These lipase homologs are lacking in the group I.1a.

Other lipophilic compounds that likely accumulated in Thaumarchaeota as a reserve material are PHAs. Polyhydroxyalkanoate synthase was found almost in all representatives of Thaumarchaeota [20,50]. The *Ca.* N. evergladensis genome encodes for class III PHA synthase (*phaC, phaE*) (Table S1 in File S2). Gene homolog for subunit PhaE of PHA synthase shares some similarity with *Ca.* N. gargensis, but it is very distantly related to representatives of group I.1a.

Isoprenoids as biomarkers for Thaumarchaeota. Archaea use isoprenoids to make phospholipids. The hydrophobic tails of the phospholipids are isoprenoid alcohols ether-linked to glycerophosphate to form monoglycerol-tetraether. Thaumarchaeota have a specific cyclopentane ring-containing dibiphytanyl glycerol tetraether membrane lipid (crenarchaeol) [51,52]. Damsté et al. [51] hypothesized that formation of cyclohexane ring in crenarchaeol may be an adaptation to cold temperatures in the marine water. However, crenarchaeol was also identified in AOA from thermophilic environments [53]. In addition to crenarchaeol, high concentrations of crenarchaeol regioisomer have been determined in *Ca.* N. gargensis, but this regioisomer is either absent or present in very low amounts in other analyzed AOA from I.1a and ThAOA groups [54]. *Ca.* N. evergladensis has the mevalonate pathway, which operates in archaea and eukaryotes [55]. This pathway is used to synthesize isopentenyl diphosphate (IPP), which is converted to different isoprenoids in the cell (quinones, hydrophobic tails of the phospholipids), using a set of enzymes present in the *Ca.* N. evegladensis genome: farnesyl pyrophosphate synthetase, and octaprenyl pyrophosphate synthetase, undecaprenyl pyrophosphate synthetase (Table S1 in File S2).

Stress response

The sequenced genome revealed the presence of multiple adaptations to survive osmotic and oxidative stress, high concentrations of heavy metals, and elevated temperatures. Moreover, it contains more diverse mechanisms than AOA from the group I.1a to resist high concentrations of heavy metals.

Osmotic stress. One of the strategies to cope with high salinity, and in some cases temperature stress, in archaea is accumulation of compatible solutes, small soluble organic molecules [56]. These solutes can be either transported into the cell or synthesized *de novo*. Several aquaporins that transport water and small uncharged molecules, and belong to the major intrinsic protein family [57], were found in the *Ca.* N. evergladensis genome as well as in other AOA genomes (Table S1 in File S2). Aquaporins identified in *Ca.* N. evergladensis are related to glycerol uptake facilitators. Glycerol is one of the uncharged compatible solutes, which can be used by AOA for osmoadaptation. Mannosyl-3-phosphoglycerate and myo-inositol-1-phosphate synthases were found in the genomes of group I.1b but not in the genomes of group I.1a. These enzymes are involved in the biosynthesis of di-myo-inositol phosphate and mannosylglycerate, two main prokaryotic compatible solutes. These compatible solutes are commonly represented in thermophilic and hyperthermophilic bacteria and archaea [58].

To mitigate oxidative damage, *Ca.* N. evergladensis encodes superoxide dismutase, peroxiredoxins, and ferritin-like proteins. The majority of enzymes involved in oxidative stress response shared similarity with *Ca.* N. gargensis and other AOA. However, some of the peroxiredoxins and ferritin-like proteins (NTE_01148, NTE_01225, NTE_01156) shared sequence similarity to Euryarchaeota and Bacteria but not to Thaumarchaeota (Table S1 in File S2).

Resistance to heavy metals. Archaea have been found in extreme environments such as mining sites with high concentration of heavy metals [59,60]. *Ca.* N. evergladensis developed mechanisms that would help it resist high external concentrations of metals with at least 21 putative heavy metal resistance proteins. Nine of these homologs were encoded only in *Nitrososphaera* genus, and not in other AOA, and six were found only in the *Ca.* N. evergladensis genome (Table S1 in File S2). Both *Ca.* N. evergladensis and *Ca.* N. gargensis are predicted to have broad tolerance to a variety of heavy metals: copper, zinc, cobalt, cadmium, arsenic, and mercury. However, AOA from group I.1a are more limited in their adaptations to high concentrations of heavy metals.

A higher tolerance of AOA than AOB to copper was shown in soil [61]. Ettema *et al.* (2006) [62] suggested a potential copper resistance gene cluster, which consists of a putative methallochaperone and P-type cation transporting ATPase. This mechanism was identified in thermoacidophilic archaeon *Sulfolobus metallicus*, *Sulfolobus solfataricus*, and *Ferroplasma acidarmanus* [63]. A similar gene cluster with encoded P-type ATPases and copper chaperones was found in *Ca.* N. evergladensis (Table S1 in File S2). This mechanism of copper tolerance was also present in *Ca.* N. gargensis genome but not in other sequenced AOA. An alternative putative mechanism of copper detoxification in AOA may involve multicopper oxidases. Multicopper oxidases play an important role in copper resistance in many bacteria [64]. Multiple putative multicopper oxidases are encoded in all known AOA, however, their role in copper tolerance remains unclear (Table S1 in File S2). Periplasmic divalent cation tolerance protein (NTE_02314) is widely represented in AOA, and may also transport copper outside the cell. Copper tolerance may also involve an inorganic polyphosphate transport system [64]. Polyphosphate kinase (PPK), which catalyzes the reversible conversion of the terminal phosphate of ATP into polyphosphates (polyP), and exopolyphosphatase (PPX), is known to hydrolyze polyP. This mechanism was described in other archaea [65]. The enzymes supporting polyP transport encoded in *Ca.* N. evergladensis showed homology to *Methanosarcina*, *Methanoregula*, and *Methanomassiliicoccus* from Euryarchaeota, but this transport is absent from other known Thaumarchaeota.

Ca. N. evergladensis has three putative nickel transporter genes, and one of these high affinity permeases (NTE_02909) is specific only for this thaumarchaeon.

Other metal resistance proteins include cobalt-zinc-cadmium resistance proteins (one is unique for *Ca.* N. evergladensis), putative tellurium resistance membrane protein, and arsenic efflux proteins. Notably, an arsenic pump was identified only in the genomes of *Nitrososphaera* species but not in AOA related to *Nitrosopumilus*.

Heat shock. Besides the altered composition of lipids in the membranes that are used to survive at elevated temperatures, Thaumarchaeota encode an entire set of proteins to cope with temperature stress [66]. The *Ca.* N. evergladensis genome harbors gene homologs of heat-shock proteins (HSP) such as small HSP, HSP60 (GroEL and Thermosomes), and chaperones such as DnaJ, DnaK and GrpE. Moreover, the copy number of these gene homologs is higher in group I.1b than in group I.1a.

Nitrogen metabolism

Ammonia assimilation. Group I.1b maybe more adapted to high concentrations of NH_3 than group I.1a. The majority of group I.1a AOA were identified in marine environments (*N. maritimus*, *Ca.* N. koreensis, *Ca.* N. limnia, *Ca.* N. sediminis, *Ca.* N. salaria) where the ammonium concentration was as low as 0.017 mg L^{-1} [23,67]. Group I.1b is mainly found in the environments at much higher ammonium concentrations (0.1–9 mg L^{-1}), such as soil, and representatives of this group are more tolerant of high ammonium levels compared to the majority of AOA isolated from the marine environments or soils with low pH [3,9,16,21,23,36].

NH_3 can be used by *Ca.* N. evergladensis not only as energy source, but also as a N source. The full set of enzymes involved in NH_3 assimilation is present (Figure 6), including glutamate dehydrogenase, glutamine synthetase, and glutamate synthase. Glutamate synthase (NTE_01407) was found only in the *Ca.* N. evergladensis genome, and not other sequenced AOA genomes.

Putative pathway for Nitrous oxide production. A cluster of genes that encode a putative multicopper oxidase related to nitrite reductase (*nirK*) and the gene homologs of the nitric oxide reductase subunits (*norD* and *norQ*) are present in the genome of *Ca.* N. evergladensis. However genes coding for the catalytic subunits (*norB* and *norC*) were not identified. Proximity of the *nirK* and *norD,Q* homologs in the genome may suggest that these genes code for the proteins involved in the same metabolic pathway [68]. A similar set of genes was found in both AOA groups (Table S1 in File S2). Nitrite reductase and nitric oxide reductase were shown to be involved in cell tolerance to NO_2^- and NO [69,70].

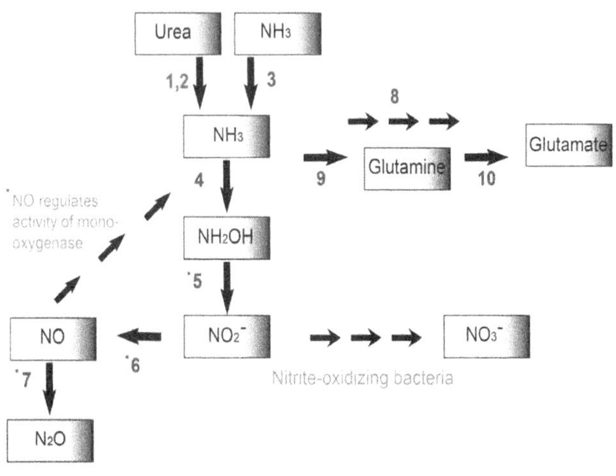

Figure 6. Components of the nitrogen metabolism of *Ca.* N. evergladensis: ammonia oxidation (4, 5), ammonia assimilation (8, 9, 10), nitrite reduction (6), nitrous oxide production (7). Reactions are mediated by the following transporters and enzymes: urea transporters, urease (1, 2), ammonia transporters (3), archaeal ammonia monooxygenase (AMO) (4), candidate enzyme: multicopper oxidase (5), nitrite reductase (NirK) (6), nitric oxide reductase (NorD, NorQ), catalytic subunit (NorB) is missing (7), glutamate dehydrogenase (8), glutamine synthetase (9), glutamate synthase (10). NO may upregulate activity of AMO. * - experimental evidences are needed.

Alternatively, as in *Nitrosomonas europaea*, *Ca.* N. evergladensis may use NO_2^- and NO as terminal electron acceptors via a putative denitrification pathway [71]. Several studies have shown that AOA cultures are able to emit N_2O and several potential pathways were suggested [15]. However, missing intermediates and missing catalytic subunits of enzymes result in an incomplete pathway for N_2O production, and further experiments are required to determine functional enzymes and intermediates for the N_2O production pathway [17,18].

Transport

In the *Ca.* N. evergladensis genome, 141 transport proteins were identified. (Table S2 in File S2), which is larger than the 89–108 found in group I.1a. Of these, 43 encoded an ATP binding cassette, 17 likely code for pores and channels, and the rest are electrochemical-potential-driven transporters, including the Twin Arginine Translocation system. Twelve of these transporters were not found in other Thaumarchaeota. Di- and tricarboxylate transporters found only in *Ca.* N. evergladensis can be involved in transport of TCA cycle intermediates, such as citrate, malate, and succinate. Thirty-one transporter genes were specific for group I.1b. They include genes coding for mechanosensitive ion channels, urea transporters, a symporter from the major facilitator superfamily (MFS), members of the cation diffusion facilitator family for transport of divalent metals, and members of solute carrier families 5 and 6-like superfamily for co-transport of Na^+ with sugars, amino acids, inorganic ions, or vitamins. Also unique for group I.1b, were proteins from the sodium bile acid symporter family. Transporters of this family were shown to be involved in sodium-dependent transport of a variety of organic molecules in plants and humans [72,73].

Motility, chemotaxis and two-component regulatory systems

At least 69 protein-coding genes related to two-component regulatory systems (TS) were found in *Ca.* N. evergladensis and 70 genes related to TS were found in *Ca.* N. gargensis (Table 1). Notably, group I.1b encodes two and a half times the number of TS genes than group I.1a (Table 1).

Motility-associated genes involved in archaeal flagella and pili assembly were clustered together with protein-coding genes for chemotaxis (Figure 5C). The operon includes genes encoding flagellins (two copies of *flaB*) followed by *fla*-associated genes (*flaG, flaH, flaJ, flaI, flaK,* and *flaD*). Structure and assembly of found flagella are closely related to type IV pili. Proteins involved in motility encoded in the *Ca.* N. evergladensis genome were observed in three other AOA, *Ca.* N. limnia, *Ca.* N. uzonensis and *Ca.* N. gargensis. These flagella-coding genes also share identity with other Thermoprotei (*Desulfurococcus kamchatkensis, Ignisphaera aggregans, Fervidicoccus fontis,* and *Sulfolobus acidocaldarius*). Adjacent to the *fla*-operon is a set of genes involved in chemotaxis.

Ammonia is the main energy and nitrogen source for AOA. The NtrB/NtrC TS, involved in the response to different NH_3 levels is present in the *Ca.* N. evergladensis genome (Table 1). NtrB senses the nitrogen levels and, under NH_3 limitation, activates NtrC by phosphorylation. NtrC activates expression of glutamine synthetase (GlnA) and it allows cells to grow under nitrogen-limited conditions.

The PhoR/PhoP (PhoB) TS is found in the *Ca.* N. evergladensis genome and in other Archaea and Bacteria [74,75]. It plays an important role for sensing inorganic phosphate levels and under phosphate-limited conditions, PhoR/PhoP activates alkaline phosphatase (Table 1).

The *Ca.* N. evergladensis genome, and other ammonia oxidizers, encode components of TS that provide a respond to different environmental stresses, such as cell-envelope stress (BaeS/BaeR) [76], osmotic pressure (EnvZ/OmpR, MtrAB, BarA/SirA), and copper resistance (CopSR) [77].

Ca. N. evergladensis encodes the sensor kinase (YpdA) of TS (YpdA/YpdB) that responds to extracellular pyruvate as a stimulus [78]. This finding supports the hypothesis that pyruvate promotes AOA growth. The response regulator of the TS ComP/ComA, which controls competence in *Bacillus subtilis* via a quorum-sensing mechanism [79], was also found in the group I.1b, but not in the group I.1a.

Information processing machinery

The information processing machinery of *Ca.* N. evergladensis is similar to other Thaumarchaeota, and it shares more homology with eukaryotes than with bacteria [52,80,81] (Table S3 in File S2). The *Ca.* N. evergladensis genome has 61 ribosomal proteins that show a phylum-specific pattern (Table S3 in File S2). One of the specific signatures of the known Thaumarchaeota is that their genomes, including *Ca.* N. evergladensis, are missing gene homologs for r-protein family LXa that is solely present in Archaea. Also, r-proteins L14e and L34e found in other archaeal phyla but not in Thaumarchaeota, are missing from *Ca.* N. evergladensis.

DNA-dependent RNA polymerase II (RNAP) is composed of 12 subunits in *Ca.* N. evergladensis as in *Ca.* N. gargensis [20]. Most of the subunits are homologous to other Archaea [80]. However, in contrast to Euryarchaeota, Crenarchaeota and Nanoarchaeota that have two genes encoding A subunit of RNAP, *Ca.* N. evergladensis and other Thaumarchaeota contain a single *rpoA* gene. This unsplit *rpoA* is common for Eykarya and it was suggested that other archaeal lineages that possess split *rpoA* branched off later in evolution than the Thaumarchaeota [52].

Archaeal RNAP requires two accessory factors: transcription factor B (TFB) (an ortholog of TFIIB), and TATA-box binding protein (TBP) [80]. *Ca.* N. evergladensis has at least nine transcription factors B (TFB), and one TATA-box-binding protein. Other representatives of Thaumarchaeota have a similar number of TFB. For example, *Ca.* N. gargensis and *Ca.* N. koreensis have 11 TFBs, *Ca.* N. limnia encodes at least 9 TFBs, *Ca.* N. sediminis and *N. maritimus* have 10 and 8 TFBs, respectively [2,20,28].

Multiprotein bridging factor 1 (MBF1) is a transcriptional cofactor that bridges the TATA box-binding protein (TBP) and regulatory DNA-binding proteins [81]. MBF1 is a conserved protein present in all eukaryotes and archaea, with exception of the *N. maritimus* and *C. symbiosum* [82]. This protein is found within the soil archaea, *Ca.* N. gargensis [20] and *Ca.* N. evergladensis. Apparently, group I.1b branched off evolutionarily earlier than group I.1a, which lost MBF1 over time.

DNA replication, repair, cell cycle. The *Ca.* N. evergladensis genome contains three *orc1/cdc6* orthologues and one of the *cdc6* orthologs is found only in group I.1b (Table S3 in File S2). *Ca.* N. evergladensis encodes small and large subunits of archaeal DNA polymerase II (pol D) and DNA polymerase type B. Also, it carries genes for the large subunit of replication factor C, both subunits of DNA primase, one copy of archaeal DNA polymerase sliding clamp, DNA ligase, RNase HII, and flap endonuclease. The gene encoding topoisomerase IB, a signature marker for Thaumarchaeota, but not for other archaeal phyla, is present in the *Ca.* N. evergladensis genome.

Cell division. Among all archaeal phyla only Thaumarchaeota shares two systems of cell division [52,82]. One is FtsZ-based, present in bacteria and most archaea, and another is CdvABC-based, which is present in Crenarchaeota, and homologous to the eukaryotic ESCRT system [10]. *Ca. N.* evergladensis codes for homologs of CdvA, CdvC (Vps4) and several homologs of CdvB-like proteins (ESCRT-III). Also, the genome has the FtsZ-based division system. *ftsZ* gene homolog encoded in the genome shares 40–63% amino-acid identity with other AOA, and less than 29% with other archaea and bacteria. Another cell division feature shared with eukaryotes and other archaea that is found in *Ca.* N. evergladensis is a homolog of pelota proteins required for meiotic cell division [83]. Pelota homologs are widely represented in archaea [83]. In *Ca.* N. evergladensis pelota homologs have 70% amino acid identity with *Ca.* N. gargensis and 42–50% identity with AOA from the group I.1a (Figure S9 in File S1). In archaea, this protein was suggested to play role in translational elongation, termination, and quality control of mRNA (mRNA surveillance) [84,85]. Archaeal pelota may be involved in the release of the stalled ribosomes and degradation of damaged mRNA [85].

DNA folding and repair. Similar to other Thaumarchaeota, *Ca.* N. evergladensis possesses the genes needed to compress and methylate DNA. The DNA repair system of *Ca.* N. evergladensis includes UvrABC endonuclease, which is common in mesophilic archaea [86]. The soil archaea possess other genes in DNA repair, such as ERCC4-type nuclease and helicase, DNA repair helicase RAD25, nucleotidyltransferase/DNA polymerase involved in DNA repair, and photolyase.

CRISPR-based system

CRISPRfinder determined only one CRISPR locus 7220 bp in the genome of *Ca.* N. evergladensis. The CRISPR region in *Ca.* N. evergladensis, which is longer than that in *Ca.* N. gargensis and many other AOA, was likely a result of more exposure to viruses in its environment [87]. The CRISPR locus consists of 99 repeat/spacer sequences, which is almost 3 times larger than that of *Ca.* N. gargensis [20]. The CRISPR spacers are 34–38 bp and equally separated with identical 37 bp direct repeats. The two genomes of soil group I.1b had higher CRISPR length and longer repeat length than the marine AOA. The five most common CRISPR-associated proteins (*cas1-4, cas7*) are adjacent to repeat/spacer sequences. Variable sequences or spacers mostly correspond to segments of captured viral sequences [88]. However, only one CRISPR spacer had significant homology to any virus, the *Helicobacter* phage phiHP33. In bacteria CRISPR-Cas system provides resistance to exogenous genetic elements and provides acquired immunity for the cell [88,89]. Most likely, *Ca.* N. evergladensis utilizes CRISPR in the similar way to maintain genome integrity.

Phylogeny and adaptations

Do the detected differences between genomes of the AOA from the group I.1a and I.1b may give us clues as to how these organisms have adapted to their environments? Many soil surveys revealed that the analyzed soils across the globe were dominated by AOA from the group I.1b (or *Nitrososphaera* cluster), while marine environments were represented by AOA from the group I.1a (or *Nitrosopumilus* cluster) [7,23]. Auguet et al. [23] studied global ecological patterns of Archaea and found that habitat classification was a strong structuring factor of the archaeal communities. Cells of isolated AOA from group I.1a are typically straight rods [3,19,90,91], whereas cultured I.1b archaea are spherically shaped [16,20,21]. Another important physiological difference between two lineages is preference of ammonia concentrations discussed above. Analysis of genomic features of *Nitrososphaera* genus may point to other physiological signatures of this group. For example, sequenced AOA representatives from the group I.1b have a larger genome size and higher G+C content than the archaea from the group I.1a.

More than 3,000 CDS are exclusively present in the genomes of Thaumarchaeota from the group I.1b but are absent in the genomes from the group I.1a AOA (Figure 3B). Coding sequences unique to the I.1b archaea included DNA repair proteins, transporters, two-component systems, and information processing machinery (Table S4 in File S2). Enzymes involved in DNA repair unique to *Ca.* Nitrososphaera included DNA repair photolyase, predicted DNA alkylation repair enzyme, an uncharacterized protein predicted to be involved in DNA repair, and replicative and repair DNA polymerase IV (family X). Also, group I.1b uniquely possessed some proteins involved in information processing machinery, such as DNA topoisomerase IA, ribosomal protein L6p, and transcriptional regulators.

The central metabolism of the AOA from the group I.1b is functionally more diverse than that of I.1a group AOA. AOA from the group I.1b have complete TCA cycle and HMP pathways. Also, unlike the I.1a group, the I.1b group seems to be capable of utilizing complex carbohydrates such as glycogen, chitin, and triacylglycerides, as suggested by the presence of genes coding for the glycogen debranching enzyme (NTE_01977), multiple chitin deacetylases, gene homologs for chitinases (NTE_00025, NTE_01408), and monoglyceride lipases. This appears appropriate given the oligotrophic nature of the environments where representatives of I.1a were isolated.

The I.1b group AOA also possess more transporters than I.1a group AOA, such as an ATPase P-type transporter, a urea transporter, a putative hydroxymethylpyrimidine transporter CytX, Di- and tricarboxylate transporters, and transporters from the solute carrier families 5 and 6-like superfamily. These transporters suggest that the compounds transported by these systems are available in the soil environment, but absent or rare in the marine environment.

Several transposases from family IS605 were encoded in the AOA genomes from the group I.1b, but not in the group I.1a. Transposable elements are widely distributed in archaeal genomes, and play an important role in the genome plasticity and response to environmental stimuli [92].

The genomes from the group I.1b also have a higher number of gene duplications compared to the genomes from the group I.1a (Figure 7). Most of the duplicated genes in the group I.1b are involved in adjusting to different environmental conditions, responses to environmental stresses, or efficient nutrient utilization.

Provisional classification and conclusion

In this study we sequenced and analyzed genome of the mesophilic AOA from the group I.1b enriched from the soil. We propose the following *Candidatus* status for this microorganism:

"Nitrososphaera evergladensis" sp. nov.

Ethymology. Nitrosus (Latin masculine adjective), nitrous, produces nitrite; sphaera (Latin feminine. n.), spherically shaped; evergladensis (Latin neutrum genitive), isolated from the Everglades.

Locality. Histosol from the Everglades agricultural area.

Diagnosis. Ammonia-oxidizing archaea phylogenetically related to the Thaumarchaeota group I.1b (*Nitrososphaera* cluster) [93]; not isolated; enriched from the agricultural soil.

Analysis of the *Ca.* N. evergladensis genome revealed many similarities of basic metabolism with the rest of AOA, including genes coding for ammonia transporters and genes for AMO subunits, genes for CO_2 fixation via modified hydroxypropionate

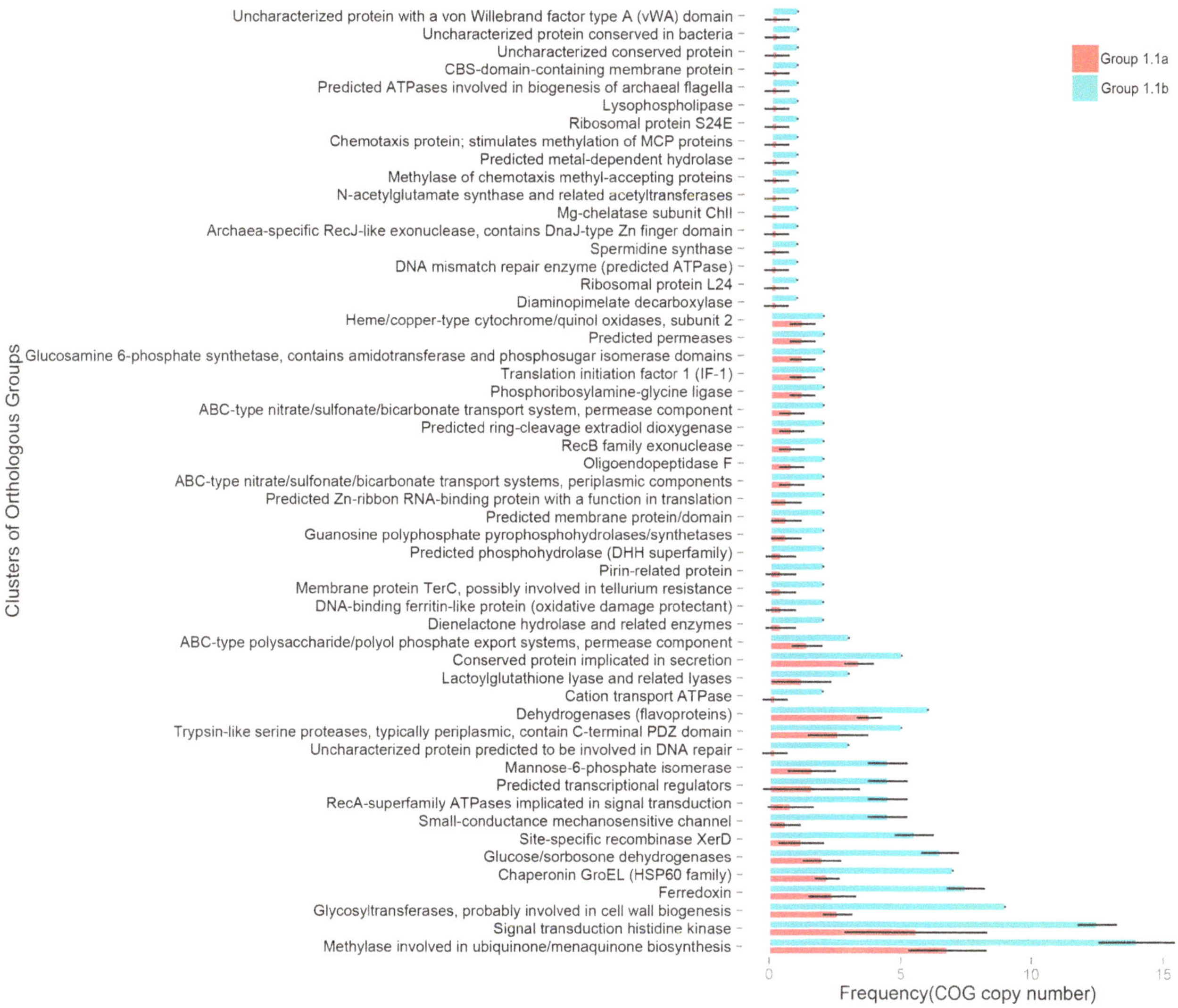

Figure 7. Distribution of Clusters of Orthologous Groups (COGs) in the AOA genomes from soil group I.1b and I.1a. COGs represented with higher copy number in the group I.1b than in the group I.1a were selected by Ttest (P_{value}<0.05). Error bars display standard deviation.

cycle, as well as the HMP, TCA and gluconeogenic pathways. This organism belongs to the group I.1b of the Thaumarchaeota, and shares most of its coding sequences with the closest sequenced relative, *Ca.* N. gargensis, isolated from hot springs. Despite the fact that *Ca.* N. evergladensis is phylogenetically closely related to *Ca.* N. gargensis, they have only 40% of whole genome homology revealing significant differences in the metabolic potential of these organisms. The majority of CDS present in *Ca.* N. evergladensis, but absent in *Ca.* N. gargensis, are hypothetical proteins (Table S5 in File S2). *Ca.* N. evergladensis is also distinct from its closest relative, *Ca.* N. gargensis in that it has a much larger CRISPR region, CRISPR-associated genes, transporters for inorganic and small organic molecules, electron carriers, steroid isomerases, chitin deacetylases, and transcriptional regulators that are completely absent in the *Ca.* N. gargensis genome.

When we compared the genetic potential of the archaeal ammonia oxidizers from group I.1b and group I.1a, the AOA from the group I.1b demonstrated a higher potential to adapt to changes in the environment, and to utilize a broad array of carbon sources compared to the AOA representatives from the group I.1a.

About half of all identified proteins were not assigned to functions and may encode completely novel pathways. Further experiments must be conducted to link novel genes to their specific functions, and determine their ecological role.

Materials and Methods

These soil samples were not collected at a national park or private land. The land is owned by the University of Florida and is within the Everglades Agricultural Area, not the Everglades National Park. No permits were required to collect the soil samples uses in this work. The field studies did not involved endangered or protected species. The GPS coordinates of the research site are: 26.667863, −80.633039.

Enrichment culture

Soil samples for the enrichment were collected from agricultural plots in the Everglades Agricultural Area planted with sugarcane. The soil from this location in the Everglades Agricultural Area is classified as a histosol with pH ~8, moisture ~123%, organic matter ~70%, nitrate concentration ~54 mg per kg soil, and ammonium concentration ~9 mg per kg soil. To enrich for *Ca.* N. evergladensis, 10 g of soil were resuspended in 0.5 L of the medium for culturing of ammonia-oxidizing archaea (AOA) [21]. The medium contained 0.5 mM NH_4Cl, and 2 ml $NaHCO_3$ (1M). The headspace above the non-shaking culture was air. One fifth of the enrichment culture was transferred to fresh medium every four weeks for over one year. To further enrich the medium, several antibiotics, including gentamicin (50 µg/ml), tetracycline (5 µg/ml), and erythromycin (10 µg/ml), were applied in order to suppress growth of co-cultured bacteria. However, the addition of antibiotics also affected archaea and did not produce a pure AOA culture. The concentrations of NH_3 and NO_2^- were determined by Griess Reagent Kit for Nitrite Determination (G-7921) (Molecular Probes, Eugene, OR, USA), and by the Ammonia Assay Kit (Sigma, St. Louis, MO, USA).

Extraction of DNA

Cells were collected by filtering 1L of culture onto 0.1 µm polycarbonate membrane (Millipore; Billerica, MA, USA). DNA was isolated from the membrane using the PowerSoil DNA Isolation Kit (MO BIO; Carlsbad, CA, USA). Extractions were performed according to the manufacturer's protocol. All genomic DNA concentration and purity were determined by NanoDrop spectrophotometry (Thermo Scientific; Wilmington, DE, USA) and by Qubit 1.27 Fluorometer (Invitrogen; Grand Island, NY, USA).

Quantification of archaeal 16S rRNA and *amoA* genes

Bacterial and archaeal 16S rRNA genes were amplified using universal prokaryotic primers 515F (5'-GTGCCAG-CAGCCGCGGTAA-3') and 806R (5'-GGACTACVSGGG-TATCTAAT-3') [94], cloned into pCR4-TOPO vector and sequenced with M13f and M13r vector primers using Sanger sequencing standard protocol. The archaeal *amoA* copy number in the culture was measured by quantitative PCR (qPCR). Primer sets Arch-amoAf and Arch-amoAr were used [95] (File S1 and S2). Bacterial *amoA* detection was carried out using primer set AmoA1f and AmoA2r [96].

DNA sequencing

Enrichment culture was sequenced using an Ion Torrent Personal Genome Machine (PGM) (Life Technologies; Grand Island, NY, USA), and the Pacific Biosciences platform (Pacific Biosciences; Menlo Park, CA, USA), according to the manufacture's protocols. Ion Torrent sequencing resulted in 2,389,864 reads with average read length 241 bp (~127X coverage). PacBio platform produced 197,138 reads with an average length 4,117 bp (~179X coverage) (Table B in File S1).

Genome assembly and annotation

Sequenced Ion Torrent reads were imported into CLC Genomics Workbench v.4.0.3 (CLC bio; Aarhus, Denmark), and quality trimmed using a minimum phred score of 20 (with a limit of 5% of low quality bases per read) and a minimum read length of 80 bp. PacBio reads were processed with BLASR mapper (http://www.pacbiodevnet.com/SMRT-Analysis/Algorithms/BLASR), and filtered by size. Ion Torrent reads were independently assembled with two *de novo* assemblers IDBA-UD [97] and Mira 3.9 [98], which resulted in 212 contigs with a length up to 41,248 bp, and 24 contigs with a maximum length 418,142 bp, respectively (Table C in File S1). PacBio reads were assembled using Mira 3.9 [98] and Celera from SMRT portal (http://www.pacbiodevnet.com/SMRT-Analysis/Software/SMRT-Pipe) assemblers, which yielded 21 contigs with maximum length 15,072 bp, and one contig 2,954,373 bp, respectively. In addition, all assembly results were compared and verified for errors using Vista [99] and Mauve [100] tools. Custom primers were designed to experimentally confirm complete genome assembly. Random regions with high fluctuations of G+C content, non-coding regions between operons, and regions with contig overlaps were verified by PCR amplifications. The assembled genome was annotated by the Rapid Annotations using Subsystems Technology (RAST) [101] and Expert Review version of the Integrated Microbial Genomes system (IMG ER) [102]. Limited inspection and clean up of coding sequences was done by comparison with the publicly available databases GenBank [103], TIGRfam [104], the database of Clusters of Orthologous Groups of proteins (COGs) [105], and Conserved Domain Database (CDD) [106]. CRISPRFinder was used to identify CRISPR loci [107]. The Conserved Domain Search tool was used for the annotation of two-component systems. The results from two different databases: TIGRfam and CDD were compared and merged together. Detailed information on the annotated genes can be found in Tables S1–S3 in File S1.

Phylogenetic analyses

Amino acid sequences of *amo* genes and nucleotide sequences of 16S rRNA were aligned using MUSCLE 3.8.31 [108]. GBLOCKS were used to select conserved sites and remove poorly aligned regions [109]. Likelihood trees were built using PhyML [110]. The optimized parameters for 16S rRNA and for AMO protein sequences are described in File S1 and S2.

Genome synteny, average nucleotide identity and whole-genome homology

The genome synteny plots were generated from pairwise alignments between the present genome and *Ca.* N. gargensis, and *N. maritimus* genomes obtained from GenBank database. The alignments were based on the six-frame amino acid translations of the compared genomes using Promer tool from MUMmer 3.0 system [111]. The JSpecies software was used to calculate average nucleotide identity between genomes based on the MUMmer ultra-rapid aligning tool [112]. Whole-genome homology was determined from alignment of whole genomes by VISTA servers [99].

Identification of unique coding sequences for group I.1a and I.1b (Venn Diagrams)

Coding sequences (CDS) of six sequenced AOA genomes were downloaded from GenBank. All CDS from the group I.1a were merged together and CDS from the group I.1b were also merged together. Redundant CDS were removed by clustering sequences from the group I.1a and I.1b at 50% identity using UCLUST v1.2.22q and choosing only unique sequence from each cluster [113]. A protein BLAST of sequences from group I.1b versus group I.1a was performed to determine shared, and unique CDS for both groups at ≥35% identity.

Sequence deposition

The genome sequence of "*Ca.* Nitrososphaera evergladensis" has been submitted to GenBank under Accession Number CP007174.

Supporting Information

File S1 Combined file containing Figures S1–S9 and Tables A–C. **Figure S1. Circular representation of the *Ca.* Nitrososphaera evergladensis genome** (A). From outside to the center: Genes on forward strand (color by COG categories); Genes on reverse strand (color by COG categories); RNA genes (tRNAs green, rRNAs red, other RNAs black); GC content; GC skew. Alignment between Mira contigs generated from Ion Torrent reads and Celera contig generated from PacBio reads (B). Vertical colored lines indicate a high alignment score and white lines indicate a low score. **Figure S2. A phylogenetic tree of ammonia-oxidizing archaea *amoA*, *amoB*, *amoC*, and *amoX* subunits of ammonia monooxygenase.** Amino-acid sequences of amo subunits of AOA were randomly selected from the National Center for Biotechnology Information databases. The multiple sequence alignment of the amino-acid sequences was used for building maximum-likelihood trees. The branching patterns are denoted by their respective bootstrap values (100 iterations). Topology is colored by the metabolic group (blue represents marine group I.1a, green represents group I.1b, red is ThAOA). **Figure S3. 3-Hydroxypropionate cycle.** Identified enzymes in *Ca.* N. evergladensis genome are in green color; missing enzymes are in red color. **Figure S4. TCA cycle.** Identified enzymes in *Ca.* N. evergladensis genome are in green color; missing enzymes are in red color. **Figure S5. Gluconeogenesis/Glycolysis.** Identified enzymes in *Ca.* N. evergladensis genome are in green color; candidates for enzymes are in red color. **Figure S6. Hexose monophosphate pathway (HMP).** Identified enzymes in *Ca.* N. evergladensis genome are in green color; missing enzymes are in red color. **Figure S7. Clustering of the *amo* genes coding for subunits of Ammonia monooxygenase (AmoA, AmoB, AmoC, AmoX) in the genomes of ammonia-oxidizing archaea (AOA) and ammonia-oxidizing bacteria (AOB). Figure S8. Electron transport chain of *Ca.* N. evergladensis.** AMO – ammonia monooxygenase; CuHAO – hydroxylamine oxidoreductase; NIR- nitrite reductase; NOR – nitric oxide reductase; PC- small blue copper-containing plastocyanin-like electron carriers; Q and QH_2 – oxidized and reduced quinone pools. Complex I - Quinone reductase; Complex III – Riske Fe-S proteins, cytochromes; Complex IV – Heme/copper-type cytochrome/quinol oxidases. * - Suggested candidate enzymes: CuHAO (multicopper oxidase), NOR (catalytic subunits NorD, Q are not found). **Figure S9. A phylogenetic tree of archaeal pelota gene homologs.** Amino-acid sequences of pelota were randomly selected from the National Center for Biotechnology Information databases. The multiple sequence alignment of the amino-acid sequences was used for building maximum-likelihood trees. **Table A**. Comparison of *Ca.* N. evergladensis with other AOA genomes that are available in the public databases. CDS were compared at amino-acid identity ≥35%. **Table B**. Sequencing reports from the Ion Torrent platform and Pacific Biosciences platform. **Table C**. Comparative results of different assembly methods and sequencing technologies.

File S2 Combined file containing Tables S1–S5. **Table S1**. Protein coding sequences of central carbon, nitrogen, lipid metabolism and genes involved in the stress response of the archaeon. **Table S2**. Transporters encoded in *Ca.* N. evergladensis genome. **Table S3**. Information processing machinery of *Ca.* N. evergladensis. **Table S4**. Protein coding sequences (COGs and TIGRfams) present only in the AOA group I.1b. **Table S5**. Coding sequences present only in *Ca.* N. evergladensis genome but missing from the genome of *Ca.* N. gargensis.

Author Contributions

Conceived and designed the experiments: KVZ RD MTL PDDQ SHD EWT. Performed the experiments: KVZ RD MTL PDDQ. Analyzed the data: KVZ RD MTL PDDQ WGF. Contributed reagents/materials/analysis tools: FAOC WGF SHD EWT. Wrote the paper: KVZ RD MTL JCD EWT.

References

1. Hallam SJ, Mincer TJ, Schleper C, Preston CM, Roberts K, et al. (2006) Pathways of carbon assimilation and ammonia oxidation suggested by environmental genomic analyses of marine *Crenarchaeota*. PLOS Biol 4: 0520–0536.
2. Walker CB, de la Torre JR, Klotz MG, Urakawa H, Pinel N, et al. (2010) *Nitrosopumilus maritimus* genome reveals unique mechanisms for nitrification and autotrophy in globally distributed marine crenarchaea. Proc Natl Acad Sci U S A 107: 8818–8823.
3. Jung MY, Park SJ, Min D, Kim JS, Rijpstra WI, et al. (2011) Enrichment and characterization of an autotrophic ammonia-oxidizing archaeon of mesophilic crenarchaeal group I.1a from an agricultural soil. Appl Environ Microbiol 77: 8635–8647.
4. Mosier AC, Allen EE, Kim M, Ferriera S, Francis CA (2012) Genome sequence of "*Candidatus* Nitrosoarchaeum limnia" BG20, a low-salinity ammonia-oxidizing archaeon from the San Francisco Bay estuary. J Bacteriol 194, 2119–2120.
5. Park SJ, Kim JG, Jung MY, Kim SJ, Cha IT, et al. (2012a) Draft genome sequence of an ammonia-oxidizing archaeon, "*Candidatus* Nitrosopumilus sediminis" AR2, from Svalbard in the Arctic Circle. J Bacteriol 194: 6948–699.
6. Park SJ, Kim JG, Jung MY, Kim SJ, Cha IT, et al. (2012b) Draft genome sequence of an ammonia-oxidizing archaeon, "*Candidatus* Nitrosopumilus koreensis" AR1, from marine sediment. J Bacteriol 194: 6940–96941.
7. Pester M, Rattei T, Fletchl S, Gröngröft A, Richter A, et al. (2012) amoA-based consensus phylogeny of ammonia-oxidizing archaea and deep sequencing of amoA genes from soils of four different geographic regions. Environ Microbiol 14: 525–539.
8. Radax R, Hoffmann F, Rapp H, Leininger S, Schleper C (2012) Ammonia-oxidizing archaea as main drivers of nitrification in cold-water sponges. Envirom Microbiol 14: 909–23.
9. Martens-Habbena W, Berube PM, Urakawa H, Torre JR, Stahl DA (2009) Ammonia oxidation kinetics determine niche separation of nitrifying Archaea and Bacteria. Nature 461: 976–979.
10. Stahl DA, de la Torre JR (2012) Physiology and diversity of ammonia-oxidizing archaea. Annu Rev Microbiol 66: 83–101.
11. Raun WR, Johnson GV (1999) Improving nitrogen use efficiency for cereal production Agron J 91: 357–363.
12. Kowalchuk GA, Stephen JA (2001) Ammonia-Oxidizing Bacteria: A Model for Molecular Microbial Ecology. Annu Rev Microbiol 55: 485–529.
13. Zhalnina K, de Quadros PD, Camargo FA, Triplett EW (2012) Drivers of archaeal ammonia-oxidizing communities in soil. Front Microbiol 3: 210.
14. Inventory of US Greenhouse Gas Emissions and Sinks: 1990–2011 Available: http://epagov/climatechange/Downloads/ghgemissions/US-GHG-Inventory-2013-Main-Textpdf.
15. Stieglmeier M, Mooshammer M, Kitzler B, Wanek W, Zechmeister-Boltenstern S, et al. (2014) Aerobic nitrous oxide production through N-nitrosating hybrid formation in ammonia-oxidizing archaea. ISME J doi:10.1038/ismej.2013.220
16. Kim JG, Jung MY, Park SJ, Rijpstra WIC, Damsté JS, et al. (2012) Cultivation of a highly enriched ammonia-oxidizing archaeon of thaumarchaeotal group I.1b from an agricultural soil. Environ Microbiol 14: 1528–1543.
17. Loscher CR, Kock A, Konneke M, LaRoche J, Bange HW, et al. (2012) Production of oceanic nitrous oxide by ammonia-oxidizing archaea Biogeosciences 9: 2419–2429.
18. Santoro AE, Buchwald C, McIlvin MR, Casciotti KL (2011) Isotopic Signature of N_2O produced by marine ammonia-oxidizing archaea. Science 333: 1282–1285.
19. Lebedeva E, Hatzenpichler R, Pelletier E, Schuster N, Hauzmayer S, et al. (2012) Enrichment and genome sequence of the group I.1a ammonia-oxidizing Archaeon "*Ca.* Nitrosotenuis uzonensis" representing a clade globally distributed in thermal habitats. PLOS ONE 8: e80835.
20. Spang A, Phoehlein A, Offre P, Zumbragel S, Haider S, et al. (2012) The genome of the ammonia-oxidizing *Candidatus* Nitrososphaera gargensis: insights into metabolic versatility and environmental adaptations. Environ Microbiol 14: 3122–3145.

21. Tourna M, Stieglmeierera M, Spang A, Könneke M, Schintlmeister A, et al. (2011) *Nitrososphaera viennensis*, an ammonia oxidizing archaeon from soil. Proc Natl Acad Sci U S A 108: 8420–8425.
22. Ochsenreiter T, Selezi D, Quaiser A, Bonch-Osmolovskaya L, Schleper C (2003) Diversity and abundance of Crenarchaeota in terrestrial habitats studied by 16S RNA surveys and real time PCR. Environ Microbiol 5: 787–97.
23. Auguet J-C, Barberan A, Casamayor E (2010) Global ecological patterns in uncultured Archaea. ISME J 4: 182–90.
24. Mosier A, Allen E, Kim M, Ferriera S, Francis C (2012) Genome sequence of "*Candidatus* Nitrosopumilus salaria" BD31, an ammonia-oxidizing archaeon from the San Francisco Bay estuary. J Bacteriol 194: 2121–2.
25. Damsté JS, Rijpstra WIC, Hopmans EC, Jung MY, Kim JG, et al. (2012) Intact polar and core glycerol dibiphytanyl glycerol tetrather lipids of Group I.1a and I.1b Thaumarchaeota in soil. Appl Environ Microbiol 78: 6866–6874.
26. Zhalnina K, de Quadros PD, Gano KA, Davis-Richardson A, Fagen JR, et al. (2013) *Ca*. Nitrososphaera and *Bradyrhizobium* are inversely correlated and related to agricultural practices in long-term field experiments. Front Microbiol 4: 104.
27. Cole JR, Wang Q, Cardenas E, Fish J, Chai B, et al. (2009) The Ribosomal Database Project: improved alignments and new tools for rRNA analysis. Nucleic Acids Res 37: D141–5.
28. Blainey PC, Mosier AC, Potanina A, Francis CA, Quake SR (2011) Genome of a low-salinity ammonia-oxidizing archaeon determined by single-cell and metagenomic analysis. PLOS ONE 6: 1–12.
29. Könneke M, Schubert DM, Brown PC, Hügler M, Standfest S, et al. (2014) Ammonia-oxidizing archaea use the most energy-efficient aerobic pathway for CO_2 fixation. Proc Natl Acad Sci (USA) early edition, doi:10.1073/pnas.1402028111
30. Zhang L, Offre PR, He JZ, Verhamme DT, Nicol GW, et al. (2010) Autotrophic ammonia oxidation by soil thaumarchaea. Proc Natl Acad Sci U S A 107: 17240–17245.
31. Zaparty M, Zaigler A, Stamme C, Soppa J, Hensel R, et al. (2008) DNA microarray analysis of central carbohydrate metabolism: glycolytic/gluconeogenic carbon switch in the hyperthermophilic crenarchaeum *Thermoproteus tenax*. J Bacteriol 190: 2231–8.
32. Siebers B, Tjaden B, Michalke K, Dörr C, Ahmed H, et al. (2004) Reconstruction of the central carbohydrate metabolism of *Thermoproteus tenax* by use of genomic and biochemical data. J Bacteriol 186: 2179–94.
33. Tang KH, Blankenship RE (2010) Both forward and reverse TCA cycles operate in green sulfur bacteria. J Biol Chem 285: 35848–35854.
34. Berg I, Kockelkorn D, Ramos-Vera W, Say RF, Zarzycki J, et al. (2010) Autotrophic carbon fixation in archaea. Nat Reviews Microbiol 8: 447–60.
35. Shan Y, Lai Y, Yan A (2012) Metabolic reprogramming under microaerobic and anaerobic conditions in bacteria. Subcell Biochem 64: 159–79.
36. Park BJ, Park SJ, Yoon DN, Schouten S, Sinninghe Damsté JS, et al. (2010) Cultivation of autotrophic ammonia-oxidizing archaea from marine sediments in coculture with sulfur-oxidizing bacteria. Appl Environ Microbiol 76: 7575–87.
37. Siebers B, Schönheit P (2005) Unusual pathways and enzymes of central carbohydrate metabolism in Archaea. Curr Opin Microbiol 8: 695–705.
38. Cheeseman P, Toms-Wood A, Wolfe RS (1972) Isolation and properties of a fluorescent compound, factor 420, from *Methanobacterium* strain MoH. J Bacteriol 112: 527–531.
39. Taylor M, Scott C, Grogan G (2013) F420-dependent enzymes - potential for applications in biotechnology. Trends Biotechnol 31: 63–4.
40. Alonso-Sáez L, Waller AS, Mende DR, Bakker K, Farnelid H, et al. (2012) Role for urea in nitrification by polar marine Archaea. Proc Natl Acad Sci U S A 109: 17989–17994.
41. Kirchman DL (2012) Marine archaea take a short cut in the nitrogen cycle. Proc Natl Acad Sci U S A 109: 17732–17733.
42. Norton JM, Alzerreca JJ, Suwa Y, Klotz MG (2002) Diversity of ammonia monooxygenase operon in autotrophic ammonia-oxidizing archaea. Arch Microbiol 177: 139–149.
43. Berube PM, Stahl DA (2012) The divergent AmoC subunit of ammonia monooxygenase functions as part of a stress response system in *Nitrosomonas europaea*. J Bacteriol 194: 3448–3456.
44. Arp D, Sayavedra-Soto L, Hommes N. (2002) Molecular biology and biochemistry of ammonia oxidation by *Nitrosomonas europaea*. Arch Microbiol 178: 250–255.
45. Klotz MG, Stein LY (2008) Nitrifier genomics and evolution of the nitrogen cycle. FEMS Microbiol Lett 278: 146–156.
46. Schleper C, Nicol GW (2010) Ammonia-oxidizing archaea-physiology, ecology and evolution. Adv Microb Physiol 57: 1–41.
47. Vajrala N, Martens-Habbena W, Sayavedra-Soto LA, Schauer A, Bottomley PJ, et al. (2012) Hydroxylamine as an intermediate in ammonia oxidation by globally abundant marine archaea. Proc Natl Acad Sci U S A 110: 1006–1011.
48. Zart D, Schmidt I, Bock E. (2000) Significance of gaseous NO for ammonia oxidation by *Nitrosomonaas eutropha*. Antonie Van Leeuwenhoek 77: 49–55.
49. Murphy DJ (2012) The dynamic roles of intracellular lipid droplets: from archaea to mammals. Protoplasma 3: 541–585.
50. Quillaguaman J, Guzman H, Van-Thuoc D, Hatti-Kaul R (2010) Synthesis and production of polyhydroxyalkanoates by halophiles: current potential and future prospects. Appl Microbiol Biotechnol 85: 1687–1696.
51. Damsté JS, Schouten S, Hopmans EC, van Duin AC, Geenevasen JA (2002) Crenarchaeol: the characteristic core glycerol dibiphytanyl glycerol tetraether membrane lipid of cosmopolitan pelagic crenarchaeota. J Lipid Res 43: 1641–1651.
52. Spang A, Hatzenpichler R, Brochier-Armanet C, Rattei T, Tischler P, et al. (2010) Distinct gene set in two different lineages of ammonia-oxidizing archaea supports the phylum Thaumarchaeota. Trends Microbiol 18: 331–340.
53. de la Torre J, Walker C, Ingalls A, Könneke M, Stahl D (2008) Cultivation of a thermophilic ammonia oxidizing archaeon synthesizing crenarchaeol. Envirom Microbiol 10: 810–8.
54. Pitcher A, Rychlik N, Hopmans E, Spieck E, Rijpstra W, et al. (2010) Crenarchaeol dominates the membrane lipids of *Candidatus* Nitrososphaera gargensis, a thermophilic group I.1b Archaeon. ISME J 4: 542–52.
55. Smit A, Mushegian A (2000) Biosynthesis of isoprenoids via mevalonate in archaea: The Lost Pathway. Genome Res 10: 1468–1484.
56. Muller V, Spanheimer R, Santos H (2005) Stress response by solute accumulation in archaea. Curr Opin Microbiol 8: 729–736.
57. Tanghe A, Van Dijck P, Thevelein JM (2006) Why do microorganisms have aquaporins? Trends Microbiol 14: 78–85.
58. Roesser M, Müller V (2001) Osmoadaptation in bacteria and archaea: common principles and differences. Envirom Microbiol 3: 743–754.
59. Maezato Y, Blum P (2012) Survival of the fittest: overcoming oxidative stress at the extremes of acid, heat and metal. Life 2: 229–242.
60. Pikuta EV, Hoover RB, Tang J (2007) Microbial extremophiles at the limits of life. Crit Rev Microbiol 33: 183–209.
61. Li X, Zhu YG, Cavagnaro TR, Chen M, Sun J, et al. (2009) Do ammonia-oxidizing archaea respond to soil Cu contamination similarly asammonia-oxidizing bacteria? Plant Soil 324: 209–217.
62. Ettema TJ, Brinkman AB, Lamers PP, Kornet NG, de Vos WM, et al. (2006) Molecular characterization of a conserved resistance (*cop*) gene cluster and its copper-responsive regulator in *Sulfolobus solfataricus* P2. Microbiology 152: 1969–1979.
63. Orell A, Remonsellez F, Arancibia R, Jerez CA (2013) Molecular characterization of copper and cadmium resistance determinants in the biomining thermoacidophilic archaeon *Sulfolobus metallicus*. *Archaea* doi:101155/2013/289236
64. Rowland JL, Niederweis M (2013) A multicopper oxidase is required for copper resistance in *Mycobacterium tuberculosis*. J Bacteriol 195: 3724–3733.
65. Remonsellez F, Orell A, Jerez CA (2006) Copper tolerance of the thermoacidophilic archaeon *Sulfolobus metallicus*: possible role of polyphosphate metabolism. Microbiology 152: 59–66.
66. Petitjean C, Moreira D, López-García P, Brochier-Armanet C (2011) Horizontal gene transfer of a chloroplast DnaJ-Fer protein to Thaumarchaeota and the evolutionary history of the DnaK chaperone system in Archaea. BMC Evol Biol 12: 226.
67. Rees A, Woodward M, Joint I (1999) Measurement of nitrate and ammonium uptake at ambient concentrations in oligotrophic waters of the North-East Atlantic Ocean. Mar Ecol Prog Ser 187: 295–300.
68. Yin Y, Zhang H, Olman V, Xu Y (2010) Genomic arrangement of bacterial operons is constrained by biological pathways encoded in the genome Proc Natl Acad Sci USA 107: 6310–5.
69. Beaumont HJE, Hommes NG, Sayavedra-Soto LA, Arp DJ, Arciero DM, et al. (2002) Nitrite reductase of *Nitrosomonas europaea* is not essential for production of gaseous nitrogen oxides and confers tolerance to nitrite. J Bacteriol 184: 2557–2560.
70. Beaumont HJE, van Schooten B, Lens SI, Westerhoff HV, van Spanning RJ (2004) *Nitrosomonas europaea* expresses a nitric oxide reductase during nitrification. J Bacteriol 186: 4417–4421.
71. Whittaker M, Bergmann D, Arciero D, Hooper AB (2000) Electron transfer during the oxidation of ammonia by the chemolithotrophic bacterium *Nitrosomonas europaea*. Biochim Biophys Acta 1459: 346–355.
72. Döring B, Lütteke T, Geyer J, Petzinger E (2012) The SLC10 carrier family: transport functions and molecular structure. Curr Top Membr 70: 105–168.
73. Furumoto T, Yamaguchi T Ohshima-Ichie Y, Nakamura M, Tsuchida-Iwata Y, et al. (2011) A plastidial sodium-dependent pyruvate transporter. Nature 476: 472–475.
74. Sola-Landa A, Moura RS, Martin JF (2003) The two-component PhoR-PhoP system controls both primary metabolism and secondary metabolite biosynthesis in *Streptomyces lividans*. Proc Natl Acad Sci U S A 100: 6133–6138.
75. Wende A, Furtwangler K, Oesterhelt D (2009) Phosphate-dependent behavior of the archaeon *Halobacterium salinarum* Strain R1. J Bacteriol 191: 3852–3860.
76. Leblanc SKD, Oates CW, Raivio TL (2011) Characterization of the induction and cellular role of the BaeSR two-component envelope stress response of *Escherichia coli*. J Bacteriol 193: 3367–3375.
77. Giner-Lamia J, López-Maury L, Reyes JC, Florencio FJ (2012) The CopRS two-component system is responsible for resistance to copper in the cyanobacterium *Synechocystis* sp. PCC 6803. Plant Physiol 159: 1806–1818.
78. Fried L, Behr S, Jung K (2013) Identification of a target gene and activating stimulus for the YpdA/YpdB histidine kinase/response regulator system in *Escherichia coli*. J Bacteriol 195: 807–815.
79. Tortosa P, Logsdon L, Kraigher B, Itoh Y, Mandic-Mulec I, et al. (2001) Specificity and genetic polymorphism of the *Bacillus* competence quorum-sensing system. J Bacteriol 183: 451–460.

80. Korkhin Y, Unligil UM, Littlefield O, Nelson PJ, Stuart DI, et al. (2009) Evolution of complex RNA polymerases: the complete archaeal RNA polymerase structure. PLOS Biol 7: 0001–0010.
81. de Koning, Blombach F, Wu H, Brouns SJJ, Oost JVD (2009) Role of multiprotein bridging factor 1 in archaea: bridging the domains? Biochem Soc Trans 37: 52–57.
82. Brochier-Armanet C, Forterre P, Gribaldo S (2011) Phylogeny and evolution of the Archaea: one hundred genomes later. Curr Opin Microbiol 14: 274–281.
83. Ragan MA, Logsdon JM, Sensen CW, Charlebois RL, Doolittle WF (1996) An archaebacterial homolog of pelota, a meiotic cell division protein in eukaryotes. FEMS Microbiol Lett 144: 151–155.
84. Kobayashi K, Kikuno I, Kuroha K, Saito K, Ito K, et al. (2010) Structural basis for mRNA surveillance by archaeal Pelota and GTP-bound EF1α complex. Proc Natl Acad Sci USA 107: 17575–9.
85. de Koning B, Blombach F, Brouns SJ, van der Oost J (2009) Fidelity in archaeal information processing. Archaea (Vancouver, BC) 2010: doi:10.1155/2010/960298
86. White MF (2003) Archaeal DNA repair: paradigms and puzzles. Biochem Soc Trans 31: 690–693.
87. Snyder JC, Bateson MM, Lavin M, Young MJ (2010) Use of cellular CRISPR (clusters of regularly interspaced short palindromic repeats) spacer-based microarrays for detection of viruses in environmental samples. Appl Environ Microbiol 76: 7251–7258.
88. Horvath P, Barrangou R (2010) CRISPR/Cas, the immune system of bacteria and archaea. Science 327: 167–170.
89. Marraffini L, Sontheimer E (2010) CRISPR interference: RNA-directed adaptive immunity in bacteria and archaea. Nat Rev Genet 11: 181–90.
90. Lehtovirta-Morley L, Stoecker K, Vilcinskas A, Prosser J, Nicol G (2011) Cultivation of an obligate acidophilic ammonia oxidizer from a nitrifying acid soil. Proc Natl Acad Sci USA 108: 15892–7.
91. Könneke M, Bernhard AE, de la Torre JR, Walker CB, Waterbury JB, et al. (2005) Isolation of an autotrophic ammonia-oxidizing marine archaeon. Nature 437: 543–546.
92. Filée J, Siguier P, Chandler M (2007) Insertion sequence diversity in archaea. Microbiology and Mol Biol Rev 71: 121–157.
93. Hatzenpichler R, Lebedeva EV, Spieck E, Stoecker K, Richter A, et al. (2008) A moderately thermophilic ammonia-oxidizing crenarchaeote from a hot spring. Proc Natl Acad Sci U S A 105: 2134–2139.
94. Caporaso GJ, Kuczynski J, Stombaugh J, Bittinger K, Bushman FD, et al. (2010) QIIME allows analysis of high-throughput community sequencing data. Nat Methods 7: 335–336.
95. Francis CA, Roberts KJ, Beman JM, Santoro AE, Oakley BB (2005) Ubiquity and diversity of ammonia-oxidizing archaea in water columns and sediments of the ocean. Proc Natl Acad Sci U S A 102: 14683–14688.
96. Rotthauwe JH, Witzel KP, Liesack W (1997) The ammonia monooxygenase structural gene *amoA* as a functional marker: molecular fine-scale analysis of natural ammonia-oxidizing populations. Appl Environ Microbiol 63: 4704–12.
97. Peng Y, Leung HC, Yiu S-M, Chin FY (2012) IDBA-UD: a de novo assembler for single-cell and metagenomic sequencing data with highly uneven depth. Bioinformatics 28: 1420–1428.
98. Chevreux B, Wetter T, Suhai Sn (1999) Genome sequence assembly using trace signals and additional sequence information. In: German conference on Bioinformatics 45: 56.
99. Frazer KA, Pachter L, Poliakov A, Rubin EM, Dubchak I (2004) VISTA: computational tools for comparative genomics. Nucleic Acids Res 32: W273–279.
100. Darling AC, Mau B, Blattner FR, Perna NT (2004) Mauve: multiple alignment of conserved genomic sequence with rearrangements. Genome Res 14: 1394–1403.
101. Aziz RK, Bartels D, Best AA, DeJongh M, Disz T, et al. (2008) The RAST Server: Rapid Annotations using Subsystems Technology. BMC Genomics 9: 75.
102. Markowitz VM, Mavromatis K, Ivanova NN, Chen I, Chu K, et al. (2009) IMG ER: a system for microbial genome annotation expert review and curation. Bioinformatics 25: 2271–2278.
103. Benson DA, Cavanaugh M, Clark K, Karsch-Mizrachi I, Lipman DJ, et al. (2013) GenBank. Nucleic Acids Res 41(D): 36–42.
104. Haft DH, Selengut JD, Richter RA, Harkins D, Basu MK, et al. (2013) TIGRFAMs and genome properties in 2013. Nucleic Acids Res 41(D): 387–395.
105. Tatusov RL, Fedorova ND, Jackson JD, Jackobs AR, Kiryutin B, et al. (2003) The COG database: an updated version includes eukaryotes. BMC Bioinformatics 4: 41.
106. Marchler-Bauer A, Lu S, Anderson JB, Chitsaz F, Deryshire MK, et al. (2011) "CDD: a conserved domain database for the functional annotation of proteins". Nucleic Acids Res 39(D): 225–229.
107. Grissa I, Vergnaud G, Poucel C (2007) CRISPRFinder: a web tool to identify clustered regularly interspaced short palindromic repeats. Nucleic Acids Res 35: W52–W57.
108. Edgar RC (2004) MUSCLE: multiple sequence alignment with high accuracy and high throughput. Nucleic Acids Res 32: 1792–1797.
109. Talavera G, Castresana J (2007) Improvement of phylogenies after removing divergent and ambiguously aligned blocks from protein sequence alignments. Syst Biol 56: 564–577.
110. Criscuolo A (2011) MorePhyML: improving the phylogenetic tree space exploration with PhyML 3. Mol Phylogenet Evol 61: 944–948.
111. Delcher AL, Salzberg SL, Phillippy AM (2003) Using MUMmer to identify similar regions in large sequence sets. Curr Protoc Bioinformatics Chapter 10: Unit 10.13.
112. Richter M, Rosselló-Móra R (2009) Shifting the genomic gold standard for the prokaryotic species definition. Proc Natl Acad Sci U S A 106: 19126–19131.
113. Edgar RC (2010) Search and clustering orders of magnitude faster than BLAST. Bioinformatics 26: 2460–2461.
114. Kurtz S, Phillippy A, Delcher AL, Smoot M, Shumway M, et al. (2004) Versatile and open software for comparing large genomes. Genome Biology 5: R12.

Changes of Soil Bacterial Diversity as a Consequence of Agricultural Land Use in a Semi-Arid Ecosystem

Guo-Chun Ding[1], Yvette M. Piceno[2], Holger Heuer[1], Nicole Weinert[1], Anja B. Dohrmann[3], Angel Carrillo[4], Gary L. Andersen[2], Thelma Castellanos[4], Christoph C. Tebbe[3], Kornelia Smalla[1]*

1 Institute for Epidemiology and Pathogen Diagnostics - Federal Research Centre for Cultivated Plants (JKI), Braunschweig, Germany, **2** Department of Ecology, Lawrence Berkeley National Laboratory, Berkeley, California, United States of America, **3** Institute for Biodiversity, Johann Heinrich von Thünen-Institut (TI), Braunschweig, Germany, **4** Centro de Investigaciones biologicas del Noroeste, S.C. La Paz, Mexico

Abstract

Natural scrublands in semi-arid deserts are increasingly being converted into fields. This results in losses of characteristic flora and fauna, and may also affect microbial diversity. In the present study, the long-term effect (50 years) of such a transition on soil bacterial communities was explored at two sites typical of semi-arid deserts. Comparisons were made between soil samples from alfalfa fields and the adjacent scrublands by two complementary methods based on 16S rRNA gene fragments amplified from total community DNA. Denaturing gradient gel electrophoresis (DGGE) analyses revealed significant effects of the transition on community composition of *Bacteria*, *Actinobacteria*, *Alpha-* and *Betaproteobacteria* at both sites. PhyloChip hybridization analysis uncovered that the transition negatively affected taxa such as *Acidobacteria*, *Chloroflexi*, *Acidimicrobiales*, *Rubrobacterales*, *Deltaproteobacteria* and *Clostridia*, while *Alpha-*, *Beta- and Gammaproteobacteria*, *Bacteroidetes* and *Actinobacteria* increased in abundance. Redundancy analysis suggested that the community composition of phyla responding to agricultural use (except for *Spirochaetes*) correlated with soil parameters that were significantly different between the agricultural and scrubland soil. The arable soils were lower in organic matter and phosphate concentration, and higher in salinity. The variation in the bacterial community composition was higher in soils from scrubland than from agriculture, as revealed by DGGE and PhyloChip analyses, suggesting reduced beta diversity due to agricultural practices. The long-term use for agriculture resulted in profound changes in the bacterial community and physicochemical characteristics of former scrublands, which may irreversibly affect the natural soil ecosystem.

Editor: A. Mark Ibekwe, U. S. Salinity Lab, United States of America

Funding: This study was financially supported by the German Federal Ministry for Education and Research, Projektträger DLR (projects MEX 06/003 and 01DN12067) and CONACYT (Concejo Nacional de Ciencia Y Tecnologia) J110.394 and the BMBF project MÄQNU 03MS642H. The funders had no role in study design, data collection and analysis, decision to publish, or preparation of the manuscript.

Competing Interests: The authors have declared that no competing interests exist.

* E-mail: kornelia.smalla@jki.bund.de

Introduction

Converting natural land into arable soils results in losses to the landscapes characterized by a typical indigenous flora and fauna. Frequently, terrestrial ecosystem diversity is being reduced by replacing indigenous flora with a few crops. The ecological consequences of such transitions have been addressed in several studies focusing on land degradation [1,2,3], losses of macro-biodiversity [4,5], nutrient exhaustion in soils [3], sustainability [6,7] and restoration [8]. Soil microorganisms, including protozoa, fungi, bacteria and archaea, are essential for the proper functioning and sustainability of ecosystems [9,10,11]. Moreover, a high microbial diversity is assumed to be critical for the stability of ecosystems by providing functional diversity and redundancy [12,13]. Changes in vegetation as well as intensive agricultural practices were shown to affect soil microbial community composition and activity [14,15] and soil physicochemical properties [3]. The influence of land use and management on soil microorganisms was addressed in several recent studies [16,17]. However, the information acquired is still not sufficient as a systematic identification of taxa responding to the transition in land use was not done.The studies investigated soils from various geographic sites in Australia, The Netherlands, and Brazil [16] although a comparison of the results might also be difficult due to the differences in the experimental designs and the resolution level of the methods used.

The Santo Domingo Valley is an agricultural area within the Southern Sonoran Desert, Baja California Sur, Mexico, which is entirely dependent on irrigation water collected from wells. Most of the farmland has been developed during the 1950's and 60's with cotton and wheat as the main crops [18]. Since then, other crops have been cultivated, e.g., oat, sorghum, chickpea, maize, and alfalfa. Due to declining water availability and increasing problems with soil salinization, yields have decreased since 1991 [18]. Most of the arable land is bordered by natural sarcocaulescent scrubland, including different crassicaulent plants, succulent cacti, woody trees and shrubs [19]. Some of these areas are utilized for grazing goats [20,21] and thus might be impacted by fecal depositions. The soils in the Santo Domingo Valley belong to the hyposodic calcisols, which are typical of many semi-arid ecosystems. The two sites selected provided almost ideal conditions to study the effects of agricultural land use on the microbial communities in these ecosystems as agricultural fields were in direct vicinity to the scrubland and the soils were typical of semi-arid deserts.

In the present study, bacterial soil communities from arable fields with alfalfa and the adjacent scrubland at two sites 50 km apart from each other were compared by denaturing gradient gel electrophoresis (DGGE) and PhyloChip analysis of16S rRNA gene fragments amplified from total community DNA to evaluate the influence of land use. Both methods are complementary but have clear differences. DGGE provides information on the relative abundance of all amplified dominant populations and thus is more suitable for comparative analysis of the community composition. In the present study the bacterial community analyses by DGGE was performed at different taxonomic levels, *Bacteria*, *Actinobacteria*, *Alpha*- and *Betaproteobacteria* in order to analyze not only dominant bacteria. The so-called PhyloChip developed by Brodie et al. [36] offers the potential to detect 8741 operational taxonomic units (OTUs) and the dataset is ideal for identifying taxa containing a high proportion of OTUs with treatment dependent significantly increased or decreased abundance. Therefore, the PhyloChip dataset was used to analyze whether, and if so which bacterial taxa responded to agricultural use, and multivariate statistics was applied to explore the relationship between discriminative soil parameters and responsive taxonomic groups.

Materials and Methods

Site and Sampling

Two typical sites in the Santo Domingo valley, Baja California, Mexico (site 1: N25°06′31.5″ W111°32′34.3″, altitude 70; site 2: N25°16′5.2″ W111° 36′ 02.5″, altitude 90; the distance between site 1 and site 2 is ca 50 km; No specific permits were required for the described field studies. The location is not protected in any way. The field studies did not involve endangered or protected species.) were selected. Soil samples were taken on 30 April 2007 from two covers, i.e., a field planted with alfalfa (*Medicago sativa*) and the adjacent natural scrubland. Four replicates per site and cover were taken respectively from plots (1 m×1 m squares) that were at least 20 m apart from each other. After removing the top 1–2 cm soil layer, the soil from 2–15 cm depth was mixed with a shovel and approx. 2 kg of soil was sampled, put into plastic bags and transferred to the laboratory. Within 24 h after sampling, the soils were sieved through a 2 mm mesh and aliquots were used for microbiological and chemical analysis.

Soil Properties

All soils were analyzed by the certified laboratory at CIBNOR (La Paz, Mexico). Briefly, soil particle size determinations were conducted with the sedimentation method [22], pH was determined in 1:1 (wt/vol) diluted water suspensions [23], electrical conductivity was determined with a CO150 conductivity meter according to the manufacturer's instructions (Hach Company, Loveland, CO, USA). Total organic matter was measured using the reduction of potassium dichromate method of Walkley and Black, as described by Nelson and Sommers [24]. Furthermore, ammonium (NH_4^+), nitrite and nitrate [25], calcium (Ca^{2+}) [26], magnesium (Mg^{2+}), phosphate, sulfate [27] were quantified. All soil parameters measured are given in Table 1.

Total Community (TC) DNA

TC DNA was extracted from 0.5 g of soil after a harsh lysis step (FastPrep FP120 bead beating system, MP Biomedicals, Santa Ana, Carlsbad, CA, USA) by means of the BIO-101 DNA spin kit for soil (Q-Biogene). The DNA was purified using the Geneclean Spin Kit (Q-Biogene) according to the manufacturer's instructions. Purified TC DNA was stored at −20°C.

PCR Amplification of 16S rRNA Gene Fragments and DGGE Analysis

Primer sets and PCR conditions employed in this study to amplify *Bacteria* [28], *Actinobacteria* [28], and *Alpha*- and *Betaproteobacteria* [29,30] and relevant information is provided in Table S1. DGGE of the 16S rRNA gene amplicons was performed according to Gomes *et al.* [31]. The gel was silver-stained according to Heuer *et al.* [32]. DGGE profiles were analyzed by GelCompar 4.5. Dendrograms were constructed by means of unweighted pair group method using arithmetic averages (UPGMA) based on pairwise Pearson correlation indices, which were also subjected to permutation tests with a modified version of PERMTEST software [33]. Box-Whisker plots were generated using R (http://www.R-project.org) based on dissimilarities (1-Pearson correlation indices) between samples within the same treatment.

PCR Amplification of 16S rRNA Genes and PhyloChip Analysis

TC DNA extracts from three replicates per site and cover were amplified using universal 16S rRNA gene primers (27f 5′-AGAGTTTGATCCTGGCTCAG-3′; 1492r 5′-GGTTACCTTGTTACGACTT-3′) and an 8-temperature gradient PCR. At each temperature, approximately 5 ng of TC DNA was used in 25 µl reactions (final concentrations were 1× Ex Taq Buffer with 2 mM $MgCl_2$, 300 nM each primer (27 f and 1492 r), 200 µM each dNTP (TaKaRa), 25 µg bovine serum albumin (Roche Applied Science, Indianapolis, IN, USA), and 0.625 U Ex Taq (TaKaRa Bio, Inc., through Fisher Scientific, Pittsburg, PA, USA)). The amplifications were performed with an iCycler (Bio-Rad, Hercules, CA, USA) as previously described by Weinert *et al.* [34]. PCR products from each annealing temperature (48–58°C) for each sample were combined, concentrated, quantified, and an amount of 500 ng product was applied to each G2 PhyloChip (Second Genome Inc., San Bruno, CA, USA) following previously described procedures [35]. The PhyloChip used in the present study contained approximately 500,000 probes (25-mer oligos) targeting 8,364 bacterial operational taxonomic units (OTUs). An OTU was considered present if more than 90 percent of the probe pairs representing this OTU showed a positive hybridization signals [36].

Statistical Analysis

An OTU-level report was produced mainly according to Brodie *et al.* [36] (background subtraction, detection and quantification criteria) except for the addition of normalizing array data by the average total array signal intensity [37]. The signal intensities of OTUs called absent were shifted to 1 to avoid errors in subsequent log transformation. Statistical analyses were done with the software package R 2.14 (http://www.r-project.org/).

Discriminative OTUs between the two different land uses were identified by multiple two-way ANOVA of log10-transformed signal intensities for each OTU (unadjusted $p<0.05$). Groups with a high proportion of OTUs significantly responding to land use were summarized at different taxonomic levels. The influence of land use on the community compositions of different taxonomic groups (from domain to family) was analyzed by a modified test based on five principal components [38]. Principal components analysis (PCA) was performed according to Weinert *et al.* [34]

Table 1. Physico-chemical characteristics (average ± standard deviation) for soils from different sites and land use.

	Site 1		Site 2	
	alfalfa	**scrubland**	**Alfalfa**	**scrubland**
pH value	8.65±0.12	8.7±0.27	8.62±0.06	8.1±0.64
Electric conductivity [mS cm^{-1}]	0.86±0.14	0.59±0.18	1.45±0.33	1.3±1.71
Ca^{2+} [mg kg^{-1}]	8.67±2.16	9.89±3.36	10.83±4.56	13.2±14.17
Mg^{2+} [mg kg^{-1}]	5.42±1.38	5±0.98	13.34±5.32	6.63±6.46
K^+ [mg kg^{-1}]	4.59±1.62	4.8±1.99	5.09±2.7	4.01±2.14
Na^+ [mg kg^{-1}]* +	33.22±6.34	13.86±9.37	44.56±8.2	26.4±32.55
Phosphate [mg kg^{-1}]*** −	5.2±1.33	41.62±19.15	7.21±3.32	30.32±17.97
NH_4^+-N [mg kg^{-1}]	10.14±4.1	7.44±1.53	10.48±2.17	10.14±3.3
Total nitrogen [mg kg^{-1}]	482.39±179.03	450.48±76.27	358.29±130.19	308.65±170.74
Cl^- [mg kg^{-1}]	37.6±5.28	11.57±7.57	60.23±15.41	65.64±110.05
NO_2^-N [mg kg^{-1}]	0.05±0.03	0.03±0.03	0.09±0.16	0.03±0.04
NO_3^-N [mg kg^{-1}]* +	2.62±0.72	0.66±0.5	1.18±1.92	0.23±0.22
Sulphate [mg kg^{-1}]*** +	8.35±1.63	4.79±1.33	29.28±6.24	2.97±0.92
Organic matter [% volumetric]* −	0.45±0.06	0.79±0.18	0.44±0.11	0.58±0.32

Two-way ANOVA was used to test for significant differences in soil parameters between land use. The significant level between land use was indicated as *$p<0.05$;
***$p<0.001$; +: significantly increased in agricultural soils; −: significantly decreased in agricultural soils.

using adjusted (log10- transformed, centered, and standardized) signal intensities of all OTUs belonging to each taxonomic group.

Soil parameters responding to land use were also identified by multiple two-way ANOVA (unadjusted $p<0.05$). Heatmap analysis was performed based on the adjusted (centered and standardized) values for different soil parameters. To analyze the influence of these discriminative soil parameters on community composition of total bacteria or responsive phyla, redundancy analysis (RDA) was performed using the R add-on package 'vegan'. A forward selection of soil parameters was applied to avoid using collinear soil parameters in the same constrained ordination model. Only those parameters contributing significantly ($p<0.05$ via 1000 times permutation tests) to community variation were added to the model.

Results

Significant Effects of Land Use on the Bacterial Community Composition Revealed by DGGE Fingerprints

To compare the bacterial community in soils under different land use, 16S rRNA gene fragments of bacteria, *Alphaproteobacteria*, *Betaproteobacteria* and *Actinobacteria* PCR-products amplified from total community DNA of alfalfa or adjacent scrubland soils sampled at two sites were analyzed by DGGE (Figures S1, S2, S3, S4).

Pairwise Pearson correlation indices were subjected to permutation tests to determine the significance of the land use effects on the bacterial community structure. A significant effect of land use was found for bacteria and all bacterial groups analyzed for both sites (Table 2). The extent of the influence was dependent on the phylogenetic group analyzed (Table 2). The strongest influence of land use was found for the *Betaproteobacteria*, especially at site 1 (Table 2). In the community profile for *Betaproteobacteria*, a strong band was observed only in all replicates in soils with alfalfa from site 1 (Figure S4). For *Alphaproteobacteria*, the dissimilarities of bacterial community fingerprints of soil under different land use were comparable between both sites. The lowest yet still significant effect of land use was observed for *Actinobacteria* (Table 2). Significant differences in community composition between the two study sites were found mainly for alfalfa growing soils as opposed to scrubland soils (Table 2). Compared to the effects of land use, the influence of different sites on community composition was smaller except for *Betaproteobacteria* (Table 2), probably still due to the strong bands for alfalfa soils from site 1 (Figure S4). For bacteria including all bacterial subgroups analyzed, Box-Whisker plots revealed that the variability of the bacterial community compositions among replicates was generally lower for alfalfa soils than the scrubland soil, except for *Betaproteobacteria* at site 2 (Figure 1). The lowest variability was found for alfalfa soils from site 1 (Figure 1). In conclusion, a significant and taxonomic group-dependent effect of land use was observed for all four targeted phylogenetic groups. Variation in community composition for soils under arable farmland use generally was lower than that from scrubland sites.

Taxa Responsive to Land Use Determined by PhyloChip Analysis

PhyloChip hybridizations were used to detect bacterial taxa with significant responses to land use. A total of 2,243 OTUs belonging to 44 phyla was detected (Table S2). The bacterial richness in terms of the number of detected OTUs was significantly ($p = 0.05$) higher for the arable field soils than for the scrubland soils.

OTUs responding to land use were identified by multiple two-way ANOVA and only taxa with a high proportion (>18% which is much higher than the unadjusted p-value) of discriminative OTUs were summarized in Table 3. Compared to the soil from the scrublands, 13% of the OTUs (295 OTUs) were more abundant in soil from alfalfa fields, while more OTUs (402 accounting for 18%) were less abundant (Table 3), suggesting few taxa probably enriched in the alfalfa fields. A large proportion (more than 25%) of the OTUs belonging to the phyla of

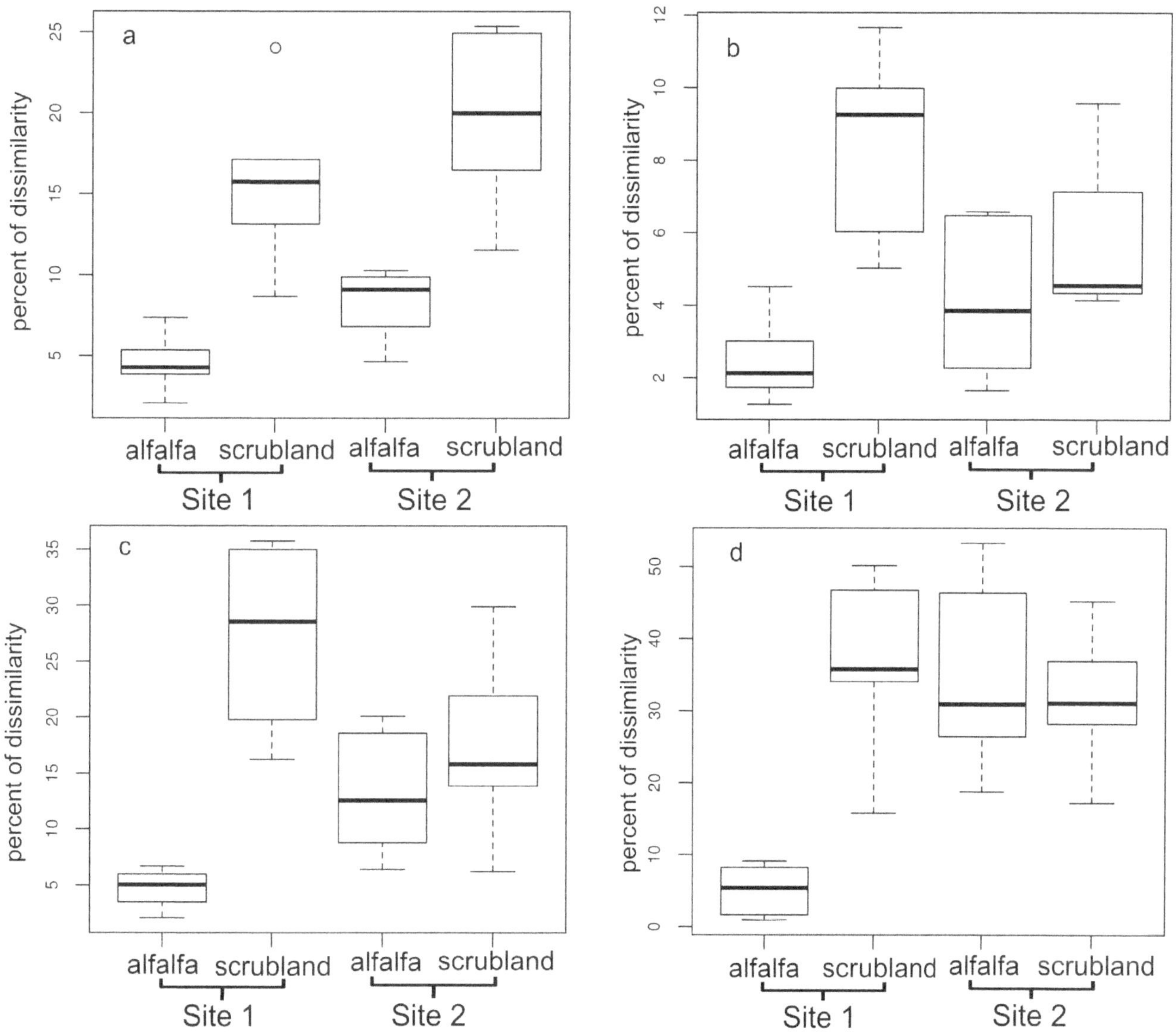

Figure 1. Boxplots of the variation of community structure under same sites and land use based on DGGE profiles for a: ***Bacteria*****; b:** ***Actinobacteria*****; c:** ***Alphaproteobacteria*****; d:** ***Betaproteobacteria*****.**

Acidobacteria, *Chloroflexi*, *Spirochaetes*, *Verrucomicrobia* and *Gemmatimonadetes* were significantly less abundant in alfalfa soils (Table 3), suggesting that the transition of scrubland soil into arable soils caused severe effects on the abundance of OTUs affiliated to these phyla. A negative effect also was observed for the class of *Clostridia* (*Firmicutes*), of which 29% of the OTUs detected had significantly lower signal intensities in the alfalfa soils (Table 3). The influence of land use on the phyla *Proteobacteria* and *Actinobacteria* was more

Table 2. Percent dissimilarity of microbial DGGE fingerprints of different taxa for soils compared between alfalfa and scrubland or between site 1 and site 2.

		Bacteria	*Alpha-proteobacteria*	*Beta-proteobacteria*	*Actinobacteria*
Land use	Site 1	25*	20.3*	68.6*	2.7*
	Site 2	13.8*	25*	31.3*	3.7*
Site	Alfalfa	10.3*	11.8*	61.5*	3.7*
	Scrubland	3.8	4.9	5.8	2.8*

*Significant ($p<0.05$) difference in community fingerprints between treatments as revealed by permutation tests.

complex as more than 40% of the OTUs belonging to three orders of the *Deltaproteobacteria* (*Desulfobacterales*, *Desulfovibrionales* and *Syntrophobacterales*) were less abundant while many taxa belonging to *Alphaproteobacteria* (*Rhizobiales*: *Rhizobiaceae*, *Phyllobacteriaceae*; *Sphingomonadales*; *Rhodobacterales*), *Betaproteobacteria* (*Burkholderiales*; *Comamonadaceae*) and *Gammaproteobacteria* (*Alteromonadales*; *Pseudomonadales*: *Pseudomonadaceae*) contained a high proportion of OTUs (>18%) that were more abundant in alfalfa soils (Table 3). An exception for *Gammaproteobacteria* was the order of *Legionellales*, of which ca. 56% of the OTUs were more abundant in the scrubland soils (Table 3). Interestingly, for *Rhodobacterales* (*Alphaproteobacteria*) 40% of the OTUs were more abundant in alfalfa soils, while 20% of the OTUs were lower compared to scrubland soils (Table 3). A high proportion (>33%) of the OTUs belonging to two orders of *Actinobacteria* (*Acidimicrobiales* and *Rubrobacterales*) had lower relative abundance in natural scrublands compared to alfalfa soils. Several families in the order of *Actinomycetales* (*Microbacteriaceae*, *Micromonosporaceae*, *Micrococcaceae*, and *Cellulomonadaceae*) had more OTUs with significantly higher abundance in alfalfa soils than in the scrubland soils (Table 3). In contrast to these families, *Mycobacteriaceae* and *Pseudonocardiaceae* decreased in relative abundance with change in land use (Table 3).

A modified test based on the first five principal components [38] was applied to study the effect of land use on bacterial community composition of the taxonomic groups listed in Table 3. It revealed and confirmed dramatic differences in the abundance of specific bacterial community members in response to land use. All the taxonomic groups listed in Table 3 had significantly different community compositions, except for *Nocardiaceae* (p = 0.07) (Table 3). As already observed with the DGGE analyses, the variation in the bacterial community composition was higher in the soils from scrubland than from alfalfa fields (Figure S5). Compared to the land use, the two different locations had less influence on the community composition. Effects of the site were only detected for a few taxonomic groups such as the *Sphingomonadales*, *Rhodobacterales*, *Rubrobacterales* and *Actinomycetales*. These orders, however, were significantly different in their community composition between the two sites (Table 3).

Table 3. Taxa and numbers (percent of all detected OTU belonging to each taxon) of OTUs significantly (unadjusted P<0.05) enriched in alfalfa or scrubland soil as identified by two-way ANOVA based on PhyloChip.

Phylum	Class	Order	Family	Alfalfa	Scrubland
Proteobacteria	*Alphaproteobacteria*	*Rhizobiales*	*Rhizobiaceae*	14 (77.8%)	0 (0.0%)
			Phyllobacteriaceae	9 (60.0%)	0 (0.0%)
		Sphingomonadales		9 (18.0%)	1 (2.0%)
		Rhodobacterales		20 (40.0%)	10 (20.0%)
	Betaproteobacteria	*Burkholderiales*	*Comamonadaceae*	36 (62.1%)	0 (0.0%)
	Gammaproteobacteria	*Alteromonadales*		15 (26.3%)	2 (3.5%)
		Pseudomonadales	*Pseudomonadaceae*	21 (53.8%)	1 (2.6%)
		Legionellales		1 (9.1%)	6 (54.5%)
	Deltaproteobacteria	*Desulfobacterales*		0 (0.0%)	19 (63.3%)
		Desulfovibrionales		0 (0.0%)	8 (50.0%)
		Syntrophobacterales		0 (0.0%)	6 (42.9%)
Firmicutes	*Clostridia*			6 (3.4%)	51 (28.8%)
Actinobacteria	***Actinobacteria***	*Acidimicrobiales*		1 (5.6%)	6 (33.3%)
		Actinomycetales	*Microbacteriaceae*	15 (65.2%)	0 (0.0%)
			Micromonosporaceae	6 (27.3%)	2 (9.1%)
			Mycobacteriaceae	0 (0.0%)	16 (76.2%)
			Micrococcaceae	13 (76.5%)	0 (0.0%)
			Nocardiaceae	2 (12.5%)	1 (6.3%)
			Pseudonocardiaceae	1 (8.3%)	4 (33.3%)
			Cellulomonadaceae	7 (63.6%)	0 (0.0%)
		Rubrobacterales		0 (0.0%)	10 (50.0%)
Acidobacteria	*Acidobacteria*	*Acidobacteriales*	*Acidobacteriaceae*	0 (0.0%)	21 (56.8%)
Bacteroidetes				22 (18.2%)	10 (8.3%)
Chloroflexi				1 (2.1%)	25 (53.2%)
Spirochaetes				0 (0.0%)	12 (30.0%)
Verrucomicrobia				0 (0.0%)	13 (48.1%)
Gemmatimonadetes				0 (0.0%)	6 (66.7%)
Bacteria (Total)				295 (13.2%)	402 (17.9%)

All listed taxa have significantly different community structure between land uses except for *Nocardiaceae*. Bold text: Taxa with significantly different community structure between sites.

Correlation of Land Use Responsive Taxa with Land Use Dependent Soil Parameters

Two-way ANOVA of soil physicochemical parameters revealed that the transition of scrubland into arable land resulted in a significantly increased concentration of sulphate, sodium, salinity, and the ratio of nitrate to total-N. The phosphate concentration and organic matter content were significantly lower in agricultural soils (Table 1).

Redundancy analysis was performed to find the correlation between discriminative physicochemical characteristics and variation in the community composition of phyla identified by PhyloChips with significant response to land use. Sulphate and phosphate (phosphate positively collinear with organic matter content) were the main factors that jointly influenced the bacterial community composition (Figure 2). In total 33% of the variation of the bacterial community could be significantly explained by sulphate and phosphate concentrations. Each factor (sulphate, phosphate and organic matter) independently could explain 17% to 21% of the variation. Sulphate and phosphate also explained a considerable amount of the variation within sub-communities of *Proteobacteria* (33%), to which most detected OTUs were affiliated (data not shown) (Figure S6). Discriminative soil parameters also explained a large amount of the variation within *Firmicutes* (19% variation explained by sulphate; Figure S7), *Actinobacteria* (40% explained by phosphate and sulphate; Figure S8), *Acidobacteria* (30% explained by nitrate; Figure S9), *Bacteroidetes* (16% explained by nitrate; Figure S10), *Chloroflexi* (30% explained by nitrate; Figure S11), *Verrucomicrobia* (24% explained by phosphate; Figure S12) and *Gemmatimonadetes* (49% explained by sulphate and sodium; Figure S13). None of these discriminative soil parameters explained the variation within *Spirochaetes,* which appeared to be linked to the pH values of the soils (explaining 32% variation; Figure S14). In general, a considerable amount of variation in the community composition of phyla responding to land use changes could be significantly explained by discriminative soil parameters, suggesting that these soil parameters and the bacterial community structure are strongly connected.

Discussion

Both DGGE and PhyloChip analysis revealed that the community composition of bacteria at various taxonomic levels differed significantly between the arable soils and the adjacent scrubland soils. The methods used to analyze the effects of the transition in land use were complementary but have clear differences which prevented the use of similar statistical analysis methods to analyze the datasets. In contrast to the DGGE fingerprints that are assumed to reflect the relative abundance of the dominant bacterial populations, the fluorescent signals detected after PhyloChip hybridizations do not necessarily reflect the relative abundance of OTUs. The strength of the PhyloChip approach is that the fluorescence signals of each OTU can be compared horizontally between treatments and thus allows to identify taxa with a high proportion of OTUs with treatment dependent changes in abundance. However, both methods used in this study to analyze PCR-amplified 16S rRNA gene fragments lead to the same conclusions: A strong and significant effect of land use and a higher variability of the community structure was observed for samples from scrubland (likely due to the absence of mixing). Furthermore, compared to the effect of land use, the effect of the site was less pronounced.

Tillage is probably one of the major forces driving shifts in the soil microbial community structure. In addition, irrigation, fertilization, and/or application of agrochemicals have been shown to affect the bacterial community structure [14,39–42]. The minor differences found between the arable soils from site 1 and site 2 may be linked to differences in their previous cropping history and agricultural management, which were most likely not identical. In general, the conversion of scrubland into agricultural land is associated with the replacement of a diverse, indigenous, highly adapted vegetation of the semi-arid desert with a few crop plants. Plant effects on soil microbial communities have been frequently observed in the rhizosphere, which refers to the soil directly influenced by root exudates [30,43]. The influence of plant species varied between different studies, in some of them it was regarded to be the major factor shaping the microbial community structure [44–49], while in others only a minor influence of plant species and vegetation composition on the soil bacterial community composition was observed [50]. However, only a few studies suggested long-term effects of plants on microbial communities in bulk soils [16]. Long-term agricultural use impacts soil physicochemical characteristics [3,16,17], and thus probably alters the composition and properties of biogeochemical interfaces in soils [51].

The present study showed that agricultural use impacted several bacterial phyla in the soils of the sites studied, e.g., *Acidobacteria, Chloroflexi, Spirochaetes, Verrucomicrobia, Gemmatimonadetes, Deltaproteobacteria* (*Proteobacteria*), *Acidimicrobiales* (*Actinobacteria*) and *Rubrobacterales* (*Actinobacteria*). *Acidobacteria* have been found to be dominant in several soils [52] though often they are difficult to cultivate. In the study by Bisette *et al.* [16], the proportion of *Acidobacteria* was found increased in grassland soils compared to soils under agricultural use at one site in Australia. Compared with agricultural soils, their relative abundance was also reported to be higher in forest, desert or prairie soils [52,53]. The proportion of *Acidobacteria* was reported to be significantly lower in nutrient-rich rhizosphere than in bulk soil, confirming their oligotrophic lifestyle [54]. Under dry conditions, the net primary productivity of plants is controlled by water [55,56]. Additional water supply in the alfalfa fields enhanced plant growth and probably elevated the extent of plant exudates into soil over a time period of about 50 years. Recently the relative abundance of *Acidobacteria* was found to be negatively correlated with the level of nitrogen input (fertilizer) [57], which in general increases the net productivity of vegetation. *Chloroflexi* also was reported to prevail in nutrient poor soils [52,58] and other oligotrophic ecosystems such as soils from high-elevation regions where vegetation is patchy [59], alpine tundra soil [60] or hyperarid polar desert soil [61]. A few bacteria belonging to *Chloroflexi* which could be retrieved from soil had very slow rates of growth and mini-colony formation [62,63] which are typical characteristics of oligotrophic organisms. In accordance with the present study, Fierer *et al.* [57] showed that the relative abundance of *Chloroflexi* was also lower in the plots with high levels of nitrogen input.

In the present study, the physicochemical analysis of the soils done for four independent replicates per site and treatment revealed that the concentration of sulphate (probably due to fertilization) was higher in alfalfa soils. Sulphate can be used as terminal electron acceptor by some anaerobic bacteria. However, a high proportion of OTUs belonging to *Desulfobacterales, Desulfovibrionales, Syntrophobacterales* or *Clostridia* had significantly lower abundances in the alfalfa soils compared to scrubland soils. By and large, bacteria belonging to these taxonomic groups are strict anaerobes [64] and play an important role in anaerobic carbon cycling in wet terrestrial ecosystems such as rice fields and wet lands [65,66].Furthermore, tilling soils also affects soil aeration [67] and can elevate the activity of aerobic microbes [14]. Notably, in the present study, a significant loss of organic matter

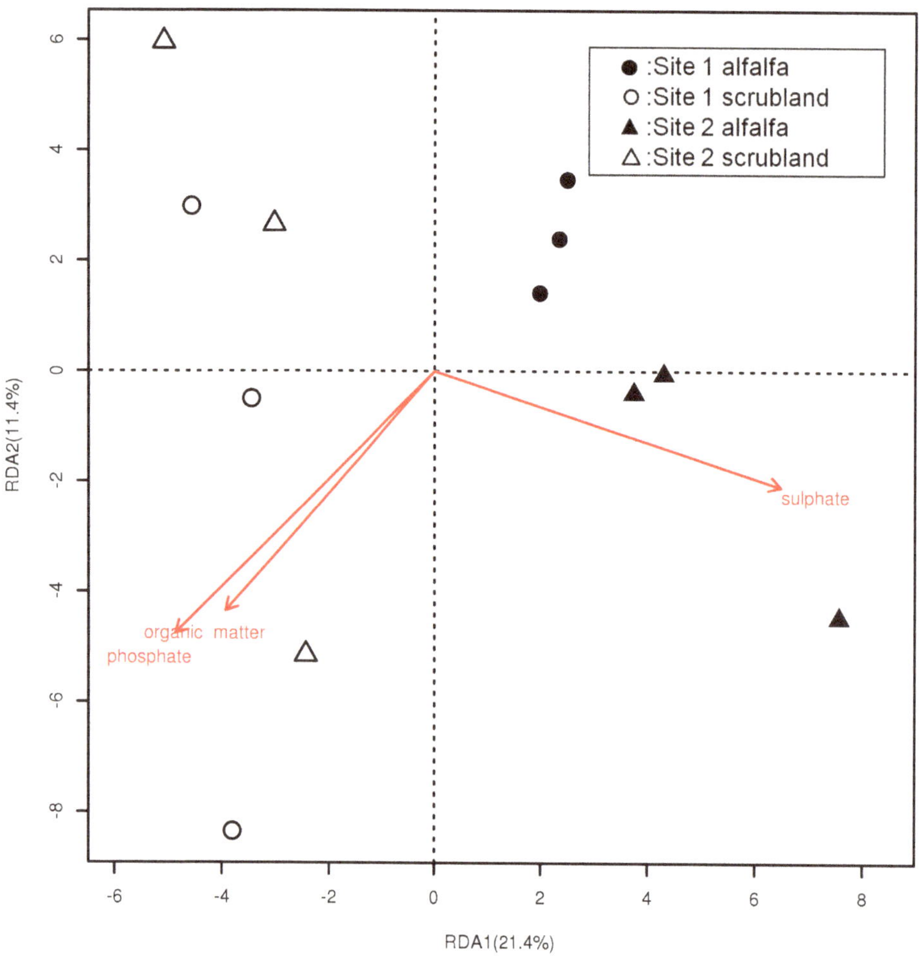

Figure 2. Redundancy analysis of the effect of discriminative soil parameters on bacterial communities using the PhyloChip data. Numbers in brackets indicate the percent of the total variance explained by each axis. Only these soil parameters which significantly ($p<0.05$ by 1000 times permutation tests) explained the bacterial community variation was shown.

also was observed in the agricultural soils and as soil organic matter is an important binding agent of soil particles into aggregates [68], presumably lowering gas diffusion, this may have contributed to lower sulphate-reducing populations in these desert agricultural soils.

High proportions of OTUs belonging to the *Bacteroidetes* or the *Alpha-*, *Beta-* and *Gammaproteobacteria* as well as several families of *Actinobacteria* were more abundant in alfalfa soils. The higher abundance of *Pseudomonas* (50% of the OTUs with significantly higher signal intensities) in the agricultural soils was also confirmed by the detection of *Pseudomonas*-specific *gacA* genes [46]. Amplicons of the *gacA* gene fragment were obtained only for TC DNA from alfalfa soils, not from scrubland soil (data not shown). *Betaproteobacteria* and *Bacteroides* were considered to contain copiotrophic taxa as their abundance has been positively correlated with carbon mineralization rate and carbon availability [69]. Several studies reported enrichments of *Alphaproteobacteria* [69], *Gammaproteobacteria* [46,70) or *Actinobacteria* [29,43) in the rhizosphere where carbon availability is increased due to root exudates. The number of OTUs detected by PhyloChip analysis was significantly lower in the soil from scrubland than alfalfa field. In general, this finding is in agreement with other studies, in which agriculture or low plant diversity (often a direct result of converting natural land into agricultural use) did not necessarily lead to a reduction of the bacterial diversity detected [48,71–73]. However, the effects of agricultural practices on soil bacterial richness remain to be explored more fully, perhaps at a finer resolution than was obtainable in this study. Firstly, total bacterial species richness in soils is difficult to assess as many populations occur at low abundance [74,75,76]. Secondly, a high proportion of OTUs belonging to *Acidobacteria*, *Chloroflexi*, *Spirochaetes*, *Verrucomicrobia*, *Gemmatimonadetes* were more abundant in the scrubland soils, suggesting that these taxa were better adapted to those soils. Compared to *Proteobacteria*, *Actinobacteria* and *Firmicutes*, many fewer OTUs belonging to these taxa were available during probe design. Therefore, bacterial richness of these phyla in the scrubland soil possibly was underestimated.

In summary, although investigated only at two sites that were assumed to be representative for this type of ecosystem, we could show that the use of scrublands for agriculture caused profound changes in the soil bacterial community structure and physico-chemical characteristics. Soil parameters that differed between land uses were highly correlated with the community composition of taxa responding to land use. Several, most likely oligotrophic or anaerobic taxa were negatively affected by the change, in contrast, populations with a potentially copiotrophic lifestyle profited and were enhanced in the agricultural soils.

Supporting Information

Figure S1 Bacterial DGGE profiles for soils from different sites and land use. *: samples not included in this study. M: bacterial standard for DGGE electrophoresis.

Figure S2 Actinobacterial DGGE profiles for soils from different sites and land use. *: samples not included in this study. M: bacterial standard for DGGE electrophoresis.

Figure S3 Alphaproteobacterial DGGE profiles for soils from different sites and land use. *: samples not included in this study. M: bacterial standard for DGGE electrophoresis.

Figure S4 Betaproteobacterial DGGE profiles for soils from different sites and land use. *: samples not included in this study. M: bacterial standard for DGGE electrophoresis.

Figure S5 Principal component analysis of PhyloChip data for soils from different sites and land use. The first and second principal components explain 31% and 16% of total variance.

Figure S6 Redundancy analysis of the effect of discriminative soil parameters on the communities of *Proteobacteria* using the PhyloChip data. Numbers in brackets indicate the percent of the total variance explained by each axis. Only these soil parameters which significantly ($p<0.05$ by 1000 times permutation tests) explained the proteobacterial community variation are shown.

Figure S7 Redundancy analysis of the effect of discriminative soil parameters on the communities of *Firmicutes* using the PhyloChip data. Numbers in brackets indicate the percent of the total variance explained by each axis. Only the soil parameter which significantly ($p<0.05$ by 1000 times permutation tests) explained the variation of *Firmicutes* community is shown.

Figure S8 Redundancy analysis of the effect of discriminative soil parameters on the communities of *Actinobacteria* using the PhyloChip data. Numbers in brackets indicate the percent of the total variance explained by each axis. Only these soil parameters which could significantly ($p<0.05$ by 1000 times permutation tests) explained the actinobacterial community variationare shown.

Figure S9 Redundancy analysis of the effect of discriminative soil parameters on the communities of *Acidobacteria* using the PhyloChip data. Numbers in brackets indicate the percent of the total variance explained by each axis. Only the soil parameters which significantly ($p<0.05$ by 1000 times permutation tests) explained the acidobacterial community variation is shown.

Figure S10 Redundancy analysis of the effect of discriminative soil parameters on the communities of *Bacteroidetes* using the PhyloChip data. Numbers in brackets indicate the percent of the total variance explained by each axis. Only the soil parameters which significantly ($p<0.05$ by 1000 times permutation tests) explained the community variation of *Bacteroidetes* is shown.

Figure S11 Redundancy analysis of the effect of discriminative soil parameters on the communities of *Chloroflexi* using the PhyloChip data. Numbers in brackets indicate the percent of the total variance explained by each axis. Only the soil parameter which significantly ($p<0.05$ by 1000 times permutation tests) explained the community variation of *Chloroflexi* is shown.

Figure S12 Redundancy analysis of the effect of discriminative soil parameters on the communities of *Verrucomicrobia* using the PhyloChip data. Numbers in brackets indicate the percent of the total variance explained by each axis. Only these soil parameters which significantly ($p<0.05$ by 1000 times permutation tests) explained the community variation of *Verrucomicrobia* are shown.

Figure S13 Redundancy analysis of the effect of discriminative soil parameters on the communities of *Gemmatimonadetes* using the PhyloChip data. Numbers in brackets indicate the percent of the total variance explained by each axis. Only these soil parameters which significantly ($p<0.05$ by 1000 times permutation tests) explained the community variation of *Gemmatimonadetes* are shown.

Figure S14 Redundancy analysis of the effect of soil parameters on the communities of *Spirochaetes* using the PhyloChip data. Numbers in brackets indicate the percent of the total variance explained by each axis. Only these soil parameters which significantly ($p<0.05$ by 1000 times permutation tests) explained the community variation of *Spirochaetes* are shown.

Table S1 Primers used for DGGE analyses in the present study.

Table S2 Numbers (average ± standard deviation)s of OTU detected for bulk soils from alfalfa field and scrubland at two sites.

Acknowledgments

We thank Professor Siegfried Kropf at the Institute for Biometry and Medical informatics, Otto von Guericke University for his help on statistical analyses and Ilse-Marie Jungkurth for proofreading the manuscript very carefully.

Author Contributions

Constructive discussion and sharp criticism to improve the manuscript: HH. Conceived and designed the experiments: KS CCT TC GLA. Performed the experiments: YMP NW ABD AC. Analyzed the data: G-CD HH. Contributed reagents/materials/analysis tools: YMP NW CCT TC G-CD. Wrote the paper: G-CD KS CCT TC GLA.

References

1. Munson SM, Belnap J, Okin GS (2011) Responses of wind erosion to climate-induced vegetation changes on the Colorado Plateau. Proc Natl Acad Sci U S A 108: 3854–3859.
2. Saygin SD, Basaran M, Ozcan AU, Dolarslan M, Timur OB, et al. (2011) Land degradation assessment by geo-spatially modeling different soil erodibility equations in a semi-arid catchment. Environ Monit Assess 180: 201–215.
3. Zhao WZ, Xiao HL, Liu ZM, Li J (2005) Soil degradation and restoration as affected by land use change in the semiarid Bashang area, northern China. Catena 59: 173–186.
4. Mahamane L, Mahamane S (2005) Biodiversity of ligneous species in semi-arid to arid zones of southwestern Niger according to anthropogenic and natural factors. Agr Ecosyst Environ 105: 267–271.
5. Underwood EC, Viers JH, Klausmeyer KR, Cox RL, Shaw MR (2009) Threats and biodiversity in the mediterranean biome. Divers Distrib 15: 188–197.
6. Foley JA, DeFries R, Asner GP, Barford C, Bonan G, et al. (2005) Global consequences of land use. Science 309: 570–574.
7. Snyman HA (1998) Dynamics and sustainable utilization of rangeland ecosystems in arid and semi-arid climates of southern Africa. J Arid Environ 39: 645–666.
8. Maestre FT, Cortina J, Vallejo R (2006) Are ecosystem composition, structure, and functional status related to restoration success? A test from semiarid Mediterranean steppes. Restor Ecol 14: 258–266.
9. Choudhary DK, Sharma KP, Gaur RK (2011) Biotechnological perspectives of microbes in agro-ecosystems. Biotechnol Lett 33: 1905–1910.
10. Miransari M (2011) Soil microbes and plant fertilization. Appl Microbiol Biotechnol 92: 875–885.
11. Uroz S, Calvaruso C, Turpault MP, Frey-Klett P (2009) Mineral weathering by bacteria: ecology, actors and mechanisms. Trends Microbiol 17: 378–387.
12. Bell T, Newman JA, Silverman BW, Turner SL, Lilley AK (2005) The contribution of species richness and composition to bacterial services. Nature 436: 1157–1160.
13. Giovannoni S (2004). Evolutionary biology: oceans of bacteria. Nature 430: 515–516.
14. Calderon FJ, Jackson LE, Scow KM, Rolston DE (2001) Short-term dynamics of nitrogen, microbial activity, and phospholipid fatty acids after tillage. Soil Sci Soc Am J 65: 118–126.
15. Zak DR, Holmes WE, White DC, Peacock AD, Tilman D (2003) Plant diversity, soil microbial communities, and ecosystem function: Are there any links? Ecology 84: 2042–2050.
16. Bissett A, Richardson AE, Baker G, Thrall PH (2011) Long-term land use effects on soil microbial community structure and function. Appl Soil Ecol 51: 66–78.
17. Peixoto RS, Chaer GM, Franco N, Reis FB, Mendes IC, et al. (2010) A decade of land use contributes to changes in the chemistry, biochemistry and bacterial community structures of soils in the Cerrado. Antonie Van Leeuwenhoek 98: 403–413.
18. Salinas-Zavala CA, Lluch-Cota SE, Fogel I (2006) Historic development of winter-wheat yields in five irrigation districts in the Sonora Desert, Mexico. Interciencia 31: 254–261.
19. Coria R, Leon de la Luz JS (1992) Flora iconografica de Baja California Sur. Centro de Investigaciones Biologicas de Baja California Sur A.C.
20. Armenta-Quintana JA, Ramirez-Orduna R, Ramirez-Lozano RG (2011) Forage utilization and diet selection by grazing goats on a sarcocaulescent scrubland in Northwest Mexico. Rev Chapingo Ser Cie 17: 163–171.
21. Ramirez-Orduna R, Ramirez RG, Romero-Vadillo E, Gonzalez-Rodriguez H, Armenta-Quintana JA, et al. (2008) Diet and nutrition of range goats on a sarcocaulescent shrubland from Baja California Sur, Mexico. Small Ruminant Res 76: 166–176.
22. Gee GW, Or D (2002) Particle size analysis. In: Dane JH, Topp GC (eds). Methods of Soil Analysis. Part 4-Physical Methods, Soil Sci. Soc., America, Madison, 255–293.
23. McLean EO (1982) Soil pH and lime requirement. In: Page, AL (ed). Methods of Soil Analysis, Part 2, American Soc. Agronomy, Madison, 199–224.
24. Nelson DE, Sommers LE (1996) Total organic carbon and organic matter. In: Page, AL (ed). Methods of Soil Analysis. Part 2, American Soc. Agronomy, Madison, 961–1010.
25. Strickland JDH, Parsons TR (1972) A practical handbook of seawater analysis. Fisheries Res. Board Canada, Ottawa, Canada.
26. Jackson ML (1958) Soil Chemical Analysis. Prentice Hall, Englewood, N.J.
27. Olsen SR, Dean LA (1965) Phosphorus. Chemical and microbiological properties. In: Black, CA (ed). Methods of Soil Analysis, Part 2. Am. Soc. Agron., Madison, WI, USA, 1035–1048.
28. Heuer H, Krsek M, Baker P, Smalla K, Wellington EM (1997) Analysis of actinomycete communities by specific amplification of genes encoding 16S rRNA and gel-electrophoretic separation in denaturing gradients. Appl Environ Microbiol 63: 3233–3241.
29. Gomes NCM, Heuer H, Schönfeld J, Costa R, Mendonça-Hagler L, et al. (2001) Bacterial diversity of the rhizosphere of maize (*Zea mays*) grown in tropical soil studied by temperature gradient gel electrophoresis. Plant Soil 232: 167–180.
30. Weinert N, Meincke R, Gottwald C, Heuer H, Gomes NC, et al. (2009) Rhizosphere communities of genetically modified zeaxanthin-accumulating potato plants and their parent cultivar differ less than those of different potato cultivars. Appl Environ Microbiol 75: 3859–3865.
31. Gomes NCM, Kosheleva IA, Abraham WR, Smalla K (2005) Effects of the inoculant strain *Pseudomonas putida* KT2442 (pNF142) and of naphthalene contamination on the soil bacterial community. FEMS Microbiol Ecol 54: 21–33.
32. Heuer H, Wieland J, Schönwälder A, Gomes NCM, Smalla K (2001) Bacterial community profiling using DGGE or TGGE analysis. In: Rouchelle, IP (ed). Environmental molecular microbiology: protocols and applications. Horizon Scientific Press: Wymondham, United Kingdom. 177–190.
33. Kropf S, Heuer H, Grüning M, Smalla K (2004) Significance test for comparing complex microbial community fingerprints using pairwise similarity measures. J Microbiol Methods 57: 187–195.
34. Weinert N, Piceno Y, Ding GC, Meincke R, Heuer H, et al. (2011) PhyloChip hybridization uncovered an enormous bacterial diversity in the rhizosphere of different potato cultivars: many common and few cultivar-dependent taxa. FEMS Microbiol Ecol 75: 497–506.
35. DeSantis TZ, Brodie EL, Moberg JP, Zubieta IX, Piceno YM, et al. (2007) High-density universal 16S rRNA microarray analysis reveals broader diversity than typical clone library when sampling the environment. Microb Ecol 53: 371–383.
36. Brodie EL, DeSantis TZ, Parker JP, Zubietta IX, Piceno YM, et al. (2007) Urban aerosols harbor diverse and dynamic bacterial populations. Proc Natl Acad Sci USA 104: 299–304.
37. Ivanov II, Atarashi K, Manel N, Brodie EL, Shima T, et al. (2009) Induction of intestinal Th17 cells by segmented filamentous bacteria. Cell 139: 485–498.
38. Ding GC, Smalla K, Heuer H, Kropf S (2012) A new proposal for a principal component-based test for high-dimensional data applied to the analysis of PhyloChip data. Biom J 54: 94–107.
39. Entry JA, Fuhrmann JJ, Sojka RE, Shewmaker GE (2004) Influence of irrigated agriculture on soil carbon and microbial community structure. Environ Manage 33: S363–S373.
40. Islam MR, Trivedi P, Palaniappan P, Reddy MS, Sa T (2009) Evaluating the effect of fertilizer application on soil microbial community structure in rice based cropping system using fatty acid methyl esters (FAME) analysis. World J Microb Biot 25: 1115–1117.
41. Lo CC (2010) Effect of pesticides on soil microbial community. J Environ Sci Health B 45: 348–359.
42. Ramirez KS, Lauber CL, Knight R, Bradford MA, Fierer N (2010) Consistent effects of nitrogen fertilization on soil bacterial communities in contrasting systems. Ecology 91: 3463–3470.
43. Smalla K, Wieland G, Buchner A, Zock A, Parzy J, et al. (2001) Bulk and rhizosphere soil bacterial communities studied by denaturing gradient gel electrophoresis: Plant-dependent enrichment and seasonal shifts revealed. Appl Environ Microb 67: 4742–4751.
44. Costa R, Götz M, Mrotzek N, Lottmann J, Berg G, et al. (2006) Effects of site and plant species on rhizosphere community structure as revealed by molecular analysis of microbial guilds. FEMS Microbiol Ecol 56: 236–249.
45. Costa R, Salles JF, Berg G, Smalla K (2006) Cultivation-independent analysis of Pseudomonas species in soil and in the rhizosphere of field-grown *Verticillium dahliae* host plants. Environ Microbiol 8: 2136–2149.
46. Costa R, Gomes NC, Krögerrecklenfort E, Opelt K, Berg G, et al. (2007) *Pseudomonas* community structure and antagonistic potential in the rhizosphere: insights gained by combining phylogenetic and functional gene-based analyses. Environ Microbiol 9: 2260–2273.
47. DeAngelis KM, Brodie EL, DeSantis TZ, Andersen GL, Lindow SE, et al. (2009) Selective progressive response of soil microbial community to wild oat roots. ISME J 3: 168–178.
48. Kowalchuk GA, Buma DS, de Boer W, Klinkhamer PG, van Veen JA (2002) Effects of above-ground plant species composition and diversity on the diversity of soil-borne microorganisms. Antonie Van Leeuwenhoek 81: 509–520.
49. Mendes R, Kruijt M, de Bruijn I, Dekkers E, van der Voort M, et al. (2011) Deciphering the rhizosphere microbiome for disease-suppressive bacteria. Science 332: 1097–1100.
50. Kielak A, Pijl AS, van Veen JA, Kowalchuk GA (2008) Differences in vegetation composition and plant species identity lead to only minor changes in soil-borne microbial communities in a former arable field. FEMS Microbiol Ecol 63: 372–382.
51. Totsche KU, Rennert T, Gerzabek MH, Kögel-Knabner I, Smalla K, et al. (2010) Biogeochemical interfaces in soil: The interdisciplinary challenge for soil science. J Plant Nutr Soil Sci 173: 88–99.
52. Janssen PH (2006) Identifying the dominant soil bacterial taxa in libraries of 16S rRNA and 16S rRNA genes. Appl Environ Microbiol 72: 1719–1728.
53. Fierer N, Jackson JA, Vilgalys R, Jackson RB (2005) Assessment of soil microbial community structure by use of taxon-specific quantitative PCR assays. Appl Environ Microbiol 71: 4117–4120.
54. Kielak A, Pijl AS, van Veen JA, Kowalchuk GA (2009) Phylogenetic diversity of Acidobacteria in a former agricultural soil. ISME J 3: 378–382.
55. Gao YZ, Chen Q, Lin S, Giese M, Brueck H (2011) Resource manipulation effects on net primary production, biomass allocation and rain-use efficiency of two semiarid grassland sites in Inner Mongolia, China. Oecologia 165: 855–864.

56. Webb W, Szarek S, Lauenroth W, Kinerson R, Smith M (1978) Primary productivity and water-use in native forest, grassland, and desert ecosystems. Ecology 59: 1239–1247.
57. Fierer N, Lauber CL, Ramirez KS, Zaneveld J, Bradford MA, et al. (2012) Comparative metagenomic, phylogenetic and physiological analyses of soil microbial communities across nitrogen gradients. ISME J 6: 1007–1017.
58. Will C, Thurmer A, Wollherr A, Nacke H, Herold N, et al. (2010) Horizon-specific bacterial community composition of German grassland soils, as revealed by pyrosequencing-based analysis of 16S rRNA genes. Appl Environ Microbiol 76: 6751–6759.
59. Freeman KR, Pescador MY, Reed SC, Costello EK, Robeson MS, et al. (2009) Soil CO2 flux and photoautotrophic community composition in high-elevation, 'barren' soil. Environ Microbiol 11: 674–686.
60. Costello EK, Schmidt SK (2006) Microbial diversity in alpine tundra wet meadow soil: novel Chloroflexi from a cold, water-saturated environment. Environ Microbiol 8: 1471–1486.
61. Pointing SB, Chan YK, Lacap DC, Lau MCY, Jurgens JA, et al. (2009) Highly specialized microbial diversity in hyper-arid polar desert. Proc Natl Acad Sci USA 106: 19964–19969.
62. Davis KE, Joseph SJ, Janssen PH (2005) Effects of growth medium, inoculum size, and incubation time on culturability and isolation of soil bacteria. Appl Environ Microbiol 71: 826–834.
63. Davis KER, Sangwan P, Janssen PH (2011) Acidobacteria, Rubrobacteridae and Chloroflexi are abundant among very slow-growing and mini-colony-forming soil bacteria. Environ Microbiol 13: 798–805.
64. Muyzer G, Stams AJ (2008) The ecology and biotechnology of sulphate-reducing bacteria. Nat Rev Microbiol 6: 441–454.
65. He JZ, Liu XZ, Zheng Y, Shen JP, Zhang LM (2010) Dynamics of sulfate reduction and sulfate-reducing prokaryotes in anaerobic paddy soil amended with rice straw. Biol Fert Soils 46: 283–291.
66. Pester M, Knorr KH, Friedrich MW, Wagner M, Loy A (2012) Sulfate-reducing microorganisms in wetlands - fameless actors in carbon cycling and climate change. Front Microbiol 3: 72.
67. Kladivko EJ (2001) Tillage systems and soil ecology. Soil Till Res 61: 61–76.
68. Bronick CJ, Lal R (2005) Soil structure and management: a review. Geoderma 124: 3–22.
69. Fierer N, Bradford MA, Jackson RB (2007) Toward an ecological classification of soil bacteria. Ecology 88: 1354–1364.
70. Rossmann B, Muller H, Smalla K, Mpiira S, Tumuhairwe JB, et al. (2012) Banana-associated microbial communities in Uganda are highly diverse but dominated by Enterobacteriaceae. Appl Environ Microbiol 78: 4933–4941.
71. Gruter D, Schmid B, Brandl H (2006) Influence of plant diversity and elevated atmospheric carbon dioxide levels on belowground bacterial diversity. BMC Microbiol 6: 68.
72. Koberl M, Muller H, Ramadan EM, Berg G (2011) Desert farming benefits from microbial potential in arid soils and promotes diversity and plant health. PLoS One 6: e24452.
73. Zul D, Denzel S, Kotz A, Overmann J (2007) Effects of plant biomass, plant diversity, and water content on bacterial communities in soil lysimeters: implications for the determinants of bacterial diversity. Appl Environ Microbiol 73: 6916–6929.
74. Curtis TP, Sloan WT, Scannell JW (2002) Estimating prokaryotic diversity and its limits. Proc Natl Acad Sci U S A 99: 10494–10499.
75. Gans J, Wolinsky M, Dunbar J (2005) Computational improvements reveal great bacterial diversity and high metal toxicity in soil. Science 309: 1387–1390.
76. Hong SH, Bunge J, Jeon SO, Epstein SS (2006) Predicting microbial species richness. Proc Natl Acad Sci U S A 103: 117–122.

18

Cross-Species Extrapolation of Prediction Model for Lead Transfer from Soil to Corn Grain under Stress of Exogenous Lead

Zhaojun Li[1]*, Hua Yang[1,2], Yupeng Li[1,3], Jian Long[2], Yongchao Liang[1]

1 Institute of Agricultural Resources and Regional Planning, Chinese Academy of Agricultural Sciences, Key Laboratory of Plant Nutrition and Fertilizer, Ministry of Agriculture, Beijing, China, **2** Guizhou Key Laboratory of Mountain Environment, Guizhou Normal University, Guiyang, China, **3** College of City and Environmental Sciences, Shanxi Normal University, Linfen, China

Abstract

There has been increasing concern in recent years regarding lead (Pb) transfer in the soil-plant system. In this study the transfer of Pb (exogenous salts) was investigated from a wide range of Chinese soils to corn grain (Zhengdan 958). Prediction models were developed with combination of the Pb bioconcentration factor (BCF) of Zhengdan 958, and soil pH, organic matter (OM) content, and cation exchange capacity (CEC) through multiple stepwise regressions. Moreover, these prediction models from Zhengdan 958 were applied to other non-model corn species through cross-species extrapolation approach. The results showed that the soil pH and OM were the major factors that controlled Pb transfer from soil to corn grain. The lower pH and OM could improve the bioaccumulation of Pb in corn grain. No significant differences were found between two prediction models derived from the different exogenous Pb contents. When the prediction models were applied to other non-model corn species, the ratio ranges between the predicted BCF values and the measured BCF values were within an interval of 2-fold and close to the solid line of 1:1 relationship. Moreover, the prediction model i.e. Log[BCF] = −0.098 pH-0.150 log[OM] −1.894 at the treatment of high Pb can effectively reduce the measured BCF intra-species variability for all non-model corn species. These suggested that this prediction model derived from the high Pb content was more adaptable to be applied to other non-model corn species to predict the Pb bioconcentration in corn grain and assess the ecological risk of Pb in different agricultural soils.

Editor: Xianglin Shi, University of Kentucky, United States of America

Funding: This work was jointly supported by the Special Fund of Public Industry in China (Agriculture, project no. 200903015, http://www.moa.gov.cn/) and the Natural Science Foundation of Institute of Agricultural Resources and Regional Planning, Chinese Academy of Agricultural Sciences (2013-9, http://www.iarrp.cn/SoisWeb/home/index.html). The funders had no role in study design, data collection and analysis, decision to publish, or preparation of the manuscript.

Competing Interests: The authors have declared that no competing interests exist.

* E-mail: lizhaojun@caas.cn

Introduction

Lead (Pb) is one of the most toxic heavy metals for people in the world [1–3]. In children, Pb has been known to cause decreases in intelligence quotient (IQ) scores, retardation of physical growth, and hearing problems. In individuals of all ages, it has been known to cause anemia brain damage, impaired function of the peripheral nervous system, high blood pressure, reproductive abnormalities, development defects, abnormal vitamin D metabolism, and in some situations death of people [4–5]. Soil-to-plant transfer of heavy metals is the major exposure pathway for humans with regards to soil contamination [6]. Therefore, the accumulation of heavy metals in plants in agricultural soils, especially in edible part of plants, is great concern regarding its posing potential threats [7–8]. The efficiency of the plants for absorbing metals is mainly evaluated by plant uptake or the metal transfer factors from soil to plant [9]. Evaluating the metal transfer factors from soil to plant especially to edible parts can further provide the information about accumulation of contaminants into the food chain [10].

The Pb transfer factor from soil to plant mainly depends on its availability in soils. Soil properties can determine the availability of Pb to crops, including soil pH, organic matter (OM), cation exchange capacity (CEC), redox potential, clay minerals, Fe and Mn oxides, and calcium carbonate content [11–13]. In addition, it has been reported that soil pH is a key factor in controlling the cation mobility and regulating the solubility of heavy metals in soils [14–15]. The heavy metals tended to be more available in acid soil [16]. The low soil pH is favorable for the transference of Pb from soil to plant [17–19]. As well as, soil OM has also been found to play an important role in influence Pb adsorption processes in soils [20–21]. Pb in soils with low OM content was difficult to be adsorbed but was easy to be uptaken by plants [22]. The importance for assessing the ecological risk of heavy metals in complex soil types has urged a lot of prediction models to be developed to predict heavy metals transfer from soil to plant [23]. The heavy metal transfer factor from soil to plant also depends on plant species. It has been observed that significant differences exist in heavy metal uptake among cultivars [24]. Due to the expenses and efforts involved, these models were limited to specific contaminated sites or single species [25–27]. In addition, most of these models were developed to describe the transfer of heavy metals from soil to plant leaves, vegetables, earthworms by using the log-transformed concentrations [28–29]. A few attempts were only made to describe heavy metals uptake by rice, wheat and barley grain directly from soil with combination of total concentrations of heavy metal and soil basic properties [30–31]. Little is reported about the prediction models developed to describe the transfer of heavy metals, especially Pb, from soils to corn grain. Intake of heavy metals from dietary sources may represent a significant exposure pathway for human populations. Prediction of heavy metals including Pb concentrations in edible part of crops based on their concentration in soil and other environmental factors are urgently required for human risk assessment and derivation of soil environmental benchmark croplands.

The species sensitivity distributions (SSD) have been proved to be successful and were increasingly used in ecological risk assessment procedures [32–33]. However, the toxicology data from the soils with different soil properties need to be normalized using prediction models to eliminate the influence of soil factors [34]. It can improve the accuracy of the sensitivity distribution of species and environmental quality standard values [35]. These models can also provide great opportunities to carrying out ecological risk assessments and establishing soil quality criteria for heavy metals in croplands. Due to the fact that significant differences existed among cultivars for heavy metal uptake, it is ideal that separate species-specific models should be developed for each single species. However, this is unrealistic due to the expenses and efforts involved. Thus the cross-species extrapolation of toxicity data models was often applied in some cases of pollutants risk assessment practice. For example, Deleebeeck et al. [36–37] have applied biotic bigand model (BLM) from *O. mykiss* to other fish species and applied BLM from D. *magna* to other cladocera species. Schlekat et al. [38] applied chronic nickel BLMs developed for the cladocera such as Daphnia magna and Ceriodaphnia dubia to predict chronic toxicity of nickel to three other invertebrates including snail *(Lymnaea stagnalis)*, insect (*Chironomus tentans*), and rotifer (*Brachionus calyciflorus*). Yang et al. [16] applied corn grain Cd BCF predication models developed for Zhengdan 958 to predict other non-model corn species and wheat grain Cd BCF. However, little effort has been put on establishing models to describe the relationship between Pb accumulation in corn grain and soil properties. The objectives of this study are: 1) to establish the prediction model for Pb accumulation in corn grain on seventeen soils with greatly different properties; 2) to assess the feasibility of applying these models to other non-model corn species, and 3) to investigate their accuracy in predicting accumulation of Pb in non-model corn species.

Materials and Methods

None of these 17 soil samples were collected from national parks or other protected areas. It is confirmed that no specific permissions were required for the soils sampling activities in the 17 locations in China. It is also can be confirmed that the field studies did not involve endangered or protected species. No tested corn species are under first- or second-class state protection, and they are not listed in the Inventory of Rare and Endangered Plants of China (http://zrbhq.forestry.gov.cn/portal/zrbh/s/3053/

Table 2. Soil Environmental Quality Standards of China (GB15618-1995) and the added contents of exogenous Pb (mg.kg^{-1}).

pH	<6.5	6.5~7.5	>7.5
Grade Two Standard	250	300	350
Low Pb	125	150	175
High Pb	250	300	350

Table 1. Selected soil properties of the 17 soils used in the lead bioconcentration factor test.

Soil NO.	Location*	pH	OM (g.kg^{-1})	CEC (cmol.kg^{-1})	TN (g.kg^{-1})	TP (mg.kg^{-1})	TK (g.kg^{-1})	Background Pb (mg.kg^{-1})
S1	Hunan	4.90	15.52	10.85	1.14	0.47	15.26	23.12
S2	Chongqing	5.74	17.48	21.34	1.00	0.55	22.61	29.72
S3	Liaoning	5.74	25.84	12.19	1.00	0.73	23.94	27.94
S4	Yunnan	5.92	34.26	11.10	2.01	0.81	4.77	24.58
S5	Jiangxi	6.01	11.69	8.70	0.51	0.52	9.96	33.48
S6	Anhui	6.25	20.04	19.08	0.99	0.35	15.41	29.34
S7	Heilongjiang	6.27	35.69	28.59	1.74	0.48	24.70	33.99
S8	Jilin	6.82	32.85	31.11	1.75	0.35	24.58	31.01
S9	Jiangsu	6.93	47.69	26.20	2.44	0.69	21.03	36.81
S10	Shaanxi	7.90	16.49	22.37	1.36	0.98	24.37	37.30
S11	Hebei	7.98	8.57	8.12	0.68	0.53	24.22	31.96
S12	Henan	8.07	17.79	16.01	1.07	0.75	19.86	35.89
S13	Xinjiang	8.12	19.43	25.25	1.32	0.78	25.49	31.73
S14	Shanxi	8.24	23.17	16.80	1.13	0.95	23.70	37.97
S15	Tianjin	8.29	22.02	24.67	1.42	0.92	24.63	34.18
S16	Gansu	8.37	19.27	11.23	1.05	0.74	23.62	35.22
S17	Shandong	8.65	11.84	13.09	0.93	0.97	21.37	34.55

*Soil sample locations are listed in the order of increasing pH.

content-457748.html), or the Key Protected Inventory of Wild Plants of China (http://zrbhq.forestry.gov.cn/uploadfile/zrbh/2010-10/file/2010-10-14-bb296addeaa047798d6b6c476aaa1da9.doc). These corn species were used for only scientific research as permitted by the Ministry of Agriculture of China.

Soil Samples

A set of 17 soils covering a wide range of soil properties was sampled from typical locations in China. In each sample location, the soils were collected from the top 20 cm of the soil profile. For analysis of their phys-chemical characteristics, the soils were air dried and sorted to pass a 2-mm sieve. Soil pH was measured in deionized water (soil: solution ratio, 1:5) [39]. CEC was determined by the unbuffered silver-thiourea method [40]. OM was measured by dry combustion [41]. The total phosphorus in soils (TP) was measured by colorimetric method [42]. The total nitrogen in soils was determined by the Kjeldahl's method [43]. The background Pb content in the soils was determined by aqua regia (1:3 fresh mixture of concentrated HNO_3 and HCl) digestion [44]. The selected properties of the 17 soil samples were shown in Table 1.

Experimental Design

Bioaccumulation factors for Pb by corn grain from seventeen soils. According to Grade Two Standard for Pb in the Soil Environmental Quality Standards of China (GB15618-1995), the tested levels of Pb added in soils were shown in Table 2. To obtain the tested levels, a certain amount of exogenous Pb i.e. $Pb(NO_3)_2$ was added to the 8 kg air-dried soils. All the soils were thoroughly mixed and placed into pots (ϕ×h, 23 cm×26.5 cm). Then the soils were moistened with deionized water to the 60% of field moisture capacity. Each tested level of Pb was replicated three times. Then the soils containing Pb at different levels were covered with plastic film with six small holes for air ventilation. All the pots were kept for 3 months for Pb aging at a greenhouse at the temperature of 25±3°C during the daytime and 20±3°C at night with a natural light photoperiod. During the period of aging, water contents of each soil sample was maintained by adding deionized water every 3 days. After 3 months' aging, corn seeds of Zhengdan 958 were sown in each soil containing Pb. During the experimental period, soil moisture content was maintained at approximately 60% of water holding capacity through weighing method and adding deionized water to the pots. The corn with grain were sampled and harvested till the corn maturity.

Bioaccumulation factors for Pb of eight corn species from two soils. Two soils with typical properties sampled from Jiangxi and Shanxi provinces were selected for this experiment. The tested levels of Pb and exogenous Pb addition were the same as those described in the experiment of bioaccumulation factors for Pb by corn from different soils (Table 2). Three replications were conducted on each tested level of Pb. After three months' aging of soils, the corn seeds, such as Jingketian 183, Chuandan 30, Liaodan 565, Tunyu 88, Zhongdan 808, and Nongda 84,were surface sterilized with 1% (v/v) NaOCl, rinsed, and soaked in distilled water for 24 h at 33°C in dark. Then these seeds were sown in each soil containing Pb. All these pots were placed in the greenhouse at the temperature of 25±3°C during the daytime and 20±3°C at night with a natural light photoperiod.

Soil Sampling and Analysis

After three months' aging, soils in the pots of the above two experiments were sampled, air-dried, passed through a <0.20-mm sieve, and then thoroughly mixed. 0.5 g of the soil sample was weighed and added to digestion tubes containing 9 ml HNO_3 and 3 ml HF.

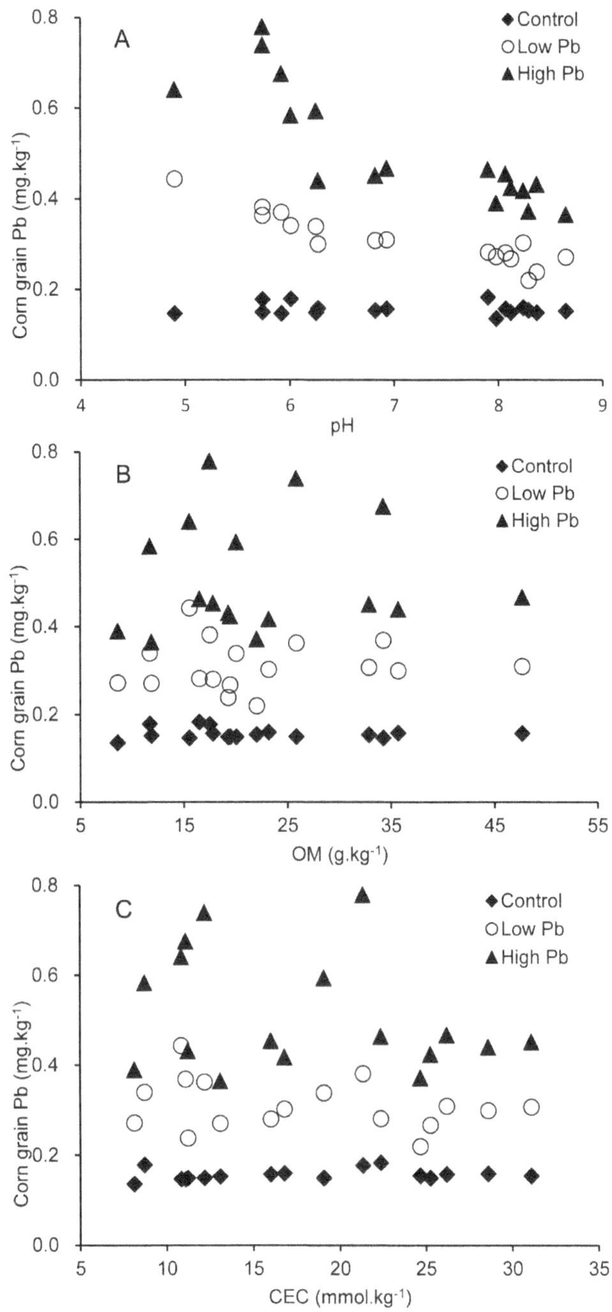

Figure 1. Relationships between Pb in corn grain and soil properties.

The corn grains were harvested and oven-dried at 105°C for 30 minutes and then kept at 70°C until the weight of the grains was stable. 0.5 g of corn grain samples oven-dried were weighed and digested in 6 ml concentrated HNO_3 (70% w/v) and 3 ml H_2O_2 in a CEM Mars X microwave oven (CEM Mars) at a pressure of 3.1 MPa. The Pb concentration in the digestion solution from soil and grain was determined by inductively coupled plasma-mass spectrometry (ICP-MS, Agilent 7500a, Agilent Technologies Co. Ltd., USA).

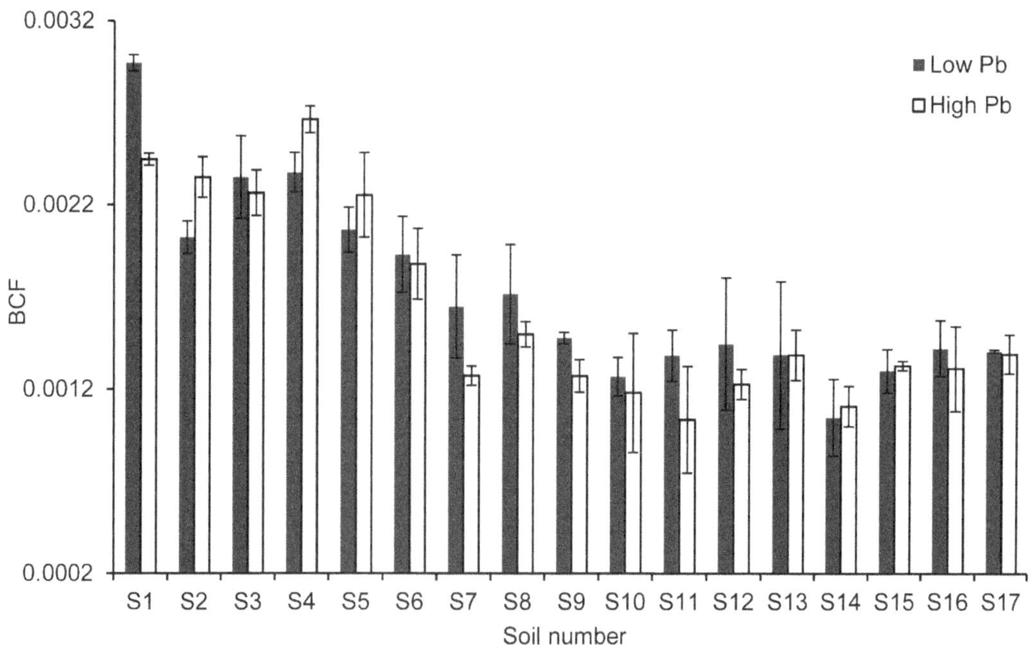

Figure 2. Effects of different soil types on the bioconcentration factor under stress of exogenous Pb.

Data Analysis

The bioconcentration factor (BCF) was calculated as the ratio of the content of Pb in the corn grain to that in the soil (equation 1) [45–46].

$$BCF = C_{grain}/C_{soil} \quad (1)$$

Where C_{grain} is the Pb concentration in the corn grain and C_{soil} is Pb concentration in the soil.

SPSS 16.0 for Windows® 107 (SPSS Inc, Chicago, IL, USA) was used for the regression analysis and statistical analysis of the significant differences. Origin 8.0 (OriginLab Co., Northampton, MA, USA) was employed for figure rendering.

Model

The prediction models for Pb transfer were established through multiple stepwise regression of the Pb BCF from Zhengdan 958 in a wide range of soil properties. The basis of these models is on the basis of equation 2 [38].

$$Log[BCF] = a \times pH + b \times \log[OM] + c \times \log[CEC] + k \quad (2)$$

Where Log [BCF], Log [OM], and Log [CEC] were the logarithm base 10 of the BCF values, the content of organic matter ($g.kg^{-1}$) in soils, and the cation exchange capacity ($cmol.kg^{-1}$) of soils, respectively? The slope of soil property parameters such as a, b, and c indicate the impacts of soil properties on Pb accumulation in corn grain. The intercept k is the intrinsic sensitivity that characterizes the ability of corn species to absorbing Pb.

Cross-species Extrapolation

In the process of cross-species extrapolation, an interim alternative was to assume that interactions between Pb accumulation in corn grain and soil pH, OM, CEC were the same among related corn species. In other words, the models stability constants including a, b, c were assumed the same among related species, and the only difference between related species was assumed to be their intrinsic sensitivity (k) [38]. The variation of intrinsic sensitivities within a species among plants reflects residual variation [34]. Under the condition of the squared error with the lowest value between predicted BCF value and the measured BCF value calculated as $\sum_{i=1}^{n} (measured\,BCF_i - predicited\,BCF_i)^2$, the intercept ($k$) for different models corresponding to various species were obtained through Excel Solver for linear optimization [47].

The accuracy of the model predictions was evaluated by comparing the measured BCF of the other non-model corn species with the predicted BCF from the model. The predictions for other non-model corn species were calculated with the prediction model derived from corn species i.e. Zhengdan 958.

Table 3. Prediction models for the different Pb levels.

Model No.	Lead levels	Prediction models	R^2	P
Model 1	Low Pb	Log[BCF] = −0.098 pH-0.150 log[OM] −1. 894	0.9085	<0.001
Model 2	High Pb	Log[BCF] = −0.108 pH-0.178 log[OM] −1. 806	0.8603	<0.001

Table 4. Intrinsic sensitivity (*k*) for non-model species fitted by models from Zhengdan 958.

Prediction Model	Intrinsic sensitivities (*k*)					
	Jingketian 183	Chuandan 30	Liaodan 565	Tunyu 88	Zhongdan 808	Nongda 84
Model 1	−1.823	−1.881	−1.924	−2.295	−2.288	−2.152
Model 2	−1.681	−1.785	−1.846	−1.857	−1.794	−1.717

Analysis of the Reduction of Intra-species Variability

The Pb BCF values of non-model corn species were normalized to the specific soil conditions through the obtained model for Zhengdan 958 and the intra-species variability was computed by equation 3 [47].

$$f=\sqrt{\frac{\sum_{i=1}^{n}(BCFsi-\overline{BCFs})^2}{(n-1)\times(\overline{BCFs})^2}} \quad (3)$$

Where the BCF of i-th conditions for specific corn species was normalized to specific soil conditions. *BCFsi* is the value that Pb BCF specific corn species were normalized to i-th soil conditions. $\overline{BCFs}$ is the mean of n BCFs. *n* is the number of different conditions for specific corn species, *f* is the intra-species variability.

The BCFs that Pb for different corn species was normalized by the prediction models to a set soil condition should be equal. Therefore, the decreases in intra-species variability indicate that normalization processing eliminated soil properties to a certain extent.

Results

Major Factors Affecting Pb Accumulation in Corn Grain in Different Soils

The relationships between Pb concentration in corn grain at different Pb levels and soil pH, OM and CEC were shown in Figure 1. At the treatment of low concentration Pb, the Pb concentration in corn grain decreased with the increasing pH of soil. The similar trend was also found at the treatment of high concentration Pb. There was no significant difference was found among Pb concentration in corn grain with the change of pH at the treatment of control. The significantly negatively direct correlations were observed between Pb accumulation in corn grain and soil pH at low concentration Pb, and high concentration Pb, respectively. However, no significantly direct correlations between Pb concentration in corn grain and soil OM, CEC were found.

Effects of Different Soil Types on Pb Transfer from Soil to Corn Grain

As shown in Figure 2, the BCF values from Zhengdan 958 under stresses of exogenous Pb significantly differed in 17 soils with a wide range of soil properties in China. The variation tendency of BCF values from Zhengdan 958 was generally consistent at different Pb levels of treatments. These BCF values decreased with

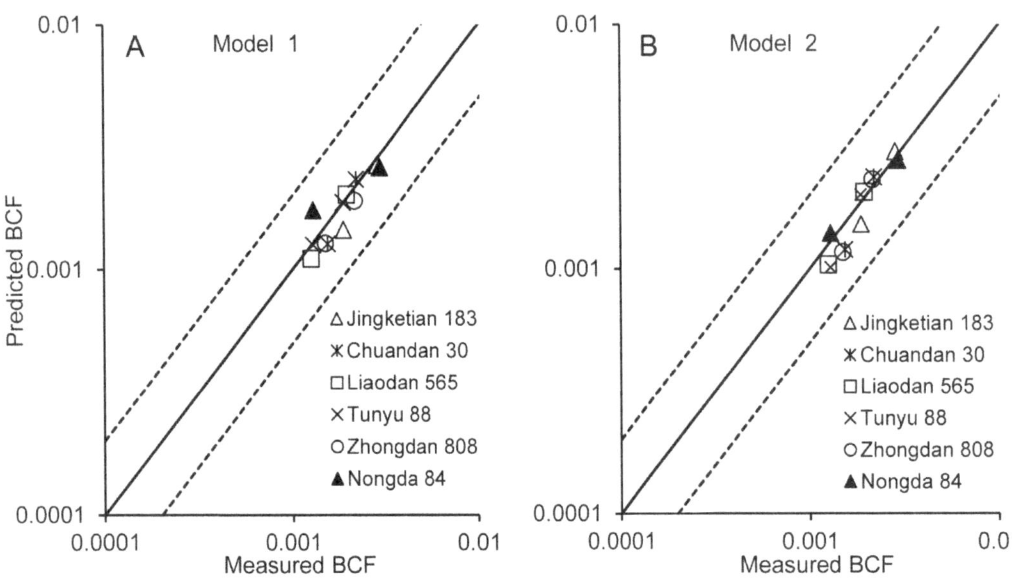

Figure 3. Relativity between measured and predicted BCF values for Pb in non-model corn grains. The predicted BCF values were estimated by model 1 and model 2 in Table 3; the solid line represents a 1:1 relationship; the interval between two dashed lines indicates a 2-fold prediction interval between the predicted and measured values.

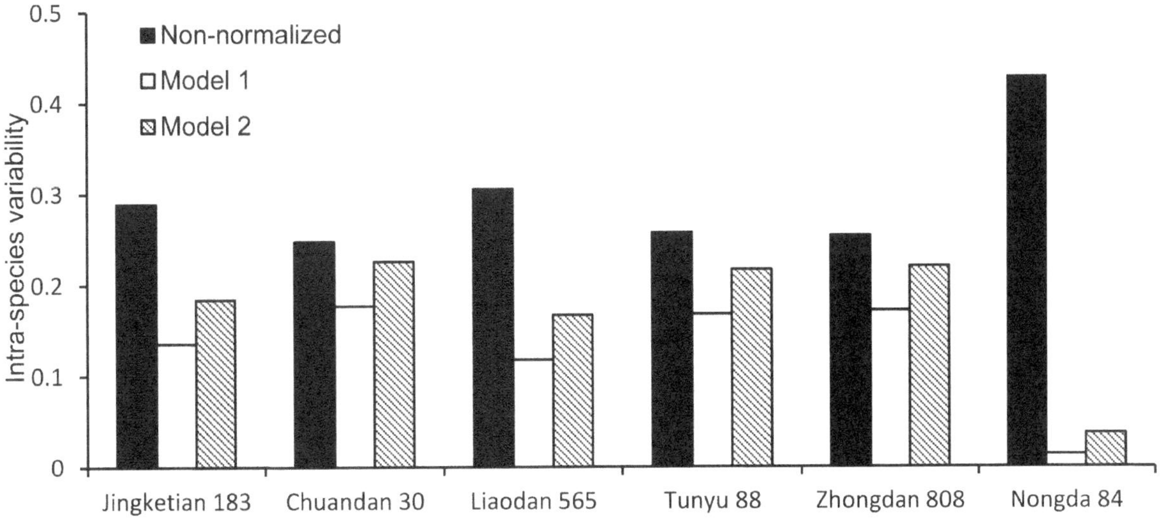

Figure 4. Intra-species variability of Pb BCF. The data from all non-model corn species were normalized with the models listed in Table 3.

the increases of soil pH. The BCF values detected from S1 to S9 (pH from 4.90 to 6.93) were significantly higher than those from S10 to S17 (pH from 7.90 to 8.80). At the treatment of low Pb, the maximum BCF (0.0030) among the 17 soils was observed in S1, and the minimum value for BCF (0.0010) was found in S14. The maximum value was 3.0 folds the minimum. At the same time, the highest BCF (0.0024) values were once again found in S1, and the lowest BCF (0.0010) was found in S11 at the treatment of high Pb, The maximum was 2.40 folds the minimum. These results indicate that Pb under acidic soil conditions is more highly bioavailable and leads to increased absorption by the plants.

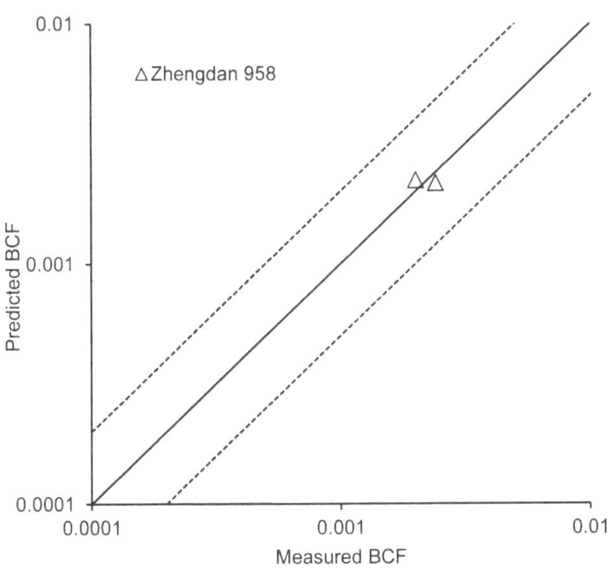

Figure 5. Relativity between measured and predicted BCF values for Pb in corn grains. The measured BCF were obtained from references. The predicted BCF values in Figure 5 were estimated by model 1. The solid line represents a 1:1 relationship; the dashed lines indicate a 2-fold prediction interval between the predicted and measured values.

Prediction Models

The prediction models established by stepwise regression were shown in Table 3. The result showed that pH and OM were the major factors affecting Pb accumulation in corn grain. A significantly negative correlation was observed between the Log[BCF] and soil pH and Log[OM] at the different treatments including low Pb, and high Pb, with the R^2 values of 0.8227, and 0.8955 ($P<0.001$), respectively. There was no significant difference between the two equations at the different treatment of Pb levels.

Cross-species Extrapolation

The intercept (*k*) can indicate the sensitivity of corn species to Pb accumulation. The intercept (*k*) for different corn species deduced by two different models were shown in Table 4. In terms of the same corn cultivar, no significant differences were observed among the *k* values for all tested non-model corn species deduced by the same model including model 1, and 2. However, the variation of *k* values among corn species deduced by the model 2 was stronger than that by the model 1. For example, the *k* values of corn species were different with the range from −1.888 to −1.736 deduced by model 1 and with range from −2.295 to −1.823 deduced by the model 2.

On the basis of the *k* values, the BCF values for non-model corn species were predicted by different models developed for Zhengdan958. The relationship between the predicted BCF values and the measured BCF values for non-model corn species was shown in Figure 3. The ratio between the predicted BCF values and the measured BCF values was within 2-fold interval and close to the solid line of 1:1 relationship. It indicates that the two models from Zhengdan 958 can be applied to predict the Pb BCF for non-model corn species.

Reduction of Intra-species Variability

Figure 4 shows the intra-species variability of Pb BCF for non-model species which were normalized with the prediction models listed in Table 3. The two prediction models can reduce the measured BCF intra-species variability for all non-model corn species. The intra-species variability of Pb BCF values fitted by model 1 was found to be significantly lower than those by model 1. It indicates that model 1 is more effective in reducing the uncertainty caused by the soil property differences than model 2.

Table 5. Soil properties and BCF values for Pb from reference.

Cultivation soil	cultivar species	pH	OM (g.kg-1)	CEC (cmol.kg^{-1})	Pb (g.kg^{-1})	BCF	Reference
Soil	Zhengdan 958	6.3	17.6	22.6	30.62	0.0020	[61]
Soil	Zhengdan 958	5.9	38.6	-	30.72	0.0024	[62]

This novel result also indicate that model 1 can be more reasonable and to be applied to other non-model corn species to establish the soil Pb SSD curves and to provide an ecological benchmark in China in the future.

Discussion

The bioavailable concentration of heavy metals is a key factor affecting its uptake and accumulation in plants [48]. In the present study, the results showed that corn grain taken up Pb from soil was mainly governed by soil pH and OM content. The Pb accumulation in the corn grain increased with the decreasing soil pH. This mainly attributed to pH played an important role in Pb uptake by plants from soil, this process through diverse reactions such as adsorption, ionic exchange, redox reactions, and precipitation-dissolution. These reactions are mainly affected by soil pH because of pH affecting Pb speciation and the charge of soil surface groups [49–50]. Christensen [51] and Naidu R et al [52] have found that the pH also influence on the solution activity of Pb and the distribution of Pb between the soil phase and the solution phase. The increased sorption of Pb at high pH values could reduce the Pb speciation in soils. In the present study, the BCF values observed in acidic soils (S1 to S9) with low pH values were significantly higher than those in alkaline soils (S10 to S17) with high pH values (Figure 2). These new results suggest that Pb is much more bioavailable under acidic conditions and is easily accumulated in corn grain. In the present paper, it is newly found that the Pb accumulation in corn grain was negatively related with OM. This is alike to the Cu accumulation in plant shoots [53], however it is different from Cd accumulation in corn grain [17]. This may be attributed to organic matter action as a primary sorbent of Pb because of its high sorption affinity [54–56].

In the present paper, no significant differences were observed between model 1 and model 2 at the different exogenous Pb concentration. In addition, there were also no significantly differences among Pb BCF values of corn grain at different exogenous Pb concentration, either. This was consistent with the findings of Rezvani and Zaefarian [57].

More importantly, we confirmed that the Pb BCF prediction models from Zhengdan 958 could be used to predict Pb BCF of other non-model corn species according to the soils pH and OM through cross-species extrapolation approach. Further, it was more accurate to apply model 2 to predict the BCF values of other corn species than model 1. Although many prediction models were developed to describe the relationship between plants uptake heavy metals and soil properties in terrestrial ecosystems, these models can't be extrapolated to non-model plant species, being only for single species [58–59]. Only a few studies were carried out to develop the prediction models for one species in aquatic ecosystems and then they are applied to other non-model species [36,38]. Few prediction models were developed for Pb accumulation in corn grain [60].

In order to further validate the feasibility of the prediction modes developed in the present study, the measured Pb BCF values of corn grains from soils were selected from the references [61,62], which were shown in Table 5. Although the corn species was also the same as that used to develop the model 1, the properties of the soils used in these references including OM, pH, and so on were different from those in the present paper. Therefore it is suitable that these data were used to test the prediction accuracy of model 1. The relativity between measured and predicted BCF values for Pb is shown in Figure 5. The result showed that the ratios between the predicted BCF values and the measured BCF values were also within 2 folds interval and close to 1:1 relationship. This indicates that the model 1 developed from Zhengdan 958 in the present study is feasible. It could be applied to other non-model corn species.

Conclusions

The soil pH and OM were the most major factors that can control Pb uptake in corn grain. The Pb accumulation in corn grain was negatively related with pH and Log [OM], respectively. The Pb BCF in corn grain can be well predicted by the model 1, which could be applied to other non-model corn species.

Acknowledgments

We are indebted to Prof. Lifang Hong for her help in soils collection, and Prof. Yibin Ma and Dr. Xiaoqing Wang for their help in the data processing and model derivation.

Author Contributions

Conceived and designed the experiments: ZL Y. Liang. Performed the experiments: Y. Li. Analyzed the data: HY. Contributed reagents/materials/analysis tools: Y. Li JL. Wrote the paper: ZL HY.

References

1. Cecchi M, Dumat C, Alric A, Felix-Faure B, Pradere P, et al (2008) Multi-metal contamination of a calcic cambisol by fallout from a lead-recycling plant. Geoderma 144: 287–298.
2. Grover P, Rekhadevi P, Danadevi K, Vuyyuri S, Mahboob M, et al (2010) Genotoxicity evaluation in workers occupationally exposed to lead. Int J Hyg Environ Health 213: 99–106.
3. Shahid M, Pinelli E, Pourrut B, Silvestre J, Dumat C (2011) Lead-induced genotoxicity to *Viciafaba L.* roots in relation with metal cell uptake and initial speciation. Ecotoxicol Environ Saf 74: 78–84.
4. Hrudey SE, Chen W, Rousseaux CG (1995) Bioavailability in environmental risk assessment. Lewis Publ, Boca Raton, FL.
5. USEPA (1992) Common chemicals found at Superfund sites. EPA540/R–94/044. U S Gov Print Office, Washington, DC.
6. Khan S, Cao Q, Zheng YM, Huang YZ, Zhu YG (2008) Health risks of heavy metals in contaminated soils and food crops irrigated with wastewater in Beijing, China. Environmental Pollution 152: 686–692.
7. Li ZJ, Xu JM, Tang CX, Wu JJ, Muhammad A, et al. (2006) Application of 16S rDNA-PCR amplification and DGGE fingerprinting for detection of shift in microbial community diversity in Cu-, Zn-, and Cd-contaminated paddy soils. Chemosphere 26: 1374–1380.
8. Chien CC, Kuo YM, Chen CC, Hung CW, Yeh CH, et al. (2008) Microbial diversity of soil bacteria in agricultural field contaminated with heavy metals. Journal of Environmental Sciences 20: 359–363.

9. Rattan RK, Datta SP, Chhonkar PK, Suribabu K, Singh AK (2005) Long-term impact of irrigation with sewage effluents on heavy metal content in soils, crops and groundwater-a case study. Agriculture Ecosystems and Environment 109: 310–322.
10. Chopin EIB, Alloway BJ (2007) Distribution and mobility of trace elements in soils and vegetation around the mining and smelting areas of Tharsis, Ríotinto and Huelva, Iberian Pyrite Belt, SW Spain. Water, Air, and Soil Pollution 182: 245–261.
11. Jung MC (2008) Heavy metal concentration in soils and factors affecting metal uptake by plants in the vicinity of a Korean Cu-W mine. Sensors 8: 2413–2423.
12. Wang J, Zhang CB, Jin ZX (2009) The distribution and phytoavailability of heavy metal fractions in rhizosphere soils of paulowniu fortune (seem) Hems near a Pb/Zn smelter in Guangdong, PR China. Geodema 148: 299–306.
13. Chapman EEV, Dave G, Murimboh JD (2013) A review of metal (Pb and Zn) sensitive and pH tolerant bioassay organisms for risk screening of metal-contaminated acidic soils. Environmental Pollution 179: 326–342.
14. Rodrıguez JA, Nanos N, Grau JM, Gil L, Lopez-Arias M (2008) Multiscale analysis of heavy metal contents in Spanish agricultural topsoils. Chemosphere 70: 1085–1096.
15. Bang J, Hesterberg D (2004) Dissolution of trace element contaminants from two coastal plain soils as affected by pH. Journal of Environmental Quality 33: 891–901.
16. Yang H, Li ZJ, Lu L, Long J, Liang YC (2013) Cross-species extrapolation of prediction models for cadmium transfer from soil to corn grain. PLoS ONE 8(12): e80855.doi: 10.1371/journal.pone.0080855.
17. Wang AS, Angle JS, Chaney RL, Delorme TA, Reeves RD (2006) Soil pH effects on uptake of Cd and Zn by Thlaspi caerulescens. Plant and Soil 281: 325–337.
18. Kirkham MB (2006) Cadmiumin plantson polluted soils: Effect of soil factors, hyperaccumulation and amendments. Geoderma 137: 19–32.
19. Zeng F, Ali S, Zhang H, Quyang Y, Qiu B, et al. (2001) The influence of pH and organic matter content in paddy soil on heavy metal availability and their uptake by rice plants. Environmental Pollution 159: 84–91.
20. Hettiarachchi GM, Ryan JA, Chaney RL, La Fleur CM (2003) Sorption and desorption of cadmium by different fractions of biosolids-amended soils. Journal of Environmental Quality 32: 1684–1693.
21. Antoniadis V, Robinson JS, Alloway BJ (2008) Effects of short-term pH fluctuations on cadmium, nickel, lead, and zinc availability to ryegrass in a sewage sludge-amended field. Chemosphere 71: 759–764.
22. Zeng FR, Shafaqat Ali, Zhang HT, Ouyang YN, Qiu BY, et al. (2011)The influence of pH and organic matter content in paddy soil on heavy metal availability and their uptake by rice plants. Environmental Pollution 159: 84–91.
23. Ge Y, Murray P, Hendershot WH (2000) Trace metal speciation and bioavailability in urban soils. Environ Pollut 107: 137–144.
24. Chen WP, Li LQ, Chang AC, Wu LS, Chaney RL, et al. (2009) Characterizing the solid-solution partitioning coefficient and plant uptake factor of As, Cd, and Pb in California croplands. Agriculture, Ecosystems and Environment 129: 212–220.
25. Zarcinas BA, Ishak CF, McLaughlin MJ, Cozens G (2004) Heavy metals in soils and crops in southeast Asia. 1. Peninsular Malaysia. Environmental Geochemistry and Health 26: 343–357.
26. Simmons RW, Noble AD, Pongsakul P, Sukreeyapongse O, Chinabut N (2008) Analysis of field-moist Cd contaminated paddy soils during rice grain fill allows reliable prediction of grain Cd levels. Plant and Soil, 302: 125–137.
27. Kuo S, Huang B, Bembenek R (2004) The availability to lettuce of zinc and cadmium in a zinc fertilizer. Soil Science, 169: 363–373.
28. Hough RL, Breward N, Young SD, Crout NMJ, Tye AM, et al. (2004) Assessing potential risk of heavy metal exposure from consumption of home-produced vegetables by urban populations. Environmental Health Perspectives, 112: 215–221.
29. Nahmani J, Hodson ME, Black S (2007) A review of studies performed to assess metal uptake by earthworms. Environmental Pollution, 145: 402–424.
30. Zhang HZ, Luo YM, Song J, Zhang HB, Xia JQ, et al. (2011) Predicting As, Cd and Pb uptake by rice and vegetables using field data from China. Journal of Environmental Sciences 23: 70–78.
31. Adamsa ML, Zhao FJ, McGratha SP, Nicholsonb FA, Chambersb BJ (2004) Predicting cadmium concentrations in wheat and barley grain using soil properties. Journal of Environmental Quality 33: 532–541.
32. Zhao XJ, Ma RC, Li ZJ (2013) Non-target organism species sensitivity distributions to oxytetracycline in water. Research Journal of Chemistry and Environment 17: 74–81.
33. Larras F, Bouchez A, Rimet FR, Montuelle B (2012) Using bioassays and species sensitivity distributions to assess herbicide toxicity towards benthic diatoms. PloS ONE 8: 1–9.
34. Van Sprang PA, Verdonck FAM, Van Assche F, Regoli L, De Schamphelaere KAC (2009) Environmental risk assessment of zinc in European freshwaters: a critical appraisal. Sci Tot Environ 407: 5373–5391.
35. Allard P, Fairbrother A, Hope BK, Hull RN, Johnson MS, et al. (2010) Recommendations for the development and application of wildlife toxicity reference values. Intergrated Environmental Assessment and Management 6: 28–37.
36. Deleebeeck NME, De Schamphelaere KAC, Janssen CR (2007a) A bioavailability model predicting the toxicity of nickel to rainbow trout (Oncorhynchus mykiss) and fathead minnow (Pimephales promelas) in synthetic and natural waters. Ecotoxicol Environ Saf 67: 1–13.
37. Deleebeeck NME, Muyssen BTA, De Laender F, Janssen CR, De Schamphelaere KAC (2007b) Comparison of nickel toxicity to cladocerans in soft versus hard surface waters. Aquat Toxicol 84: 223–35.
38. Schlekat CE, Genderen EV, Schamphelaere KACD, Antunes PMC, Rogevich EC, et al. (2010). Cross-species extrapolation of chronic nickel biotic ligand models. Science of the Total Environment 408: 6148–6157.
39. Sparks DL, Page AL, Helmke PA, Loeppert RH, Soltanpour PN, et al. (1996) Methods of soil analysis. Part 3 - chemical methods. Soil Science Society of America, Inc, Madison. WI.
40. Dohrmann R (2006) Cation exchange capacity methodology II: A modified silver-thiourea method. Applied Clay Science 34: 38–46.
41. Ball DF (1964) Loss-on-ignition as an estimate of organic matter and organic carbon in noncalcareous soils. Journal of Soil Science 15: 84–92.
42. Bray RH, Kurtz LT (1945) Determination of total, organic, and available forms of phosphorus in soils. Soil Science 59: 39–46.
43. Bremner JM (1996) Nitrogen total Methods of soil analysis. Part 3: Chemical methods, American Society of Agronomy-Soil Science Society of America, 1996, PP 1085–1121.
44. Zarcinas BA, McLaughlin MJ, Smart MK (1996) The effect of acid digestion technique on the performance of nebulisation systems used in inductively coupled plasma spectrometry. Commun Soil Sci Plant Anal 27: 1331–1354.
45. Baker AJM, Reeves RD, Hajar ASM (1994) Heavy metal accumulation and tolerance in British populations of metallophyte Thlaspi caerulescens. New Phytol 127: 61–68.
46. Raskin I, Kumar PBAN, Dushenkov S, Salt DE (1994) Bioconcentrationof heavy metals by plants. Curr Opin Biotechnol 5: 285–290.
47. Wang XQ (2012) Ecological Thresholds for Copper and Nickel in Chinese Agricultural Soils, Beijing, China Mining University 1–50.
48. Sarwar N, Saifullah, Malhi SS, Zia MH, Naeem A, et al. (2010) Role of mineral nutrition in minimizing cadmium accumulation by plants. J Sci Food Agric 90: 925–937.
49. Martınez-Villegas N, Flores-Velez LM, Domınguez O (2004) Sorption of lead in soil as a function of pH:a study case in Mexico. Chemosphere 57: 1537–1542.
50. Smical AI, Hotea V, Oros V, Juhasz J, Pop I (2008) Studies on transfer and bioaccumulation of heavy metals from soil into lettuce. Environmental Engineering and Management Journal 7: 609–615.
51. Christensen TH (1984) Cadmium soil sorption at low concentrations: I. Effect of time, cadmium load, pH, and calcium. Water. Air. Soil Pollut 21: 105–114.
52. Naidu R, Bolan NS, Kookana RS, Tiller KG (1994) Ionic-strength and pH effects on the sorption of cadmium and the surface charge of soils. European Journal of Soil Science 45: 419–429.
53. Moreno JL, Garcia C, Hernandez T, Pascual JA (1996) Transference of heavy metals from a calcareous soil amended with sewage sludge compost to barley plants. Bioresource Technology 55: 251–258.
54. Ge Y, Hendershot W (2005) Modeling sorption of Cd, Hg and Pb in soils by the NICA (non-ideal competitive adsorption)-Donnan model. Soil Sediment Contam 14: 53–69.
55. Sauve S, Manna S, Turmel MC, Roy AG, Courchesne F (2003) Solid solution partitioning of Cd, Cu, Ni, Pb, and Zn in the organic horizons of a forest soil. Environ Sci Technol 37: 5191–5196.
56. Puschenreiter M, Horak O, Friesl Wand Hartl W (2005) Low-cost agricultural measures to reduce heavy metal transfer into the food chain – a review. Plant Soil Environ 51: 1–11.
57. Rezvani M, Zaefarian F (2011) Bioaccumulation and translocation factors of cadmium and lead in Aeluropus littoralis. Australian Journal Agricultural Engineering 2: 114–119.
58. Gandoisa L, Probstb A, Dumata C (2010) Modelling trace metal extractability and solubility in French forest soils by using soil properties. European Journal of Soil Science 61: 271–286.
59. Wang XP, Shan XQ, Zhang SZ, Wen B (2004) A model for evaluation of the phytoavailability of trace elements to vegetables under the field conditions. Chemosphere 55: 811–822.
60. Rainbow PS (2007) Trace metal bioaccumulation: Models, metabolic availability and toxicity. Environment international 33: 576–582.
61. Cao Y, Liu Y, Wang GJ, Han Y, Wang SB (2009). Differences in Lead and Cadmium Concentrations among Plant Tissues of Maize Cultivars under the Combined Stress of Lead and Cadmium. Journal of Maize Sciences 17: 80–85.
62. Li YF, Liu L, Chen X, Hu LL, Tang JJ (2010) Plant Growth,Lead Uptake and Partitioning of Maize (*Zea mays L.*) Under Simulated Mild/Moderate Lead Pollution Stress. Journal of Agro-Environment Science 29: 2260–2267.

19

Modeling Soil Organic Carbon Change across Australian Wheat Growing Areas, 1960–2010

Guocheng Wang[1,3], Yao Huang[1,2*], Enli Wang[4], Yongqiang Yu[1], Wen Zhang[1]

1 State Key Laboratory of Atmospheric Boundary Layer Physics and Atmospheric Chemistry, Institute of Atmospheric Physics, Chinese Academy of Sciences, Beijing, China, 2 State Key Laboratory of Vegetation and Environmental Change, Institute of Botany, Chinese Academy of Sciences, Beijing, China, 3 University of the Chinese Academy of Sciences, Beijing, China, 4 Land and Water, Commonwealth Scientific and Industrial Research Organisation, Canberra, Australia

Abstract

Soil organic carbon (SOC) dynamics in Australian wheat-growing areas were simulated from 1960 to 2010 using Agro-C, a calibrated and validated biogeophysical model. Previously published data from field measurements were used to parameterize the Agro-C model. Model simulations show a decreasing trend in SOC over the last 50 years, mainly attributable to relatively low organic carbon (C) inputs. The rate of decrease in SOC tended to slow in the last two decades due primarily to an increase in wheat yields, which resulted in an increase in C input. Overall, we estimate that Australian wheat-growing areas, covering an area of 15.09 million hectares (Mha), lost 156 (86–222, 95% confidence interval) Tg C in the topsoil (to 30 cm depth) from 1960 to 2010. Approximately 80% of the SOC loss occurred in the period between the 1960s and the 1980s. Spatially, the SOC loss in areas with relatively high temperature and low precipitation, such as Queensland, the northern part of New South Wales and Western Australia, was more significant than that in other areas. We suggest that the loss of SOC could be halted, or even reversed, with an additional input of organic C into the soil at a minimum rate of 0.4 Mg ha^{-1} yr^{-1}.

Editor: Ben Bond-Lamberty, DOE Pacific Northwest National Laboratory, United States of America

Funding: This work was jointly supported by the National Basic Research Program of China on Global Change (2010CB50604) and the National Natural Science Foundation of China (41075108). The funders had no role in study design, data collection and analysis, decision to publish, or preparation of the manuscript.

Competing Interests: The authors have declared that no competing interests exist.

* E-mail: huangy@mail.iap.ac.cn

Introduction

Cultivation of natural soils generally leads to a reduction of soil organic carbon (SOC) because cultivation enhances the rates of carbon (C) mineralization, reduces the amount of biomass C returned to the soil, and accelerates SOC erosion and leaching processes [1,2]. SOC loss can be slowed, or even reversed, by optimizing agricultural management (e.g., stubble retention, fertilization and conservation tillage), thereby not only mitigating climate change but also improving soil fertility [2,3].

Australian cropland has experienced tremendous loss of SOC since the European settlement began, due largely to the loss of above-ground biomass C after the conversion of native land for agriculture [4]. Recently, a national meta-analysis showed that SOC stores in the top meter of Australian agricultural soils have decreased by 40 to 60% compared with pre-clearing levels [5]. Based on a regional meta-analysis, Dalal and Chan [6] suggested that, in the Australian wheat belt, soil would sequester a large amount of atmospheric CO_2 20 years after the adoption of improved management techniques. However, cropland SOC dynamics depend on a balance between C production and decomposition and are regulated not only by management but also by highly variable climate and soil conditions [5,7]. A regional survey, based on field experiments, reported large uncertainties in SOC estimates due to the spatiotemporal variability of the study sites. Accurately assessing regional SOC dynamics is difficult, particularly in Australian croplands, due to the lack of detailed climate, soil, and management-related information.

Wheat-farming lands constitute approximately 40% of all Australian agricultural areas [8] and are distributed across several natural resource management (NRM) regions, primarily on the mainland, in a narrow crescent known as the wheat belt (Figure 1). Due to the adoption of modern cultivars and improved management practices, Australian wheat yields per unit area have increased significantly during the past several decades, especially from 1960 to 2010, during which time yields approximately doubled [9]. This increased wheat production has resulted in an enhanced amount of wheat residues [10]. Since the 1960s, improved management techniques (such as conservation tillage and stubble retention) have been widely adopted across many parts of the wheat belt to reduce the risk of soil and water erosion [11,12,13,14]. Both the improvement of wheat production and the promotion of stubble retention are likely to cause an increase in SOC accumulation or a slow-down of SOC loss. Nevertheless, to the best of our knowledge, no comprehensive study has been conducted on large-scale cropland SOC change across the Australian wheat-growing areas.

It is widely recognized that a modeling approach has advantages when estimating spatiotemporal changes in SOC [1]. Dynamics of agricultural SOC are generally well captured by process-based models such as DNDC [15], RothC [16] and CENTURY [17], which have already been widely used to estimate changes in SOC at both national and continental scales [18,19,20]. More recently, Huang *et al.* [7] developed a biogeophysical model, Agro-C, which has been used to assess the national long-term agricultural SOC

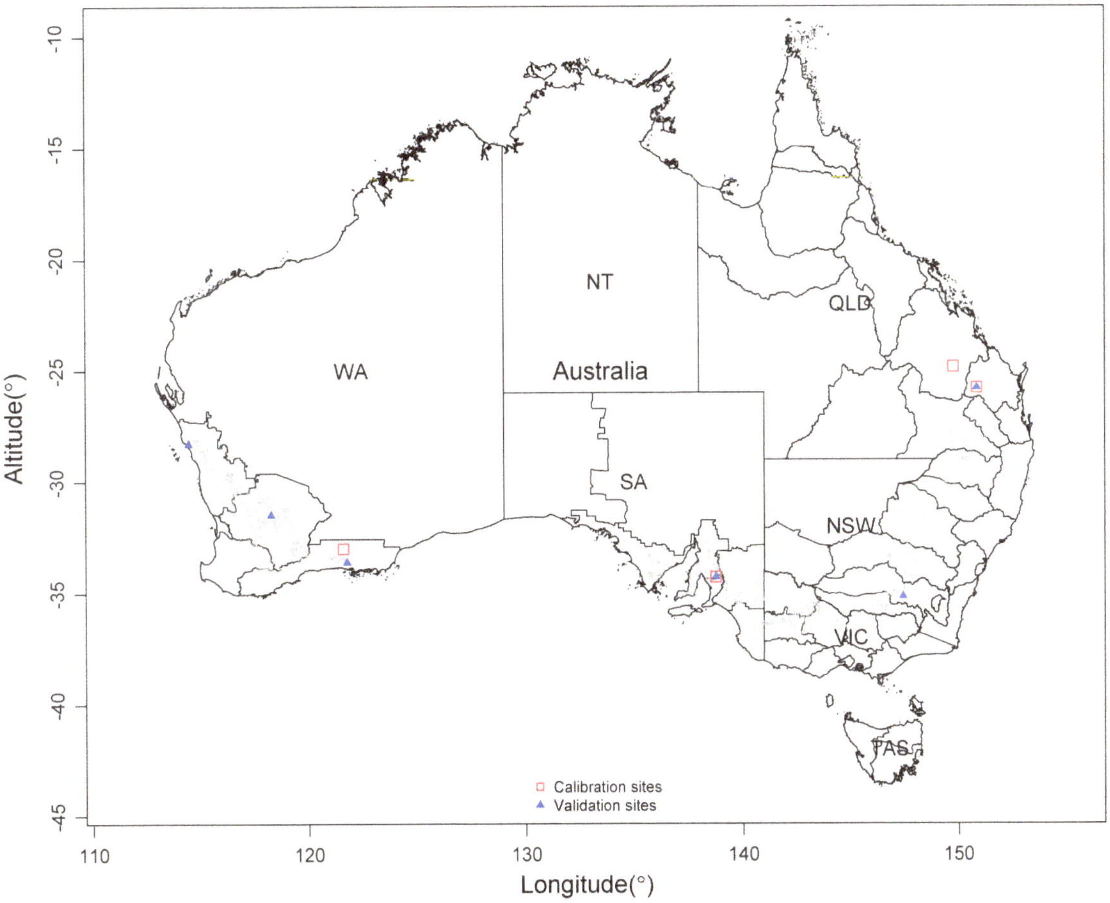

Figure 1. Spatial distribution of wheat-growing areas across the Australian wheat belt and locations of long-term experiments. Wheat is planted in the grey areas, red open squares show the sites for model calibration and blue solid triangles show the sites for model validation. NT, QLD, NSW, VIC, SA, WA and TAS refer to Northern Territory, Queensland, New South Wales, Victoria, Southern Australia, Western Australia and Tasmania, respectively.

change in China [1]. The model consists of two sub-models: Crop-C for simulating crop net primary production, and Soil-C for computing SOC change. Previous validation studies indicate that the Soil-C sub-model properly simulates the observed changes in SOC in most cases across Chinese croplands [1,7,21,22,23].

Determining the regional cropland SOC dynamics would not only provide a better understanding of the factors and processes regulating C cycling and balance in the agro-ecosystems but would also provide insight into the effectiveness of methods for enhancing soil quality and plant production, as well as reducing greenhouse gas emissions. Our objective in this study is to quantitatively estimate regional-scale spatiotemporal changes in SOC across Australian wheat-growing areas between 1960 and 2010, so that policy-makers can make sensible, region-oriented decisions regarding SOC management.

Materials and Methods

1.1 Model Calibration and Validation

1.1.1 Model calibration. Agro-C is a biogeophysical model for simulating regional carbon budgets of agro-ecosystems on a large scale initially developed by Huang *et al.* [7]. The model consists of two submodels, Crop-C and Soil-C. Crop-C simulates processes involved with crop photosynthesis, autotrophic respiration, and net primary production (NPP). Soil-C simulates soil heterotrophic respiration via the decomposition of both input organic C (e.g., crop residues, roots and manure) and soil organic carbon. Changes in SOC are then determined by balancing the loss of soil carbon with the gain of input organic carbon. A detailed description of the Agro-C model was shown in Huang *et al.* [7]. The Soil-C submodel was recently modified by Yu *et al.* [1], who split the previous SOC pool into two sub-pools named light-C and heavy-C, with specific decomposition rates for each sub-pool. The light-C sub-pool is considered more biologically reactive, with turnover times ranging from a few months to a few years, while the heavy-C sub-pool is much more resistant to decomposition and can remain in the soil for decades or centuries [1]. Meanwhile, the C flow between different pools was also modified based on the following assumptions: (i) the decomposition of both the labile-C and the resistant-C converts a fraction of the C into the light-C pool, (ii) a fraction of the decomposed light-C is transferred into the heavy-C pool, and (iii) the decomposition of the heavy-C pool only produces CO_2. Detailed structural modification of the Soil-C model was shown in Yu *et al.* [1].

In Soil-C, the parameters that need to be determined include the first-order reaction rates of light-C (K_{LC}) and heavy-C (K_{HC}), the fractions of the decomposed labile-C and resistant-C entering into the light-C pool (F_{LL} and F_{RL}, respectively), and the fraction of the decomposed light-C entering into the heavy-C pool (F_{LH}). Other parameters, such as the first-order reaction rates of labile-C

Table 1. Parameters in the original and modified Soil-C submodel.

Parameter	Original [a]	Modified [a]	Description
K_{LC} (d^{-1})	2.5×10^{-4} (7.6)	2.5×10^{-4} (7.6)	First-order reaction rate of light-C pool
K_{HC} (d^{-1})	1.8×10^{-5} (105.4)	2.6×10^{-5} (73)	First-order reaction rate of heavy-C pool
F_{LL}	0.3	0.3	Fraction of decomposed labile-C to light-C
F_{RL}	0.45	0.45	Fraction of decomposed resistant-C to light-C
F_{LH}	0.45	0.3	Fraction of decomposed light-C to heavy-C

[a]Values in parentheses are the half-life residence time in years.

(K_L) and resistant-C (K_R), are taken from the original version of Soil-C [7]. The values of the five key parameters were initially set within a given range of intervals [1]. The final values of these parameters were determined by minimizing the mean deviation (see Eq. 5) between the simulated and observed SOC after running each combination of the five parameters on Soil-C using a parameter-space search method.

Measurements of SOC from long-term field experiments at four sites, including Brigalow, Tarlee, Warra and Salmon Gums (Figure 1 and Table S1), were used to calibrate Soil-C. The field experiment data used for this study were initially published by the Queensland Department of Primary Industry, South Australian Research and Development Institute, Queensland Department of Natural Resources and Mining, and CSIRO Land and Water. None of the field studies mentioned in this study involved endangered or protected species. In the four sites, annual mean temperature and precipitation ranged from 16.8 to 21.4°C and from 467 to 685 mm, respectively. At Brigalow, the original land use type was native forest; however, this site was cleared and converted to cropland in 1982. Following a 2-year fallow period, sorghum and wheat were planted in alternating years. Two cropping treatments, the continuous wheat system (CW) and the fallow-wheat system (FW), were adopted at the Tarlee site. The FW treatment was selected to for Soil-C submodel calibration, whereas the CW treatment was used for model validation. The site has been used for cereal production since 1935. In 1986, long-term field experiments were established covering various combinations of cropping systems and fertilization management. The wheat-lucerne (1 year, 2 years, respectively) rotation (WL, the first crop being wheat) was selected for model parameterization, and the lucerne-wheat (2 years, 1 year, respectively) rotation (LW, the first crop was lucerne) was selected for model validation. The Salmon Gums site, established in 1979, was parameterized for the Soil-C model using a continuous wheat treatment.

Table 1 shows the values of K_{LC}, K_{HC}, F_{LL}, F_{RL} and F_{LH} in the calibrated Soil-C that resulted in a mean deviation of 0.002 MgC ha^{-1} between the simulated and the observed SOC at the four experimental sites (Figure S1). Generally, the above five parameters are within reasonable ranges. For example, the value of K_{LC} is smaller than the reaction rate of BIO (Microbial Biomass), and the value of K_{HC} is lower than the reaction rate of HUM (Humified Organic Matter) and higher than that of IOM (Inert Organic Matter) in the Roth-C model [16].

1.1.2 Model validation. The calibrated Soil-C submodel was validated against independent datasets from 6 long-term experimental sites (Figure 1 and Table S1). These experimental data were derived from the National Carbon Accounting System Technical Report (*Integrated soils modeling for the National Carbon Accounting System No. 36*, available: http://www.pandora.nla.gov.au/tep/23322). At each site, crop biomass production was recorded at harvest every year, and SOC at a given profile depth was measured at both the start of the trials and after the harvest in some years. Table S1 shows the initial soil information at the start of each experiment.

For some experimental sites where SOC density was only reported in the top 10 cm ($SOC_{0-10\ cm}$) or 20 cm ($SOC_{0-20\ cm}$), the SOC density in the top 30 cm soil layer ($SOC_{0-30\ cm}$) was calculated using the SOC vertical distribution [24,25,26]:

$$SOC_{0-20cm} = SOC_{0-10cm} + SOC_{10-20cm} \quad (1)$$

$$SOC_{0-30cm} = 1.32 \times SOC_{0-20cm} \quad (2)$$

$$SOC_{0-30cm} = 2.35 \times SOC_{0-10cm} \quad (3)$$

where 1.32 and 2.35 are the conversion coefficients.

Table 2. Main input data for running the Agro-C model.

Category	Item (Unit)	Source (Reference)
Climate	Daily maximum and minimum temperature (°C), precipitation (mm), solar radiation (MJ m^{-2} d^{-1}), wind speed (m s^{-1}) and relative humidity (%)	SILO Patched Point Dataset (http://www.bom.gov.au/silo/)
Soil properties	Concentrations of organic carbon and total nitrogen (g kg^{-1}), bulk density (g cm^{-3}), sand and clay fraction (kg kg^{-1}) and pH	Australian Soil Resource Information System (ASRIS) (http://www.asris.csiro.au/)
Farming management and crop yield	Dates of planting, heading and harvest (dd/mm), timing (dd/mm) and rate of fertilizer N application (kg ha^{-1}) and grain yield (kg ha^{-1})	Year Book of Australia (1960–2010, http://www.abs.gov.au/) Scott *et al.* [41]; Llewellyn and D'Emden [33]
Atmosphere	Atmospheric CO2 concentration	WMO World Data Centre for Greenhouse Gases (http://gaw.kishou.go.jp/wdcgg/)

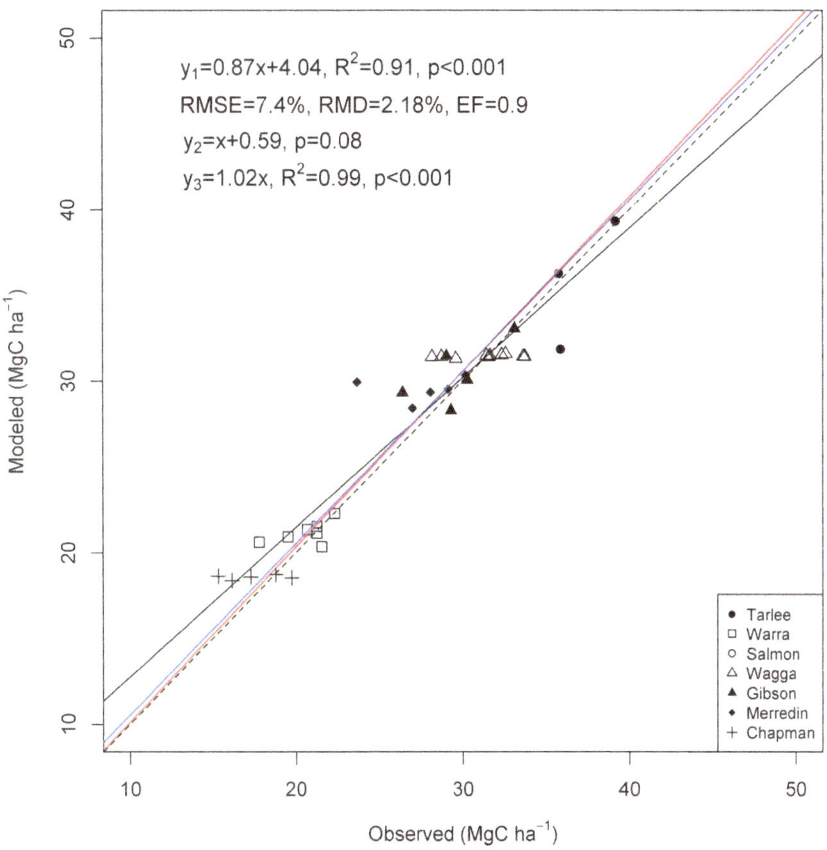

Figure 2. Modeled vs. observed SOC at the validation sites. Dashed line is 1:1. Black solid line (y_1) shows the linear regression analysis, blue solid line (y_2) shows the statistical tests on the intercept, and red solid line (y_3) shows the statistical tests on the slope.

Following Yu *et al.* [1], three statistical criteria were used to evaluate the model performance: the root mean squared error (*RMSE*, Eq. 4) was calculated to measure the coincidence between the observed and the simulated SOC, the relative mean deviation (*RMD*, Eq. 5) was computed to evaluate the systematic bias of the model, and the model efficiency (*EF*, Eq. 6) was calculated to estimate model performance in relation to the observed mean. Linear regression analysis between simulated and observed SOC was also used to evaluate the model performance.

$$RMSE = \frac{100}{\overline{O}}\sqrt{\sum_{i=1}^{n}\frac{(P_i - O_i)^2}{n}} \quad (4)$$

$$RMD = \frac{100}{\overline{O}}\sum_{i=1}^{n}\frac{(P_i - O_i)}{n} \quad (5)$$

$$EF = 1 - \frac{\sum_{i=1}^{n}(P_i - O_i)^2}{\sum_{i=1}^{n}(\overline{O} - O_i)^2} \quad (6)$$

where P and O represent the model estimates and the field measurements, respectively, $\overline{O}$ is the mean observed SOC, and n is the total number of observations.

1.2 Simulation of Changes in SOC

The SOC change was simulated using the Agro-C model because the soil moisture that affects SOC dynamics in the Soil-C submodel is computed via the Crop-C submodel in Agro-C [1,7]. The changes in SOC were simulated with a daily step from 1960 to 2010. Up-scaling of the Agro-C model was accomplished by first computing the model inputs within each interested NRM region across the wheat-belt and then running the model in each of the NRM regions.

1.2.1 Compilation of model inputs. In this study, the main input data used for the Agro-C model include climate, soil, farming management and crop yield (Table 2).

Gridded daily climate data (i.e., maximum and minimum temperature, precipitation, relative humidity and radiation) from 1960 to 2010 with a spatial resolution of 0.05° × 0.05° were obtained from the Queensland Climate Change Centre of Excellence (http://chp.qld.gov.au/index.html). These datasets are constructed by applying spatial interpolation algorithms to historical climate data from approximately 4,600 observations across Australia [27]. For each studied NRM region, the daily climate input data were calculated using the mean of the gridded climate data located within that specific NRM region.

Input soil properties include the concentrations of organic carbon and total nitrogen, bulk density, clay and sand fraction, and pH in the topsoil to 30 cm depth. These soil properties, with a 0.01° × 0.01° spatial resolution, were obtained from Australian Soil Resources Information System (ASRIS) [28]. These datasets represent some 164,030 soil profile measurements made since the

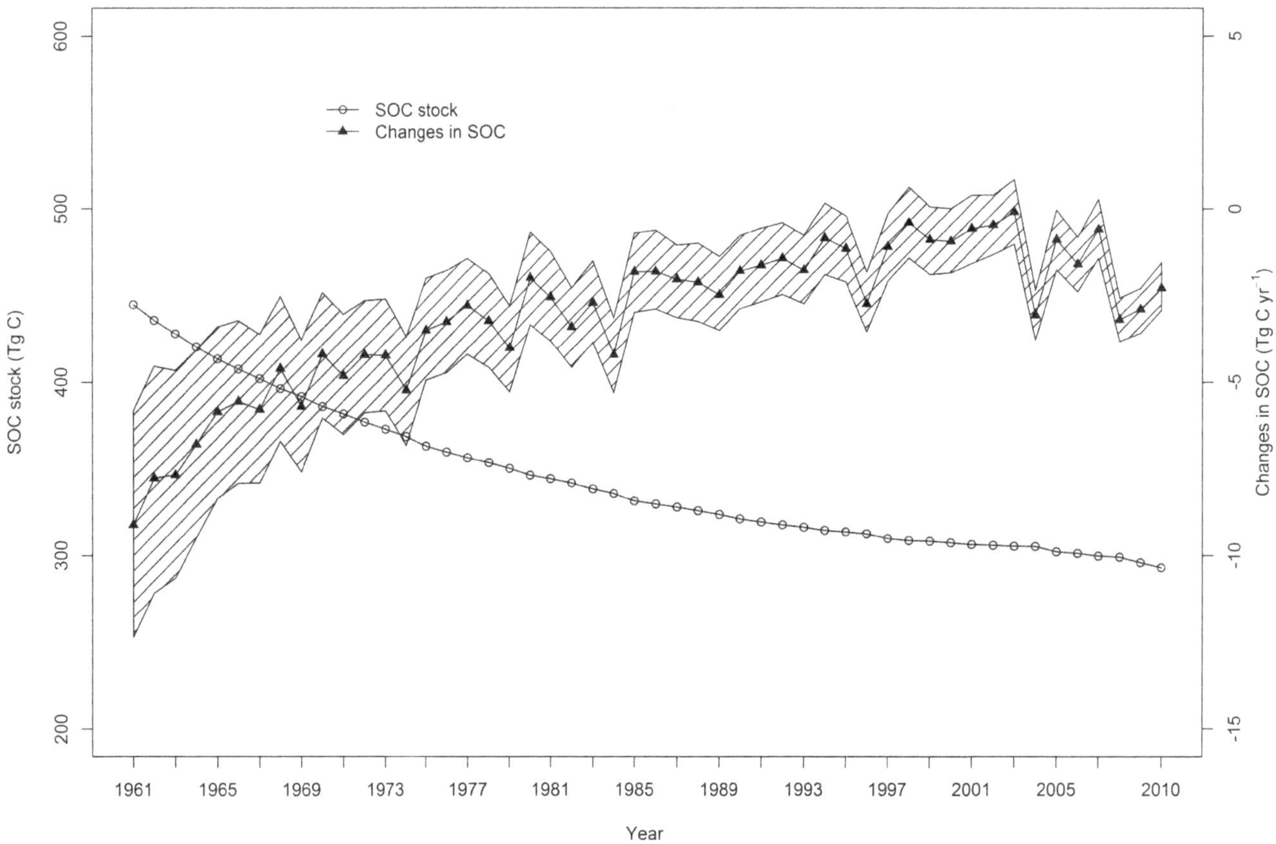

Figure 3. Simulated SOC levels and annual change in SOC of Australian wheat-growing land from 1960 to 2010. The open circles show the SOC levels and the solid triangles show the annual SOC change. The grey and shaded areas show the 95% confidence interval of SOC levels and annual SOC change, respectively.

1960s across intensively used agricultural zones in Australia and are currently the best continental-scale soil profile data available to the public [29].

The spatial distribution of cropland was computed based on a national-scale land use map of Australia (Version 3 for 2005/2006), obtained from the Australian Bureau of Rural Sciences [30]. Due to a lack of yearly cropland data with sufficient spatial distribution from 1960 to 2010, we assumed that the area of cropland in the wheat-growing regions did not significantly change over this time period. The annual wheat yield at the NRM regional level from 1960 to 2010 was obtained from the Australian Bureau of Statistics (http://www.abs.gov.au/).

Carbon enters the soil through crop residue retention, root mass, and organic amendments (e.g., animal manure). Following Huang *et al.* [31], annual above-ground residue mass was calculated via the economic yield of the crop and the ratio of above-ground residue production to economic yield. The contribution from root mass was estimated as 40% of the crop residue mass, and all above-ground and root residues were assumed to have a C content of 45% [32]. The proportion of the above-ground residue that is retained in the soil across different regions was estimated based on the work of Llewellyn and D'Emden [33]. The rate of residue retention in different cropping Australian croplands over the study period has been uneven. In Queensland, the burning of crop residues has been almost abandoned, with approximately 95% of stubble being incorporated into soils or retained in the fields. In WA, approximately 17% of stubble has traditionally been burned. In NSW, Victoria and SA, however, burning was more common (24–55% of stubble burned), especially in NSW and Victoria. Due to a lack of yearly stubble retention rate data, we assume the stubble retention rate across different regions did not change significantly over time. Additionally, because the areas that have adopted animal manure application constitutes only 0.4% of the whole wheat belt [6] and there is no detailed information on the spatial distribution of these areas, we ignore the impacts of animal manure application on the regional cropland SOC change.

1.2.2 Model initialization. In Agro-C, the half-life residence time for labile-C, resistant-C, light-C and heavy-C were 0.1 y, 2.3 y, 7.6 y, and 105.4 y [1], respectively. To obtain initial values for the different C pools and to initialize the Agro-C model, a 'spin-up' procedure was performed by running the model for one hundred years (i.e., 1900–2000). The initial fractions of light-C and heavy-C were assigned to be 0.25 and 0.75 [1], respectively. The input values for organic carbon were derived from annual crop yield. Climatic data from the same time period were used during the spin-up procedure.

1.2.3 Uncertainty assessment. The two main factors that can influence SOC change are soil environment and carbon input [34], both of which can result in major uncertainties in the estimated SOC change.

For each initial soil-related model input (i.e., initial soil organic carbon and nitrogen concentrations, clay fraction and pH), the probability distribution function (PDF) for each NRM region was calculated based on the grid values within that region. We also used a Monte Carlo analysis to develop the PDF for carbon input

Table 3. Estimated changes in SOC levels (0–30 cm) in different regions of Australian wheat-growing areas over the last 50 years.

			1960s			1970s		
Region	Area (Mha)	Initial SOC density (Mg C ha^{-1})	Average SOC density (Mg C ha^{-1})	Average C input (Mg C ha^{-1} yr^{-1})	Changes in SOC (Tg C)	Average SOC density (Mg C ha^{-1})	Average C input (Mg C ha^{-1} yr^{-1})	Changes in SOC (Tg C)
QLD	0.86	33 (22–43)	30 (21–39)	1.18	−4 (−6– −2)	26 (19–33)	1.13	−3 (−4– −1)
NSW	3.83	31 (21–42)	28 (19–38)	0.88	−19 (−28– −9)	25 (17–32)	0.93	−11 (−17– −6)
WA	5.28	29 (24–34)	26 (22–30)	0.86	−23 (−30– −17)	23 (20–26)	0.96	−13 (−17– −9)
SA	2.87	29 (23–34)	26 (21–31)	0.98	−11 (−14– −6)	24 (20–28)	1.01	−6 (−9– −4)
VIC	2.24	28 (24–33)	26 (22–30)	1.15	−7 (−9– −5)	24 (21–27)	1.31	−4 (−5– −2)
Wheat belt	15.08	30 (23–36)	27 (21–33)	0.95	−64 (−87– −39)	24 (19–28)	1.02	−37 (−52– −22)

1980s			1990s			2000s			1960–2010
Average SOC density (Mg C ha^{-1})	Average C input (Mg C ha^{-1} yr^{-1})	Changes in SOC (Tg C)	Average SOC density (Mg C ha^{-1})	Average C input (Mg C ha^{-1} yr^{-1})	Changes in SOC (Tg C)	Average SOC density (Mg C ha^{-1})	Average C input (Mg C ha^{-1} yr^{-1})	Changes in SOC (Tg C)	Changes in SOC (Tg C)
23 (17–29)	1.27	−2 (−3– −1)	21 (16–26)	1.14	−2 (−3– −1)	20 (15–23)	1.29	−1 (−2– −1)	−12 (−18– −6)
22 (16–28)	0.99	−7 (−12– −3)	21 (16–26)	1.31	−5 (−8– −1)	20 (16–24)	1.02	−6 (−9– −3)	−48 (−74– −22)
21 (18–24)	1.03	−10 (−13– −6)	20 (17–22)	1.48	−3 (−6–0)	19 (17–21)	1.40	−4 (−5– −2)	−53 (−71– −34)
22 (18–25)	1.10	−4 (−6– −3)	21 (18–24)	1.43	−2 (−3–0)	21 (18–23)	1.35	−3 (−4– −1)	−26 (−36– −14)
23 (21–25)	1.41	−2 (−4– −1)	22 (20–24)	1.40	−2 (−2– −1)	21 (20–23)	1.34	−2 (−3– −1)	−17 (−23– −10)
22 (18–26)	1.10	−25 (−38– −14)	21 (17–24)	1.40	−14(−22– −3)	20 (17–23)	1.28	−16 (−23– −8)	−156 (−222– −86)

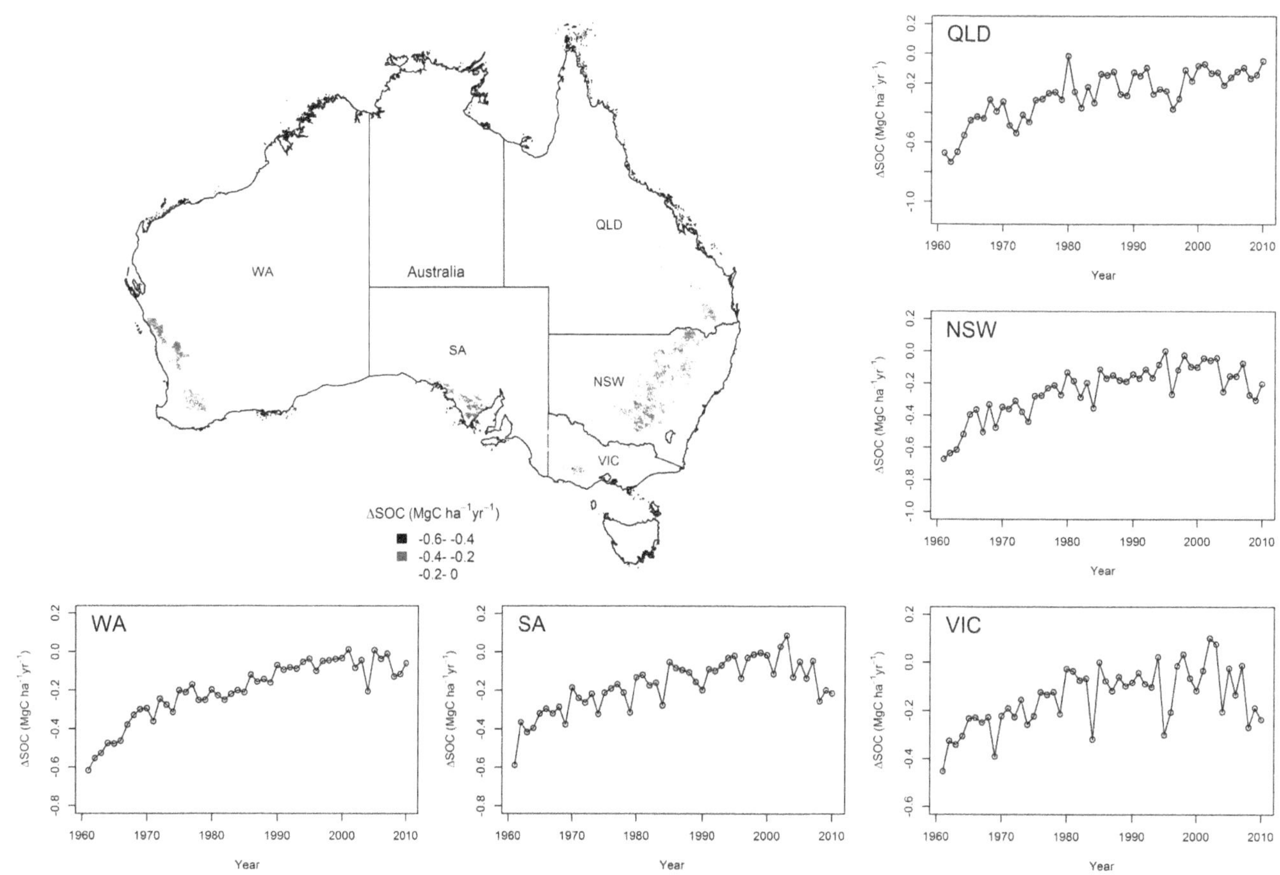

Figure 4. Estimated changes in SOC density (△SOC, Mg C ha^{-1} yr^{-1}) in different regions of Australian wheat-growing areas. The shaded areas represent the lower and upper bounds of the estimates at 95% confidence level. QLD, NSW, VIC, SA and WA refer to Queensland, New South Wales, Victoria, Southern Australia and Western Australia, respectively.

within the ranges of mean±10% for each NRM region. The calibrated and validated Agro-C model was then run 300 times in each NRM region with the initial soil properties and carbon input randomly assigned from the PDFs of soil-related covariates and carbon input, respectively. Finally, the model output was analyzed to produce an empirical distribution of SOC changes on a regional level [1,20]. A 95% confidence interval of SOC change was also estimated based on the 500 simulations in each NRM region. All the above analyses were conducted using the R version 2.15.1 software package [35].

Results

2.1 Model Performance

The calibrated Agro-C model, in general, captured SOC changes better than the original version of Agro-C at all of the calibration sites except Tarlee (Figure S1). The simulated trends in SOC change in the top 30 cm of soil at different validation sites generally agreed well with the observed data during the selected study period (Figure 2 and Figure S1), except some exceptional variations (e.g., observed SOC in 1993 at Warra and in 1986 at Merredin). In general, Agro-C simulations could explain more than 90% of the variation in the SOC when data from all validation sites were pulled together (Figure 2), indicating an overall good performance of the calibrated Agro-C model. Statistical tests on both the slope and the intercept indicated that, in general, Agro-C modeled SOC agreed well with corresponding observations (Figure 2).

2.2 Estimated Changes in SOC from 1960 to 2010

Our Agro-C simulation results suggest that SOC in Australian wheat-growing areas has decreased over the last five decades (Figure 3). On average, the modeled SOC densities of the entire Australian wheat-growing areas in each decade were 27 (±22%, 95% confidence interval) Mg ha^{-1} during the 1960s, 24 (±17%) Mg ha^{-1} during the 1970s, 22 (±18%) Mg ha^{-1} during the 1980s, 21 (±14%) Mg ha^{-1} during the 1990s, and 20 (±15%) Mg ha^{-1} during the 2000s, respectively (Table 3). Although average SOC densities decreased over the study period, the rate of SOC loss is declining, with more SOC lost in the first 30 years than in the last 20 years (Table 3 and Figure 3).

Averaging for each decade, the rate of SOC loss was estimated to be approximately 6.4, 3.7, 2.5, 1.4 and 1.6 Tg yr^{-1} in the 1960s, 1970s, 1980s, 1990s and 2000s, respectively. Summing up the yearly changes in SOC over the period from 1960 to 2010, the loss of SOC in Australian wheat-growing areas was estimated to be 156 Tg C, with a range from 86 to 222 Tg C at the 95% confidence level (Figure 3). It is noteworthy that after the decline in the first few years, SOC seemed to reach a new, steady state in the mid-2000s, followed by a decrease in the following several years (Figure 3).

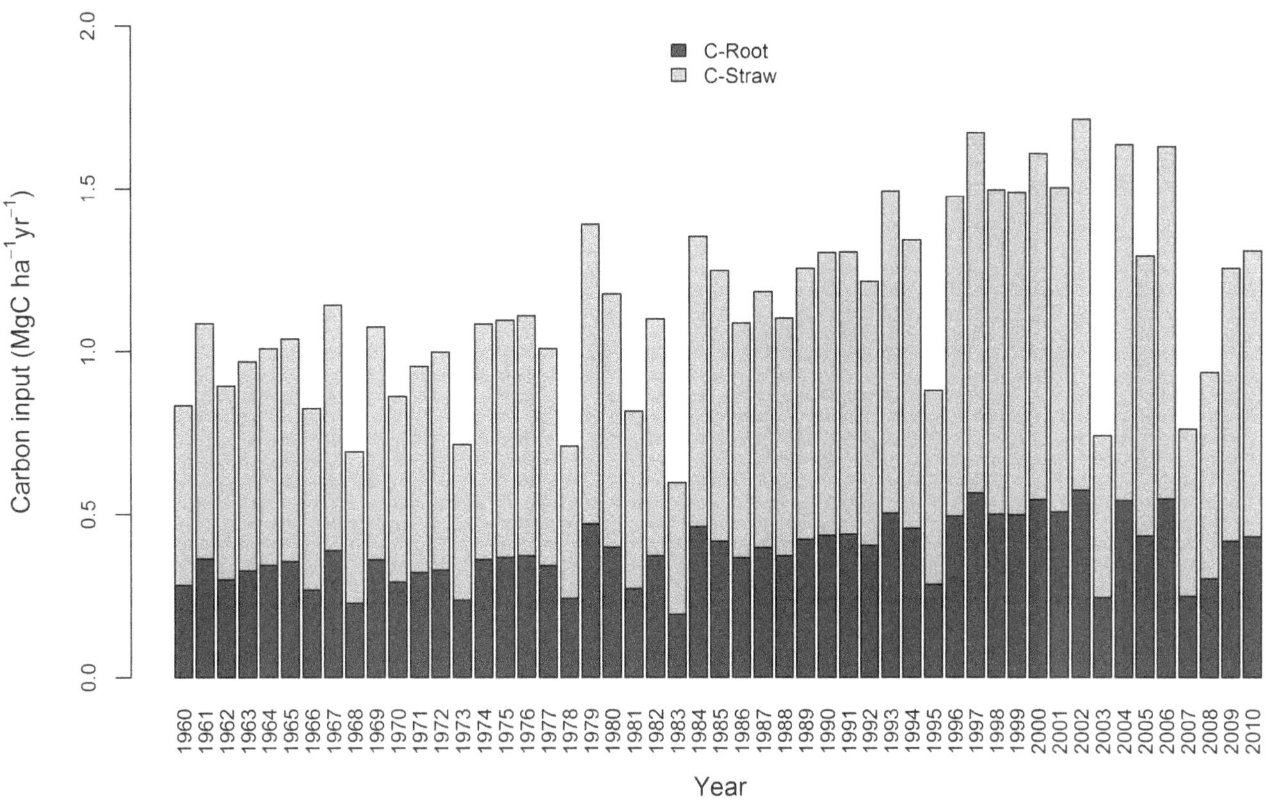

Figure 5. Average rates of carbon input from 1960 to 2010.

2.3 Spatial Characteristics of Estimated SOC Change

In general, changes in SOC density from 1960 to 2010 within different states are consistent with that of the entire Australian wheat-growing area. Nevertheless, there are differences in SOC density changes across different states. On average, compared with the initial status, SOC density during the 2000s decreased by 13 Mg ha^{-1} in Queensland, 11 Mg ha^{-1} in New South Wales, 10 Mg ha^{-1} in Western Australia, 8 Mg ha^{-1} in Southern Australia, and 7 Mg ha^{-1} in Victoria, respectively (Table 3).

Overall, SOC stocks in the topsoil to 30 cm depth were estimated to have decreased by 12 (±50%) Tg in Queensland, 48 (±54%) Tg in New South Wales, 53 (±34%) Tg in Western Australia, 26 (±38%) Tg in Southern Australia and 17 (±35%) Tg in Victoria (Figure 4 and Table 3).

Discussion

Our simulation results showed that although the Australian wheat-growing areas have in general experienced SOC loss from 1960 to 2010, the rate of SOC loss has declined over time (Figure 3 and Table3). Simulations indicated that approximately 80% of the loss of SOC across Australian wheat-growing areas occurred in the first 30 years, with a lower amount of SOC loss in the later 20 years. SOC loss appeared to reach a new steady-state during the first few years of the 2000s (Table 3 and Figure 3). The achievement of steady-state SOC level is attributed to two primary factors. First, enhanced crop production increased the amount of residue and root input into the soil (Figure 5), thus leading to a reduction in SOC loss over time. Second, following the initial rapid decline, SOC would reach a new steady-state after about 50 years of cultivation, therefore resulting in a slower rate of SOC loss during the later years [2,36].

Model estimates suggested that SOC loss was slowed during the 1990s and 2000s, particularly between 1998 and 2003 when SOC loss almost halted (Figure 3). This cessation in SOC loss was likely due to relatively high C input rates, averaging 1.56 Mg C ha^{-1} yr^{-1}, between 1997 and 2002 (Figure 5). However, SOC decreased significantly from 2003 to 2010 (Figure 3), when crop yields were low and C input averaged only 1.20 Mg C ha^{-1} y^{-1} (Figure 5). We suggest that the SOC loss across most of the Australian wheat-growing areas could be halted or even reversed with an additional input of organic carbon into the soil at a minimum rate of 0.4 Mg ha^{-1} yr^{-1}, relative to the mean rate from 2003 to 2010.

Recently, using Agro-C, Yu et al. [1] modeled the SOC changes across Chinese agro-ecosystems during the last three decades and found that Chinese cropland SOC has generally increased. The differences in cropland SOC patterns in the two countries during the past several decades are mainly due to, first, the different initial SOC values. In generally, Chinese croplands have experienced a long history of intense cultivation, which led to a very low level of SOC [18], whereas the Australian soils have not been cultivated as intensively for as long. Second, the annual amount of organic C input in China [1] is generally higher than that in Australia, allowing for more carbon to be sequestrated in the soil. For example, SOC densities increased by 0.35 Mg C ha^{-1} yr^{-1} in south and central China but decreased by 0.18 Mg C ha^{-1} yr^{-1} in Victoria during the first 30 years of the simulations, even though their initial SOC densities were similar (approximately 28 Mg C ha^{-1}). In contrast, the annual C input in

Victoria was only 1.29 Mg ha^{-1} yr^{-1} (Table 3), approximately 38% of that in south and central China [1].

On average, the rate of SOC loss across Australian wheat-growing areas from 1960 to 2010 is 3.12 Tg y^{-1}, approximately 4.8% of the total Australian fossil-fuel carbon emissions during the same period (http://cdiac.ornl.gov/trends/emis/aus.html). These results imply that although Australian agricultural soil acted as a net carbon source, its contribution to increasing atmospheric carbon dioxide is much smaller than that of fossil fuel production and consumption.

The net change in SOC is determined by the balance of C inputs by crop production and outputs by decomposition, both of which are regulated by both management and environmental factors. In this study, higher rates of SOC decreases were found in regions with relatively low annual precipitation and high annual temperature (climatic data not shown), such as Queensland, northern parts of New South Wales, and Western Australia (Figure 4 and Table 3). This can be attributed to the following factors: precipitation provides water to support crop growth and leads to more C input into soil through stubble retention, whereas lower temperatures can inhibit soil respiration, thereby reducing SOC decomposition rates [37]. Furthermore, historical land use also impacts long-term SOC change. For example, SOC declined more in Queensland, which has a relatively higher initial SOC level (Table 3) than other regions. This difference in initial conditions is due to the relatively short history of cultivation in Queensland compared to the other regions [38]. Consequently, a higher C input rate is needed in Queensland to balance SOC loss.

Another implication of our results is that during the last five decades, C input through retention of crop residues has failed to compensate for C output across Australian wheat-growing areas. Thus, soil has acted as a net C source and might therefore be contributing to global warming. For the purpose of accumulating SOC and reducing greenhouse gas emissions, future agricultural management methods need to be improved. For example, increasing rotation complexity (e.g., increasing crop diversity or introducing legumes into rotation) has been reported to significantly promote SOC accumulation when compared to single wheat farming systems [39]. Additionally, in many regions, nitrogen is a dominant limiting factor for crop growth [40]. Adopting nitrogen fertilization would inevitably enhance both crop production and SOC accumulation. Finally, addition of animal manure would not only supply nutrients for crop growth but would also directly add C into the soil [5].

Supporting Information

Figure S1 Comparison of simulated and observed SOC at different calibration sites. Open circles show the observed values, dashed lines show the simulated values before model calibration, and solid lines show the simulated values after calibration.

Figure S2 Comparison of simulated and observed SOC at different validation sites. Open circles show the observed values and solid lines show the simulated values.

Table S1 Locations and initial soil properties of the calibration and validation sites.

Acknowledgments

We greatly thank David Jacquier of CSIRO Land and Water, Canberra, for providing regional soil profile data and land-use information.

Author Contributions

Conceived and designed the experiments: YH. Performed the experiments: GW. Analyzed the data: GW. Contributed reagents/materials/analysis tools: EW YY WZ. Wrote the paper: GW. Revised the manuscript: YH.

References

1. Yu Y, Huang Y, Zhang W (2012) Modelling soil organic carbon change in croplands of China, 1980–2009. Global and Planetary Change: 115–128.
2. Lal R (2004) Soil carbon sequestration impacts on global climate change and food security. Science 304: 1623–1627.
3. Smith P, Martino D, Cai Z, Gwary D, Janzen H, et al. (2008) Greenhouse gas mitigation in agriculture. Philosophical Transactions of the Royal Society B: Biological Sciences 363: 789–813.
4. Barrett DJ, Galbally IE, Graetz RD (2001) Quantifying uncertainty in estimates of C emissions from above-ground biomass due to historic land-use change to cropping in Australia. Global Change Biology 7: 883–902.
5. Luo Z, Wang E, Sun OJ (2010) Soil carbon change and its responses to agricultural practices in Australian agro-ecosystems: A review and synthesis. Geoderma 155: 211–223.
6. Dalal R, Chan K (2001) Soil organic matter in rainfed cropping systems of the Australian cereal belt. Soil Research 39: 435–464.
7. Huang Y, Yu Y, Zhang W, Sun W, Liu S, et al. (2009) Agro-C: A biogeophysical model for simulating the carbon budget of agroecosystems. Agricultural and Forest Meteorology 149: 106–129.
8. Sanderman J, Farquharson R, Baldock J (2009) Soil carbon sequestration potential: a review for Australian agriculture. Soil carbon sequestration potential: a review for Australian agriculture. Available: http://www.csiro.au/en/Portals/Publications/Research-Reports/Soil-Carbon-Sequestration-Potential-Report.aspx. Accessed 20 August 2012.
9. Brain Pink (2012)Year Book Australia. Canberra: Australian Bureau of Statistics. 885 p.
10. Anderson G (2009) The impact of tillage practices and crop residue (stubble) retention in the cropping system of Western Australia. Perth: Department of Agriculture and Food. 93 p.
11. Steed GR, Ellington A, Pratley JE, Carter M (1994) Conservation tillage in the southeastern Australian wheat-sheep belt. Conservation tillage in temperate agroecosystems: 231–251.
12. Mermut A, Eswaran H (2001) Some major developments in soil science since the mid-1960s. Geoderma 100: 403–426.
13. Tullberg J, Ziebarth P, Li Y (2001) Tillage and traffic effects on runoff. Soil Research 39: 249–257.
14. Ugalde D, Brungs A, Kaebernick M, McGregor A, Slattery B (2007) Implications of climate change for tillage practice in Australia. Soil and Tillage Research 97: 318–330.
15. Li C, Frolking S, Harriss R (1994) Modeling carbon biogeochemistry in agricultural soils. Global biogeochemical cycles 8: 237–254.
16. Jenkinson D, Andrew S, Lynch J, Goss M, Tinker P (1990) The Turnover of Organic Carbon and Nitrogen in Soil [and Discussion]. Philosophical Transactions: Biological Sciences 329: 361–368.
17. Parton W, Scurlock J, Ojima D, Gilmanov T, Scholes R, et al. (1993) Observations and modeling of biomass and soil organic matter dynamics for the grassland biome worldwide. Global biogeochemical cycles 7: 785–809.
18. Li CS, Zhuang Y, Frolking S, Galloway J, Harriss R, et al. (2003) Modeling soil organic carbon change in croplands of China. Ecological Applications 13: 327–336.
19. Falloon P, Smith P, Bradley R, Milne R, Tomlinson R, et al. (2006) RothCUK–a dynamic modelling system for estimating changes in soil C from mineral soils at 1-km resolution in the UK. Soil Use and Management 22: 274–288.
20. Ogle SM, Breidt F, Easter M, Williams S, Killian K, et al. (2010) Scale and uncertainty in modeled soil organic carbon stock changes for US croplands using a process-based model. Global Change Biology 16: 810–822.
21. Yu S, Yao H (2003) Simulation and prediction of soil organic carbon dynamics in Jiangsu province based on models and GIS techniques. Scientia Agriculatura Sinica 2: 1007–1013.
22. Yu YQ, Huang Y, Zhang W (2006) Modeling farmland soil organic carbon dynamics in eastern China: model validation and sensitivity analysis. Geogr Geo-Inf Sci 22: 83–88 (in Chinese).
23. Shi XZ, Yang RW, Weindorf DC, Wang HJ, Yu DS, et al. (2010) Simulation of organic carbon dynamics at regional scale for paddy soils in China. Climatic Change 102: 579–593.
24. Jobbagy EG, Jackson RB (2000) The vertical distribution of soil organic carbon and its relation to climate and vegetation. Ecological Applications 10: 423–436.

25. Wang S, Huang M, Shao X, Mickler RA, Li K, et al. (2004) Vertical distribution of soil organic carbon in China. Environmental management 33: 200–209.
26. Qin ZC, Huang Y (2010) Quantification of soil organic carbon sequestration potential in cropland: a model approach. SCIENCE CHINA Life Sciences 53: 868–884.
27. Jeffrey SJ, Carter JO, Moodie KB, Beswick AR (2001) Using spatial interpolation to construct a comprehensive archive of Australian climate data. Environmental Modelling & Software 16: 309–330.
28. Henderson B (2001) ASRIS: Continental-scale soil property predictions from point data. Canberra: CSIRO Land and Water. 107 p.
29. Johnston R, Barry S, Bleys E, Bui E, Moran C, et al. (2003) ASRIS: the database. Soil Research 41: 1021–1036.
30. ACLUMP (2010) Land Use and Land Management Information for Australia: Workplan of the Australian Collaborative Land Use and Management Program. Available: http://adl.brs.gov.au/landuse/. Accessed 30 August 2012.
31. Huang Y, Zhang W, Sun W, Zheng X (2007) Net primary production of Chinese croplands from 1950 to 1999. Ecological Applications 17: 692–701.
32. Skjemstad J, Spouncer L, Cowie B, Swift R (2004) Calibration of the Rothamsted organic carbon turnover model (RothC ver. 26.3), using measurable soil organic carbon pools. Soil Research 42: 79–88.
33. Llewellyn D'Emden (2009) Adoption of no-till and conservation farming practices in Australian grain growing regions: current status and trends. In "Report for SA No-till Farmers Association and CAAANZ". Available: http://www.canfa.com.au/articles.html. Acessed 30 August 2012.
34. Brady NC, Weil RR (2008) The nature and properties of soils (14 th ed). New Jersey: Pearson Education Publication. 975 p.
35. R Development Core Team (2012) *R: A language and environment for statistical computing.* R Foundation for Statistical Computing, Vienna, Austria.
36. Davidson EA, Ackerman IL (1993) Changes in soil carbon inventories following cultivation of previously untilled soils. Biogeochemistry 20: 161–193.
37. Post W, Emanuel W, Zinke P, Stangenberger A (1982) Soil Carbon Pools and World Life Zones. Nature 298: 156–158.
38. Luo Z, Wang E, Sun OJ, Smith CJ, Probert ME (2011) Modeling long-term soil carbon dynamics and sequestration potential in semi-arid agro-ecosystems. Agricultural and Forest Meteorology 151: 1529–1544.
39. West TO, Post WM (2002) Soil organic carbon sequestration rates by tillage and crop rotation: A global data analysis. Soil Science Society of America Journal 66: 1930–1946.
40. Wang E, Xu JH, Smith CJ (2008) Value of historical climate knowledge, SOI-based seasonal climate forecasting and stored soil moisture at sowing in crop nitrogen management in south eastern Australia. Agricultural and Forest Meteorology 148: 1743–1753.
41. Scott BJ, Eberbach P, Evans J, Wade L (2010) EH Graham Centre Monograph No. 1: Stubble Retention in Cropping Systems in Southern Australia: Benefits and Challenges. Ed by EH Clayton and HM Burns. Industry & Investment NSW, Orange. Available: http://www.csu.edu.au/research/grahamcentre/. Accessed 20 August 2012.

20

Effect of Land Use History and Pattern on Soil Carbon Storage in Arid Region of Central Asia

Xiaoyu Li, Yugang Wang*, Lijuan Liu, Geping Luo, Yan Li, Xi Chen*

State Key Lab of Desert and Oasis Ecology, Xinjiang Institute of Ecology and Geography, Chinese Academy of Sciences, Urumqi, Xinjiang, China

Abstract

The purpose of this study is to investigate variations in soil organic carbon (SOC) in arid areas due to differences in the cultivation history, land use, and soil salinization. The study area is the lower Sangong River basin on the piedmont of the northern TianShan mountains, which experiences heavy land-use activities. In 1982 and 2005,127(152) and 74 (161) samples in old (new) oasis were collected from each site at the surface soil (i.e., 0–20 cm). The data reveal that the mean value of the surface soil organic carbon content of the old oasis was higher than that of the new oasis by 4.01 g/kg in 1982 and 3.79 g/kg in 2005. Additionally, the soil organic carbon content decreased more rapidly in the newly reclaimed oasis than in the old oasis from 1982 to 2005. The spatial pattern of the SOC content was correlated with the exploitation time in the new oasis, the agricultural land use history, and the SOC content. The decreasing trend is clearer in the high SOC content area than in the low SOC content area. Farmland is the largest carbon pool in both the new and old oases. The carbon density of the old oasis was higher than that of the new oasis by 4.01 and 3.79 g/kg in 1982 and 2005 respectively. The loss of SOC in the agricultural watershed of the arid region in NW China is obvious. Improvements of land management practices, such as no tillage, straw returning to soil, and balanced fertilization techniques, should be adopted to increase the SOC content.

Editor: Peter Shaw, Roehampton university, United Kingdom

Funding: This study was financed by 973 Program of China (No. 2009CB825102), National Natural Science Foundation of China (No.41271202 and 41101179), and the Knowledge Innovation Program of the Chinese Academy of Sciences (No.KZCX2-EW-QN-316). The funders had no role in study design, data collection and analysis, decision to publish, or preparation of the manuscript.

Competing Interests: The authors have declared that no competing interests exist.

* E-mail: wangyg@ms.xjb.ac.cn (YW); chenxi@ms.xjb.ac.cn (XC)

Introduction

Changes in land use and land cover are among the main human activities affecting the surface of the earth [1], with the expansion of agriculture posited as one of the main causes of land cover change globally. The role of terrestrial ecosystems as sources and sinks of C has been highlighted, underscoring the impact of land use and land cover changes on the global climate [2–5]. Globally, the soil organic carbon (SOC) stock is 1550 Pg (1 Pg = 10^{15} g) at a1-m depth [6]; it is as much as three times greater than the atmospheric carbon pool and double that in the biota [7–8]. Soils are the largest organic carbon sink on Earth. The abundance of organic C in the soil affects and is affected by land use and land cover changes, and organic carbon's role as a key determinant of soil fertility and vegetation production has been documented in recent years [9–13]. The effects of land use change on the carbon cycle are complex and are likely to vary among land use types on different landscape scales [14]. Understanding the carbon cycle across different landscape scales, including patch, field, watershed, and global scales, is important [15–17]. Therefore, understanding the changes in the organic carbon storage space distribution in soil is crucial for assessing current regional, continental, and global soil C stores and predicting and ameliorating the consequences of global change [18].

The magnitude of the change in C storage depends on how physical, chemical, or biological processes are altered over time under different land uses [19]. The extent to which soils and vegetation act as CO_2 sinks or sources depends largely on land-use management [20]. Appropriate land use management practices can improve SOC, whereas unreasonable practices reduce SOC [21]. Some of the land use activities that affect carbon fluxes in soils are deforestation, afforestation, biomass burning, cultivation, crop residue management, and the application of inorganic fertilizers and organic manure [20]. Generally, soil organic carbon (SOC) stocks under cropland are lower than those under pasture or forest, and forest SOC stocks tend to be higher than pasture SOC stocks. The conversion of forest to pasture or to cropland or the conversion of pasture to cropland decreases SOC stocks, whereas the opposite conversions usually lead to increased SOC stocks [22–24]. Historically, when croplands have been established on land previously used for native vegetation, the soil C pool has been a major source of atmospheric CO_2, contributing approximately 180–200 Pg C over the last two centuries [25], which is approximately 40% of the total anthropogenic CO_2 emissions [26–27]. At a global scale, land use and land cover changes (LULCC) are estimated to contribute 25% of the anthropogenic flux of carbon dioxide to the atmosphere, the second highest contribution after that of fossil fuels [28]. The amount of C released to the atmosphere after LULCC in the 1980s and 1990s is estimated as 1.41 and 1.6$PgCyr^{-1}$ [29]; however, reducing CO_2 emissions and increasing C sequestration by vegetation and soils can contribute to decreasing this rate [5,30–31]. Davidson and Ackerman [32] suggested that nearly all C lost from soil occurs within 20 years and that most occurs within 5 years, after initial cultivation. Therefore, decreasing the atmospheric CO_2 contraction is necessary, which requires understanding the variations in regional SOC stocks by land use history and the resulting patterns.

Arid and semi-arid lands cover approximately 45% of the global terrestrial area and contain 16% of the global soil carbon pool [33]. Overgrazing, human activities, and climate variation have contributed to the desertification and degradation of more than two-thirds of these fragile ecosystems [34]. Changes in the vegetation and the release of carbon from these drier regions must be considered when discussing carbon balance on a global scale [35–36]. Desertification is estimated to affect approximately 1.137 Bha of soils and an additional 2.576 Bha of rangeland vegetation in drylands around the world. Furthermore, the total historic loss of C due to desertification may be 18–28 Pg [37], indicating that land desertification characterized by soil degradation and the diminution or destruction of the biological potential of ecosystems have played an important role in the atmospheric CO_2enrichment [38].

China is among the countries seriously affected by desertification. It is estimated that dryland and susceptible dryland (excluding hyperarid areas) comprise357.05 and 331.70 million ha of China, respectively, and in 2004, 263.62 million ha had suffered or was suffering from desertification [39]. Feng et al. [40] estimated that desert lands in north China caused a net emission of 2.17 Pg of C from the 1950s to the 1990s.

Xinjiang, the Uighur Autonomous Region, lies in the hinterland of the Eurasian continent, far from the ocean, in what was known in China as the Western Territory in ancient times. Its 1.66 million sq km represents approximately one-sixth of the total territory of China, making it the largest of China's regions and provinces. Surrounded by mountains, Xinjiang features a temperate continental climate with low precipitation. It has semi-arid and arid climates, with a long-term average annual precipitation of 130 mm and a long-term average of 55 rainy days per year. It is characterized by widespread desert and a fragile ecosystem. Over the past 50 years, the local vegetation and soil have been changed or modified by large-scale land reclamation, irrigation, and cultivation, as well as the application of fertilizers [41]. The farmland area and population increased by 63.5% and 80.1% during 1949–2011 [42]. In particular, intensive soil cultivation was initiated in the1950s. An overuse of water resources, expansion of agricultural lands, and environmental degradation have led to severe problems, such as soil organic carbon loss in Xinjiang oases [43]. The knowledge of the spatial variation of SOC and itsrelation to land use and soil properties is thus crucial for enacting more rational management measures at regional levels [44].

Thus, the objective of this work was to estimate SOC in arid areas as a function of cultivation history, land use, and soil salinization. The study site was the lower Sangong River basin on the piedmont of the northern TianShan mountains, which features heavy land-use activities (Fig. 1). The goal was to identify land use and soil management options to help achieve the full SOC sequestration potential.

Materials and Methods

Study Area

For the Sangong river cachment, where the 308 surface soil samples (0–20 cm) were collected in October 2005, no specific permissions were required for these locations. Thus the soil sampling in the agricultural region was permitted and no specific permissions were required.

We confirmed that the field studies did not involve any endangered or protected species.

The Sangong River basin, a typical inland river basin with spring and autumn flooding, is located in the northernTianShan mountains and the southern Guerbantonggute Desert in central Asia, extending between 87°47′-88°17′ E and 44°09′-44°29′ N (Fig. 1). The Sangong River originates from the northern slope of the TianShan mountains. The elevation along the basin descends from 750 m (the southern region) to 450 m (the northern region) above sea level. Passing by plain oases in the middle, it reaches the southern edge of the Gurbantunggut Desert, a total distance of less than 80 km. The basin is a mountain-oasis-desert landscape pattern typical of the region. Oases, more productive than the surrounding deserts, are the primary sites for human settlement because of the availability of fertile soil, fresh groundwater and surface runoff from the nearby mountains [41].

The southern part of the oasis region is an old oasis with thousands of years of agricultural history; however, the northern part has featured agricultural production for fewer than 60 years and is thus a new oasis. The boundary of new oasis is straight-sided because of this area was reclaimed from desert land by XinJiang Production and Construction Group in 1950s under unified planning. And the boundary of old oasis is curled naturally because of this area was developed from natural land for hundreds years. A reservoir was constructed in the 1970s at the end of the Sangong River to supply water for agricultural irrigation. These hydrological and land-use changes have promoted soil salinization and land degradation in the entire lower reaches of the Sangong River catchment [45].

The climate is an arid continental climate with an annual precipitation of 163 mm and an annual pan evaporation of 1780–2460 mm. The main soil type is Solonchak, covering approximately 37% of the study area; less frequent soil types include Haplic calcisols and Aquert on substrates of varied fertility. The natural vegetation of the region is characterized by different types of xeric or halophyte communities, dominated by the desert shrubs *Tamarix ramosissima* and *Haloxylon ammodendron* as well as *Reaumuria soongorica* scrubs. Crops in the oasis include cotton, wheat, hops, grapes, and corn.

Soil Sampling and Measurements

In 1982, soil samples were collected in the Sangong river watershed as part of China's National Soil Inventory of the 1980s [46]. In all, 127 surface soil (0–20 cm) samples from the old oasis and 152 samples from the new oasis were positioned according to the recorded geographic location and land cover information. The soil organic matter for each soil sample was converted to soil organic carbon (SOC) by multiplying by 0.58 [47–49].

In 2005, a total of 308 surface soil samples (0–20 cm) were collected in different landscape types throughout the catchment oasis in October. Among these samples, 74 samples were obtained from the old oasis and 161 samples represented the new oasis. The sample sites covered various land use types. Soil samples from these sites were obtained by hand soil auger and analyzed in the laboratory. Samples were air-dried and crushed to pass through a 2-mm mesh. All of the soil samples were analyzed for soil salt content and soil organic matter, and the soil organic matter were converted to soil organic carbon (SOC) by multiplying by 0.58. Similarly to those in 1982.

An equation was developed to quantify and characterize the soil organic carbon storage state in different land-use types based on the distribution of soil organic carbon content by land-use type. For an individual land-use type with k patches, the equations below were used to calculate the soil organic carbon storage (g):

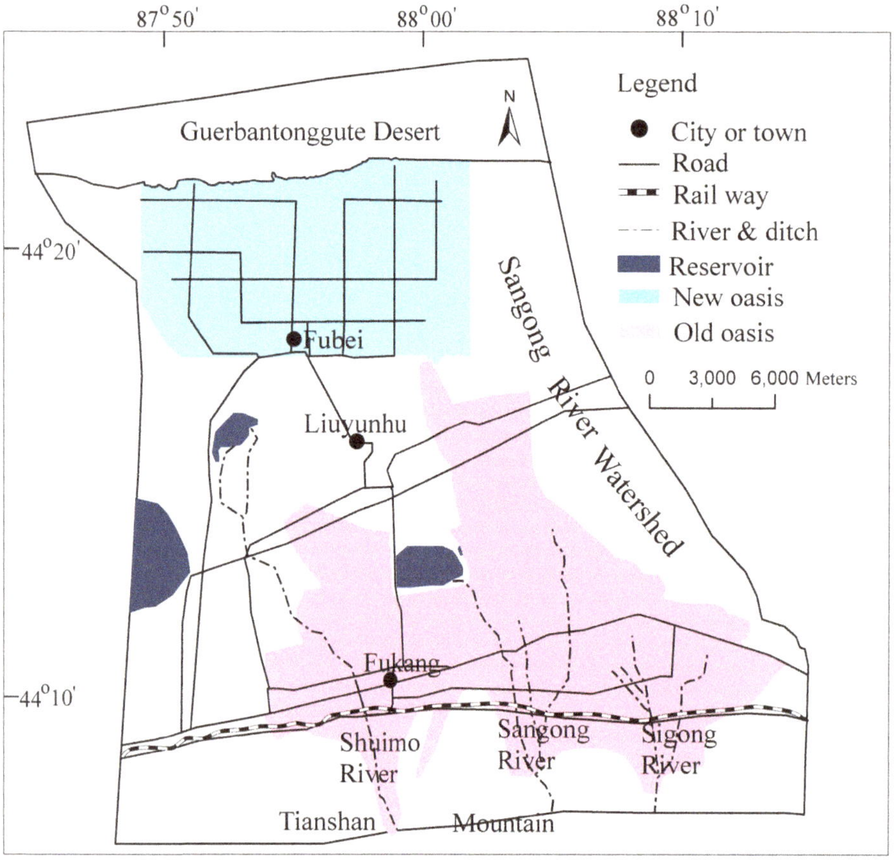

Figure 1. Study area of the Sangong river basin and its oases.

$$SOCA_d = \sum_{i=1}^{k} SOCA_i = \sum_{i=1}^{k} \rho_i \times S_i \times D_i \times (1 - V_i) \times A_i \quad (1)$$

$$SOCA_i = \rho_i \times S_i \times D_i \times (1 - V_i) \times A_i \quad (2)$$

where k is the number of patches, ρ_i is the bulk density (Mg m^{-3}), S_i is the distribution area of soil organic carbon content (m^2) for a degree of soil organic content i, D_i is the thickness of soil layer (m), V_i is the volume fraction of fragments >2 mm, and A_i is the mean of the soil organic content (g kg^{-1}). The 1982 soil bulk density data [46] and the 2005 soil bulk density under the same conditions and land-use type were determined using a soil corer (100-cm^{3}stainless steel cylinder).

Land Use Data

The 1982 and 2005 land use data were obtained from remote sensing images, topography maps, and field investigations. A topographic map at the scale of 1:50,000 was used to create a 1:50,000image. In 2004, spot imagery (June, with 10×10 m resolution) was geo-referenced to the Universal Transverse Mercator projection with a root mean square (RMS) error of 0.45 pixels. To obtain the map of land uses in 2005, the land use types were investigated in June 2005 according to the maps of spot imagery, topography, and field. Using the rectified 2004 image as a basis, the 1982 image was rectified with an RMS error of 0.5 pixels (15 m). Visual interpretations of Landsat imagery have been demonstrated to be useful for land cover and vegetation mapping [50]. Land use types in the study area were divided into eight types: farmland, planted forest, shrub, grassland, construction land, reservoir/flood land, saline alkali land, and bald land (Fig. 2).

Results and Discussion

Land Use Change

As shown in Fig. 2 and Table 1, farmland is the main type of land use in both the new and old oases of the Sangong River watershed, accounting for 42.37% of the new oasis in 1982 and 50.49% in 2005 and accounting for 50.22% of the old oasis in 1982 and 58.56% in 2005. The farmland area increased by 8% from 1982 to 2005 in both the new and old oases. At the same time, grassland, the second largest land use type, decreased by 11.85% in the new oasis and 9.76% in the old oasis. Much of the grassland was converted into farmland over the last 25 years. The contribution of planted forest increased by 6.22% in the new oasis and was almost constant in the old oasis because land exploitation in the new oasis has been accompanied by the planting of shelter forest since 2000 due to land integrated management (Wang & Li, 2013). The shrub land increased by 3.98% in the new oasis and decreased by 5.03% in the old oasis. The flood land in the new oasis was disappeared in 2005 because of the runoff was dammed to the reservoir, which located at the upper stream of the new oasis, and the flood land was reclaimed to extend the farmland. In both the new and old oases, the landscape patch changed dramatically from 1982 to 2005 (Fig. 2). These remarkable

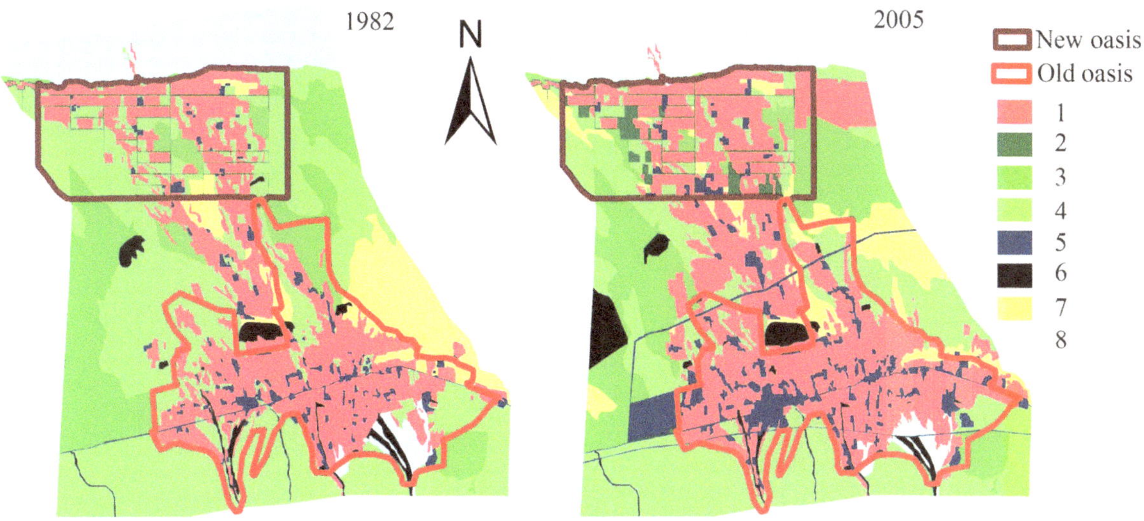

Figure 2. Land use map of the Sangong river basin and its oases in 1982 and 2005 (1, Farmland; 2, planted forest; 3, shrub land; 4, grassland; 5, construction land; 6, reservoir/flood land; 7, saline alkali land; 8, bald land).

transformations were responsible for the intensification of human activities in the oases. Socioeconomic factors created by human activities are the main driver of land-use change in shaping the agricultural land-use pattern [51–53].

Descriptive Characteristics of Soil Organic Carbon

In the old oasis, the average surface soil organic carbon (SOC) was 11.44 g/kg in 1982, with a range of 4.93 to 19.89 g/kg, and 9.812 g/kg in 2005, with a range of 1.33 to 18.30 g/kg (Table 2). The average SOC content decreased by 14.2% from 1982 to 2005.

In the new oasis, the average SOC content was 7.428 g/kg in 1982, with a range of 1.91 to 16.00 g/kg, and 6.026 g/kg in 2005, with a range of 0.58 to 14.82 g/kg (Table 2). The average SOC content decreased by 18.9% from 1982 to 2005.

The mean SOC content in the old oasis is higher than that of the new oasis by 4.01 g/kg in 1982 and 3.79 g/kg in 2005. This difference may be the attributed to the fact that the old oasis was cultivated for over 500 years and is located in the middle river of the watershed, with good soil and geographical conditions. However, in both the old and new oases, the surface SOC decreased over the past 25years. The decrease is more rapid for the newly reclaimed oasis than the old oasis, indicating that the SOC content decrease improved by human activity is stronger in the surface soil, especially in the new oasis region. The result of land reclamation converted natural ecosystems to agricultural ecosystems, which generally depletes the SOC pool because of the lower C input and higher decomposition rates of agricultural ecosystems [54–55]. Furthermore, the change in SOC content varies between the old and new oasis due to differences in soil moisture and temperature regimes caused by plowing, drainage, biomass burning, and residue removal [56–57].

Spatial Pattern of Soil Organic Carbon (SOC)

To describe the time-space distribution of SOC precisely and simply, spatial distribution maps of SOC in new and old oases in 1982 and 2005 were compiled by geostatistics (Table 3, Fig. 3). According to semivariogram calculation (Table 3), the experi-

Table 1. Area and percentage of land use types in the new and old oases in 1982 and 2005.

Land use type	New oasis				Old oasis			
	1982		2005		1982		2005	
	Area(hm)	%	Area(hm)	%	Area(hm)	%	Area(hm)	%
Farmland	6313.42	42.37%	7523.07	50.49%	12058.27	50.22%	14061.12	58.56%
Planted forest	465.55	3.12%	1391.07	9.34%	2.94	0.01%	9.18	0.04%
Shrub land	1298.33	8.71%	1891.07	12.69%	1462.87	6.09%	255.36	1.06%
Grassland	6090.79	40.88%	2835.01	19.03%	6025.54	25.10%	3682.24	15.34%
Construction land	278.24	1.87%	548.78	3.68%	1465.14	6.10%	3363.02	14.01%
Reservoir/flood land	29.02	0.19%	0	0.00%	869.31	3.62%	667.24	2.78%
Saline alkali land	411.16	2.76%	677.52	4.55%	1176.24	4.90%	1150.26	4.79%
Bald land	13.33	0.09%	33.06	0.22%	950.46	3.96%	822.24	3.42%
Total	14900	100%	14900	100%	24010.78	100%	24010.68	100%

Table 2. Soil organic carbon (SOC) content (g/kg) in the Sangong river basin and its oases in 1982 and 2005.

	Minimum	Maximum	Mean	Median	SD	Skewness	Kurtosis
Whole basin 1982	1.91	19.89	9.25	8.70	3.87	0.23	−0.87
Whole basin 2005	0.58	21.71	7.15	6.55	3.64	0.66	0.32
New oasis 1982	1.91	16.00	7.43	6.61	3.39	0.70	−0.35
New oasis 2005	.58	14.82	6.03	5.39	2.92	0.76	−0.02
Old oasis 1982	4.93	19.89	11.44	11.60	3.23	0.01	−0.60
Old oasis 2005	1.33	18.30	9.81	10.21	3.25	0.01	0.52

mental semivariogram of SOC content could be fitted with the exponential model for the old oasis in 1982, the spherical model for the old oasis in 2005, and the Gaussian model for new oasis in 1982 and 2005. The theoretical variation function and experimental variation function exhibits a better fitting (Table 3). The coefficients of determination (R^2) were >0.55. The values of R were significant at the 0.01 level by F test, which shows that the semivariogram models well reflect the spatial structural characteristics of the SOC content. The ratio of nugget and sill (C0/C0+C) was between 0.25 and 0.75, except for in old oasis in 2005, suggesting moderate spatial autocorrelation [58]. However, the nugget to sill ratio for the SOC content in the old oasis in 2005 was <0.25, strong suggesting that the spatial dependence of SOC content was mainly due to structural factors in this case. The spatial variability of SOC content might be directly or indirectly caused by natural and human action on soil processes in the region, e.g., crop sequences, irrigation, precipitation, evaporation, and residue management [45].

In the new oasis (Fig. 3), the high SOC content was distributed in the eastern part of the center region the SOC content generally decreased from the center region to the eastern and western regions. The SOC content was clearly correlated with the exploitation time. The exploitation of the new oasis began in the eastern part of the center region in the 1950s and then expanded westward. The shorter the agricultural land use history, the lower the SOC content. The SOC content decreased dramatically from 1982 to 2005. The decreasing trend is clearer in the high SOC content area than in the low SOC content area, indicating that land-use change in the form of land reclamation was accompanied by SOC content loss in surface soil between 1982 and 2005. This

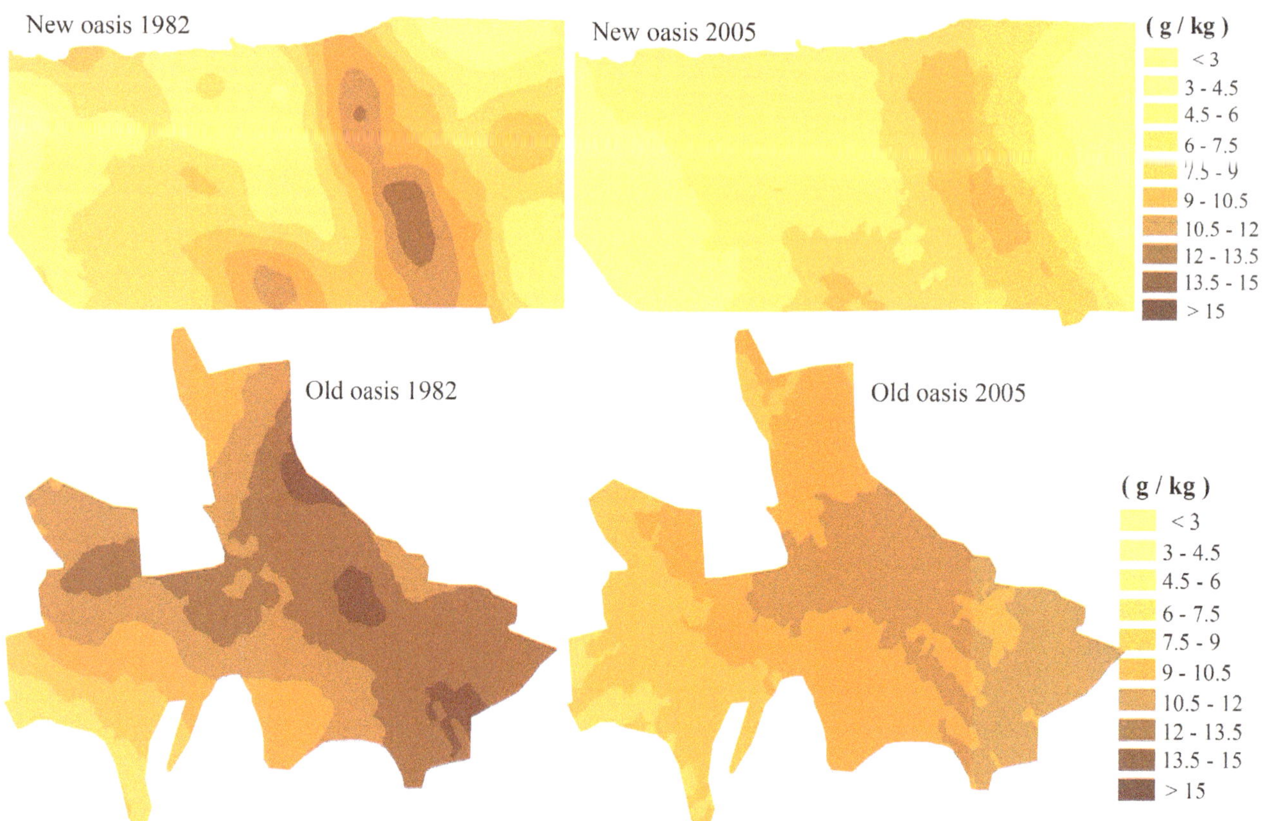

Figure 3. Estimated SOC content maps in the new and old oasis in 1982 and 2005 (mg/kg).

Table 3. Parameters and F-test of fitted semivariogram models for soil organic carbon in oases.

	Theoretical model	Nugget (C_0)	Sill (C_0+C)	C/(C_0+C)	Range	R^2	RSS	F test
New oasis 1982	Gaussian model	5.21	12.49	0.583	4156.9	0.982	1.30	545.56**
New oasis 2005	Gaussian model	2.88	8.492	0.661	2303.6	0.862	4.15	62.46**
Old oasis 1982	Exponential model	5.50	12.45	0.558	19290	0.845	6.04	54.51**
Old oasis 2005	Spherical model	0.17	10.37	0.984	1550	0.547	10.50	12.07**

**Significance at $\alpha = 0.01$ level of F test.

inference was also consistent with the conclusion that conversion of land-use types induces SOC content change [59].

The area of the new oasis with SOC contents above 10.5 g/kg (Table 4) decreased from 1434.34 ha (9.63%) in 1982, 76.15% of which was distributed in farmland, to 0 in 2005. The area with

Table 4. Area percent (%) of SOC content groups in different land use types in the new and old oases in 1982 and 2005.

			SOC content groups (g/kg)									
Land use type	Oasis	Period	<3	3–4.5	4.5–6	6–7.5	7.5–9	9–10.5	10.5–12	12–13.5	>13.5	Total
1	New	1982	–	2.38	10.93	9.32	6.39	6.02	5.53	1.80	–	42.37
		2005	0.78	3.47	21.37	12.76	10.22	1.88	–	–	–	50.49
	Old	1982	–	–	–	2.65	2.12	8.21	14.81	19.80	2.65	50.22
		2005	–	–	–	2.17	12.85	21.85	21.70	–	–	58.56
2	New	1982	–	0.04	1.03	0.91	0.38	0.47	0.20	0.10	–	3.12
		2005	–	0.66	5.18	1.83	1.47	0.21	–	–	–	9.34
	Old	1982	–	–	–	–	–	0.01	–	–	–	0.01
		2005	–	–	–	0.02	0.01	0.01	–	–	–	0.04
3	New	1982	–	–	3.12	3.37	2.00	0.22	–	–	–	8.71
		2005	0.03	5.77	5.31	1.52	0.06	–	–	–	–	12.69
	Old	1982	–	–	–	0.08	0.03	1.50	2.32	1.29	0.87	6.09
		2005	–	–	–	–	0.21	0.31	0.55	–	–	1.06
4	New	1982	–	2.45	11.44	17.89	5.06	2.50	1.38	0.18	–	40.88
		2005	–	7.99	6.57	3.15	1.14	0.18	–	–	–	19.03
	Old	1982	–	–	–	1.62	3.22	4.06	6.52	8.95	0.72	25.09
		2005	–	–	–	1.08	4.50	5.85	3.91	–	–	15.34
5	New	1982	–	–	0.36	0.32	0.38	0.47	0.21	0.12	–	1.87
		2005	–	–	1.08	1.67	0.90	0.03	–	–	–	3.68
	Old	1982	–	–	–	0.19	0.20	1.46	1.43	2.44	0.38	6.10
		2005	–	–	–	0.94	4.53	4.62	3.93	–	–	14.01
7	New	1982	–	–	0.01	0.11	0.07	–	–	–	–	0.19
		2005	–	–	–	–	–	–	–	–	–	–
	Old	1982	–	–	–	0.38	0.62	0.35	0.74	1.44	0.09	3.62
		2005	–	–	–	0.14	0.66	1.11	0.86	–	–	2.78
8	New	1982	–	0.54	0.58	0.02	0.81	0.70	0.11	–	–	2.76
		2005	–	1.97	1.18	1.13	0.28	–	–	–	–	4.55
	Old	1982	–	–	–	–	–	0.29	1.74	2.51	0.37	4.90
		2005	–	–	–	–	0.45	2.85	1.49	–	–	4.79
9	New	1982	–	0.03	0.05	0.01	–	–	–	–	–	0.09
		2005	–	–	0.17	0.06	–	–	–	–	–	0.22
	Old	1982	–	–	–	–	–	0.21	0.62	2.91	0.22	3.96
		2005	–	–	–	–	–	1.37	2.05	–	–	3.42

1, Farmland; 2, planted forest; 3, shrub land; 4, grassland; 5, construction land; 6, reservoir/flood land; 7, saline alkali land; 8, bald land.

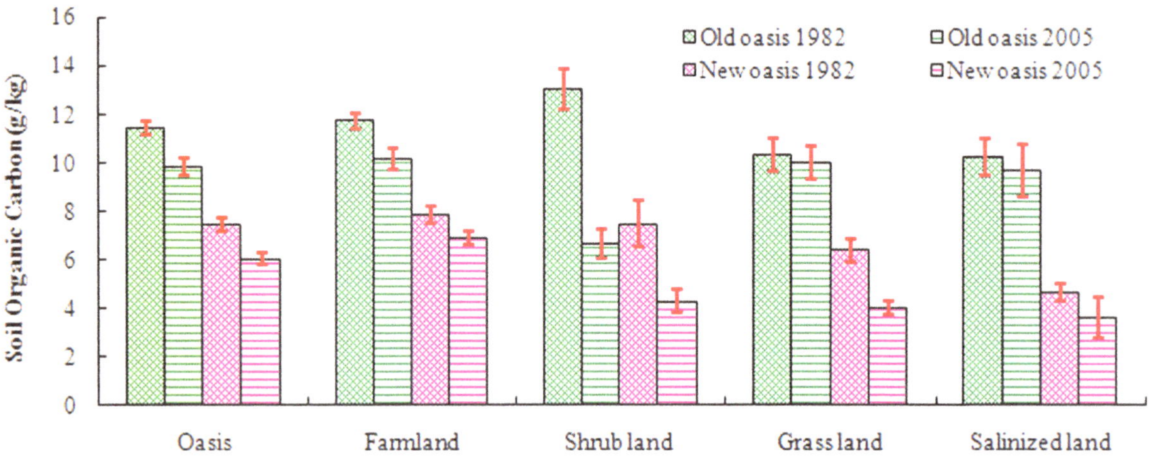

Figure 4. Mean SOC content and standard error in the oasis and main land use types.

SOC contents of 7.5–10.5 g/kg decreased by 1357.54 ha (9.11%), 68.41% of which was from grassland. The area with SOC contents below 4.5 g/kg increased from 808.72 ha (5.43%) in 1982 to 3080.98 ha (20.68%) in 2005, a change of2272.26 ha (15.25%), of which 38.03% was shrub land and 36.36% was grassland. The area with an SOC content of 4.5–7.5 g/kg reached 8860.41 ha (59.47%) in 1982 and 9379.78 ha (62.95%) in 2005, of which the farmland area increased by 2067.49 ha and the grassland area decreased by 2921.75 ha from 1982 to 2005. Over the past 25 years, the SOC content in the new oasis decreased dramatically in the farmland, shrub land, and grassland.

In the old oasis (Fig. 3), the spatial distribution of SOC content in 1982 and 2005 shows that the SOC exhibited an obvious spatial distribution pattern, gradually decreasing from east to west. Temporally, the SOC content in the old oasis also decreased substantially from 1982 to 2005.

The area with an SOC content above 12 g/kg, which was mostly distributed in the eastern region of the oasis, accounted for 44.64% (10717.55 ha) of the total area in 1982, of which farmland and grassland occupied 50.28% and 21.66%, respectively. However, this area decreased to 0 in 2005. At the same time, the areas with SOC contents of 9–12 g/kg and below 9 g/kg increased from 44.26% (10627.37 ha) and 11.10% (2665.86 ha) in 1982 to 72.44% (17394.36 ha) and 27.56% (6616.32 ha) in 2005, respectively. Among the increases in low SOC content areas, the loss of SOC from the surface soil of farmland was most remarkable (68.98%).

Effect of Land Use History and Land Use Type on Soil Organic Carbon (SOC)

Fig. 4 shows the SOC content by land use. The representative SOC content of the 4 main land use types in arid areas ranged from 4.640 g/kg in salinized land to12.998 g/kg in shrub land in 1982 and from 3.608 g/kg in salinized land to 10.125 g/kg in farmland in 2005.

In the old oasis, relative to the 1982 levels, the SOC decreased by 1.597 g/kg in farmland, 6.365 g/kg in shrub land, 0.331 g/kg in grassland, and 0.588 g/kg in salinized land in 2005, accounting for 13.62%, 48.97%, 3.21%, and 5.74% of the 1982 values, respectively. The SOC content of shrub land was higher than that of any other land use type in 1982 and decreased most rapidly over the last 20 years.

In both 1982 and 2005, the SOC of farmland was highest among the land use types in the new oasis. The decrease in SOC from 1982 to 2005 varied from 0.952 g/kg in farmland to 3.225 g/kg in shrub land. The contribution to the decrease in SOC was 12.19% for farmland, 43.18% for shrub land, 37.69% for grassland, and 22.23% for salinized land in 1982. Just as in the old oasis, the largest decrease occurred for shrub land. The SOC in grassland also decreased significantly.

As shown in Table 4, in the farmland in the new oasis, the area with SOC contents above 9 g/kg decreased from 13.35% of the new oasis area in 1982 to 1.88% in 2005, and the area with SOC contents below 9 g/kg increased from 29.02% in 1982 to 48.60% in 2005. In the farmland in the old oasis, the area with SOC contents above 12 g/kg decreased from 22.45% in 1982 to 0% in 2005, and the area with SOC contents below 12 g/kg increased from 27.79% in 1982 to 58.57% in 2005. This finding clearly shows the loss of SOC in farmland from 1982 to 2005 in both the new and old oasis. Grassland is the second most important land use type in the study area. The area of grassland in the new oasis with SOC content above 4.5 g/kg decreased from 38.45% of the new oasis area in 1982 to 11.04% in 2005, and the area with SOC contents below 4.5 g/kg increased from 2.445% in 1982 to 7.99% in 2005. In the old oasis, the grassland area with SOC contents above 10.5 g/kg decreased from 16.19% of the old oasis area in 1982 to 3.91% in 2005, and the area with SOC below 10.5 g/kg increased from 8.9% in 1982 to 11.43% in 2005.

Dynamics of Land Use and Carbon Storage

In the new oasis, the total SOC storage was 306373.43 t (20.56 t/ha) in 1982 and 251234.78 t (16.86 t/ha) in 2005, a decrease of 55138.65 t (18.00%). Farmland is the largest carbon pool, accounting for 49.44% and 58.46% of the total SOC storage in 1982 and 2005, respectively. Although the farmland area increased by 1209.65 ha and 8.12% of the total new oasis area, the carbon storage in farmland decreased from 151469.56 t in 1982 to 146861.21 t in 2005, which is a decrease of4608.35 t or 3.04% of that in 1982, because of the decrease in the carbon density from23.99 t/ha to 19.52 t/ha. In grassland, the second largest carbon pool, the SOC storage decreased from 109317.91 t in 1982 to 38678.44 t in 2005, a decrease of70639.48 t or 64.62% of that in 1982. The grassland area and carbon density also decreased from 6090.79 ha (40.88%) and 17.95 t/ha in 1982 to 2835.01 ha (19.03%) and 13.64 t/ha in 2005, respectively. Shrub land also contributes substantially to SOC storage. With the shrub land area increased by 4% of the oasis area, the SOC storage increased by 895.92 t and the carbon density decreased from 15.66 t/ha to 11.22 t/ha.

In the old oasis, the total SOC storage and mean carbon density were 805616.15 t and 33.55 t/ha in 1982 and 722257.97 t and 30.08 t/ha in 2005, respectively. The carbon storage and density decreased by 10.35% over the past 25 years. The carbon density of the old oasis was much higher than that of the new oasis at the same time, reaching 1.63 and 1.78 times that of the new oasis in 1982 and 2005, respectively. Farmland is the dominant land use type, and 50.22% of the land area accounted for 53.26% of the SOC storage in 1982 and 58.56% of the land area accounted for 59.92% of the SOC storage in 2005. The carbon density of the farmland decreased from 35.58 t/ha in 1982 to 30.78 t/ha in 2005 (13.49%). Grassland is the second largest land use type in the old oasis; 25.10% of the land area accounted for 21.97% of the SOC storage in 1982, and 15.34% of the land area accounted for 12.95% of the SOC storage in 2005. The carbon density of grassland decreased from 29.37 t/ha in 1982 to 25.41 t/ha in 2005 (13.50%). Due to the decreases in the area and carbon density of grassland, the SOC storage in grassland decreased by 47.14%. However, the change of the carbon density value of the old oasis is not significantly different than that of new oasis between 1982 and 2005. This result indicated that the soil carbon storage ability is the same throughout the region, regardless of the age of the oasis, between 1982 and 2005. The loss of SOC in each land-use type was <8 Mg C/ha, and the amount of SOC lost varies by land-use type in the past 25 years (Table 5). The loss of SOC may reach20–50 Mg C/ha within the first 5–50 years after conversion [60]. In the topsoil, the carbon process in the oasis was mainly dominated by consumption and agricultural cultivation, which may produce unexpectedly high carbon stocks in deep soil but carbon loss in the topsoil [61]. Obviously, the dynamics of land use in the oasis can cause SOC loss and thereby reduce the carbon storage in the surface soil.

Effects of Soil Salinity on SOC Content

In recent years, the study area has experienced severe soil salinization as a result of agricultural land exploitation. Agricultural land exploitation requires irrigation, which directly increases the groundwater table, and the resultant higher evaporation leads to soil salinization [45]. The correlations between SOC content and soil salinity were analyzed for the new oasis (Fig. 5). We grouped the samples (152 samples in 1982 and 161 samples in 2005) by soil salinity to make the relationship between SOC and soil salinity more clear (Fig. 5).

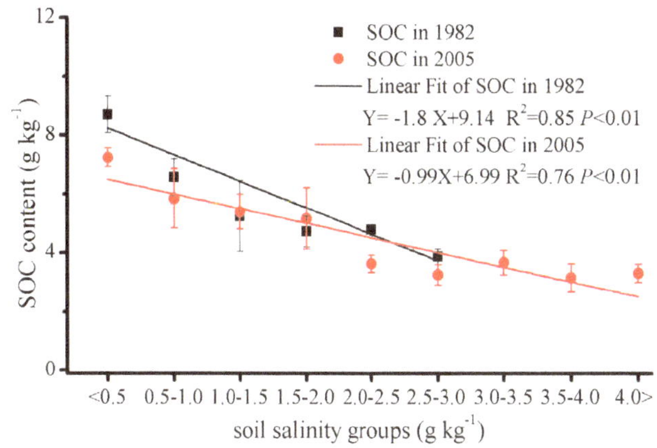

Figure 5. Relationship between SOC content and soil salinity.

The SOC content was negatively correlated with soil salinity in both 1982 and 2005 ($p<0.01$), the analysis of covariance showed that the mean difference is significant at the 0.05 level. This suggests that soil salinization can affect SOC content. Because soil salinity decreases plant productivity and hence the input of C into the soil, it also affects microbial activity and therefore SOC decomposition rates [62]. The dramatic salt accumulation was a result of agricultural land exploitation, which requires the oasis to be irrigated. Water consumption by agricultural irrigates may leach SOC into the subsoil. The SOC is easily transported by irrigation water because of its relatively low density (<1.8 Mg/m^3) [59] and is therefore concentrated near the soil surface. Thus, the SOC content was decreased in the artificial landscape of both the new and old oasis, including farmland, planted forest, etc. Therefore, much more effort is needed to decrease soil salinity and then increase SOC content by adjusting land-use structure as an effective strategy for both the new and old oases.

Conclusion

Long-term loss of soil organic matter is a major threat to sustainable agriculture in China. This study investigated the SOC dynamics under different land use histories and patterns in a typical agricultural watershed in an arid area of NW China. The farmland area increased by 8% from 1982 to 2005 in both the new oasis and the old oasis, whereas grassland, the second largest land

Table 5. SOC storage and density of different land use types in the new and old oases in 1982 and 2005.

			1	2	3	4	5	7	8	9	Total or Mean
	1982	Carbon storage (t)	151469.56	9345.29	20327.22	109317.91	8540.69	697.44	6479.70	195.63	306373.43
New		Carbon density(t/ha)	23.99	20.07	15.66	17.95	30.70	24.03	15.76	14.67	20.35
Oasis	2005	Carbon storage (t)	146861.21	22898.98	21223.14	38678.44	13390.58	0	7613.31	569.13	251234.78
		Carbon density(t/ha)	19.52	16.46	11.22	13.64	24.40		11.24	17.21	16.24
	1982	Carbon storage (t)	429035.88	78.57	39995.96	176975.59	61104.50	30986.22	31212.32	36227.11	805616.15
Old		Carbon density(t/ha)	35.58	26.72	27.34	29.37	41.71	35.64	26.54	38.12	32.63
Oasis	2005	Carbon storage (t)	432812.28	195.96	6164.93	93553.02	116109.36	21366.45	25263.57	26792.40	722257.97
		Carbon density(t/ha)	30.78	21.343	24.14	25.41	34.53	32.02	21.96	32.58	27.84

1, Farmland; 2, planted forest; 3, shrub land; 4, grassland; 5, construction land; 6, reservoir/flood land; 7, saline alkali land; 8, bald land.

use type, decreased by 11.85% in the new oasis and 9.76% in the old oasis. Farmland expansion at the expense of shrub land and grassland is a common pattern in the arid region.

The mean surface soil organic carbon (SOC) was higher in the old oasis than in the new oasis by 4.01 g/kg in 1982 and 3.79 g/kg in 2005. Additionally, the SOC decrease in the newly reclaimed oasis is more rapid than that in the old oasis. Spatially, the SOC content pattern was clearly correlated with the exploitation time in the new oasis: the shorter the agricultural land use history, the lower the SOC content. The decreasing trend in the high SOC content area is clearer than that in the low SOC content area. In the new oasis, the total SOC storage was 306373.43 t (20.56 t/ha) in 1982 and 251234.78 t (16.86 t/ha) in 2005, a decrease of 55138.65 t (18.00%). Farmland is the largest carbon pool in the new oasis, with the highest SOC content, accounting for 49.44% and 58.46% of the total SOC storage in 1982 and 2005, respectively. In the old oasis, the total SOC storage was 805616.15 t in 1982 and 722257.97 t in 2005 and decreased by 10.35% over the past25 years. The carbon density of the old oasis was much higher than that of the new oasis at the same time, reaching 1.63 and 1.78 times that of the new oasis in 1982 and 2005, respectively. The loss of SOC in the agricultural watershed of the arid region in NW China is substantial, and the change in carbon density does not differ significantly between the old oasis and new oasis but does vary among land use types.

The loss of SOC in the agricultural watershed of the arid region in NW China is high. Best management practices are likely to stabilize and reduce carbon release from soils, as well as increase further the storage of carbon in terrestrial ecosystems [20,63]. Land management practices should be improved by the implementation of such approaches as no tillage, straw returning to soil, and balanced fertilization techniques to increase the SOC content.

Acknowledgments

We would like to thank the entire staff at the Fukang Station of Desert Ecology for their technical and field help.

Author Contributions

Conceived and designed the experiments: YL XC. Performed the experiments: YW LL. Analyzed the data: XL YW GL. Contributed reagents/materials/analysis tools: YW. Wrote the paper: XL YW GL.

References

1. Lambin EF, Turner BL, Geist HJ (2001) The causes of land use and land-cover change: moving beyond the myths. Global Environ Chang 11: 261–69.
2. Houghton RA, Boone RD, Melillo JM, Palm CA, Woodwell GM, et al. (1985) Net flux of carbon dioxide from tropical forest in 1980. Nature 316: 617–620.
3. Freibauer A, Rounsevell MDA, Smith P, Verhagen J (2004) Carbon sequestration in the agricultural soils of Europe. Geoderma 122: 1–23.
4. Lal R (2004) Soil carbon sequestration impacts on global climate change and food security. Science 304: 1623–1627.
5. Muñoz-Rojas M, Rosa DD, Zavala LM, Jordán A, Anaya-Romero M (2011) Changes in land cover and vegetation carbon stocks in Andalusia, Southern Spain (1956–2007). Sci Total Environ 409: 2796–2806.
6. Singh SK, Pandey CB, Sidhu GS, Sarkar D, Sagar R (2011) Concentration and stock of carbon in the soils affected by land uses and climates in the western Himalaya, India. Catena 87: 78–89.
7. Lal R. (2008) Carbon sequestration. Philos T R Soc B 63: 815–830.
8. Watson RT, Noble IR, Bolin B, Ravindranmath NH, Verardo DJ, et al. (2000) Land Use Change, and Forestry. Cambridge: Cambridge University Press.
9. Houghton RA, Hackler JL, Lawrence KT (1999) The U.S. carbon budget: Contributions from land-use change. Science 285: 574–578.
10. Jobbágy EG, Jackson RB (2000) The vertical distribution of soil organic carbon and its relation to climate and vegetation. Ecol Appl 10: 423–436.
11. Halvorson AD, Wienhold BJ, Black AL (2002) Tillage, nitrogen, and cropping system effects on soil carbon sequestration. Soil Sci Soc Am J 66: 906–912.
12. Tiessen H, Cuevas E, Chacon P (1994) The role of soil organic matter in sustaining soil fertility. Nature 371: 783–785.
13. Yoo K, Amundson R, Heimsath AM, Dietrich WE (2006) Spatial patterns of soil organic carbon on hill slopes: integrating geomorphic processes and the biological C cycle. Geoderma 130: 47–65.
14. Brye KR, Gower ST, Norman JM, Bundy LG (2002) Carbon budgets for a prairie and agroecosystems: effects of land use and interannual variability. Ecol Appl 12: 962–979.
15. Ehleringer JR, Buchmann N, Flanagan LB (2000) Carbon isotope rations in belowground carbon cycle processes. Ecol Appl 10: 412–422.
16. Landi A, Mermut AR, Anderson DW 2004. Carbon distribution in a hummocky landscape from Saskatchewan, Canada. Soil Sci Soc Am J 68: 175–184.
17. Singh SK, Singh AK, Sharma BK, Tarafdar JC (2007) Carbon stock and organic carbon dynamics in soils of Rajasthan, India.J Arid Environ 68: 408–421.
18. Wang Y, Li Y, Ye X, Chu Y, Wang X (2010) Profile storage of organic/inorganic carbon in soil: from forest to desert. Sci Total Environ 408: 1925–1931.
19. IPCC (2000) Land use, land use change, and forestry. In: Watson RT, Noble IR, Bolin B, Ravindranath N, Verardo D, et al. (Eds.), Special Report of the Intergovernmental Panel on Climate Change. Cambridge: Cambridge University Press.
20. Lal R, Fausey NR, Eckert DJ (1995) Land use and soil management effects on emissions of radiatively active gases from two soils in Ohio. In: Lal R, Kimble J, Levine E, (Eds). Soil Management and Greenhouse Effect. Boca Raton: CRC Press.
21. Lal R, Mokma D, Lowery B (1999) Relation between soil quality and erosion. In: Lal R (Ed.), Soil Quality and Soil Erosion. Washington: CRC Press.
22. Guo LB, Gifford RM (2002) Soil carbon stocks and land use change: a meta analysis. Global Change Biol 8: 345–360.
23. Falloon P, Smith P, Bradley RI, Milne R, Tomlinson RW, et al. (2006) RothCUK–a dynamic modelling system for estimating changes in soil C from mineral soils at 1-km resolution in the UK. Soil Use Manage 22: 274–288.
24. Schulp CJE, Nabuurs G, Verburg PH (2008) Future carbon sequestration in Europe–Effects of land use change. Agr Ecosyst Environ 127: 251–264.
25. DeFries RS, Field CB, Fung I, Collatz GJ, Bounoua L (1999) Combining satellite data and biogeochemical models to estimate global effects of human-induced land cover change on carbon emissions and primary productivity. Global Biogeochem Cy 13: 803–815.
26. Marland G, Boden TA, Ndres RJ (2000) Global, Regional, and National CO2 Emissions, in Trends: A Compendium of Data on Global Change. Oak Ridge: U.S. Department of Energy.
27. Poeplau C, Don A (2013) Sensitivity of soil organic carbon stocks and fractions to different land-use changes across Europe. Geoderma 192: 189–201.
28. Houghton JT, Ding Y, Griggs DJ, Noguer M, Van der Linden, et al. (2001) Climate change 2001: the scientific basis. Cambridge: Cambridge University Press.
29. IPCC (2007) Climate change 2007: the physical science basis. Cambridge: Cambridge University Press.
30. Cruickshank MM, Tomlinson RW, Trew S (2000) Application of CORINE land-cover mapping to estimate carbon stored in the vegetation of Ireland. J Environ Manage 58: 269–287.
31. Robinson DT, Brown DG, Currie WS (2009) Modelling carbon storage in highly fragmented and human-dominated landscapes: Linking land-cover patterns and ecosystem models. Ecol Model 220: 1325–1338.
32. Davidson EA, Ackerman IL, (1993) Changes in soil carbon inventories following cultivation of previously untilled soils. Biogeochemistry 20, 161–193.
33. Nosetto MD, Jobbágy EG, Paruelo JM (2006) Carbon sequestration in semi-arid rangelands: comparison of Pinus ponderosa plantations and grazing exclusion in NW Patagonia,J Arid Environ 67: 142–156.
34. Li YQ, Zhou XH, Brandle JR, Zhang TH, Chen YP, et al. (2012) Temporal progress in improving carbon and nitrogen storage by grazing exclosure practice in a degraded land area of China's Horqin Sandy Grassland. Agr Ecosyst Environ 159: 55–61.
35. San Jose JJ, Montes RA (2001). Management effects on carbon stocks and fluxes across the Orinoco savannas. Forest Ecol Manag 150: 293–311.
36. Woomer PL, Toure A, Sall M (2004) Carbon stocks in Senegal's Sahel transition zone.J Arid Environ 59: 499–510.
37. Lal R. (2001) Potential of desertification control to sequester carbon and mitigation the greenhouse effect. Climatic Change 51: 35–72.
38. Su YZ, Wang XF, Yang R, Lee J (2010) Effects of sandy desertified land rehabilitation on soil carbon sequestration and aggregation in an arid region in China. J Environ Manage 91: 2109–2116.
39. Ci LJ, Yang XH (2010) Desertification and Its Control in China. New York: Springer.
40. Feng Q, Cheng GD, Kunihiko E (2001) Carbon storage in desertified lands: a case study from North China. GeoJournal 51: 181–189.
41. Luo GP, Zhou CH, Chen X, Li Y (2008) A methodology of characterizing status and trend of land changes in oases: A case study of Sangong River watershed, Xinjiang, China. J Environ Manage 88: 775–783.
42. Ling HB, Xu HL, Fu JY, Zhang QQ, Xu XW (2012) Analysis of temporal-spatial variation characteristics of extreme air temperature in Xinjiang, China. Quatern Int 282: 14–26.

43. Wang Y, Li Y (2013). Land exploitation resulting in soil salinization in a desert-oasis ecotone. Catena 100: 50–56.
44. Cerri CEP, Cerri CC, Paustian K, Bernoux M, Mellilo JM (2004) Combining soil C and N spatial variability and modeling approaches for measuring and monitoring soil carbon sequestration. Environ Manage 33: 274–88.
45. Wang Y, Xiao D, Li Y, Li X (2008) Soil salinity evolution and its relationship with dynamics of groundwater in the oasis of inland river basins: case study from the Fubei region of Xinjiang Province, China. Environ Monit Assess 140: 291–302.
46. Land Resource Administration of Fukang (1983) Land use and soil inventory report in 1982.
47. Post WM, Emanuel WR, Zinke PJ, Stangenberger AG (1982) Soil carbon pools and world life zones. Nature 298: 156–159.
48. Xie ZB, Zhu JG, Liu G, Cadisch G, Hasegawa T, et al. (2007) Soil organic carbon stocks in China and changes from 1980s to 2000s. Global Change Biol 13: 1989–2007.
49. Yang YH, Mohammat A, Feng JM, Zhou R, Fang JY (2007) Storage, patterns and environmental controls of soil organic carbon in China. Biogeochemistry 84: 131–141.
50. Lillesand TM, Kiefer RW (2000) Remote sensing and image interpretation, 4th edn. New York: John Wiley & Sons Inc.
51. Viglizzo EF, Paruelo JM, Laterra P, Jobbágy EG (2012) Ecosystem service evaluation to support land-use policy. Agr Ecosyst Environ 154: 78–84.
52. Fu B, Hu C, Chen L, Hommay O, Gulinck H (2006) Evaluating change in agricultural landscape pattern between 1980 and 2000 in the Loess hilly region of Ansai County, China. Agr Ecosyst Environ 114: 387–396.
53. Li XY, Xiao DN, He XY, Chen W (2007) Factors associated with farmland area changes in arid regions: a case study of the Shiyang River basin, northwestern China. Front Ecol Environ 5: 139–144.
54. Voroney RP, van Veen JA, Paul EA (1981) Organic C dynamics in grassland soils: 2. Model validation and simulation of the long-term effects of cultivation and rainfall erosion. Can J Soil Sci 61: 211–224.
55. Gregorich EG, Anderson DW (1985) The effects of cultivation and erosion on soils of four toposequences in the Canadian prairies. Geoderma 36: 343–354.
56. Gregorich EG, Greer KJ, Anderson DW, Liang BC (1998) Carbon distribution and losses: erosion and deposition effects. Soil Till Res 47: 291–302.
57. Lal R (2000) World cropland soils as a source or sink for atmospheric C. Adv Agron 71: 145–91.
58. Cambardella CA, Moorman TB, Novak JM, Parkin TB, Karlen DL, et al. (1994) Field-scale variability of soil properties in central Iowa soils. Soil Sci Soc Am J 58: 1501–1511.
59. Lal R (2003) Soil erosion and global carbon budget. Environment International 29: 437–450.
60. Davidson EA, Ackerman IL (1993) Changes in soil carbon inventories following cultivation of previously untilled soils. Biogeochemistry 20: 161–193.
61. Li C, Tang L, Li Y (2010) Soil organic carbon stock and carbon efflux in deep soils of desert and oasis. Environ Earth Sci 60: 549–557.
62. Setia R, Gottschalk P, Smith P, Marschne P, Baldock J, Setia D, Smith J (2012) Soil salinity decreases global soil organic carbon stocks. Sci Total Environ, http://dx.doi.org/10.1016/j.scitotenv.2012.08.028.
63. Li ZP, Han FX, Su Y, Zhang TL, Sun B, et al. (2007) Assessment of soil organic and carbonate carbon storage in China. Geoderma 138: 119–126.

21

Soil Carbon and Nitrogen Changes following Afforestation of Marginal Cropland across a Precipitation Gradient in Loess Plateau of China

Ruiying Chang[1,2]ᶦ, Tiantian Jin[2,3]ᶦ, Yihe Lü[2], Guohua Liu[2], Bojie Fu[2]*

1 Key Laboratory of Mountain Surface Processes and Ecological Regulation, Institute of Mountain Hazards and Environment, Chinese Academy of Sciences, Chengdu, Sichuan, China, **2** State Key Laboratory of Urban and Regional Ecology, Research Center for Eco-Environmental Sciences, Chinese Academy of Sciences, Beijing, China, **3** China Institute of Water Resources and Hydropower Research, Beijing, China

Abstract

Cropland afforestation has been widely found to increase soil organic carbon (SOC) and soil total nitrogen (STN); however, the magnitudes of SOC and STN accumulation and regulating factors are less studied in dry, marginal lands, and therein the interaction between soil carbon and nitrogen is not well understood. We examined the changes in SOC and STN in younger (5–9-year-old) and older (25–30-year-old) black locust (*Robinia pseudoacacia* L., an N-fixing species) plantations that were established on former cropland along a precipitation gradient (380 to 650 mm) in the semi-arid Loess Plateau of China. The SOC and STN stocks of cropland and plantations increased linearly with precipitation increase, respectively, accompanying an increase in the plantation net primary productivity and the soil clay content along the increasing precipitation gradient. The SOC stock of cropland decreased in younger plantations and increased in older plantations after afforestation, and the amount of the initial loss of SOC during the younger plantations' establishment increased with precipitation increasing. By contrast, the STN stock of cropland showed no decrease in the initial afforestation while tending to increase with plantation age, and the changes in STN were not related to precipitation. The changes in STN and SOC showed correlated and were precipitation-dependent following afforestation, displaying a higher relative gain of SOC to STN as precipitation decreased. Our results suggest that the afforestation of marginal cropland in Loess Plateau can have a significant effect on the accumulation of SOC and STN, and that precipitation has a significant effect on SOC accumulation but little effect on STN retention. The limitation effect of soil nitrogen on soil carbon accumulation is more limited in the drier area rather than in the wetter sites.

Editor: Xiujun Wang, University of Maryland, United States of America

Funding: This research was supported the National Natural Science Foundation of China (41230745), the Chinese Academy of Sciences (No. XDA05060000) and the International Partnership Program for Creative Research Teams of "Ecosystem Processes and Services." The funders had no role in study design, data collection and analysis, decision to publish, or preparation of the manuscript.

Competing Interests: The authors have declared that no competing interests exist.

* E-mail: bfu@rcees.ac.cn

ᶦ These authors contributed equally to this work.

Introduction

Clearing trees to create cropland can reduce soil organic carbon (SOC) significantly by up to 50% of the SOC in the initial forest [1,2,3,4]. In contrast, the afforestation of previously arable land is generally found to sequester carbon (C) in the soil and play an important role in climate change mitigation [5,6,7], which was proposed as an effective method of C sequestration in Article 3.3 of the Kyoto Protocol in 1997. Subsequently, large-scale afforestation has been practiced worldwide in recent decades, especially in marginal land [8]. Thus, it is meaningful and necessary to study the process and mechanism of soil C changes and their causes during afforestation, especially when considering large-area plantation activities conducted in marginal land to mitigate CO_2 emissions.

Soil C sequestration following afforestation has been the subject of a substantial body of research, which suggests that the direction and magnitude of SOC changes are determined by many factors and processes, such as climate, stand age and soil depth [6,7,9,10,11]. However, the effects of these factors on SOC sequestration are still controversial and unclear. Some reviews suggest that climate factors have a weak effect on soil C accumulation on a global scale [6,11], whereas other meta-analyses indicate that climate plays an important role in influencing SOC during afforestation [9,12]. Along a precipitation gradient (650 to 1450 mm), Berthrong *et al.* [10] found that SOC sequestration was significantly related to precipitation. In arid areas, precipitation is the most important ecological limiting factor and determines the net primary productivity (NPP); it may therefore have a significant effect on SOC, but such evidence is limited. The changes in soil C with stand age can vary, particularly in the initial stage of afforestation (<10 years). Soil C can decrease [7,9,10,11], increase [10,11], show no trend [13,14], or produce a Covington curve [11] with stand age. It has been suggested that the different initial trend may depend on mean annual precipitation (MAP) [10], with soil C decreasing in wetter regions (MAP>1150 mm) but increasing in drier regions (MAP<1150 mm).

In contrast to SOC, the changes in soil nitrogen (N) following afforestation are poorly documented, even though soil N is strongly correlated with SOC, and SOC sequestration is controlled by the accumulation of soil N as a result of a stoichiometric relationship [15,16]. In their meta-analysis, Li *et al.* [7] and Yang *et al.* [11] found a linear relationship between soil C and N following afforestation, and a 1 g N accretion was associated with a 7 to 13 g C accumulation. However, the interactions between soil C and N are less studied and less understood at the regional scale [17].

Plantations of N-fixing trees accumulate more soil C than non N-fixing forest [18,19,20]. An increase of 12 to 15 g C in the soils was found to be associated with an increase of a 1 g N in 19 case studies on N-fixing tree plantations [18]. The mechanisms behind the greater accumulation of soil C in N-fixing tree plantations may be the reduced decomposition of older soil C, increased formation of soil C, and a higher C input from vegetation, which is associated with the supply of N derived from fixation [18,19,20]. Yet the processes affecting soil C and N change once an N-fixing forest is established, and their relationship with stand age, remain unclear, especially in water-limited marginal lands. The N supply from decomposition is severely limited on dry land because of water limitations. As a result, increases in the external N supply from N-fixing tree plantations in arid areas may result in greater soil C than similar increases in moist regions.

Globally, most of the modern afforestation activities have occurred in China [8]. From 2000 to 2010, the area of afforestation and reforestation was approximately 4.9×10^6 ha globally, and over 60% of the new forest area was in China [8]. Large areas of new forest have expanded during the modern period in China due to a series of ecological restoration programs, such as the Natural Forest Conservation Program and the Grain for Green project (GFG). The GFG is one of the most ambitious programs, initiated in 1999 in some areas and expanded in 2000 to the whole country. It aims to convert marginal, low-yield cropland slopes into woodland and grassland. The Loess Plateau, located in northwestern China, has a semi-arid continental climate (MAP ranges from approximately 650 mm in the southern Loess Plateau to approximately 250 mm in the north) and is the most important target for the GFGP due to its tremendous soil erosion. Between 2000 and 2008, the total area of the marginal cropland involved in the project was approximately 4.83×10^6 ha, of which 3.84×10^5 ha was converted into forest [21]. An N-fixing tree species, black locust (*Robinia pseudoacacia* L.), was one of the most commonly planted trees. Such large-scale plantation activity provides suitable conditions for studying the effects of afforestation on SOC and soil total nitrogen (STN) in marginal lands at a regional scale. The objectives of this study are to (1) detect the effects of MAP on soil C and N changes in younger (<10 years) and older (approximately 30 years) black locust stands in marginal cropland and (2) analyze the interaction between SOC accumulation and soil N accretion at the regional scale.

Materials and Methods

Site Description

All our field activities were conducted with the permission of the farmers who owned the land we tested, and we confirmed that the field studies did not involve any endangered or protected species.

Six sites (S1 to S6) were selected across the Loess Plateau (Fig. S1). The MAP increases from 380 mm in the northern site (S1) to 650 mm in the southern site (S6), with most precipitation occurring between June and August at all sites. The mean annual temperature is similar among the sites (Table 1). In all sites, the landform is typical loess and hilly, and the soil is a calcareous loamy soil (classified as an Entisol in USA soil taxonomy) and well drained [22]. The clay content of the soils increased significantly and linearly from 21.3% in the driest site (S1) to 38.2% in the wettest site (S6), whereas the soil sand content decreased significantly and linearly from 27.2% to 8.5% along this gradient (Fig. S2).

In July and August of 2008, 2009, and 2010, a total of 17 younger (5 to 9 years at sample time) and 16 older (25 to 30 years at sample time) black locust plantation stands were chosen in the six sites, and in each site nearby, steep cropland stands also were chosen for comparison (Table 1). The aspect and angle of the cropland sites were similar to those of the corresponding black locust plantation stands. Most of the sloping croplands had been converted to other land use, especially since 1999 to 2000, so only a small amount of cropland existed, and only one or two cropland sites were selected in each location. Therefore, a few additional cropland data were cited and used from previous studies conducted at the same site (Table 1), which were believed to be reliable because the management history of cropland was similar and the soil C or N content of sloping cropland were found to be quite consistent in the same site in the Loess Plateau (e.g., Sun *et al.*, [23], Table 1).

We assumed there were no differences in the soil physicochemical properties between the black locust plantations and cropland sites before afforestation, and the changes in soil C and N afterward were primarily due to the plantation establishment. There are three lines of evidence to support these assumptions. First, all plantations were converted from long-term (at least 30 years) cultivated cropland, and soil C and N are generally thought to reach a constant value after long-term cultivation [3]. Second, the forest and cropland sites had similar histories and management before afforestation. Third, in each site, the soil characteristics, such as soil type (loess soil) and texture, were largely uniform [22,24]. Furthermore, our assumption that there were no differences in soil C and N between pre-plantation and cropland was able to be strongly supported by the evidence that the soil C and N content of sloping cropland were consistent in the same site as shown above.

In each forest stand, black locust trees were planted in a regular pattern, and the same pre-planting site preparation, including disc trenching, was used across all the sites. After plantation, but prior to 2000, the older stands were occasionally thinned or used for livestock-grazing activities. Since 2000, there has been little human activity in the older plantations, and there has never been much in the younger stands. None of the stands has ever been fertilized. The annual fertilization application to cropland (farm manure or urea) was approximately 300 to 500 kg ha^{-1} across all sites. There has never been irrigation in the plantations or cropland.

Field Measurements, Soil Sampling and Analysis

Generally, the area of the plantation stands was less than 0.5 ha, and only one plot (10 m×10 m or 20 m×10 m plot, depending on the area of the stand) was established in the center of each. The diameter at breast height (DBH), tree height, and tree density of the black locust trees in each plot were recorded (Table 1).

In each plot, locust trees were separated into three or four (dependent on the DBH values distribution) classes based on DBH, and one sample tree with the mean DBH value was selected per class. A 1 m×1 m subplot was established using the sample tree as a vertex, with three or four sample tree subplots in each plot. Sometimes, in addition to the sample tree subplots, another 1 m×1 m subplot (sometimes two) was randomly established in some plots. In total, three to six subplots were selected and were

Table 1. Basic Descriptive Statistics of Six Plantation Sites Along a Precipitation Gradient.

Sites	Longitude/Latitude	Temperature(°C)	Precipitation(mm)	[a]Category	Plantation age (years)	Mean DBH (cm)	Mean tree height (m)	Tree density (trees·ha^{-1})	[b]Fine roots	[c]Description
S1	110.22°/37.47°	9.5	380	Cropland (4)	0					
				Cropland (3)	0					
				Y ([illegible])	7	6.5	5.2	900		
				A (3)	30	15.0	8.0	350		
				A (4)	30	14.0	11.0	700		
S2	108.11°/37.11°	8.5	420	Cropland (3)	0				Y (6)	
				Y ([illegible])	8	5.0	3.5	1000		
				Y ([illegible])	8	7.8	7.0	625	Y (7)	
				A (3)	30	15.0	7.0	800		
				A (5)	25	10.6	6.0	625	Y (9)	
S3	109.30°/36.92°	9.0	460	Cropland	0					[62]
				Cropland	0					[63]
				Cropland (4)	0				Y (8)	
				Y (5)	5	2.4	3.0	1200	Y (16)	
				Y (4)	8	6.1	6.5	1950	Y (16)	
				Y (3)	9	4.2	2.5	1100		
				A (3)	30	13.5	8.0	400		
				A (3)	30	11.0	8.0	1800		
				A (5)	30	16.1	10.0	500	Y (14)	
S4	109.53°/36.52°	8.7	530	Cropland	0					[64]
				Cropland	0					[23]
				Cropland	0					[65]
				Y (3)	5	6.3	7.0	2200		
				Y (3)	9	5.0	6.2	2700		
				Y (3)	9	5.1	5.8	2300		
				A (3)	25	9.0	6.8	1800		
				A (3)	30	10.5	6.5	2000		
				A (3)	30	18.0	7.0	1500		
S5	109.18°/36.07°	9.2	580	Cropland	0					[66]
				Cropland (5)	0				Y (10)	
				Y (3)	5	2.0	2.5	1400	Y (8)	
				Y (3)	6	5.0	7.0	5200		
				Y (3)	8	–	–	–		Tree census not detected
				Y (4)	8	5.6	7.0	2700	Y (16)	
				A (3)	26	12.0	12.0	1000		

Table 1. Cont.

Sites	Longitude/Latitude	Temperature(°C)	Precipitation(mm)	[a]Category	Plantation age (years)	Mean DBH (cm)	Mean tree height (m)	Tree density (trees·ha^{-1})	[b]Fine roots	[c]Description
				A (4)	28	13.2	15.0	1300		
				A (5)	30	7.1	7.5	1700	Y (17)	
S6	109.12°/35.33°	8.5	650	Cropland (4)	0					
				Cropland (5)	0				Y (10)	
				Y (3)	5	5.3	6.0	3000		
				Y (3)	6	5.2	6.5	3800		
				Y (6)	8	5.6	8.0	3200	Y (24)	
				Y (3)	9	5.8	8.0	4000		
				A (3)	25	9.2	12.0	2200		
				A (3)	30	14.3	13.0	1000		
				A (5)	30	13.8	13.0	1550	Y (20)	

[a]Y and A indicate younger and older plantations, respectively. The number in the parentheses is the soil sample size in each stand.
[b]Y (Yes) indicates that the fine root biomass was collected in this stand. The number in the parentheses is the sample size.
[c]The data on soil organic carbon, total nitrogen content, and soil bulk density of some croplands were cited from other studies in the same site, and the data in the other cropland and forest stands were collected and analyzed in this paper. There was no tree census of the S5 forest stand, but the soil properties were measured.

evenly distributed in each plot, and the sample size of each stand is shown in Table 1. In each subplot, an auger (3.3 cm diameter, 85 cm^3) was used to collect soil samples for SOC and STN concentrations, and another soil auger (5.0 cm diameter, 100 cm^3) was used to collect the soil samples for bulk density (BD). In all, there were three to six soil samples for SOC, STN, and BD taken from each plot, and all the samples were collected in increments of 0–10 cm and 10–20 cm. In some of the plots, soil cores (3.3 cm diameter, 85 cm^3) were also collected to investigate the fine root (≤2 mm in diameter) biomass in each subplot, for a total of 6 to 24 soil cores in each plot (Table 1). Fine root samples were also partitioned into 0–10 cm and 10–20 cm increments. A detailed description of the fine root collection protocol is provided in Chang *et al.* [25].

In each cropland site, a 30 to 50 m transect was established at the center of the stand. Three to five 1 m×1 m subplots were established at approximately 10 m intervals along each transect. In each subplot, one soil core (3.3 cm diameter, 85 cm^3) was collected to analyze SOC and STN concentrations. Another soil core was collected using a 100 cm^3 cylindrical auger (5.0 cm diameter) for soil BD. In some of the cropland stands, two soil samples were collected in each subplot to assess the fine root biomass (Table 1). The soil samples for both physicochemical properties and fine root biomass were partitioned into 0–10 cm and 10–20 cm increments.

The SOC was measured using potassium-dichromate oxidation (titration of dichromate with ammonium ferrous sulfate), and the C content of the soil organic matter was assumed to be 0.58 [26]. The STN concentration was determined by the dry combustion method (1150°C) using a CN analyzer (Elementar Vario EL)(principle with thermal-conductivity detection for N_2). Fine root samples were washed to collect the fine roots. Roots that were less than 2 mm in diameter were dried at 65°C to a constant mass in order to estimate the biomass. The soil samples for soil BD were dried at 105°C to a constant mass and weighed to determine the soil BD.

Data and Statistical Analysis

Plantation aboveground biomass and fine root biomass. The aboveground biomass at the stand level in the plantations was calculated by multiplying the mean stem biomass by the tree density. The mean stem biomass was determined from the mean DBH based on the allometric relationship (between stem biomass and DBH) established in a previous study on black locust plantations [27]. The fine root biomass at stand level was calculated using the sample value.

SOC and STN stock estimation. The equivalent mass method (SOC stock calculated based on a reference soil mass) was not used in this study because there was a significant difference in soil BD in only a few plantation stands (four stands, 12.9% of total stands) and the corresponding cropland across the six sites. Thus the SOC or STN stock of each stand was calculated at a fixed depth as follows:

$$SCT_i = SC_i \times BD_i \times H_i \times 10 \times (1 - \eta_i) \quad (1)$$

where SCT_i represents either the SOC or STN stock in layer i (in the top 10 cm layer or the 10–20 cm layer, Mg C·ha^{-1}); SCi represents the SOC or STN concentration (g·kg^{-1}) in layer i; BDi is the soil bulk density (g·cm^{-3}) in layer i; and Hi is the thickness (cm) of layer i. The term ηi represents the volumetric percentage of particles with a size fraction >2 mm (rock fragments, %) in layer i, and this term was zero in each soil layer. The parameter 10 is a conversion factor.

SOC and STN stock changes. For each site, the mean SOC and STN of the cropland were calculated, and the absolute changes in SOC and STN stock (Mg C·ha^{-1}) after planting were estimated as the difference between the SOC or STN stock in the tree plantation soil and the mean value of cropland soil, as in the following:

$$Absolute\ change = SCT_{forest} - SCT_{cropland} \quad (2)$$

where SCT_{forest} is the SOC or STN stock in the plantation (Mg C·ha^{-1}), and $SCT_{cropland}$ is the mean SOC or STN stock of the corresponding cropland for each site (Mg C·ha^{-1}).

The relative SOC or STN stock change (%) after the establishment of tree plantations was calculated as follows:

$$Relative\ change = (SCT_{forest} - SCT_{cropland})/SCT_{cropland} \times 100 \quad (3)$$

where the mean SCT_{forest} and $SCT_{cropland}$ are shown above.

The C:N ratio change was used to represent the relative gain of SOC to STN following afforestation, and it was calculated as follows:

$$C:N\ ratio\ change = C:N\ ratio\ of\ plantations\ /C:N\ ratio\ of\ cropland \quad (4)$$

where C:N ratio means SOC:STN.

Statistical Analysis

A simple linear regression was used to test whether the tree density, tree height, and the plantation aboveground biomass increased with MAP and MAP had a significant effect on the SOC and STN stock and the C:N ratio (SOC to STN ratio) of the cropland and plantation stands and whether the MAP were significant predictors of the effects of afforestation on the SOC stock, STN stock and C:N ratio. In addition, the relationships between the SOC and STN stock changes during afforestation were tested using linear regression. The difference in the slope of the linear regression for different soil layers was tested (Z-test) using the following formula proposed in Paternoster *et al.* [28]:

$$Z = (b_1 - b_2)/\sqrt{SEb_1^2 + SEB_2^2} \quad (5)$$

where b_1 and b_2 are the regression coefficient, and SEb_1 and SEb_2 are the coefficient variances of the two regressions, respectively.

A one-way ANOVA was used (Scheffé's test was used if the data had equal variances and, if not, Tamhane's T2 test was used) to compare the fine root biomass among the cropland and the younger and older plantations within each site, as well as among the sites for younger and older plantations. All analyses were performed in SPSS 11.0, based on a significance level of 0.05.

Results

Tree Census along the Precipitation Gradient

Tree density increased significantly linearly with MAP increasing ($R^2 = 0.42$, $p<0.001$), and tree height tended to increase along the gradient but with a low R^2 ($R^2 = 0.12$, $p = 0.054$). As the trend in tree density and height increased, the black locust plantation aboveground biomass also increased along the MAP gradient ($R^2 = 0.18$, $p = 0.015$). In contrast, the fine root biomass decreased from the drier sites to wetter sites for younger and older plantations (Fig. 1).

With stand age, the fine root biomass of the younger plantations increased significantly in all four sites (Fig. 1), although the average age difference between younger and older plantations was approximately 22 years. In comparison to cropland, the fine root biomass of the younger plantations was significantly greater in the two drier sites (S2, MAP = 420 mm and S3, MAP = 460 mm), but not in the two wetter sites (S5, MAP = 580 mm and S6, MAP = 650 mm), and the fine root biomass of the older plantations was pronounced greater in all sites (Fig. 1).

SOC, STN Stocks and C:N Ratio along the Precipitation Gradient

With an increase in MAP, the SOC stock of the top 20 cm layer of the cropland increased linearly from 5.2 Mg C·ha^{-1} in the driest site (S1) to 19.6 Mg C·ha^{-1} in the wettest site (S6; about a fourfold increase; $R^2 = 0.92$, $p<0.001$), and the STN stock increased linearly from 0.8 Mg N·ha^{-1} at S1 to 2.2 Mg N·ha^{-1} at S6 (about a threefold increase; $R^2 = 0.83$, $p<0.001$, Fig. 2). The greater sensitivity of SOC to MAP (compared with STN) led to a positive relationship between the C:N ratio of cropland and MAP ($R^2 = 0.50$, $p = 0.01$, Fig. 2). For all plantations, the SOC and STN stocks in the top 20 cm layers also increased linearly with MAP ($R^2 = 0.75$, $p<0.001$ for SOC of younger plantation; $R^2 = 0.80$, $p<0.001$ for STN of younger plantation; $R^2 = 0.44$, $p = 0.005$ for SOC of older plantation; and $R^2 = 0.48$, $p = 0.004$ for STN of older plantation; Fig. 2). However, the regression slope was similar between SOC and STN for both younger and older plantations, resulting in a poor relationship between the C:N ratio of the plantations and MAP (Fig. 2).

SOC and STN Stock Changes Following Afforestation along the Precipitation Gradient

The changes in SOC stock following the afforestation of cropland were dependent on soil depth and plantation age (Table 2, Fig. 3). For all the sites, the average absolute and relative SOC remained unchanged in the uppermost 10 cm and for the total 0–20 cm layer following afforestation, but decreased significantly (by 1.45 Mg C·ha^{-1} or 20.9%) in the lower 10–20 cm of soil ($p<0.05$, Table 2). In addition, the size of the decrease in absolute SOC in the 10–20 cm soil layer had a positive relationship with MAP ($R^2 = 0.39$, $p<0.01$, Fig. 3), indicating that SOC stock decreased more in the wetter sites relative to the drier sites in the first plantation period.

With an increase in plantation age, absolute and relative SOC increased significantly, by 6.70 Mg C ha^{-1} or 161.1% in the top 10 cm layer, and by 7.57 Mg C ha^{-1} or 93.8% in the total 0–20 cm layer ($p<0.05$ for both, Table 2). However, the absolute SOC showed little increase in the lower 10–20 cm soil layer ($p>0.05$, Table 2), suggesting that the initial decrease of SOC stock in this layer recovered to the level of cropland approximately 30 years after planting. For older plantations, the SOC stock in both the uppermost 10 cm layer and lower 10–20 cm increased more quickly in the drier sites than in the wetter sites, as shown by the negative linear relationship between MAP and the relative SOC changes ($R^2 = 0.57$, $p<0.01$ for the top 10 cm, and $R^2 = 0.35$, $p = 0.02$ for the 10–20 cm layer, Fig. 3). Moreover, the slope of the linear regression was greater for the top 10 cm than for the 10–20 cm layer ($p<0.01$, Fig. 3), implying that the relative SOC increased more rapidly in surface layers (0–10 cm) than in the subsurface layer (10–20 cm) as precipitation decreased.

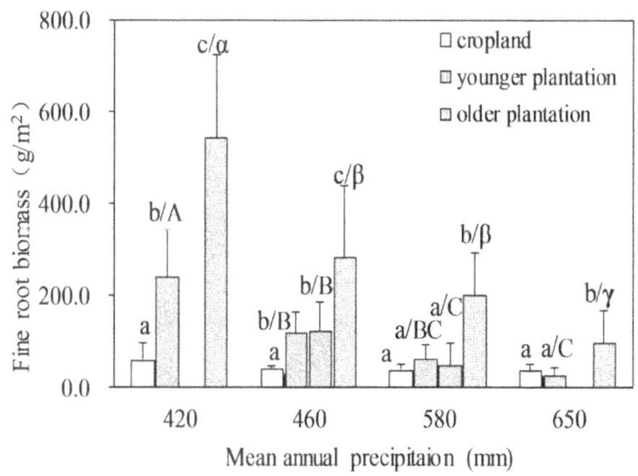

Figure 1. Comparison of fine root biomass among cropland and younger and older plantations in each site as well as among the younger and older plantations in four sites. Younger plantations are 5- to 9-year- old and older plantations are 25- to 30-year-old forests, respectively. The error bars represent the standard deviations of the means. The sample numbers are shown in Table 1. A different lowercase English letter is used for cropland and for younger and older plantations in the same site, and a different uppercase English letter represents younger plantations across the sites, whereas a different Greek letter is used for older plantations to indicate a significant difference at the 0.05 level.

In contrast to SOC, neither absolute nor relative STN were related to MAP at any depth (Fig. 4). For all values of MAP, the absolute STN increased in the top 10 layer and remained constant in the lower 10–20 cm following afforestation, whereas there was no trend or a decrease in the absolute SOC in the upper and lower soil layers, respectively (Table 2). In older plantations, the absolute and relative STN had increased significantly in each soil layer, compared with the lack of change in absolute SOC in the lower 10–20 cm in younger plantations (Table 2). These results suggested that STN increased more quickly than SOC following cropland afforestation.

Relationship between SOC and STN Changes Following Afforestation

Absolute and relative STN were found to increase significantly and linearly with increasing SOC (Figs. 5A, 5B), indicating that there is a close relationship between SOC and STN during afforestation at the regional scale. According to the relationship between absolute SOC and STN, an average gain of 1 g N was associated with a 9.1 g gain in C during afforestation (Fig. 5A). The slope (1.27) of the relationship between the relative SOC and STN was significantly greater than one (Fig. 5B), which implies that the relative STN exceeded SOC across the sites, confirming the relationship in average gains. However, the changes in SOC and STN during afforestation were dependent on precipitation (Figs. 5C, 5D). The negative linear relationship between the C:N ratio and MAP suggests that the relative gain of SOC to STN following afforestation was higher in drier lands and lower in wetter areas.

Discussion

SOC and STN Stocks along the Precipitation in the Loess Plateau

The SOC and STN stocks of younger and older plantations increased linearly as MAP increased from the north to the south of the Loess Plateau. These findings were consistent with the measured trend of SOC content in *Pinus tabulaeformis* plantations along the increasing precipitation gradient of the Loess Plateau [29] and the SOC stock's tendency in the Nordic forest soils [30]. The positive relationship between precipitation and SOC and STN stocks could be related to an increase in NPP with precipitation [30,31]. Also, the accompanying increase in tree density with precipitation is generally found to lead to a higher aboveground NPP and total belowground carbon allocation in forest [32]; however, the tree density has a nonsignificant effect on SOC stocks [32] because of the synchronous increase in soil CO_2 efflux with tree density [33]. In addition, the greater SOC and STN stocks in the wetter Loess Plateau is associated with the higher clay content, which, it is suggested, plays a positive role in accumulating SOC and STN resulting from the chemical protection of SOC by bonds with clay surfaces and from physical protection through occlusion with aggregates [34,35].

SOC Changes Following Afforestation

Soil C is generally found to decrease in the first few years (>10 years) following afforestation, after which it increases with plantation age to pre-afforestation levels and then to net gains, which is indicated as a Covington curve [7,9,11]. Our findings showed a similar temporal pattern in absolute SOC, with an initial decrease in 8-year-old plantations followed by a return to the initial cropland SOC level in 30-year-old plantations in the 10–20 cm soil layer. The balance between soil C input and output determined the soil C pool. Accordingly, the initial decrease in SOC following afforestation indicates that the C input was too low to match the decomposition of SOC in younger plantations. A study using a natural ^{13}C abundance approach indicated that when cane (a C_4 photosynthetic pathway) is planted with *Eucalyptus* (a C_3 photosynthetic pathway), the new SOC_3 (derived from the *Eucalyptus* planting) in the 10–20 cm layer increased by 1.8 Mg C/ha, whereas the older SOC_4 (derived from cane) was lost at a rate of 3.2 Mg C/ha, resulting in a net C loss of 1.4 Mg C/ha in 10 to 13 years [14].

The initial accelerated SOC decomposition may be due in part to the intensive pre-planting site preparation, which included disc trenching in our study sites. In two reviews, Johnson [36] suggested that site preparation results in a net loss of soil C, although there is an accompanying increase in productivity, and Laganière *et al.* [6] suggested that low-intensity preparation (e.g., hand planting) may lead to a greater accumulation of SOC, and that high-intensity preparation (e.g., disc trenching) stops affecting the SOC level only 21 years after afforestation of cropland. Besides, the decomposition of older SOC is frequently found to be stimulated by the input of new liable C to soil in a short time scale, which is called a priming effect (PE) [37,38,39]. PE is suggested to be a global phenomenal, including in terrestrial and aquatic ecosystems [40], and the rhizosphere priming effect on SOM decomposition is from −70% to more than 380% [37,39]. Thus the priming effect associated with the new C input, especially derived from fine roots during forest establishment, may in part induce the initial loss of old C from the arable soil.

The lose in absolute SOC in the 10–20 cm layer in younger stands increased with rising MAP, even though the NPP of vegetation is generally found to increase along the MAP gradient across the Loess Plateau [41]. This result provided evidence supporting the idea that the initial SOC loss during afforestation is more likely to occur in wetter soils other than in drier soils [10]. This condition may be ascribed to four causes. First, as predicted by the functional balance theory [42,43], the fine root biomass of plantation stands was found to be greater in the drier sites where

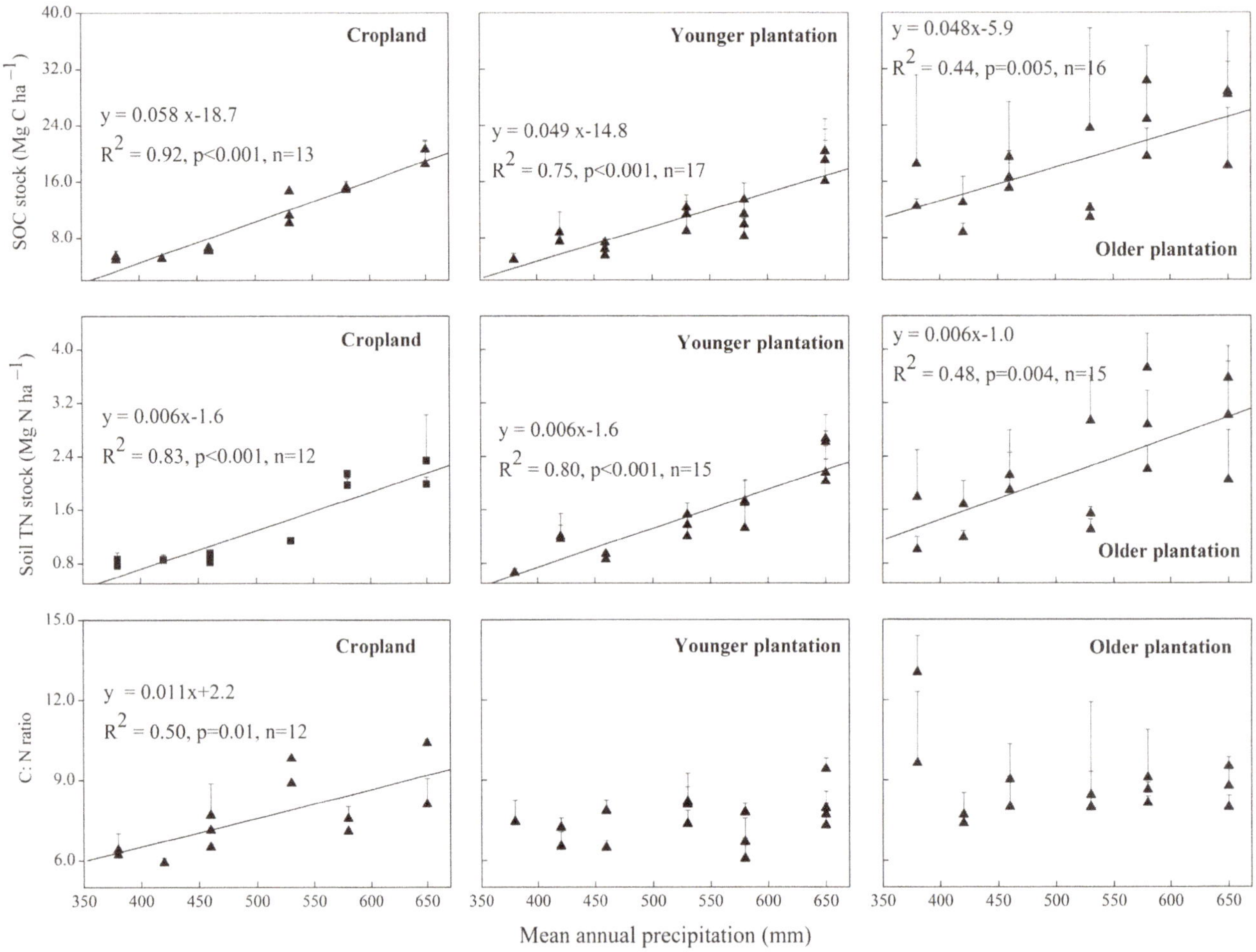

Figure 2. Relationship between mean annual precipitation and soil organic carbon (SOC), soil total nitrogen (STN) stock, and C:N ratio in cropland and younger and older plantations. Younger and older plantations are 5- to 9-year-old and 25- to 30-year-old forests, respectively. The error bars represent the standard deviations of the means (only shown in positive, n = three to six for each stand). The data for some cropland stands were cited from other studies (Table 1). Regressions were conducted for the top 20 cm of soil, and only significant regression models are displayed.

Table 2. Changes in Soil Organic Carbon (SOC) and Soil Total Nitrogen (STN) Stocks in Younger (n = 17 and 15 for SOC and STN, Respectively) and Older Plantations (n = 16 and 15 for SOC and STN, Respectively) Established on Croplands.

Afforestation stages	Absolute SOC change ($Mg\ C{\cdot}ha^{-1}$)			Relative SOC change (%)			Absolute STN change ($Mg\ N{\cdot}ha^{-1}$)			Relative STN change (%)		
	0–10 cm	10–20 cm	0–20 cm	0–10 cm	10–20 cm	0–20 cm	0–10 cm	10–20 cm	0–20 cm	0–10 cm	10–20 cm	0–20 cm
Younger plantation	0.43	−1.45**	−1.01	16.6	−20.9**	−2.3	0.12**	−0.08	0.04	21.5**	−9.8**	5.5
	(−0.44/1.31)	(−2.07/ −0.82)	(−2.41/0.38)	(−4.7/37.8)	(−30.9/ −10.9)	(−16.8/12.3)	(0.01/0.23)	(−0.17/0.01)	(−0.16/0.23)	(3.8/39.1)	(−19.5/ −0.2)	(−7.1/ 18.2)
Older plantation	6.70**	0.85	7.57**	161.1**	27.1**	93.8**	0.66**	0.11**	0.77**	109.4**	18.4**	63.5**
	(4.79/8.65)	(−0.23/1.93)	(4.85/10.29)	(97.5/224.4)	(3.9/50.4)	(52.3/135.4)	(0.41/0.91)	(0.01/0.21)	(0.45/1.10)	(63.8/154.9)	(5.0/31.8)	(35.8/ 91.2)

**Significant at 0.05 level compared with the value of zero (one-sample T test).
The numbers in parentheses indicate the 95% confidence interval.
Younger and older plantations mean 5- to 9-year-old and 25- to 30-year-old plantations, respectively.

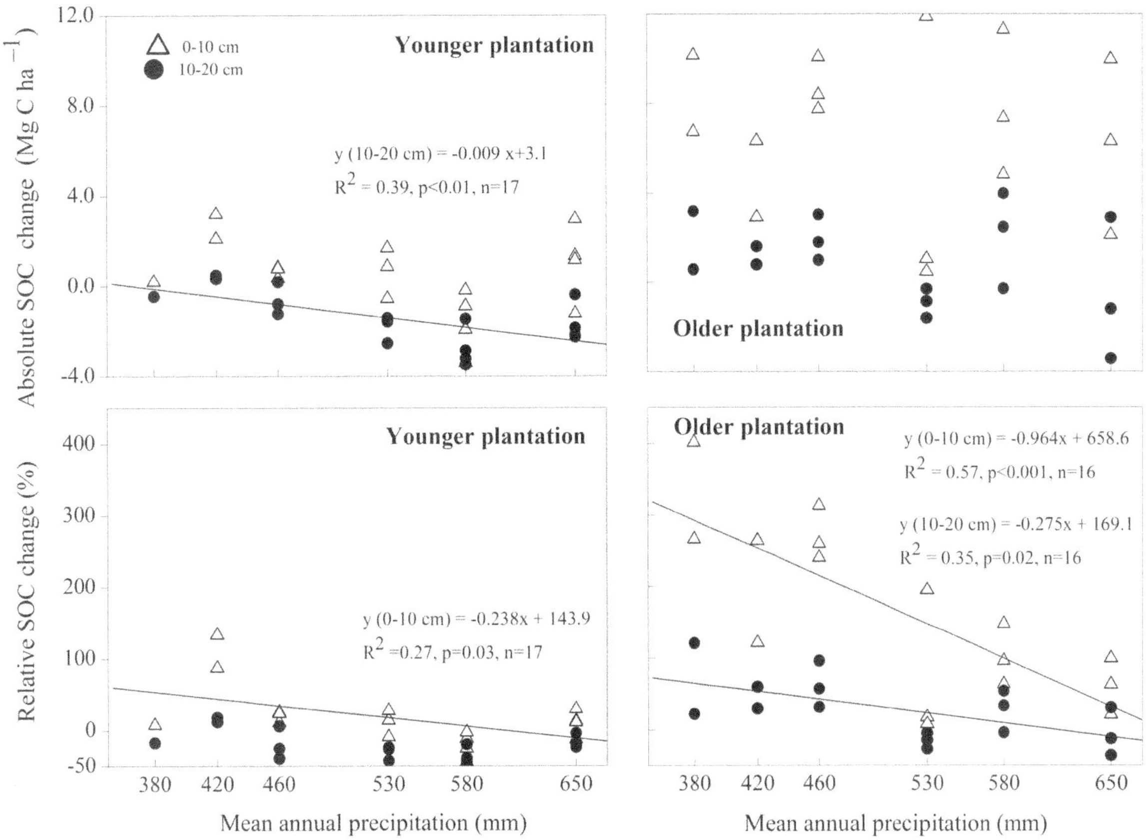

Figure 3. Relationship between mean annual precipitation and the absolute and relative change in soil organic carbon (SOC) in younger and older plantations. Younger and older plantations are 5- to 9-year-old and 25- to 30-year-old forests, respectively. Regressions were conducted separately for different soil layers, and only significant regression models are displayed. Positive values indicate an increase in the SOC stock due to afforestation, and negative values indicate a decrease.

soil water availability was limited, unlike aboveground biomass. Fine roots have been suggested to play a more important role in soil C accumulation than the aboveground parts [44,45]; therefore, the increased fine root biomass along the increasing MAP gradient may drive the trend in SOC change across the Loess Plateau. Second, the tree density increased with increasing MAP. Although tree density is suggested to have a limited effect on SOC changes during afforestation [6], higher tree density could entail an increase in the associated plantation disturbance. Thus the SOC loss resulting from the pre-planting preparations may be higher in the wetter sites. Third, microbial activity is suggested to be limited by the amount of soil water available [46], and soil respiration is generally found to be positively correlated with precipitation [47]. Thus the loss of soil C derived from decomposition would be less in the drier sites. Fourth, in the drier, infertile sites, the PE may also be constrained by limited soil water availability and decomposition substrate [38,48], which also could reduce the SOC loss. Therefore, our results suggest that less SOC decomposition and higher belowground C input determined the initial SOC changes along the decreasing MAP gradient, rather than the NPP of afforestation.

After the initial decrease in SOC in younger plantations, soil C is usually found to increase with plantation age and to reach net gains in approximately 30 years [6,7,9,10]. Based on global afforestation meta-data, Laganière *et al.* [6] found that the SOC stock decreased by 5.6% in young plantations (<10 years) but increased by 6.1% and 18.6% in medium-aged (10 to 30 years) and older (>30 years) plantations, respectively. Our findings showed a greater gain in SOC stock (161.1% and 93.8% in the top 10 cm and total 20 cm layers, respectively) in the 30-year-old plantations, which may be because the plantations were primarily N-fixing tree species.

In older plantations, there was a weak relationship between the change in absolute SOC stock and MAP. However, the negative relationship between the relative SOC changes and MAP suggests that the relative SOC gain was higher in drier sites than in the wetter sites. Berthrong et al. [10] also showed that more SOC accumulated in drier lands (MAP<1150 mm) than in wetter lands (MAP>1150 mm) following the afforestation of grassland. This current finding, in contrast to the trend of SOC during forest cultivation, indicates that SOC loss is greater in wetter regions than in drier regions [49,50]. The greater relative increase in SOC in the drier sites was associated with a lower initial value of soil C in cropland. Thus, afforestation can increase SOC more rapidly and improve soil quality more in infertile arable lands than in fertile land, although it may have a similar effect on the accumulation capacity for SOC across the sites.

STN Changes Following Afforestation

As with SOC, STN following afforestation also can show various temporal patterns [7,10,11,13,51]. In a global meta-analysis, STN stocks were generally found to follow a Covington curve, similar to the general trends in SOC, although the duration of the initial decrease in STN (50 years) is longer than that of SOC

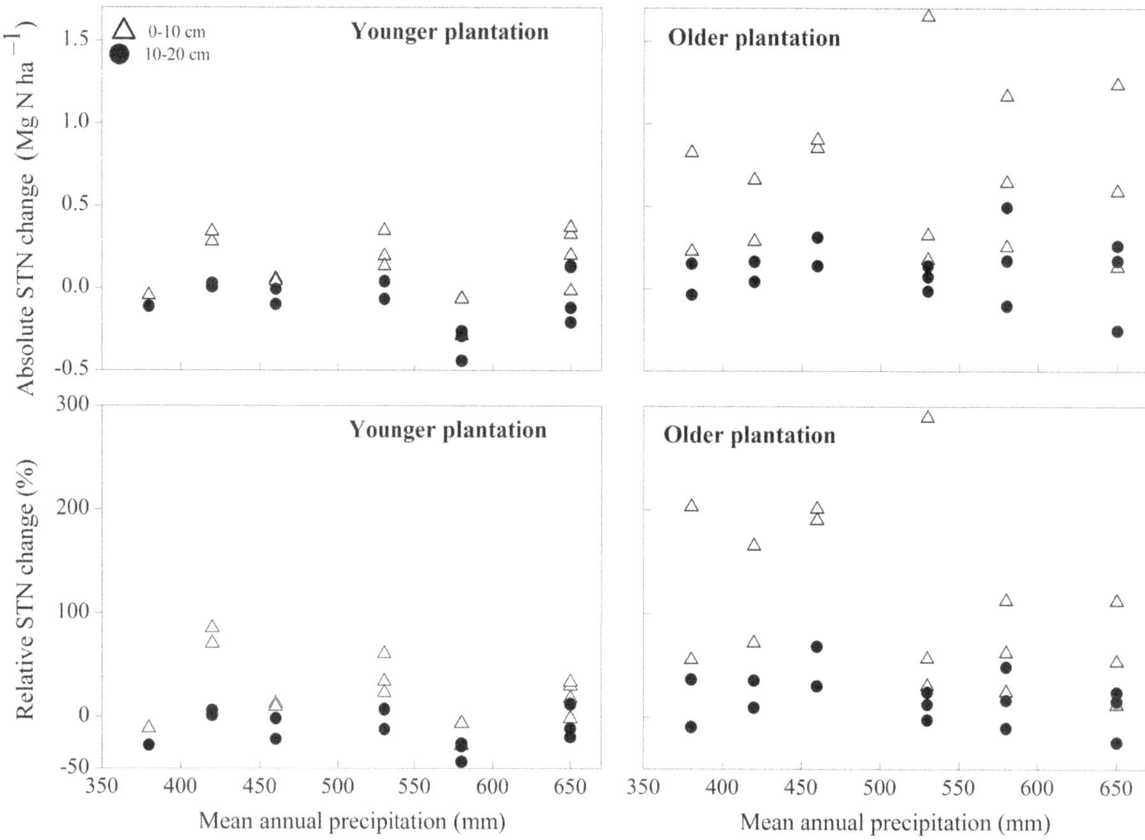

Figure 4. Association of absolute and relative change in soil total nitrogen (STN) in younger and older plantations with mean annual precipitation. Younger and older plantations mean 5- to 9-year-old and 25- to 30-year-old forests, respectively. Positive values indicate an increase in the STN stock due to afforestation, and negative values indicate a decrease. No significant association was found.

(30 years) [7]. However, our results showed no significant decrease in STN during early afforestation, whereas STN stocks tended to increase with plantation stand age. These findings were consistent with other studies of black locust plantations in the Loess Plateau [52,53]. This increase in STN may, to a large extent, be due to the effect of biological nitrogen fixation by the black locust. Boring and Swank [54] reported that a 4-year-old black locust stand fixed 30 kg N/ha/yr during a one-year study period, and Rice *et al.* [55] found that there was 86 kg N/ha/yr of leaf litter returned to the soil by a 20- to 35-year-old black locust stand. These estimates exceed the average observed annual N accretion in our sites (4 to 9 kg N/ha/yr for younger stands and 26 to 31 kg N/ha/yr for older stands, respectively). Other N sources and mechanisms also may affect STN accumulation. For example, atmospheric N deposition (including dry and wet deposition) is often used to explain increasing STN during afforestation [7,11]. However, few studies have examined the difference in atmospheric N deposition between plantation and cropland, and the N retained in the soil through deposition has rarely been quantified after afforestation. In addition, plantation areas may be less susceptible to soil erosion than croplands in the Loess Plateau [56]; therefore, the lower N loss from soil erosion may result in higher soil N after afforestation.

N mainly enters soil in three ways: atmospheric deposition, fertilizer, and N fixation. There was no N input from fertilizer in the plantations of the Loess Plateau. N inputs from atmospheric deposition are generally found to increase with MAP increase from the northern to southern areas of the Loess Plateau [57]. Meanwhile, biological N fixation is predicted to increase with tree density increase from the drier to wetter sites, and N-fixation is also suggested to be limited by drought conditions in the drier areas [58,59]. Nevertheless, the changes in STN stock after afforestation were not found to be correlated with MAP in any age stand, which is in contrast to Berthrong *et al.*'s [10] finding that STN changes following afforestation were significantly related to MAP. This condition may result from higher N loss from plant uptake and mineralization and from less N gain from soil-erosion control in the wetter southern Loess Plateau. First, the aboveground biomass of plantation stands was positively correlated with MAP, and this higher aboveground biomass in the wetter sites may consume more soil N. Evidence shows that available inorganic soil N (NH_4^+ and NO_3^-) is lower with higher tree density (higher aboveground biomass) [32] and which may offset, in part, the benefit of greater N-fixation resulting from higher tree density. Second, stronger microbial activity and a more pronounced PE in the wetter environment (as discussed above) may lead to higher SOM mineralization and consequently higher N loss from gas (e.g., NH_3, N_2O). However, McCulley *et al.* [60] suggested the N cycling is tighter in the wetter grassland sites along a precipitation gradient in the Central Great Plains, and the net soil N mineralization did not increase along the precipitation gradient. Thus a more precise measurement of N mineralization in the forests across the Loess Plateau is needed to understand the

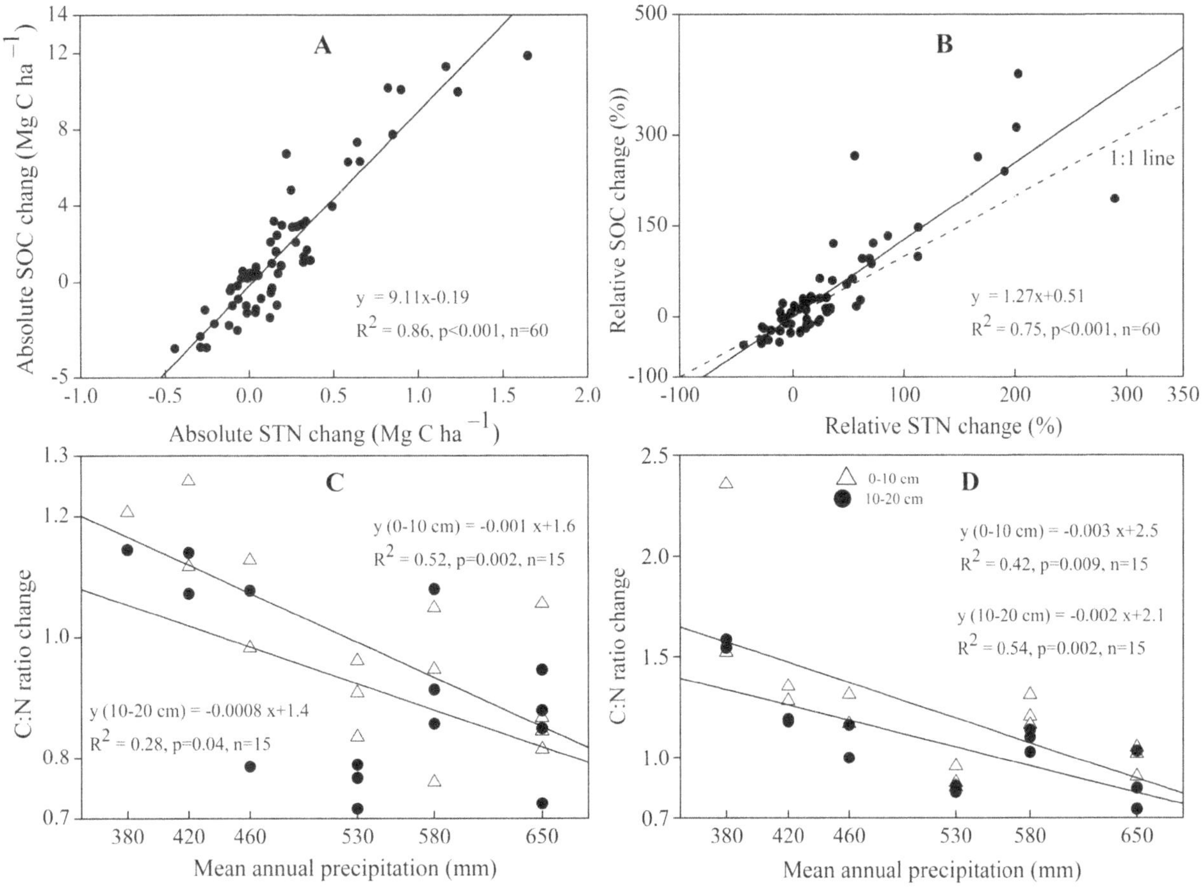

Figure 5. Relationship of changes in absolute soil organic carbon (SOC) and soil total nitrogen (STN) (A), changes in relative SOC and STN (B), and changes in the C:N ratio in younger (C) and older (D) plantations with mean annual precipitation. Younger and older plantations mean 5- to 9-year-old and 25- to 30-year-old forests, respectively. The dashed line in panel B indicates the 1:1 line. Values greater than 1.0 in panels C and D indicate a relative gain of SOC to STN following afforestation, and lower values indicate the inverse.

changes in soil N. Third, soil erosion in cropland is more serious in the northern, drier sites relative to the southern, wetter areas [61], and the N gain derived from the reduced soil erosion in afforested sites may be more pronounced in the northern, more eroded sites.

Soil C and N Interaction and the Implications

The loss of SOC is greater than that of STN when forests were converted into cropland, resulting in a lower C:N ratio in the cropland [1,4]. In contrast, the accumulation rate of SOC is suggested to be higher than that of soil N after cropland afforestation [4,7,51]. In the present study, however, STN was found to increase more quickly than SOC following afforestation, as indicated by higher relative changes in STN. Nevertheless, the relative gains of SOC and STN also were precipitation dependent, with higher C gain (compared with N gain) in the drier lands and lower C gain in the wetter sites. Furthermore, SOC changes following afforestation were related to MAP, whereas STN changes were not. Therefore, C and N were more likely to be decoupled following marginal cropland afforestation in this semi-arid environment, and the changes in soil C and N were not synchronous at the regional scale, which is a proxy for precipitation. This finding implied that the mechanisms and factors controlling the accumulation rates were different for C and N, as suggested by McLauchlan [4].

As a result of the stoichiometric relationship between C and N, the terrestrial C (including soil C) sequestration is suggested to be determined by the availability of N and may be down-regulated as the supply of N is limited, as argued in the progressive N limitation theory [15]. In our study, however, the relative soil C gain to N retention increased when precipitation decreased, indicating a limited effect of N limitation on SOC accumulation in the drier area, at least in a relatively short plantation stage (during about 30 years). It seemed that an increase in soil N derived from N-fixing tree plantation in such N-limited arid lands could have a significant effect on soil C accumulation, and this effect was greater in the drier areas rather than in the wetter ones. Nevertheless, these results are based on a relatively short period of afforestation (30 years), and they may not hold up in the long term, although some studies suggest that forest ecosystems have an intrinsic ability to regulate the soil C and N accumulation and reduce N limitation in a relatively longer time [11]. So further investigations focused on the older stands in such arid areas are needed.

Conclusion

Our study suggests that afforestation of marginal cropland could lead to a significant increase in SOC and STN stocks in a relatively short time (about 30 years) and underlines the effect of precipitation on SOC dynamics during afforestation in an arid environment. The results show that, along the increasing precipitation gradient, the initial absolute loss of SOC increased but the relative increase in SOC in the older ages decreased. This phenomenon is suggested to be associated with the greater soil C decomposition loss and the less belowground C input in the wetter sites, rather than the NPP of afforestation.

Unlike the SOC dynamics, there was no significant loss in STN during initial afforestation, and the changes in STN were not correlated with precipitation. This decoupled relationship between STN and SOC implies that a mechanism may have been regulating the effect of soil N retention on soil C accumulation to prevent progressive N limitation during forest development.

Supporting Information

Figure S1 Geographic locations of the study sites and croplands involved in Grain for Green project in China's Loess Plateau.

Figure S2 Relationship between soil clay content of croplands and mean annual precipitation. The error bars represent the standard deviations of the means (only shown in positive, n = three to five for each stand). The datum of one stand in S4 (with MAP of 530 mm) was cited from Li *et al.* [66].

Author Contributions

Conceived and designed the experiments: RC TJ BF. Performed the experiments: RC TJ YL. Analyzed the data: RC. Contributed reagents/materials/analysis tools: RC TJ GL. Wrote the paper: RC YL BF.

References

1. Murty D, Kirschbaum MUF, McMurtrie RE, McGilvray H (2002) Does conversion of forest to agricultural land change soil carbon and nitrogen? A review of the literature. Global Change Biology 8: 105–123.
2. Liu SG, Bliss N, Sundquist E, Huntington TG (2003) Modeling carbon dynamics in vegetation and soil under the impact of soil erosion and deposition. Global Biogeochemical Cycles 17: 1074.
3. Davidson E, Ackerman I (1993) Changes in soil carbon inventories following cultivation of previously untilled soils. Biogeochemistry 20: 161–193.
4. McLauchlan K (2006) The nature and longevity of agricultural impacts on soil carbon and nutrients: a review. Ecosystems 9: 1364–1382.
5. Post WM, Kwon KC (2000) Soil carbon sequestration and land use change: processes and potential. Global Change Biology 6: 317–327.
6. Laganière J, Angers DA, Par D (2010) Carbon accumulation in agricultural soils after afforestation: a meta analysis. Global Change Biology 16: 439–453.
7. Li D, Niu S, Luo Y (2012) Global patterns of the dynamics of soil carbon and nitrogen stocks following afforestation: a meta-analysis. New Phytologist 195: 172–181.
8. FAO (2010) The state of forest resources – a regional analysis. In: FAO, editor. The State of the World's Forests 2011.
9. Paul KI, Polglase PJ, Nyakuengama JG, Khanna PK (2002) Change in soil carbon following afforestation. Forest ecology and management 168: 241–257.
10. Berthrong ST, Piñeiro G, Jobbágy EG, Jackson RB (2012) Soil C and N changes with afforestation of grasslands across gradients of precipitation and plantation age. Ecological Applications 22: 76–86.
11. Yang Y, Luo Y, Finzi AC (2011) Carbon and nitrogen dynamics during forest stand development: a global synthesis. New Phytologist 190: 977–989.
12. Guo LB, Gifford RM (2002) Soil carbon stocks and land use change: a meta analysis. Global Change Biology 8: 345–360.
13. Sartori F, Lal R, Ebinger MH, Eaton JA (2007) Changes in soil carbon and nutrient pools along a chronosequence of poplar plantations in the Columbia Plateau, Oregon, USA. Agriculture, Ecosystems & Environment 122: 325–339.
14. Bashkin MA, Binkley D (1998) Changes in soil carbon following afforestation in Hawaii. Ecology 79: 828–833.
15. Luo Y, Su BO, Currie WS, Dukes JS, Finzi A, et al. (2004) Progressive Nitrogen Limitation of Ecosystem Responses to Rising Atmospheric Carbon Dioxide. BioScience 54: 731–739.
16. Kirkby CA, Richardson AE, Wade LJ, Batten GD, Blanchard C, et al. (2013) Carbon-nutrient stoichiometry to increase soil carbon sequestration. Soil Biology and Biochemistry 60: 77–86.
17. Gärdenäs AI, Ågren GI, Bird JA, Clarholm M, Hallin S, et al. (2011) Knowledgegaps in soilcarbon and nitrogeninteractions – From molecular to global scale to global scale. Soil Biology and Biochemistry 43: 702–717.
18. Binkley D (2005) How nitrogen-fixing trees change soil carbon. In: Binkley D, Menyailo O, editors. Tree species effects on soils: implications for global change. Dordrecht: Kluwer Academic Publishers. pp. 155–164.
19. Forrester DI, Pares A, O'Hara C, Khanna PK, Bauhus J (2013) Soil Organic Carbon is Increased in Mixed-Species Plantations of Eucalyptus and Nitrogen-Fixing Acacia. Ecosystems 16: 123–132.
20. Resh SC, Binkley D, Parrotta JA (2002) Greater soil carbon sequestration under nitrogen-fixing trees compared with Eucalyptus species. Ecosystems 5: 217–231.
21. Lü YH, Fu BJ, Feng XM, Zeng Y, Liu Y, et al. (2012) A Policy-Driven Large Scale Ecological Restoration: Quantifying Ecosystem Services Changes in the Loess Plateau of China. PLoS ONE 7: e31782.
22. Wen QZ, editor (1989) Geochemistry in Chinese Loess. Beijing: Science Press.
23. Sun WY, Guo SL, Zhou XG (2010) Effects of Topographies and Land Uses on Soil Organic Carbon in Subsurface in Hilly Region of Loess Plateau. Environmental Science 31: 2740–2747.
24. Guo ZY (1992) Shaanxi Soils. Beijing: Science Press.
25. Chang RY, Fu BJ, Liu GH, Yao XL, Wang S (2012) Effects of soil physicochemical properties and stand age on fine root biomass and vertical distribution of plantation forests in the Loess Plateau of China. Ecological Research 27: 827–836.
26. Lu R (1999) Analytical Methods of Soil Agrochemistry. Beijing: Science and Technology Press.
27. Tian QF, Du LH, Li XJ (1997) Study on Biomass of *RobiniaPseudoacacia* Plantation in the Beijing Xishan National Forest Park. Journal of Beijing Forestry University 19: 104–107.
28. Paternoster R, Brame R, Mazerolle P, Piquero A (1998) Using the correct statistical test for the equality of regression coefficients. Criminology 36: 859–866.
29. Wei XR, Shao MA, Fu XL, Horton R, Li Y, et al. (2009) Distribution of soil organic C, N and P in three adjacent land use patterns in the northern Loess Plateau, China. Biogeochemistry 96: 149–162.
30. Callesen I, Liski J, Raulund-Rasmussen K, Olsson MT, Tau-Strand L, et al. (2003) Soil carbon stores in Nordic well-drained forest soils–relationships with climate and texture class. Global Change Biology 9: 358–370.
31. Schimel DS, Braswell BH, Holland EA, McKeown R, Ojima DS, et al. (1994) Climatic, edaphic, and biotic controls over storage and turnover of carbon in soils. Global Biogeochemical Cycles 8: 279–293.
32. Litton CM, Ryan MG, Knight DH (2004) Effects of tree density and stand age on carbon allocation patterns in postfire lodgepole pine. Ecological Applications 14: 460–475.
33. Litton CM, Ryan MG, Knight DH, Stahl PD (2003) Soil-surface carbon dioxide efflux and microbial biomass in relation to tree density 13 years after a stand replacing fire in a lodgepole pine ecosystem. Global Change Biology 9: 680–696.
34. Six J, Conant RT, Paul EA, Paustian K (2002) Stabilization mechanisms of soil organic matter: Implications for C-saturation of soils. Plant and Soil 241: 155–176.
35. Krull ES, Baldock JA, Skjemstad JO (2003) Importance of mechanisms and processes of the stabilisation of soil organic matter for modelling carbon turnover. Functional Plant Biology 30: 207–222.
36. Johnson DW (1992) Effects of forest management on soil carbon storage. Water, Air, & Soil Pollution 64: 83–120.
37. Cheng WX (2009) Rhizosphere priming effect: Its functional relationships with microbial turnover, evapotranspiration, and C–N budgets. Soil Biology and Biochemistry 41: 1795–1801.
38. Kuzyakov Y, Friedel JK, Stahr K (2000) Review of mechanisms and quantification of priming effects. Soil Biology and Biochemistry 32: 1485–1498.
39. Cheng WX, Kuzyakov Y (2005) Root effects on soil organic matter decomposition. In: Zobel RW, Wright, SF., editor. Roots and Soil Management: Interactions between Roots and the Soil. Madison Wisconsin: ASA-SSSA.
40. Guenet B, Danger M, Abbadie L, Lacroix G (2010) Priming effect: bridging the gap between terrestrial and aquatic ecology. Ecology 91: 2850–2861.
41. Li Z, Yan F, Fan X. The variability of NDVI over Northwest China and its relation to temperature and precipitation; 2003. IEEE. pp. 2275–2277.
42. Hendricks JJ, Hendrick RL, Wilson CA, Mitchell RJ, Pecot SD, et al. (2006) Assessing the patterns and controls of fine root dynamics: an empirical test and methodological review. Journal of Ecology 94: 40–57.
43. Farrar JF, Jones DL (2000) The control of carbon acquisition by roots. New Phytologist 147: 43–53.
44. Rasse DP, Rumpel C, Dignac MF (2005) Is soil carbon mostly root carbon? Mechanisms for a specific stabilisation. Plant and Soil 269: 341–356.

45. Kuzyakov Y, Schneckenberger K (2004) Review of estimation of plant rhizodeposition and their contribution to soil organic matter formation. Archives of Agronomy and Soil Science 50: 115–132.
46. Broughton LC, Gross KL (2000) Patterns of diversity in plant and soil microbial communities along a productivity gradient in a Michigan old-field. Oecologia 125: 420–427.
47. Raich JW, Schlesinger WH (1992) The global carbon dioxide flux in soil respiration and its relationship to vegetation and climate. Tellus B 44: 81–99.
48. Dijkstra FA, Cheng W (2007) Moisture modulates rhizosphere effects on C decomposition in two different soil types. Soil Biology and Biochemistry 39: 2264–2274.
49. Ogle SM, Breidt FJ, Paustian K (2005) Agricultural management impacts on soil organic carbon storage under moist and dry climatic conditions of temperate and tropical regions. Biogeochemistry 72: 87–121.
50. Lugo A, Sanchez M, Brown S (1986) Land use and organic carbon content of some subtropical soils. Plant and Soil 96: 185–196.
51. Hooker TD, Compton JE (2003) Forest ecosystem carbon and nitrogen accumulation during the first century after agricultural abandonment. Ecological Applications 13: 299–313.
52. Wang B, Liu G, Xue S (2012) Effect of black locust (Robinia pseudoacacia) on soil chemical and microbiological properties in the eroded hilly area of China's Loess Plateau. Environmental Earth Sciences 65: 597–607.
53. Liu D, Huang YM, An SS (2012) Changes in Soil Nitrogen and Microbial Activity During *Robinia Pseudoacacia* Recovery Period in the Loess Hilly-Gully Region. Chinese Journal of Eco-Agriculture 20: 322–329.
54. Boring LR, Swank WT (1984) Symbiotic Nitrogen Fixation in Regenerating Black Locust (Robinia pseudoacacia L.) Stands. Forest Science 30: 528–537.
55. Rice S, Westerman B, Federici R (2004) Impacts of the exotic, nitrogen-fixing black locust (Robinia pseudoacacia) on nitrogen-cycling in a pine–oak ecosystem. Plant Ecology 174: 97–107.
56. Fu BJ, Wang YF, Lu YH, He CS, Chen LD, et al. (2009) The effects of land-use combinations on soil erosion: a case study in the Loess Plateau of China. Progress in Physical Geography 33: 793–804.
57. Peng L, Yu CZ, Wang JZ (1995) The Nutrient Content and Nutrient Supply of Dry-farming Soil in Loess Plateau. Journal of Northwest University (Natural Science Edition) 25: 117–122.
58. Zahran HH (1999) Rhizobium-Legume Symbiosis and Nitrogen Fixation under Severe Conditions and in an Arid Climate. Microbiology and Molecular Biology Reviews 63: 968–989.
59. Gálvez L, González EM, Arrese-Igor C (2005) Evidence for carbon flux shortage and strong carbon/nitrogen interactions in pea nodules at early stages of water stress. Journal of Experimental Botany 56: 2551–2561.
60. McCulley R, Burke I, Lauenroth W (2009) Conservation of nitrogen increases with precipitation across a major grassland gradient in the Central Great Plains of North America. Oecologia 159: 571–581.
61. Wang WZ, Jiao JY (2002) Temporal and Spatial Variation Features of Sediment Yield Intensity on Loess Plateau. Acta Geographica Sinica 57: 210–217.
62. Wang B, Liu GB, Xue S, Zhu BB (2011) Changes in soil physico-chemical and microbiological properties during natural succession on abandoned farmland in the Loess Plateau. Environmental Earth Sciences 62: 915–925.
63. Jiao JY, Zhang ZG, Bai WJ, Jia YF, Wang N (2010) Assessing the Ecological Success of Restoration by Afforestation on the Chinese Loess Plateau. Restoration Ecology 20: 240–249.
64. Wang XL, Guo SL, Ma YH, Huang GY, Wu JS (2007) Effects of land use type on soilorganic C and total N in a small watershed in loess hilly-gully region. Chinese Journal of Applied Ecology 18: 1281–1285.
65. Liu Y, Zheng FL, An SS, Guo M (2007) The responses of soil organic carbon, total nitrogen and enzymatic activity to vegetation restoration in abandoned lands in Yangou watershed. Agricultural Research in the Arid Areas 25: 220–225.
66. Li YY, Shao MA, Zheng JY, Zhang XC (2005) Spatial-temporal changes of soil organic carbon during vegetation recovery at Ziwuling, China. Pedosphere 15: 601–610.

22

Lead Tolerance and Accumulation in *Hirschfeldia incana*, a Mediterranean *Brassicaceae* from Metalliferous Mine Spoils

Florence Auguy[1,2,3], Mouna Fahr[2,3,4], Patricia Moulin[3,5], Anaïs Brugel[1], Laurent Laplaze[1,6], Mohamed El Mzibri[2,3], Abdelkarim Filali-Maltouf[3,7], Patrick Doumas[1,2,3,8]*, Abdelaziz Smouni[2,4]

1 Institut de Recherche pour le Développement, Unité Mixte de Recherche Diversité Adaptation et Développement des Plantes, Montpellier, France, **2** Centre National de l'Energie, des Sciences et des Techniques Nucléaires, Laboratoire de Biotechnologie des Plantes, Rabat, Maroc, **3** Laboratoire Mixte International, Université Mohammed V - Agdal, Rabat, Maroc, **4** Laboratoire de Physiologie et Biotechnologie Végétale, Université Mohammed V - Agdal, Rabat, Maroc, **5** Institut de Recherche pour le Développement, Unité de Service Instrumentation, Moyens Analytiques, Observatoires en Géophysique et Océanographie, Laboratoire de Microbiologie et Biologie Moléculaire, Université Mohammed V - Agdal, Rabat, Maroc, **6** Institut de Recherche pour le Développement, Laboratoire mixte international Adaptation des Plantes et microorganismes associés aux Stress Environnementaux, Laboratoire Commun de Microbiologie, Dakar, Sénégal, **7** Laboratoire de Microbiologie et Biologie Moléculaire, Université Mohammed V - Agdal, Rabat, Maroc, **8** Institut National de la Recherche Agronomique, Unité Mixte de Recherche Biochimie et Physiologie Moléculaire des Plantes, Montpellier, France

Abstract

Lead is a heavy metal of particular concern with respect to environmental quality and health. The lack of plant species that accumulate and tolerate Pb is a limiting factor to understand the molecular mechanisms involved in Pb tolerance. In this study we identified *Hirschfeldia incana*, a *Brassicaceae* collected from metalliferous mine spoils in Morocco, as a Pb accumulator plant. *H. incana* exhibited high Pb accumulation in mine soils and in hydroponic cultures. Major Pb accumulation occurred in the roots and a part of Pb translocated from the roots to the shoots, even to the siliques. These findings demonstrated that *H. incana* is a Pb accumulator species. The expression of several candidate genes after Pb-exposure was measured by quantitative PCR and two of them, *HiHMA4* and *HiMT2a*, coding respectively for a P1B-type ATPase and a metallothionein, were particularly induced by Pb-exposure in both roots and leaves. The functional characterization of *HiHMA4* and *HiMT2a* was achieved using Arabidopsis T-DNA insertional mutants. Pb content and primary root growth analysis confirmed the role of these two genes in Pb tolerance and accumulation. *H. incana* could be considered as a good experimental model to identify genes involved in lead tolerance and accumulation in plants.

Editor: Abidur Rahman, Iwate University, Japan

Funding: This research was supported by grants from the International Joint Laboratory "Laboratoire de Biotechnologie Microbienne et Végétale", Centre National de l'Energie, des Sciences et Techniques Nucléaires (CNESTEN)/Institut de Recherche pour le Développement (IRD)/Université Mohamed V Agdal, Rabat, Morocco. Abdelaziz Smouni gratefully acknowledges the support from IRD through a BESCD grant. The funders had no role in study design, data collection and analysis, decision to publish, or preparation of the manuscript.

Competing Interests: The authors have declared that no competing interests exist.

* E-mail: patrick.doumas@supagro.inra.fr

Introduction

Lead (Pb) is a heavy metal of particular concern with respect to environmental quality and health [1]. As a non-essential trace metal for biological functions, it is highly toxic to plants and animals. Sources of anthropogenic soil contamination by Pb include industrial and agricultural activities such as mining and smelting of metalliferous ores, battery-engine waste, wastewater irrigation, and overuse of chemical fertilizers and pesticides [2]. Therefore, the cleanup of Pb-contaminated soils is imperative. In this context, phytoremediation can be considered as a potent tool in the near future. Phytoextraction is based on the genetic and physiological capacity of specialized plants to tolerate high amounts of metal, to translocate from roots to shoots, and to accumulate in shoots [3]. The idea of using plants to remove metals from soils came from the discovery of different wild plants that accumulate high concentrations of metals in their foliage [4]. Currently, phytoremediation of Pb-polluted soils presents two major drawbacks namely, on the one hand, the limited number of species which hyperaccumulate Pb and, on the other hand, the insufficient knowledge of the molecular mechanisms implicated in Pb tolerance in plants.

Over 450 species of metal hyperaccumulator and tolerant plants are known, in both tropical and temperate zones, and most of them are nickel hyperaccumulators [5,6]. Plant species which hyperaccumulate cobalt, copper or zinc are in second rank, although in much smaller numbers. Hyperaccumulation of arsenic, cadmium, gold, lead, manganese and thallium occur in a limited number of species [7,8]. Lead accumulating species are rather exceptional. It was well established by Baker *et al.* [4] that *Noccaea caerulescens* can accumulate Pb to high concentrations from nutrient solutions with low concentration of added Pb, mostly fixed in the root with limited translocation to the leaves. Few other species have been described as Pb hyperaccumulators such as *Agrostis tenuis*, *Festuca ovina*, *Rumex acetosa* or *Thlaspi cepaeifolium* [9].

Concerning the molecular elements implicated in lead-tolerance mechanisms, the number of identified genes is quite small. The tobacco plasma membrane protein NtCBP4 and the Arabidopsis gene *CNGC1* were reported to be components of a transport pathway responsible for Pb entry into plant cells [10]. An Arabidopsis P-type ATPase, HMA3, was described to improve tolerance by sequestrating Pb in the vacuole [11,12]. HMA4, another P-type ATPase, from *Noccaea caerulescens* has also been suggested as having a potential role in Pb efflux transport in yeast [13]. In Arabidopsis, three members of ABC (ATPase-binding cassette) transporters family AtATM3, AtPDR12 and AtPDR8 contribute to Pb resistance [1,14,15]. Recently, ACBP1, an acyl-CoA-binding protein, was found to be involved in mediating Pb tolerance through accumulation of Pb in shoots [16] and *AtMRP3* transcription was also shown to be strongly induced by Pb treatment [17] in *A. thaliana*. An important mechanism controlling heavy metal tolerance is chelation that involves small molecules such as metallothioneins, phytochelatins and glutathione [18,19].

Here, we identified *Hirschfeldia incana*, a member of the *Brassicaceae* family, collected in abandoned lead mining sites in the east of Morocco, as a Pb accumulator plant and we demonstrated that this species, with a close genetic proximity to Arabidopsis, is a good experimental model to identify genes involved in lead tolerance and accumulation in higher plants.

Material and Methods

Ethics Statement

No specific permits were required for the described field studies because sample collection does not involve any endangered or protected plant species or privately-owned locations.

Plant Material and Growth Conditions

Seeds of *H. incana* were harvested from plants growing in abandoned metalliferous mine spoils in the mine district of Oued El Himer, south of Oujda city in eastern Morocco (34°26′88″N, 1°54′03″W; Smouni *et al.*, 2010). Seeds were surface sterilized, sown on one half MS medium [20] with 1.2% agar and grown vertically in a growth chamber at 22°C with a photoperiod of 16 h of light (0.1 $mMol.m^{-2}.s^{-1}$) from fluorescent lamps. After 2 weeks, seedlings were transferred to hydroponic culture on BD medium [21] containing 5 mM KNO_3, 1 mM $CaCl_2$, 0.5 mM KH_2PO_4, 0.25 mM MgSO4, 0.25 mM K2SO4, 1 µM MnSO4, 50 µM FeEDTA, 2 µM H_3BO_3, 0.5 µM $ZNSO_4$, 0.2 µM $CuSO_4$, 0.1 µM $CoSO_4$ and 0.1 µM Na_2MoO_4. Lead treatment was done by adding 50, 100 or 300 µM $Pb(NO_3)_2$ to a fresh BD medium without phosphate to avoid Pb precipitation. For the cultivation of *H. incana* on different soils with various lead concentrations, seedlings were grown on sterile compost for 20 days and transferred to four different soils from metalliferous mine spoils containing various amounts of lead (soil 1:6972 $mg.kg^{-1}$ DW, soil 2:18626 $mg.kg^{-1}$ DW, soil 3:7531 $mg.kg^{-1}$ DW and soil 4:1577 $mg.kg^{-1}$ DW; table 1).

Arabidopsis thaliana ecotype Columbia (Col-0) and T-DNA mutant seeds were obtained from the Nottingham Arabidopsis Stock Centre [22]. Seeds were surface-sterilized and sown on square Petri dishes containing BD medium without phosphate, with 1.2% agar, with or without 40 µM $Pb(NO_3)_2$. After sowing on either standard or Pb-treated medium, seeds were cold-treated at 4°C for 48 h in darkness to promote and synchronize germination, subsequently transferred in a vertical position to a growth chamber at 22°C with a photoperiod of 16 h of light (0.1 $mMol.m^{-2}.s^{-1}$). In hydroponic cultures, *A. thaliana* Col-0 seeds were prepared as described previously for *H. incana* seeds and seedlings were transferred after two weeks on liquid BD medium. The Pb-treatment was done by adding 40 µM $Pb(NO_3)_2$ to a fresh BD medium without phosphate.

Lead Quantification

Shoots or roots (three independent replicates per sample) were washed twice in cold 0.2 mM $CaSO_4$ and rinsed with cold distilled water. For the roots, rinsing with water and $CaSO_4$ does not remove adsorbed Pb, and hence all root tissue Pb measurements reported in this study include both the Pb taken up by the plant and the Pb adsorbed at the root surface. Samples were dried at 72°C for 48 h and a maximum of 200 mg of dried tissues were treated according to the acid hydrolysis protocol described by Temminghoff and Houba [23]. Soil samples were air-dried at 70°C to constant weight. Dried soils were sieved through a 2 mm mesh and ground in a porcelain pestle and mortar. To estimate total lead amount from each prepared soil, 0.5 g (three replicates per sample) was treated as described by Smouni *et al.* [24]. The mobile and mobilisable fractions in the soils were estimated by using 10 mM $CaCl_2$ and 50 mM EDTA at pH 7, respectively. Two grams of soil samples (three replicates per sample) were suspending in 20 mL of 10 mM $CaCl_2$ or 50 mM EDTA and shacked for two hours. The suspensions were centrifuged at 8000 rpm for 12 min and lead concentration was analyzed in the supernatants. Lead concentration was determined by ICP-AES (Inductively Coupled Plasma-Atomic Emission Spectrometry; Ultima2 JY) at a wavelength of 220.353 nm in accordance with the method devised by Margui *et al.* [25].

DNA Extraction

Hundred mg of fresh leaf tissue of *H. incana* were harvested directly in 500 µL of extraction buffer (100 mM TrisHCl, 500 mM NaCl, 50 mM EDTA, pH 8.0) and 70 µL of SDS 10% (w/v) were added. Samples were homogenised, incubated 10 min at 65°C and 130 µl of 5M potassium acetate were added. After homogenisation, samples were centrifuged 15 min at 13 000 rpm and 500 µL of the supernatant were added to 500 µL of propanol 2. The precipated DNA was recovered by centrifugation

Table 1. Pb levels of the metalliferous mine soils in the plant cultivation.

Fractions	Soil 1	Soil 2	Soil 3	Soil 4
Total Pb (mg/Kg)	6972.8±59.4	18626.2±125.1	7531.1±86.2	1577.2±36.7
10 mM CaCl2 extractable Pb (mg/Kg)	5.8±0.1	2.8±0.2	1.2±0.1	1.5±0.3
50 mM EDTA extractable Pb (mg/Kg)	6665.9±195.4	18315.8±268.1	7166.8±822.3	1445.6±72.2

Pb-polluted soils were collected in different zones from abandoned metalliferous mines of Oued El Himer, south of Oujda city in eastern Morocco. Data are the average (± SE) of three independent measurements.

5 min at 13 000 rpm. The pellet was dried and resuspended in 30 µL of ultra pure water. DNA extraction was done on three independent replicate experiments.

RNA Extraction

Total RNA was extracted from 100 mg of frozen samples, using the RNeasy Plant mini kit (Qiagen, USA) for roots and the SV Total RNA isolation system (Promega, USA) for shoots. All samples were DNase treated using the Turbo DNA-free (Ambion, USA) in accordance with the manufacturer's protocol. The RNA quality was confirmed by non-denaturating electrophoresis. RNA extraction was done on three independent replicate experiments.

Quantitative Real-time Reverse-transcription PCR

First-strand synthesis was carried out using 1 µg of total RNA with the reverse transcription system (Promega, USA) and oligo$(dT)_{20}$ primers. For the three biological replicates, two independent reverse transcriptions were done and pooled to minimize variation in reverse transcription yield. An equal amount of cDNA was used for each reaction, corresponding to 1.5 ng of total RNA. A reaction contained 7.5 µL of Brillant II Sybr Green QPCR master mix (Agilent, USA), 2.25 pmol of each primer in a total volume of 15 µL. The reactions were done on the MX3005P apparatus (Agilent, USA) under the following thermal profile: 5 min at 95°C, 40 repeats of 10 s at 95°C, and 30 s at 60°C and a final stage to determine dissociation curves of 1 min at 95°C, 30 s at 55°C and after, constant temperature increasing to 95°C with the fluorescence reading in continuous. The primer set was designed with the Primer 3 software and primers are listed in table S1. The efficiency of each primer pair was determined by amplifying serial dilutions of cDNA. Tubulin gene was chosen as reference gene because there was less than 1 threshold cycle (Ct) difference among the different samples and conditions. Relative expression ratios were calculated using the comparative ΔCt method and efficiencies of each gene were taken into account. The calibration was done against roots or leaves from plants growing in standard conditions (without lead). The RNA level from the three biological samples was measured in triplicate.

Internal Transcribed Spacer (ITS) Sequence Isolation

The complete ITS region (including ITS 1 and 2 and the 5.8 S r DNA gene) was amplified with the primers ITS4 and ITS5 [26]. PCR products of 722 bp were cloned into pGEM-T easy vector (Promega) and sequenced on both sides with the universal M13 primers. Three ITS regions were sequenced from three PCR reactions done on the three independent DNA extractions. DNA sequences were aligned with the software Clustal. Percentage identity of the *H. incana* ITS consensus sequence and *A. thaliana* ITS sequence was determined with the software Blast.

H. incana Gene Cloning

PCR was achieved on roots or shoots of *H. incana* cDNA treated with 100 µM $Pb(NO_3)_2$ for 3 days. Primers were designed directly from the literature or from *A. thaliana* sequences or by alignment of multiple sequences (Table S2). PCR fragments were purified with the Wizard SV PCR Clean up system (Promega) and ligated in the pGEM-T easy vector (Promega) before transforming *E. coli* competent cells. PCR products were sequenced by automated DNA sequencing (Eurofin MWG Operon) and gene identities were confirmed by comparison with the BLAST algorithm against the Genbank sequences.

Identification of Arabidopsis T-DNA Insertion Mutants

Multiple alignments of nucleotide sequences reveal high sequence identity with respectively HiHMA4 and At HMA4, and HiMT2a and AtMT2a. In addition, phylogenic trees point to HiHMA4 and HiMT2a as putative orthologs of respectively AtHMA4 and AtMT2a (Figure S1). Consequently, homozygote plants of the T-DNA insertion lines Salk_093482 and Salk_059712 for the genes At2g19110 (*AtHMA4*) and At3g09390 (*AtMT2a*) respectively were identified by PCR using 3 different primers. In the case of Salk_093482, a T-DNA specific primer (LBb1, 5′-GCGTGGACCGCTTGCTCAACT-3′) and two AtHMA4 specific primers (AtHMA4-F 5′-CACTTGACGGCGT-TAAAGAA-3′ and AtHMA4-R 5′-AACCATGACGCAAAAC-CACT-3′) were used. In the case of Salk_059712, the same LBb1 T-DNA specific primer and two AtMT2a specific primers (AtMt-F 5′- CCATAACACACGGAACATCG-3′ and AtMt-R 5′-AGATCCACATCCGCAGTTT-3′) were used. The genotype of the F3 individuals was checked by PCR using gene-specific primers and T-DNA primers. Individual homozygous mutants were backcrossed twice with the wild-type Col-0.

Root Growth Analysis

Images of the root system were acquired using a desktop scanner (with a resolution of 450 dpi) directly from Arabidopsis plants growing in Petri dishes after 13 days of culture. Images were analyzed using Optimas software version 6.1 (Media Cybernetics, MD, USA). Data were exported to an Excel worksheet for final processing.

Statistical Analysis

All data are expressed as arithmetic means +/− SD of replicate plants within an experiment. All data shown are from one experiment representative of a total of two or three independent biological experiments. All results were statistically analyzed using Statistica software version 7.1 (Statsoft, Tulsa, OK, USA). For the analysis of variance, 2 factors ANOVA with a LSD Post Hoc test were used to measure differences ($p<0.01$).

Results

H. incana Accumulates Pb in Both Leaves and Roots

An exploration of the flora from mining sites in Oued El Himer region, located in the south of Oujda city (Eastern Morocco), was conducted in order to identify new species able to accumulate heavy metals in their shoots for a future phytoremediation project. The sites we studied have been heavily affected by lead mining and smelting activities, and soils are subject of a polymetallic contamination [24]. In these areas, lead concentrations in soils varied from 26 to 9479 $mg.kg^{-1}$ [24]. Plants were collected and leaf samples were analyzed by ICP-AES in order to quantify heavy metal contents. The *Brassicaceae H. incana* was particularly interesting because of its accumulation profile of toxic heavy metals in leaf tissue (Figure 1). In natural conditions, this species presented a high level of Pb in leaves ranging from 0.53 to 1.43 $mg.g^{-1}$ DW with an average of 0.79 $mg.g^{-1}$ DW (Figure 1). The other metals analyzed were present at lower concentration such as Cd or Zn with an average of 0.04 and 0.11 $mg.g^{-1}$ DW respectively even if the concentrations of these metals were high in the different sampled soils [24]. These results show that *H. incana* seems to be highly specific for Pb accumulation in accordance to the contents of the different heavy metals analyzed. In order to confirm the accumulator trait observed in natural conditions and to avoid air-borne contamination such as dust deposits, *H. incana* plants were grown under controlled growth chamber conditions on four different soils collected in metalliferous mine spoils. These soils contained various amounts of total Pb (soil 1:6973 $mg.kg^{-1}$ DW; soil 2:18626 $mg.kg^{-1}$ DW; soil 3:7531 $mg.kg^{-1}$ DW; soil 4:1577 $mg.kg^{-1}$ DW, table 1). After 2 months of culture, Pb content was quantified in the shoots (Figure 2). The concentration of Pb in leaves was variable from a minimum of 0.43 $mg.g^{-1}$ DW for plants grown on soil 1 to a maximum of 3.58 $mg.g^{-1}$ DW for those grown on soil 3. These results confirmed the lead hyperaccumulator trait for *H. incana.* On the other hand, no correlation was observed between total Pb content in the different soils and Pb level in the plant tissues suggesting variations in the amount of bioavailable lead. CaCl2 and EDTA extractible fractions corresponding respectively to the mobile and the mobilisable fractions were quantified (Table 1). No correlation could be demonstrated between the Pb levels of the extractible fractions and those of the *H. incana* assimilated fraction.

In the light of these results and according to the standard criteria on heavy metal hyperaccumulation (i.e. >0.1% DW of Pb; [4,8]), *H. incana* could be considered as a lead hyperaccumulator plant.

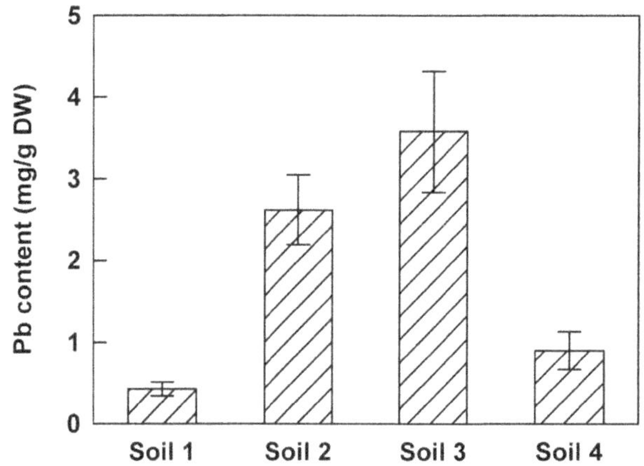

Figure 2. Lead concentration in shoots of *H. incana* growing on different polluted soils. Seedlings were grown on sterile compost for 20 days and transferred for 2 months on 4 different soils containing various amounts of total lead (soil 1:6972 $mg.kg^{-1}$ DW, soil 2:18626 $mg.kg^{-1}$ DW, soil 3:7531 $mg.kg^{-1}$ DW and soil 4:1577 $mg.kg^{-1}$ DW; Table 1). Data are the average (± SE) of three independent measurements.

H. incana Translocates Pb from Roots to Shoots

A four-day kinetic study, in hydroponic conditions, was achieved to confirm the transport of lead from roots to shoots in *H. incana* (Figure 3). Three Pb concentrations were tested (50, 100 and 300 µM $Pb(NO_3)_2$). In the study period and with these concentrations of Pb, no visual damage could be observed neither in leaves and roots. The profiles of lead accumulation in roots or in shoots were similar for the different concentrations with a proportional response to the Pb concentration in the medium. In hydroponic conditions, lead accumulation in roots and shoots was found to be dose and time dependent: roots showed more accumulation than shoots at the same concentration and exposure periods. In the shoots, a slow but constant increase of lead content after a latency period of 24 hours was observed for the three concentrations (Figure 3A). In the roots, the levels of Pb gradually increased with time of exposure for expected values of 42, 69 and 121 $mg.g^{-1}$ DW for respectively 50, 100 and 300 µM of Pb in the medium. A slight decrease of the curve was observed for the last points of kinetics for the 50 µM concentration and a plateau for the 100 µM concentration (Figure 3B). Finally a maximum of translocation was observed at day-4 for 300 µM of Pb in the medium where 3% of total Pb in the plant were localized in the shoots.

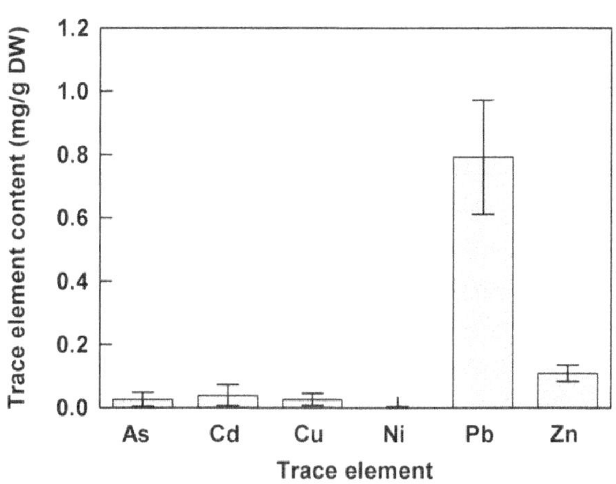

Figure 1. Amount of metallic trace elements (As, Cd, Cu, Ni, Pb and Zn) in shoots of *H. incana.* Plants were collected from different zones in a heavy metal-polluted area. Data are the average (± SE) of six independent measurements. Amount (average ± SE) of metallic trace elements in the soil (mg/Kg): As = 298±80; Cd = 10±4; Cu = 459±88; Ni = 13±6, Pb = 7157±746; Zn = 14164±1460.

In order to evaluate the distribution of Pb in different parts of the plant, *H. incana* was grown in hydroponic conditions supplied weekly with fresh medium containing 100 µM $Pb(NO_3)_2$. After 2 months of Pb exposure, *H incana* was able to accumulate up to 106 and 77 mg $Pb.g^{-1}$ DW in roots at the vegetative stage and in roots at the floral stage respectively (Figure 4A). The aerial plant organ with the highest Pb concentration was the rosette leaves reaching up to 0.95 $mg.g^{-1}$ DW following by the siliques where the Pb content may reach 0.65 $mg.g^{-1}$ DW. When data were expressed per organ (Figure 4B), strong accumulation of Pb was also observed in the roots (19 and 23.7 mg Pb in roots at the vegetative stage and in roots at the floral stage respectively) but a significant presence of Pb has also been noted in the shoots (0.54 mg Pb in the rosette leaves). These results confirm that the highest amount of lead accumulated in studied plants stays at the root level as described above but an important translocation of Pb can be observed from the roots to the leaves and even to the siliques.

Identification of Genes Involved in Lead Tolerance in *H. incana*

In order to characterize the molecular mechanisms involved in Pb tolerance and accumulation in *H. incana*, we identified *H. incana* homologues of several genes previously described in the literature as involved in heavy metal tolerance but not necessarily to Pb, such as *ATM3* (ATP-binding cassette transporter of mitochondrial protein), *CNGC1* (cyclic nucleotide-gated channel), *GS2* (glutathione synthetase), *HMA4* (heavy metal ATPase), *MRP3* (multidrug resistance-associated protein), *MT2a* (metallothionein) and *PCS1* (phytochelatin synthase). A tubulin (*TUB*) gene was identified and used as a reference gene for analysis of gene expression. Partial cDNA sequences were obtained by RT-PCR on leaves or roots total RNA from *H. incana* using primer sequences chosen in the literature or in consensus regions of the corresponding proteins from *A. thaliana* (Table S2). Identification of *H. incana* cDNA

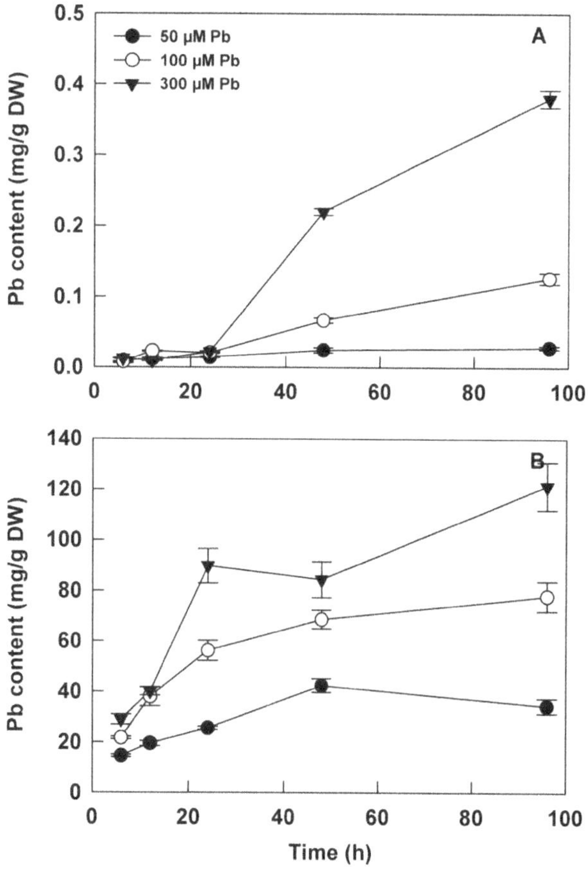

Figure 3. Kinetic effect of lead exposure on lead accumulation in (A) shoots and (B) roots of *H. incana*. Seedlings were grown on hydroponic system and lead treatments were done after 2 weeks by adding 50, 100 and 300 μM $Pb(NO_3)_2$ in fresh medium. Data are the average (± SE) of three independent measurements.

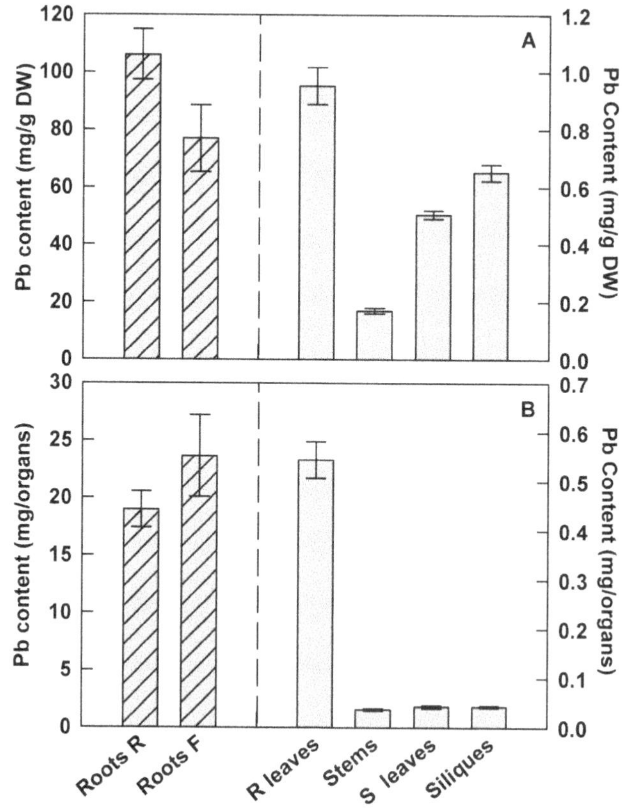

Figure 4. Lead distribution in different parts of *H. incana*. Lead contents are expressed as (A) mg $Pb.g^{-1}$ DW or (B) mg $Pb.organ^{-1}$. Seedlings were grown for 2 months in hydroponic conditions supplied weekly with fresh media containing 100 μM $Pb(NO_3)_2$. Roots R = roots at the rosette stage; Roots F = roots at the floral stage; R leaves = rosette leaves; S leaves = stem leaves. Data are the average (± SE) of three independent measurements.

sequences was based on the analysis of deduced amino acid sequences, using the BLASTX program. This analysis revealed that all isolated cDNA sequences from *H.incana*, which contained one or more motifs characterizing the corresponding proteins, were highly similar to equivalent sequences from *A. thaliana* (Table 2). On the basis of these 8 genes, the percentage of sequence identities found between *H. incana* and *A. thaliana* was 89% in average with a maximum of 93% for the *PCS1* gene and a minimum of 84% for the *CNCG1* gene in the coding regions. The intergenic transcribed spacer (ITS) region from *H. incana* was cloned and sequenced. The percentage of identity in the ITS region between *H. incana* and *A. thaliana* was 85%. These results verify a close genetic proximity between *H. incana* and *A. thaliana*. This does not mean that the nearest species of *H. incana* was *A. thaliana*. For each gene studied, comparison with available sequences shows that H. incana is also close to *Brassica rapa*, *Noccaea caerulescens*, *Brassica juncea* and *Arabidopsis lyrata* (Figure S2).

The expression profile of the selected genes in response to Pb treatment was determined in *H. incana* and in the non-tolerant *A. thaliana* (Figure 5). Seedlings previously grown on the standard medium were transferred to Pb medium (100 and 40 μM $Pb(NO_3)_2$ respectively for *H. incana* and *A. thaliana*) for a further 3 days. For the selected genes, basic gene expression levels were low in both plants and organs (data not shown). In roots of *H. incana*, an important increase of transcript levels (fold change >2) associated to Pb-treatment was observed for five out of the seven target genes studied: *HiATM3*, *HiGS2*, *HiHMA4*, *HiMRP3* and *HiMT2a* but, among these, only *ATM3*, *GS2* and *HMA4* have a significant expression greater than that obtained in *A. thaliana* (Figure 5A). In leaves of *H. incana*, Pb treatment enhanced approximately by two-fold the expression of *HiHMA4* and more than eight-fold the expression of *HiMT2a* (Figure 5B). The expression of these genes was not significantly increased in Pb-treated plants of *A. thaliana*. These two genes, *HiHMA4* and *HiMT2a*, were particularly affected in both roots and leaves suggesting putative roles in lead tolerance and accumulation in *H. incana*.

If *HiHMA4* and *HiMT2a* are important for lead tolerance, as suggested by their overexpression in the presence of Pb, we expect the corresponding Arabidopsis T-DNA mutants to be more sensitive to heavy metals than wild type plants. To test the physiological function of HMA4 and MT2a in *A. thaliana*, two homozygote lines that contain T-DNA insertion in the *HMA4* gene (Salk_093482, Figure S3) and in the *MT2a* gene (Salk_059712, Figure S4) respectively were isolated from the T-DNA insertion collection generated at the Salk Institute (http://signal.salk.edu). To evaluate the consequences of *HMA4 and MT2a* gene disruption on root growth responses to Pb treatment, we examined the primary root length of mutant plants grown under Pb-treatment and control conditions and found them to be distinguishable from the wild type plants. Wild-type, *hma4* and

Table 2. Description of the cDNA sequences isolated from *H. incana*.

Gene name	Length (bp)	Accession[a]	Similarity to *A. thaliana*[b]
HiATM3	407	HQ398196.1	91% to At5g58270
HiCNGC1	866	HQ398199.1	84% to At5g53130
HiGS2	531	HQ398198.1	88% to At5g27380
HiHMA4	863	HQ398195.1	86% to At2g19110
HiMRP3	633	HQ398194.1	89% to At3g13080
HiMT2a	184	HQ398197.1	88% to At3g09390
HiPCS1	118	JF288760.1	93% to At5g44070
HiTUB	1066	HQ398200.1	92% to At1g50010

[a]GeneBank accession number (http://www.ncbi.nlm.nih.gov/genbank/).
[b]AGI gene code (http://www.arabidopsis.org/).

mt2a seeds were germinated directly on standard medium or on medium with 40 µM $Pb(NO_3)_2$. The lead concentration used in this experiment corresponds to the threshold of tolerance that Arabidopsis can withstand in our experimental conditions (data not shown). After 13 days of culture, Pb-treated wild-type plants did not seem to be affected by the Pb treatment compared to the control plants. Similarly when grown on standard medium, both *hma4* and *mt2a* mutants and wild-type plants showed a similar root phenotype. Pb treatment highly significantly reduced primary root length in both *hma4* (45%) and in *mt2a* (48%) mutants (Figure 6). This reduction in primary root growth reflects an increased sensitivity to Pb in the Arabidopsis T-DNA insertion mutants and it suggests that the *HMA4* and *MT2a* genes contribute to lead tolerance. We measured the Pb contents in roots and shoots of the 2 week-old wild-type and *hma4*- and *mt2a*-mutants (Figure 7). Roots of *hma4* plants contained 1.8-fold more Pb than those of the wild-type plants ($p<0.01$). Although the Pb content of *mt2a*-plants seems slightly higher than that of wild-type plants in $Pb(NO_3)_2$-containing medium, the difference was not statistically significant. No effect was observed in Pb content in the shoots. The Pb accumulation observed in roots of *hma4* plants is compatible with an interruption of the translocation resulting from *HMA4* gene disruption.

Discussion

Unlike Zn and Cu which are essential micronutrients, Pb is non-essential and is detrimental to plant development. Its concentration through the food chain adversely affects biological functions of all living organisms [27]. The identification of genes associated with Pb tolerance and accumulation in plants is the first step towards their application in phytoremediation. Genes conferring Pb tolerance are rare [28] and rarer still are those that confer both Pb tolerance and Pb accumulation.

H. incana is a Lead Accumulator Plant

High tolerance to heavy metals has evolved in a number of plant species leading to a class of rare plants named hyperaccumulators. In the case of Pb the number of plant species able to accumulate this heavy metal in any aboveground tissue in their natural habitat is surprisingly low with a poor degree of diversity [5,9]. This fact may become an obstacle for the development of phytoremediation projects. The first implementation of phytoremediation strategies is the identification of plant species adapted to environmental conditions and able to accumulate large amount of at least one

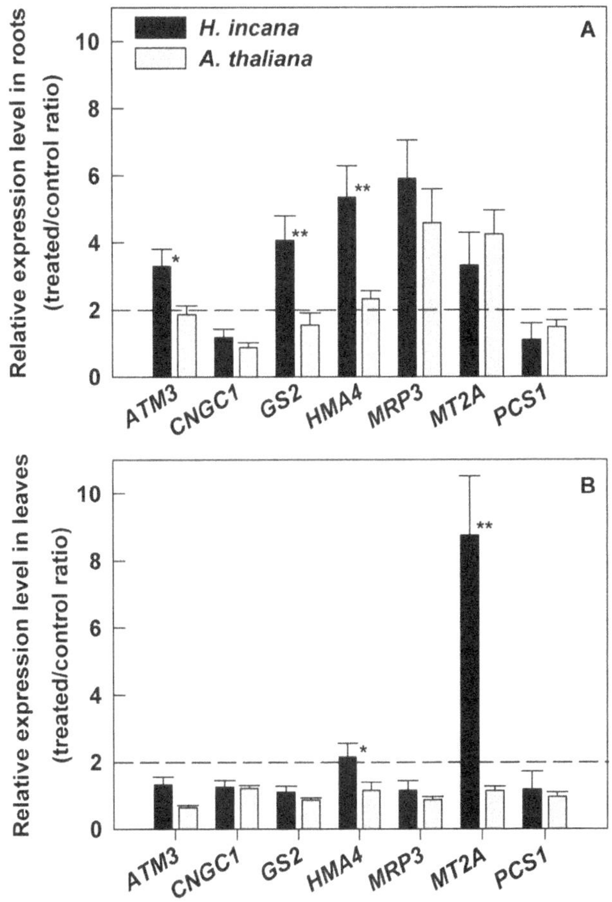

Figure 5. Quantitative analysis of the expression levels of *ATM3, CNGC1, GS2, HMA4, MRP3, MT2a and PCS1* genes. Expression profiles were obtained from (A) roots and (B) leaves of *H. incana* and *A. thaliana*, Pb-treated or not for 3 days. *ATM3*: ATP-binding cassette transporter of mitochondrial protein, *CNGC1*: cyclic nucleotide-gated channel, *GS2*: glutathione synthetase, *HMA4*: heavy metal ATPase, *MRP3*: multidrug resistance-associated protein, *MT2a*: metallothionein and *PCS1*: phytochelatin synthase. The tubulin gene was used as reference gene and the calibration was done against the roots or the leaves of plants cultivated without Pb. Data are expressed as relative expression level (treated/control ratio) and are the average (± SE) of three independent replicates composed by nine seedlings. * and **Significant difference of values at $p<0.05$ and $p<0.01$, respectively, by Student's t-test in comparison between *H. incana* and *A. thaliana*. Dashed line indicates a fold change of 2.

trace element. For Pb, the most common definition of a hyperaccumulator plant meets the following requirement: the concentration of metal in the shoot must be higher than 1 $mg.g^{-1}$ DW of Pb [4,8]. Considering this definition, our data obtained from plants collected in tailing mine areas have shown that *H. incana* is a Pb hyperaccumulator (Figure 1). *H. incana* was previously described to accumulate Cu [29], Tl [30] and Zn [31]. Nevertheless, on the polluted site we studied, *H. incana* seems to accumulate preferentially Pb even if Cd, Cu and Zn are present in large amounts in the soil (Figure 1; [24]). The absence of pronounced accumulation of other heavy metals by *H. incana* may be due to competitive interactions between metal ions in these multi-contaminated soils, inhibiting uptake of heavy metals, relative to medium having elevated levels of single metals [31,32]. In controlled growth conditions on different Pb contaminated soils with no possibility of air-borne contaminations, *H.*

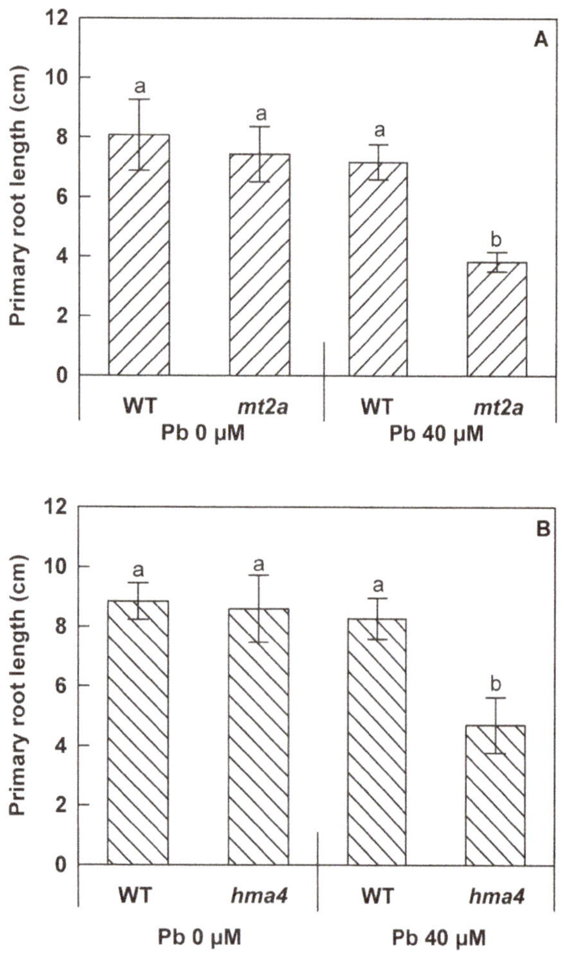

Figure 6. Effects of lead treatment on primary root length in wild-type and knockout mutant seedlings of *A. thaliana*. Seedlings were grown for 13 days on vertically orientated agar plates with or without 40 μM of $Pb(NO_3)_2$. (A) *Atmt2a* T-DNA mutant analysis. (B) *Athma4* T-DNA mutant analysis. All results are the average value (± SE) of 16 seedlings. The letters represent statistically homogenous subgroups using LSD post hoc test at a $\alpha = 0.01$ significance level.

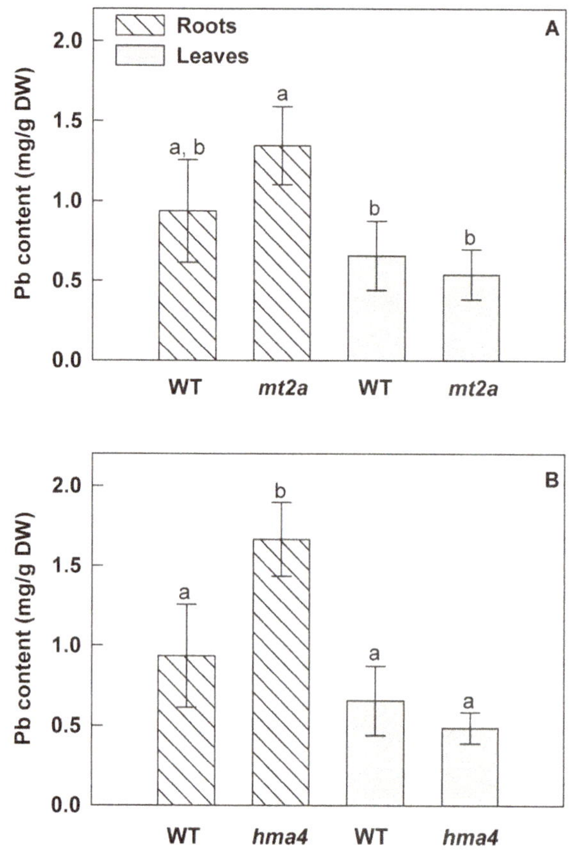

Figure 7. Lead accumulation in roots and shoots of wild-type and knockout mutant seedlings of *A. thaliana*. Seedlings were collected after 13 days of culture on agar plates with or without 40 μM of $Pb(NO_3)_2$. (A) *Atmt2a* T-DNA mutant analysis. (B) *Athma4* T-DNA mutant analysis. All results are the average value (± SE) of three independent replicates. For the roots and leaves separately, the letters represent statistically homogenous subgroups using LSD post hoc test at a $\alpha = 0.01$ significance level.

incana could accumulate concentration until 3.58 $mg.g^{-1}$ DW of Pb in the aboveground part (Figure 2), confirming the hyperaccumulator trait of *H. incana*. Hydroponic cultures using the bioavailable $Pb(NO_3)_2$ in phosphate free medium is an interesting way to estimate Pb accumulation in both roots and leaves in order to confirm Pb translocation from root to leaves (Figure 3). The amount of lead in leaves after 4 days in hydroponic culture with medium containing 300 μM of $Pb(NO_3)_2$ reaches 0.4 $mg.g^{-1}$ DW. With a longer Pb-exposition, we find lead not only in the stems, the rosette and the stem leaves but also in the siliques (Figure 4). Large differences of lead concentration between root and leaf may indicate an important restriction of the internal transport of this metal from the roots towards the stem, leaves and siliques. However our results rather showed that, although *H. incana* accumulated high concentrations of Pb in the shoots (>1$mg.g^{-1}$ DW of Pb), Pb was mainly stored but not completely immobilized in its roots. Highest concentrations of Pb in roots than in shoots seem to be a general pattern of lead distribution in plants [27,33,34]. Taken together, these different experiments demonstrate that *H. incana* could be classified as a Pb accumulator species because it tolerates by far more than 1 $mg.g^{-1}$ DW of Pb in the shoots. However the second criteria defining a hyperaccumulating species, which consists in a higher concentration of the considered heavy metal in the shoots than in the roots, was not fulfilled in *H. incana*. Nevertheless this species implements all the mechanisms that are necessary to translocate Pb from the roots to the aerial parts of the plant and accumulate Pb in different organs.

H. incana is a Powerful Tool to Identify Genes Involved in Pb Tolerance and Accumulation

Several genes previously described in the literature as being involved in heavy metal tolerance and accumulation were selected in order to measure their expression pattern in Pb-treated plants compared to control plants from *H. incana* and *A. thaliana* (Figure 5). Thus among these genes, ATM3 is a mitochondrial transporter that is essential for Fe homeostasis in *A. thaliana* [35] but Kim *et al.* [14] provided several lines of evidence to implicate ATM3 in heavy metal resistance. They showed that the expression of *ATM3* gene is induced either by Cd or Pb and demonstrated that *ATM3*-overexpressing plants grow better than wild-type plants on either Cd- or Pb-containing medium. The activity of *MRP3* promoter from *A. thaliana*, a gene encoding an ABC-transporter, is induced by As, Cd, Cu, Ni, and Pb, but not by Zn and Fe [17]. Disruption by T-DNA insertion mutagenesis of the Arabidopsis gene *CNGC1* conferred plant tolerance to Pb [11]. HMA4 is a Zn/Cd transporter that presented plant tolerance to Cd by loading it into the xylem, thus increasing translocation to the shoot where it might have less damaging effects [13,36,37]. Enhanced expression of *HMA4* gene has been shown to be essential for Cd tolerance in Cd-accumulator *A. halleri* [38]. Yeast functional complementation assays revealed that the expression of *MT1a*, *MT2a* and *MT3a* genes in Arabidopsis could increase the tolerance of yeast mutants to Cu and Cd [39,40,41]. However, with regard to Pb, little is known about the relationship between the induction of plant MT genes and lead tolerance [42,43,44]. It is also the same for phytochelatin that were identified as heavy metal-binding peptides involved in the accumulation, detoxification and metabolism of metal ions (for review [18]). But an alternative to phytochelatins could be glutathione that can also bind lead [19].

In our experiments, gene expression profiles in *H. incana* showed a significant up-regulation (FC>2) by Pb-treatment for some genes such as *HiATM3, HiGS2, HiHMA4, HiMRP3* and *HiMT2a* in roots (Figure 5A) and *HiHMA4* and *HiMT2a* in aerial parts (Figure 5B). The analysis of the expression of these genes in *A. thaliana*, a non Pb tolerant species, showed that only *ATM3, GS2* and *HMA4* were specifically over-expressed in roots and only *HMA4* and *MT2a* were specifically over-expressed in shoots by Pb-treatment in *H incana* (Figure 5). In agreement with previous report, an overexpression of *ATM3* was observed in *H. incana* roots by Pb treatment [14].

From this expression profiling experiment (Fig. 5), two genes, *HMA4* and *MT2a*, retained our attention because they were both specifically over-expressed in roots and/or shoots of *H. incana* compared to *A. thaliana.* The expression of *HiHMA4* and *HiMT2a* was induced two-fold and eight-fold respectively by Pb-exposure in the leaves as well as five-fold and three-fold respectively in the roots. In previous reports *HMA4* gene expression was shown to be modulated by Cd, Mn and Zn treatments [45,46], and, in the present work, by Pb treatment. The involvement of *HiHMA4* and *HiMT2a* genes in the tolerance process was indirectly studied using T-DNA insertional mutants from Arabidopsis. Two *A. thaliana* mutant lines for respectively *HMA4* and *MT2a* genes were isolated and compared with Col0 wild-type for root growth and Pb content. A significant primary root length decrease was observed under Pb-treatment for both mutant lines, which comforted the implication of these two genes in lead tolerance mechanisms in *A. thaliana* (Figure 6). Similar approaches were used for the *Athma4* mutant in presence of Zn, Cd and Co but no significant phenotype was found [36,47] whilst in our case AtHMA4 confers resistance to Pb. In parallel a significant increase of Pb content was observed in the roots and a slight but not significant decrease in shoots of *Athma4* mutants under Pb-treatment (Figure 7). This result can be interpreted as the blocking of translocation leading to an accumulation of Pb in roots and a reduction in shoots. This hypothesis is in concordance with the role of HMA4 in cytosolic metal efflux [6,13,48]. In the Cd/Zn hyperaccumulators, *A. halleri* and *N. caerulescens*, HMA4 would be involved in the detoxification of roots by translocated Cd or Zn to the shoots [13,6]. HMA4 seems to be implicated in lead accumulation in *H. incana* by functioning as metal xylem loading.

For the *MT2a* gene both the up-regulation of the expression by lead exposure in *H. incana* and the phenotype of *A. thaliana mt2a* mutant under Pb-treatment confirmed, for the first time, the implication of a metallothionein in lead tolerance. Several lines of evidence suggest that, as in mammals, plant metallothioneins are involved in the response to oxidative stress and metal toxicity as well as in the regulation of metal homeostasis [49,50,51]. In *Festuca rubra*, functional complementation studies using yeast mutant confirmed the functional implication of *MT1* gene in sequestering both essential (Cu, Zn) and non-essential metals (Cd, Pb, Cr) [52]. Recently it was demonstrated that the expression of *AtMT2a* was induced under H_2O_2 and low temperature stresses, and AtMT2a functions as an ROS scavenger in the cytosol under abiotic stress conditions [53]. A similar role of MT2a in response to lead exposure can also be suggested.

Taken together, our results demonstrated that *H. incana* is an interesting experimental model to identify new genes involved in the molecular mechanisms of Pb accumulation and tolerance in plants. Indeed its close genetic proximity to *A. thaliana* allows the use of its powerful genetic resources: complete sequencing and annotated genome to identify target genes, T-DNA insertion mutants for functional analysis of these genes.

Supporting Information

Figure S1 Phylogenic tree, based on the unweighted pair group method with arithmetic mean, showing the relationships (A) between HiHMA1, AtHMA1 (AT4G37270), AtHAM2 (AT4G30110), AtHMA3 (AT4G30120) and AtHMA4 (AT2G19110), (B) between HiMT2a, AtMT2a (AT3G09390), AtMT2b (AT5G02380), AtMT3 (AT3G15353). Bootstrap values are indicated in percentage (100 replicates). Multiple sequence alignments were made from the coding DNA sequence using Clustal software.

Figure S2 Phylogenetic trees based on the Neighbor-joining method showing the relationships between *Hirschfeldia incana* and the others *brassicaceae*. (A) ATM3, *H. incana*: HQ398196, *Arabidopsis thaliana*: NM_125212, *Thellungiella halophila*: AK353402, *Arabidopsis lyrata*: XM_002864513, *Brassica rapa*: AC189355, *Noccaea caerulescens*: AJ746246 and *Populus trichocarpa*: XM_002325589. (B) CNGC1, *H. incana*: HQ398199, *A. thaliana*: NM_124692, *A. lyrata*: XM_002865908 and *P. trichocarpa*: XM_002317724. (C) GS2, *H. incana*: HQ398198, *A. thaliana*: NM_122620, *A. lyrata*: XM_002872228, *B. rapa*: GQ996584, *Oryza sativa*: EU267952. (D) HMA4, *H. incana*: HQ398195, *A. thaliana*: AF412407, *A. halleri*: AY960757, *A. lyrata*: XM_002886195, *N. caerulescens*: JQ904707, *Brassica juncea*: EU418580, *Ricinus communis*: XM_002532190. (E) MRP3, *H.incana*: HQ398194, *A. thaliana*: NM_202570, *A. lyrata*: XM_002884895, *N. caerulescens*: AB162907, *P. trichocarpa*: XM_002300326. (F) MT2A, *H. incana*: HQ398197, *A. thaliana*: NM_111773, *A. lyrata*: XM_002884686, *B. juncea*: Y10850, *B. rapa*: GQ996588, *Brassica oleracea*: AF200712, *N. caerulescens*: FJ439656, *Nicotiana tabacum*: DQ132853. (G) PCS1, *H. incana*: JF288760, *A. thaliana*: NM_123774, *A. lyrata*: XM_002865338, *A. halleri*: AY463694, *B. juncea*: AJ278627, *N. caerulescens*: AY540104, *Triticum aestivum*: AF093752. (H) TUB, *H. incana*: HQ398200, *A. thaliana*: AY149922, *A. lyrata*: XM_002863549, *B. rapa*: DQ414683, *O. sativa*: DQ683569. Multiple sequence alignments were made from the coding DNA sequence using Clustal software.

Figure S3 Isolation of the *mha4* T-DNA insertional mutant. (A) Intron-exon organization of the *Arabidopsis HMA4* gene (At2g19110) and T-DNA location. Solid black boxes and the solid line indicate coding regions and introns, respectively. The position of the T-DNA insertion in the *hma4* allele is indicated by triangle (not to scale). (B) PCR analysis of *HMA4* transcript in wild-type (Col-0) and mutant allele. Expression of *tubulin* (At1g50010) was analyzed as a control. (C) Phenotype of *hma4* mutant seedlings with or without Pb treatment.

Figure S4 Isolation of the *mt2a* T-DNA insertional mutant. (A) Intron-exon organization of the *Arabidopsis mt2a* gene (at3g09390) and T-DNA location. Solid black boxes and the solid line indicate coding regions and introns, respectively. The position of the T-DNA insertion in the *mt2a* allele is indicated by triangle (not to scale). (B) PCR analysis of *MT2A* transcript in wild-type (Col-0) and mutant allele. Expression of *tubulin* (At1g50010) was analyzed as a control. (C) Phenotype of *mt2a* mutant seedlings with or without Pb treatment.

Acknowledgments

We are thankful to Sophia El Hassouni for technical assistance and Pr Mohamed Ater (Laboratoire Diversité et Conservation des Systèmes Biologiques, Faculté des Sciences de Tétouan, Morocco) for his help and advice in the field.

Author Contributions

Conceived and designed the experiments: AS FA PD LL. Performed the experiments: AB FA MF PM. Analyzed the data: AS FA PD. Contributed reagents/materials/analysis tools: MEM AFM. Wrote the paper: AS FA PD LL.

References

1. Lee M, Lee K, Lee J, Noh EW, Lee Y (2005) AtPDR12 contributes to lead resistance in Arabidopsis. Plant Physiology 138: 827–836.
2. Sharma P, Dubey RS (2005) Lead toxicity in plants. Brazilian Journal of Plant Physiology 17: 35–52.
3. Van Nevel L, Mertens J, Oorts K, Verheyen K (2007) Phytoextraction of metals from soils: How far from practice? Environmental Pollution 150: 34–40.
4. Baker AJM, Reeves RD, Hajar ASM (1994) Heavy metal accumulation and tolerance in British populations of the metallophyte *Thlaspi caerulescens* J. & C. Presl (Brassicaceae). New Phytologist 127: 61–68.
5. Kramer U (2010) Metal hyperaccumulation in plants. Annual Review of Plant Biology 61: 517–534.
6. Verbruggen N, Hermans C, Schat H (2009) Molecular mechanisms of metal hyperaccumulation in plants. New Phytolologist 181: 759–776.
7. Anderson CWN, Brooks RR, Chiarucci A, LaCoste CJ, Leblanc M, et al. (1999) Phytomining for nickel, thallium and gold. Journal of Geochemical Exploration 67: 407–415.
8. Baker AJM, Brooks RR (1989) Terrestrial higher plants which hyper-accumulate metallic elements - A review of their distribution, ecology and phytochemistry. Biorecovery 1: 81–126.
9. Reeves RD (2006) Hyperaccumulation of trace elements by plants. In: Morel JL, Echevarria G, Goncharova N, editors. Phytoremediation of Metal-contaminated Soils. NATO Sciences Series 68. Springer, New York. 25–52.
10. Sunkar R, Kaplan B, Bouche N, Arazi T, Dolev D, et al. (2000) Expression of a truncated tobacco NtCBP4 channel in transgenic plants and disruption of the homologous Arabidopsis CNGC1 gene confer Pb tolerance. Plant Journal 24: 533–542.
11. Gravot A, Lieutaud A, Verret F, Auroy P, Vavasseur A, et al. (2004) AtHMA3, a plant P1B-ATPase, functions as a Cd/Pb transporter in yeast. FEBS Letters 561: 22–28.
12. Morel M, Crouzet J, Gravot A, Auroy P, Leonhardt N, et al. (2009) AtHMA3, a P1B-ATPase allowing Cd/Zn/Co/Pb vacuolar storage in Arabidopsis. Plant Physiology 149: 894–904.
13. Papoyan A, Kochian V (2004) Identification of *Thlaspi caerulescens* genes that may be involved in heavy metal hyperaccumulation and tolerance. Characterization of a novel heavy metal transporting ATPase. Plant Physiology 145: 3814–3823.
14. Kim DY, Bovet L, Kushnir S, Noh EW, Martinoia E, et al. (2006) AtATM3 is involved in heavy metal resistance in Arabidopsis. Plant Physiology 140: 922–932.
15. Kim DY, Bovet L, Maeshima M, Martinoia E, Lee Y (2007) The ABC transporter AtPDR8 is a cadmium extrusion pump conferring heavy metal resistance. Plant Journal 50: 207–218.
16. Xiao S, Gao W, Chen QF, Ramalingam S, Chye ML (2008) Overexpression of membrane-associated acyl-CoA-binding protein ACBP1 enhances lead tolerance in Arabidopsis. The Plant Journal 54: 141–151.
17. Zientara K, Wawrzynska A, Lukomska J, Lopez-Moya JR, Liszewska F, et al. (2009) Activity of the AtMRP3 promoter in transgenic *Arabidopsis thaliana* and *Nicotiana tabacum* plants is increased by cadmium, nickel, arsenic, cobalt and lead but not by zinc and iron. Journal of Biotechnology 139: 258–263.
18. Cobbett C, Goldsbrough P (2002) Phytochelatins and metallothioneins: roles in heavy metal detoxification and homeostasis. Annual review of Plant Biology 53: 159–182.
19. Gupta DK, Huang HG, Yang XE, Razafindrabe BH, Inouhe M (2010) The detoxification of lead in *Sedum alfredii* H. is not related to phytochelatins but the glutathione. Journal of Hazardous Materials 177: 437–444.
20. Murashige T, Skoog F (1962) A revised medium for rapid growth and bio assays with tobacco tissue cultures. Physiologia Plantarum 15: 473–497.
21. Broughton WJ, Dilworth MJ (1971) Control of Leghaemoglobin synthesis in snake beans. Biochemical journal 125: 1075–1080.
22. Scholl RL, May ST, Ware DH (2000) Seed and molecular resources for Arabidopsis. Plant Physiology 124: 1477–1480.
23. Temminghoff EJM, Houba VJG (2004) Digestion with HNO3-H2O2-HF. In: Temminghoff EJM, Houba VJG, editors. Plant Analysis Procedures (second edition). Kluwer, Dordrecht. 16–19.
24. Smouni A, Ater M, Auguy F, Laplaze L, El Mzibri M, et al. (2010) Assessment of contamination by metallic trace elements in a mining area of eastern Morocco. Cahiers Agriculture 19:, 26–76.
25. Marguí E, Queralt I, Carvalho ML, Hidalgo M (2005) Comparison of EDXRF and ICP-OES after microwave digestion for element determination in plant specimens from an abandoned mining area. Analytica Chimica Acta 549: 197–204.
26. White TJ, Bruns T, Lee S, Taylor J (1990) Amplification and direct sequencing of fungal ribosomal RNA genes for phylogenetics. In: Innis MA, Gelfand DH, Sninsky JJ, White TJ, editors. PCR protocols. A guide to methods and applications. Academic Press, San Diego, Calif. 315–322.
27. Pourrut B, Shahid M, Dumat C, Winterton P, et al. (2011) Lead uptake, toxicity, and detoxification in plants. In: Whitacre DM, editor. Reviews of Environmental Contamination and Toxicology 213. Springer, New York. 113–136.
28. Xiao S, Chye ML (2008) Arabidopsis ACBP1 overexpressors are Pb(II)-tolerant and accumulate Pb(II). Plant Signaling & Behavior 3: 693–94.
29. Poschenrieder P, Bech J, Llugany M, Pace A, Fenés E, et al. (2001) Copper in plant species in a copper gradient in Catalonia (North East Spain) and their potential for phytoremediation. Plant and Soil 2: 247–256.
30. Madejón P, Murillo JM, Marañón T, Lepp NW (2007) Factors affecting accumulation of thallium and other trace elements in two wild Brassicaceae spontaneously growing on soils contaminated by tailings dam waste. Chemosphere 67: 20–28.
31. Gisbert C, Clemente R, Navarro-Aviñó J, Baixauli C, Ginér A, et al. (2006) Tolerance and accumulation of heavy metals by Brassicaceae species grown in contaminated soils from Mediterranean regions of Spain. Environmental and Experimental Botany 56: 19–27.
32. Walker DJ, Bernal MP (2004) The effects of copper and lead on growth and zinc accumulation of *Thlaspi caerulescens* J. and C. presl: implications for phytoremediation of contaminated soils. Water Air Soil Pollution 151: 361–372.
33. Yanqun Z, Yuan L, Schvartz C, Langlade L, Fan L (2004) Accumulation of Pb, Cd, Cu and Zn in plants and hyperaccumulator choice in Lanping lead–zinc mine area, China. Environment International 30: 567–576.
34. Deng DM, Deng JC, Li JT, Zhang J, Hu M, et al. (2008) Accumulation of zinc, cadmium, and lead in four populations of *Sedum alfredii* growing on lead/zinc mine spoils. Journal of Integrative Plant Biology 50: 691–698.
35. Kushnir S, Babiychuk E, Storozhenko S, Davey MW, Papenbrock J, et al. (2001) A mutation of the mitochondrial ABC transporter Sta1 leads to dwarfism and chlorosis in the Arabidopsis mutant starik. Plant Cell 13: 89–100.
36. Verret F, Gravot A, Auroy P, Leonhardt N, David P, et al. (2004) Overexpression of AtHMA4 enhances root-to-shoot translocation of zinc and cadmium and plant metal tolerance. FEBS Letters 576: 306–312.
37. Mills RF, Francini A, Ferreira da Rocha PS, Baccarini PJ, Aylett M, et al. (2005) The plant P1B-type ATPase AtHMA4 transports Zn and Cd and plays a role in detoxification of transition metals supplied at elevated levels. FEBS Letters 579: 783–791.
38. Hanikenne M, Talke IN, Haydon MJ, Lanz C, Nolte A, et al. (2008) Evolution of metal hyperaccumulation required cis-regulatory changes and triplication of HMA4. Nature 453: 391–395.
39. Roosens NH, Bernard C, Leplae R (2004) Evidence for copper homeostasis function of metallothionein (MT3) in the hyperaccumulator *Thlaspi caerulescens*. FEBS Letters 577: 9–16.
40. Lee J, Shim D, Song WY, Hwang I, Lee Y (2004) Arabidopsis metallothioneins 2a and 3 enhance resistance to cadmium when expressed in *Vicia faba* guard cells. Plant Molecular Biology 54: 805–815.

41. Guo WJ, Meetam M, Goldsbrough PB (2008) Examining the specific contributions of individual Arabidopsis Metallothioneins to copper distribution and metal tolerance. Plant Physiology 146: 1697–1706.
42. Yu LH, Umeda M, Liu JY, Zhao NM, Uchimiya H (1998) A novel MT gene of rice plants is strongly expressed in the node portion of the stem. Gene 206: 29–35.
43. Xu YF, Zhou GK, Zhou L, Li YQ, Liu JY (2007) Expression patterns of the rice class I metallothionein gene family in response to lead stress in rice seedlings and functional complementation of its members in lead-sensitive yeast cells. Chinese Science Bulletin 52: 2203–2209.
44. Dong CJ, Wang Y, Yu SS, Liu JY (2010) Characterization of a novel rice metallothionein gene promoter: its tissue specificity and heavy metal responsiveness. Journal of Integrative Plant Biology 52: 914–924.
45. Mills RF, Krijger GC, Baccarini PJ, Hall JL, Williams LE (2003) Functional expression of AtHMA4, a P-1B-type ATPase of the Zn/Co/Cd/Pb subclass. Plant Journal 35: 164–176.
46. Bernard C, Roosens N, Czernic P, Lebrun M, Verbruggen N (2004) A novel CPx-ATPase from the cadmium accumulator *Thlaspi caerulescens*. FEBS letters 569: 140–148.
47. Hussain D, Haydon MJ, Wang Y, Wong E, Sherson SM, et al. (2004) P-type ATPase heavy metal transporters with roles in essential zinc homeostasis in Arabidopsis. Plant Cell 16: 1327–1339.
48. Wong CKE, Cobbett C (2009) HMA P-type ATPases are the major mechanism for root-to-shoot Cd translocation in Arabidopsis. New Phytologist 181: 71–78.
49. Murphy A, Taiz L 1(995) Comparison of metallothionein gene expression and nonprotein thiols in ten Arabidopsis ecotypes (correlation with copper tolerance). Plant Physiology 109: 945–954.
50. Jin S, Cheng Y, Guan Q, Liu D, Takano T, et al. (2006) A metallothionein-like protein of rice (rgMT) functions in *E. coli* and its gene expression is induced by abiotic stresses. Biotechnology Letters 28: 1749–1753.
51. Huang GY, Wang YS (2009) Expression analysis of type 2 metallothionein gene in mangrove species (*Bruguiera gymnorrhiza*) under heavy metal stress. Chemosphere 77: 1026–1029.
52. Ma M, Lau PS, Jia YT, Tsang WK, Lam SKS, et al. (2003) The isolation and characterization of Type 1 metallothionein (MT) DNA from a heavy-metal-tolerant plant, *Festuca rubra* cv. Merlin. Plant Science 164: 51–60.
53. Zhu W, Zhao D-X, Miao Q, Xue T-T, Li X-Z, et al. (2009) Arabidopsis thaliana metallothionein, AtMT2a, mediates ROS balance during oxidative stress. Journal of Plant Biology 52: 585–592.

23

Sulfonamide-Resistant Bacteria and Their Resistance Genes in Soils Fertilized with Manures from Jiangsu Province, Southeastern China

Na Wang[1,2]❡, Xiaohong Yang[3]❡, Shaojun Jiao[2], Jun Zhang[3], Boping Ye[3*¶], Shixiang Gao[1*¶]

1 State Key Laboratory of Pollution Control and Resource Reuse, School of the Environment, Nanjing University, Nanjing, 210093, China, 2 Nanjing Institute of Environmental Science, Ministry of Environmental Protection of China, Nanjing, 210042, China, 3 School of Life Science and Technology, China Pharmaceutical University, Nanjing, 210009, China

Abstract

Antibiotic-resistant bacteria and genes are recognized as new environmental pollutants that warrant special concern. There were few reports on veterinary antibiotic-resistant bacteria and genes in China. This work systematically analyzed the prevalence and distribution of sulfonamide resistance genes in soils from the environments around poultry and livestock farms in Jiangsu Province, Southeastern China. The results showed that the animal manure application made the spread and abundance of antibiotic resistance genes (ARGs) increasingly in the soil. The frequency of sulfonamide resistance genes was *sul*1 > *sul*2 > *sul*3 in pig-manured soil DNA and *sul*2 > *sul*1 > *sul*3 in chicken-manured soil DNA. Further analysis suggested that the frequency distribution of the *sul* genes in the genomic DNA and plasmids of the SR isolates from manured soil was *sul2* > *sul1* > *sul3* overall ($p<0.05$). The combination of *sul*1 and *sul*2 was the most frequent, and the co-existence of *sul1* and *sul3* was not found either in the genomic DNA or plasmids. The sample type, animal type and sampling time can influence the prevalence and distribution pattern of sulfonamide resistance genes. The present study also indicated that *Bacillus, Pseudomonas* and *Shigella* were the most prevalent *sul*-positive genera in the soil, suggesting a potential human health risk. The above results could be important in the evaluation of antibiotic-resistant bacteria and genes from manure as sources of agricultural soil pollution; the results also demonstrate the necessity and urgency of the regulation and supervision of veterinary antibiotics in China.

Editor: Jose Luis Balcazar, Catalan Institute for Water Research (ICRA), Spain

Funding: The funding source of this study are the Commonwealth and Environmental Protection project granted by the MEP "The health risk assessment and management technology of veterinary medicine" (201109038), "Study of determination method, pollution level and pollution control strategy of antibiotic-resistant gene in China" (201309031). The funders had no role in study design, data collection and analysis, decision to publish, or preparation of the manuscript.

Competing Interests: The authors have declared that no competing interests exist.

* Email: yebp@cpu.edu.cn (BY); ecsxg@nju.edu.cn (SG)

❡ These authors contributed equally to this work.

¶ These authors also contributed equally to this work.

Introduction

In the past few decades, veterinary antibiotics have been widely used in many countries to treat disease and promote animal growth. However, this release together with antibiotic-resistant bacteria (ARB) is a great concern recently [1], primarily because the land application of antibiotic-polluted manure in agricultural practice not only introduced bacteria carrying antibiotic resistance genes (ARGs) into the soil but also had a significant effect on the ARB promotion and selection. In the soil, antibiotics provide a positive selective pressure for these bacteria [2]. The horizontal transfer of ARGs between bacteria is an important factor in resistance dissemination [3]. It is worth noting that some ARB in soil and manure are phylogenetically close to human pathogens, making genetic exchange more likely [3]. Evidence from the last 35 years demonstrates that there was consistent correlation between the use of antibiotic-contaminated manure on farms and the transfer of ARGs in human pathogens, as well as the direct shift of ARB from animals to humans [4]. Therefore, ARGs are recognized as new environmental pollutants, and special concern is warranted due to their potential environmental and human health risks.

The used amount of veterinary medicines in China is more than that of other countries. According to a 2007 survey, the usage of antibiotics in livestock was almost half of the total antibiotics produced in China, which was 210,000 tons [5]. It was approximately 10-fold higher than in the USA and approximately 300-fold higher than in the UK [6]. It would be a good chance to analyze the impact of livestock practices on ARGs in the environment in China, where the animal farm was large-scale and the antibiotics usage was great [7]. However, there are few reports on veterinary ARGs in China.

Sulfonamides are synthetic veterinary antibiotics that are the most widely used veterinary antibiotics in China, the European Union and some developing countries due to their low costs [8,9].

However, sulfonamides were ranged as "High priority" of veterinary medicines, due to the high potential to reach the environment [10]. Sulfonamide resistance is primarily mediated by the *sul*1, *sul*2 and *sul*3 genes encoding dihydropteroate synthetase (DHPS) with a low affinity for sulfonamides [11–13]. A wide range of bacterial species harbor these genes, which are located in transposons and in self-transferable or mobilizable plasmids with a broad host range; these genes manifest multiple antibiotic resistance that is co-selected by sulfonamides [14–16].

Numerous recent studies have focused solely on the prevalence of sulfonamide resistance genes in bacterial isolates from manured agricultural soils or on the quantification of the total ARGs from environmental soil media to reflect the resistance reservoir. Few studies have systematically covered the identity of sulfonamide-resistant (SR) bacteria and the distribution patterns of sulfonamide ARGs in the total soil DNA and in sulfonamide-resistant bacteria.

The objectives of this study were (i) to determine the influence of the fertilization with antibiotic-polluted manure on the selection of sulfonamide ARB and ARGs and (ii) to investigate the distribution pattern of the *sul*1, *sul*2 and *sul*3 genes in the total soil DNA and the identified SR bacteria. Furthermore, (iii) the identification of the SR bacteria genera and description of the genotypes in each genus were also conducted to identify resistant opportunistic pathogens that increased the risk of ARGs affecting public health. To the best of our knowledge, this is the first comprehensive study of sulfonamide ARB and ARGs in livestock and poultry farms in China. The present study could be important in the evaluation of the pollution of soils used for agriculture by ARB and ARGs from manure; this study also demonstrates the necessity and urgency for the regulation and supervision of veterinary antibiotics in China.

Materials and Methods

Sampling

Soil samples from 10 sites were studied, including four pig farms, four chicken farms, one non-arable agricultural area and one mountain forest. The animal feeding farms of different sizes and scales were selected (detailed information about the sampling sites and the person in charge of sampling are given in Table S1 in File S1). The study was permitted and approved by the Ministry of Environmental Protection, China. The land accessed was not privately owned or protected. No protected species were sampled. There were vegetable cultivation area and grain planting area, which were all fertilized with animal manure, in each animal feeding farm. Therefore, two replicates of 1 kg soil samples for each type in every animal feeding farm were collected from depth of 10 to 15 cm, loaded into sterile glass flasks. The soil samples of the same type in different animal feeding farms were mixed (50 g from each source) to processed within 1 to 2 days after collection. The following description was the name rule of samples: (i) samples from the vegetable region of pig farms collected in the winter, the mixture of which was marked as PVW; (ii) samples from the agricultural region of pig farms collected in winter, the mixture of which was marked as PAW; (iii) samples from the vegetable region of pig farms collected in the summer, the mixture of which was marked as PVS; (iv) samples from the agricultural region of pig farms collected in the summer, the mixture of which was marked as PAS; (v) samples from the vegetable region of chicken farms, the mixture of which was marked as CV; (vi) samples from the agricultural region of chicken farms, the mixture of which was marked as CA; (vii) non-arable soils (marked as NA) where manure was not used for a few years near a Nanjing chicken farm; and (viii) forest soil collected from the Fangshan mountain in the Jiangning district of Nanjing (manure and/or antibiotics were not used), which was marked as F. Soil P represents the mixture of soil samples from a pig farm in winter, and soil C is the mixture of soil samples from a chicken farm. The manure (M) was obtained from chickens that were treated with sulfonamides.

For each sample, 100 g was taken for the isolation of SR bacteria and the measurement of sulfonamide residues, and the remainder was stored at 4°C for DNA extraction. Meanwhile, the concentration of sulfonamides in the samples was analyzed in this study using a previously published method [17].

Viable plate counts

The isolation of SR bacteria from the soil or manure was performed by cultivating bacteria on nutrient broth agar plates containing 60 μg/ml sulfadiazine (SDZ) [15] followed by the spread plate technique [17]. Total bacteria from samples M, F, NA, P and C were cultivated on nutrient broth agar plates without SDZ. In brief, 1.0 ml of each soil sample solution, which was prepared by dissolving 5 g of soil in 45 ml of sterile physiological saline (0.9% NaCl), was mixed with 9 ml of sterile physiological saline. The process was repeated to make additional serial 10-fold dilutions, i.e., 10^{-3}, 10^{-4}, 10^{-5} and 10^{-6}. After 2–5 days of incubation at 37°C, the number of resistant bacteria on the agar plates were counted to calculate the colony-forming units (CFUs) per gram of soil with the following formula: CFU/g soil = 45 × average colony number × dilution factor. For subsequent analyses, SR isolates were randomly picked from the plates of each soil sample, with a total of 237 SR bacterial isolates, including 6 isolates from M; 1 isolate from F; 2 isolates from NA; 65, 57, 25 and 25 isolates from PVW, PAW, PVS, and PAS, respectively; and 20 and 36 isolates from CV and CA, respectively. All bacterial strains were stored at −80°C in nutrient broth medium containing 15% glycerol.

DNA extraction

Total soil DNA was extracted from 0.5 g of soil using a PowerSoil® DNA Isolation Kit (MoBio Laboratories, Carlsbad, California, USA) following the manufacturer's instructions. SR isolates were cultured at 37°C overnight with constant shaking at 200 rpm/min in 5 ml of LB supplemented with 60 μg/ml SDZ. DNA extraction was performed with 3.0 ml of cultured SR isolates using the TIANamp bacteria DNA kit (Tiangen, Beijing, China). The plasmids were extracted with the Biomiga EZgene™ Plasmid Miniprep kit (Biomiga, USA) following the manufacturer's protocol. The genomic DNA and plasmids were examined by 1% and 1.5% agarose gel electrophoresis, respectively. Moreover, the λDNA and DNA5000 were used as the marker of genomic DNA and pasmid, respectively. Usually, the molecular weight of genomic DNA was greater than that of the plasmid.

The detection of the *sul*1, *sul*2, and *sul*3 genes in the SR isolates

The prevalence of the *sul*1, *sul*2, and *sul*3 genes in the genomic DNA and plasmids of the isolates was examined via PCR with gene-specific primers (Table S2 in File S1). The amplification conditions for the *sul*1 and *sul*2 genes were as follows: 94°C for 5 min; 30 cycles of 94°C for 30 s, 69°C for 30 s and 72°C for 45 s; and one cycle of 72°C for 7 min. The amplification conditions for the *sul*3 gene were 94°C for 5 min, 30 cycles of 94°C for 30 s, 52°C for 30 s and 72°C for 60 s, and one cycle of 72°C for 7 min. Gel electrophoresis was performed on 1.5% agarose gels. The CA01 (a bacteria from soil CA) plasmid containing the *sul*1 gene was used as the positive control for the detection of the *sul*1 gene; the M01 (bacteria from chicken manure) plasmid containing the *sul*2 and *sul*3 genes was used as the positive control for the detection of the *sul*2 or *sul*3 genes. *E. coli* DH5α cells were used as the negative control. When the PCR product appeared as a single clear band with the same migration profile as the corresponding gene control, the isolate was counted as positive for that gene.

Quantitative PCR

The relative abundances of the *sul*1, *sul*2, and *sul*3 genes in the soil DNA were determined in triplicate via SYBR Green-based real-time PCR on a CFX96 Touch Real-Time PCR Detection System. The primer sequences are listed in Table S3 in File S1. Each 10-μl reaction mixture contained 5 μl of SYBR Premix (Cwbio, China), 1 μl of 2 μM forward and reverse primer mix, 1 μl of template, and 3 μl of ddH_2O. The PCR conditions were 95°C for 10 min, followed by 39 cycles of 95°C for 15 s and 60°C for 60 s. The samples were assessed via $2^{-\Delta\Delta Ct}$ relative quantitative analysis to compare the relative abundance of the *sul* genes among samples. All samples were analyzed in triplicate. The CA01 (a bacteria from soil CA) plasmid containing the *sul*1 gene was used as the positive control for the detection of the *sul*1 gene; the M01 (bacteria from chicken manure) plasmid containing the *sul*2 and *sul*3 genes was used as the positive control for the detection of the *sul*2 or *sul*3 genes. *E. coli* DH5α cells were used as the negative control.

16S rRNA sequencing of SR isolates

The complete 16S rRNA gene was used to identify the genera present in the bacterial isolates. Genomic DNA was used as the template for the PCR amplification of the 16S rRNA gene using the universal bacterial 16S rRNA primers 27F and 1492R (Table S2 in File S1). Each 50-μl reaction mixture consisted of 1 to 4 μl of genomic DNA, Taq plus polymerase buffer containing 1.5 mM $MgCl_2$, 0.2 mM each of the 4 deoxynucleoside triphosphates (dNTPs), 1 mM each of the 27F and 1492R primers, and 1 U of Taq plus polymerase (Tiangen). PCR was performed using a Bio-Rad thermal cycler under the following conditions: 94°C for 5 min, followed by 30 cycles of 94°C for 30 s, 58°C for 30 s, and 72°C for 1.5 min, and 1 cycle of 72°C for 10 min. The PCR products were separated via electrophoresis on 1.0% agarose gels. The PCR amplicons were sequenced by Sangon (Shanghai, China). A pair-wise 16S rRNA gene sequence similarity was performed using the EzTaxon server (http://www.eztaxon.org/) [18] and NCBI BLAST (http://blast.ncbi.nlm.nih.gov/blast.cgi). A bacterial genus was considered present when a sample 16S rRNA gene sequence was ≥97% identical to the reference sequence of the bacteria in that genus.

Statistical analysis

The statistical analysis was performed using SAS 9.1. The group mean levels were analyzed via a one-way Analysis of Variance (ANOVA). Statistical significance was defined as a p-value≤0.05. This p-value was chosen because the standard error associated with CFU plating and qPCR results are generally approximately 5% of the mean. The mean and standard error (SE) displayed in the figures were generated using the means procedure without transformation.

Results and Discussion

Enumeration of the total culturable microbial populations and SR Bacteria in the soil

The number of total culturable microbial populations on the nutrient agar ranged from 1.96×10^7 to 9.75×10^7 CFU/g soil and that of the SR isolates on the nutrient agar ranged from 4.5×10^5 to 9.0×10^7 CFU/g soil (Figure 1), which were higher than those of the reported aquaculture-agriculture ponds (3.0×10^4 to 1.6×10^6 and 3.0×10^2 to 4.1×10^4, respectively) [19]. The higher numbers of total bacteria and SR isolates were found in chicken manure (9.75×10^7 and 9.00×10^7, respectively), which was most likely due to the amount of easily accessible nutrients in the manure that stimulated the growth of bacteria [20]. The number of SR bacteria from the soils affected by pig or chicken manure (3.02×10^6 to 9.40×10^6 CFU/g soil) was higher than that from non-arable soil (1.96×10^6 CFU/g soil) or forest soil (4.5×10^5 CFU/g soil). This difference was most likely due to the application of manure to the soil. Previous studies reported that manure from treated pigs was rich in antibiotics and bacteria carrying ARGs, which were both transferred to the soil via fertilization [3,10]. Furthermore, the number of SR isolates from the vegetable soils was significantly higher than that from the agricultural soils (5.96×10^6 and 3.02×10^6 CFU/g soil for PVW and PAW, respectively; 9.40×10^6 and 4.98×10^6 CFU/g soil for PVS and PAS, respectively; 7.50×10^6 and 4.11×10^6 CFU/g soil for CV and CA, respectively). Because liquid manure or wastewater was frequently used to irrigate the vegetable region, manure was more frequently applied to the vegetable soils than to the agricultural soils, and the repeated application of manure to the vegetable soils may have increased bacterial resistance. Additionally, the mean number of SR isolates from the winter soils (4.49×10^6 CFU/g soil for PW) was lower than that from the summer soils (7.19×10^6 CFU/g soil for PS). This difference most likely occurred because the temperature in the summer is more suitable for the growth of bacteria than that in the winter.

The concentration sums of sulfadiazine, sulfamerazine, sulfathiazole, sulfamethazine, sulfadimethazine and sulfamethoxazole were 4503, 0, 0.536, 35.6, 25.9, 15.8, 12.6, 239 and 193 μg/kg in the mixed samples of M, F, NA, PVW, PAW, PVS, PAS, CV and CA, respectively. The number of cultivable bacteria was not consistent with the concentration of antibiotic sulfonamides in the soil. The pollution level of sulfonamides was found to be significantly higher in chicken farms than in pig farms, but there was no significant difference among the numbers of cultivable bacteria.

Characterization of SR bacteria

All 237 SR isolates that were identified via 16S r RNA belonged to 26 typical soil bacteria genera, including *Achromobacter*, *Arthrobacter*, *Bacillus*, *Brevibacterium*, *Chryseobacterium*, *Citrobacter*, *Cupriavidus*, *Escherichia*, *Flavobacterium*, *Hydrogenophaga*, *Klebsiella*, *Lysinibacillus*, *Massilia*, *Microbacterium*, *Microvirga*, *Pseudomonas*, *Pseudoxanthomonas*, *Rhizobium*, *Rhodococcus*, *Shigella*, *Sphingobacterium*, *Sphingopyxis*, *Staphylococcus*, *Stenotrophomonas*, *Streptococcus*, and *Streptomyces*. *Bacillus* was the most prevalent genus in all 9 environmental samples with a frequency of 43.88%, followed by *Pseudomonas* and *Shigella* (11.39% and 8.02%, respectively; Figure 2). However, it is reported that *Acinetobacter* was abundant in pig wastewater in Vietnam [21]. Both pig- and chicken-manured soil samples were rich in bacteria species; for example, 12 genera were found in PVW and CA (see Figure S1).

Relative abundance of the *sul* genes in the soils

A qPCR analysis of sulfonamide resistance genes was performed on the total DNA extracted directly from the soil. There was significant variation in the relative quantities of the *sul*1, *sul*2, and *sul*3 genes in the DNA extracted from the eight types of soils (see Figure 3). The DNA from the pig-manured soils (PVW, PAW, PVS and PAS) contained relatively higher copy numbers of *sul*1

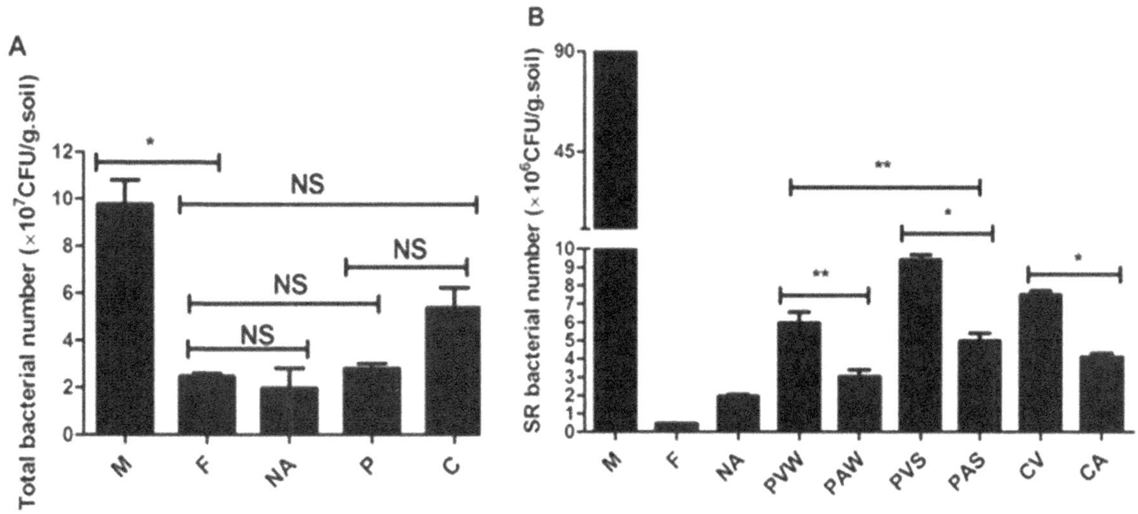

Figure 1. Numbers of cultivable bacteria. (M = Manure, F = Forest, NA = non-arable fied, P = Pig, C = Chicken, W = winter, V = vegetable garden soil, A = agricultural soil; *p≤0.05, **p≤0.01, n = 3; NS, not significant).

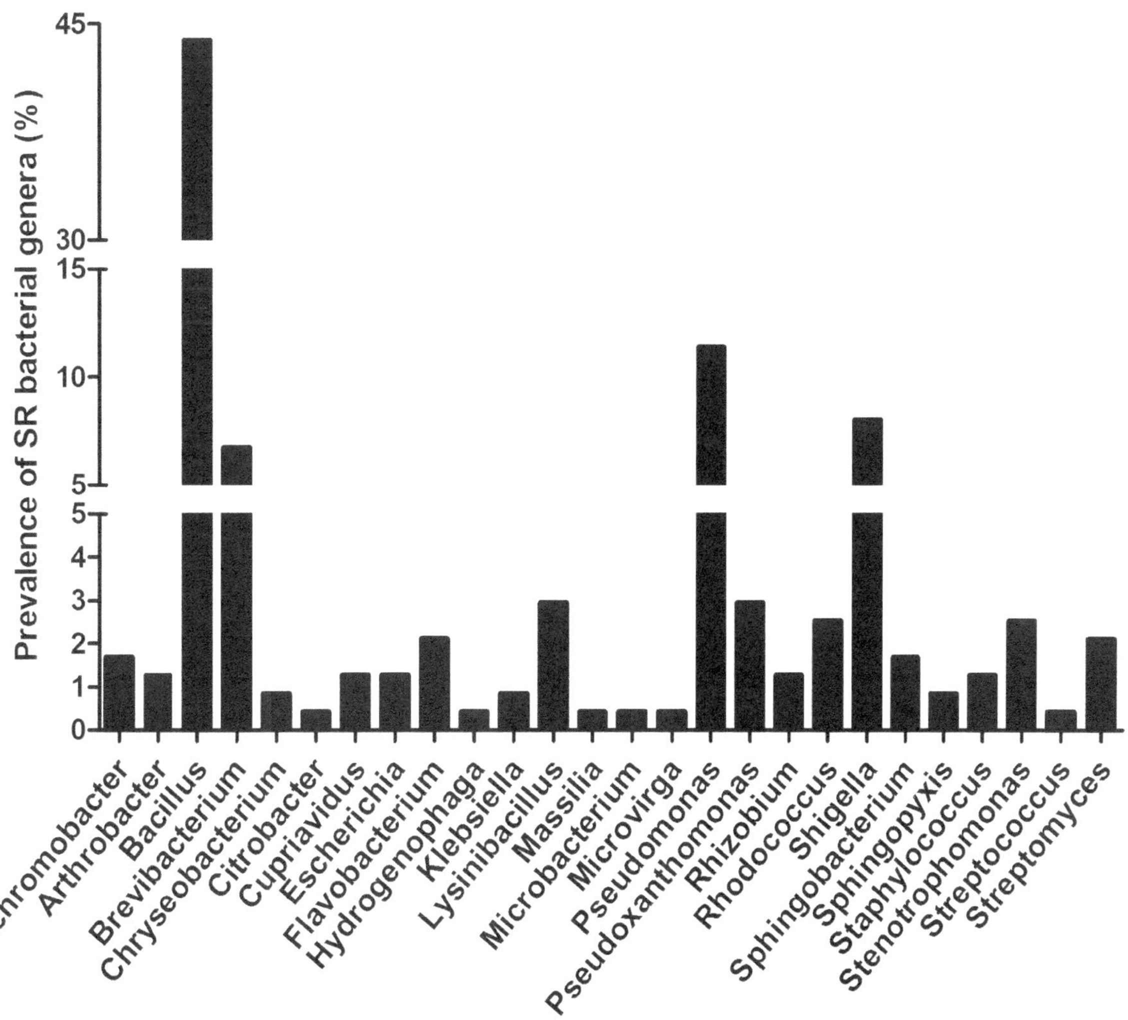

Figure 2. The genera of SR bacteria and their detected frequency in all sampling sites.

than *sul*2. Comparatively, the relative quantity of the *sul*1 gene in the chicken-manured soils was lower than that of the *sul*2 gene. Additionally, the *sul*3 genes were detected at low relative quantities in the DNA extracted from the eight soils but were not detected via PCR in bacteria isolated from forest and pig-manured agricultural (summer) soils. The results of our study were consistent with other reports that demonstrated that the repeated application of manure from pigs or chickens treated with SDZ increased the transfer and abundance of ARGs in the soil [3,10,20]. Furthermore, good positive linear correlations were observed between the relative abundance of the *sul2* genes and the number of culturable SR isolates in the soil. For the *sul2* gene and sum of the three *sul* genes, the correlation coefficients (R^2) were 0.95 and 0.65, respectively ($p<0.05$). However, the abundance of *sul1* and *sul3* showed no significant correlation with the numbers of culturable SR isolates in the soil ($R^2 = 0.44$, $p>0.05$ for *sul1* and $R^2 = 0.39$, $p>0.05$ for *sul3*). This lack of a correlation could be attributed to the fact that the viable plate counts method only sampled microbes that were culturable and expressed their ARGs under those conditions, so most of the microbes carrying *sul1* and *sul3* genes may not be culturable. The other probable reason was that some "silent" or unexpressed *sul1* and *sul3* genes may be existed in the isolates of soils, which could be horizontally transferred or expressed under other conditions.

In brief, the number of culturable SR isolates in the soil can reflect the total relative abundance of the three *sul* genes, showing that the plate count method was effective in assessing the antibiotic resistance risk of the soil. Therefore, the diversity of ARGs enriched at the farm level should be the focus of more attention.

Distribution of *sul* genes in SR isolates

The number and percentage of isolates carrying the *sul* genes in their genomic DNA and plasmids are summarized in Table 1 and Table 2. The distribution and spread of SR genes in the soil microbes are sufficiently frequent to warrant special concern. The *sul1*, *sul2*, and *sul3* genes were all detected at a frequency of 100% in the genomic DNA and plasmids of the SR isolates from the manure sample, indicating that ARGs were extensively harbored in the chromosome and mobile genetic elements of the bacteria in manure, leading to the high potential of horizontal gene transfer of ARGs in soil. Interestingly, the *sul2* genes were only present in the genomic DNA of the isolates collected from forest soil and non-arable soil, which had no history of manure application. This finding may be attributed to the notion that the *sul1* and *sul3* genotype in genomic DNA maybe associate with the amended manure. However, the *sul1*, *sul2* and *sul3* genes were all located in the plasmids of the isolates from non-arable soil but were absent from the plasmids of the isolates from the forest soil; a potential explanation for this difference could be that the bacteria carrying *sul* genes in the manured soil may transfer to the nearby region by aerosolization or runoff, then horizontal transfer occurred in close bacteria via plasmids.

For the manured soil, the frequency distribution of the *sul* genes in the genomic DNA and plasmids of the SR isolates investigated overall followed a trend of *sul2* > *sul1* > *sul3* ($p<0.05$). This result was in contrast to several previous studies showing that the *sul1* gene was more prevalent than the *sul2* gene in the DNA from manure and manured soils [10,15] due to different conditions in various countries. The *sul3* gene was found at low frequencies in

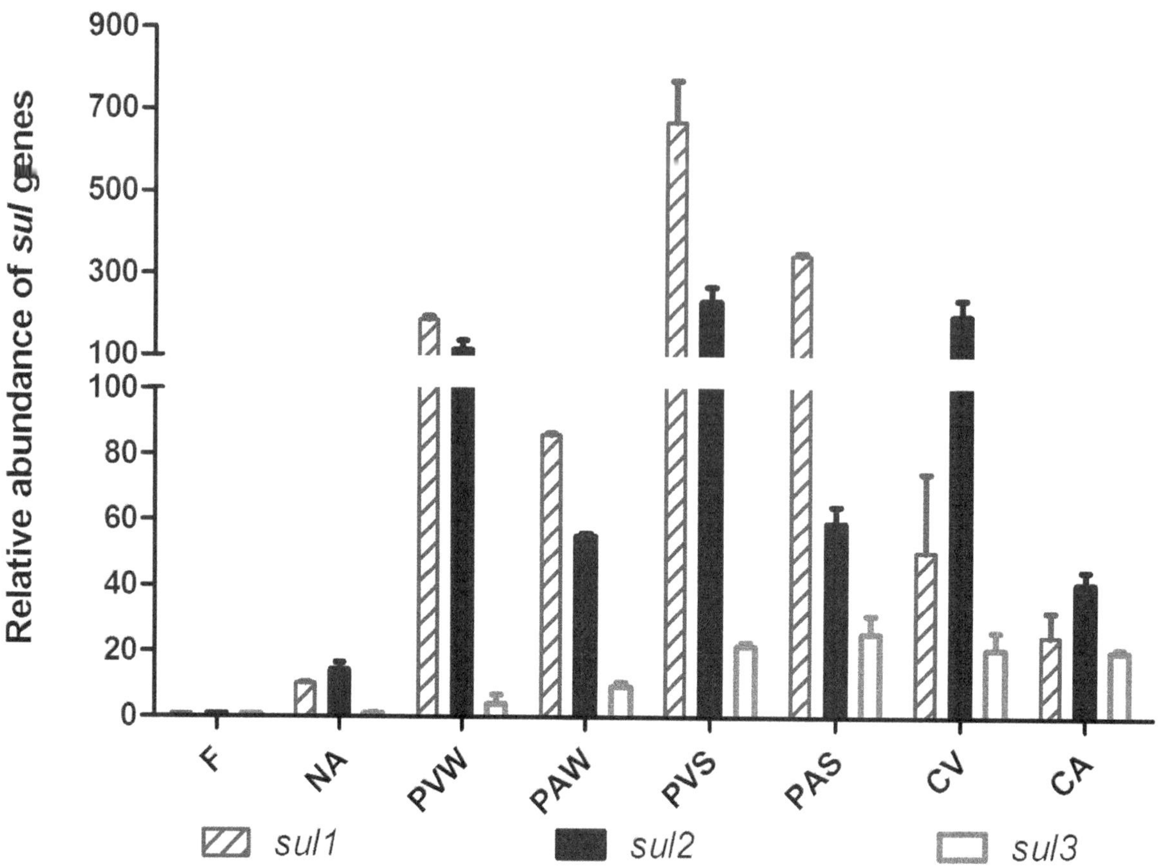

Figure 3. Relative quantity of sulfonamides resistant genes in soils with and without manure treatment.

Table 1. Distribution of *sul*1, *sul*2 and *sul*3 genes in genomic DNA and plasmid of SR isolates (in samples M, F, NA, CV and CA).

sul gene combination		M (n = 6[a]/6[b])		F (n = 1/0)		NA (n = 2/2)		CV (n = 20/20)		CA (n = 36/36)	
		NO. of isolates (%)		NO. of isolates (%)		NO. of isolates (%)		NO. of isolates (%)		NO. of isolates (%)	
		Genomic DNA	Plasmid DNA	Genomic DNA	Plasmid DNA	Genomic DNA	Plasmid DNA	Genomic DNA	Plasmid DNA	Genomi c DNA	Plasmid DNA
Single genes	*sul*1	0 (0.0)	0 (0.0)	0 (0.0)	0 (0.0)	0 (0.0)	0 (0.0)	0 (0.0)	1 (5.0)	2 (5.6)	0 (0.0)
	*sul*2	0 (0.0)	0 (0.0)	1 (100.0)	0 (0.0)	2 (100.0)	0 (0.0)	2 (10.0)	3 (15.0)	3 (8.3)	2 (5.6)
	*sul*3	0 (0.0)	0 (0.0)	0 (0.0)	0 (0.0)	0 (0.0)	0 (0.0)	0 (0.0)	0 (0.0)	1 (2.8)	0 (0.0)
Two genes	*sul*1+*sul*2	0 (0.0)	0 (0.0)	0 (0.0)	0 (0.0)	0 (0.0)	0 (0.0)	4 (20.0)	7 (35.0)	10 (27.8)	2 (5.6)
	*sul*1+*sul*3	0 (0.0)	0 (0.0)	0 (0.0)	0 (0.0)	0 (0.0)	0 (0.0)	0 (0.0)	0 (0.0)	0 (0.0)	0 (0.0)
	*sul*2+*sul*3	0 (0.0)	0 (0.0)	0 (0.0)	0 (0.0)	0 (0.0)	1 (50.0)	0 (0.0)	0 (0.0)	1 (2.8)	9 (25.0)
Three genes	*sul*1+*sul*2 +*sul*3	6 (100.0)	6 (100.0)	0 (0.0)	0 (0.0)	0 (0.0)	1 (50.0)	12 (60.0)	9 (45.0)	19 (52.8)	23 (63.9)
None		0 (0.0)	0 (0.0)	0 (0.0)	0 (0.0)	0 (0.0)	0 (0.0)	0 (0.0)	0 (0.0)	0 (0.0)	0 (0.0)
Total	*sul*1	6 (100.0)	6 (100.0)	0 (0.0)	0 (0.0)	0 (0.0)	1 (50.0)	16 (80.00	17 (85.0)	27 (75.0)	25 (69.4)
	*sul*2	6 (100.0)	6 (100.0)	1 (100.0)	0 (0.0)	2 (100)	2 (100.0)	19 (95.0)	19 (95.0)	29 (80.6)	36 (100.0)
	*sul*3	6 (100.0)	6 (100.0)	0 (0.0)	0 (0.0)	0 (0.0)	2 (100.0)	13 (65.0)	9 (45.0)	21 (58.3)	32 (88.9)
Total of SR isolate positive for *sul* genes		6 (100.0)	6 (100.0)	1 (100.0)	0 (0.0)	2 (100.0)	2 (100.0)	19 (95.0)	20 (100.0)	36 (100.0)	36 (100.0)

a = genomic DNA, b = plasmid.

Table 2. Distribution of *sul*1, *sul*2 and *sul*3 genes in genomic DNA and plasmid of SR isolates (in samples PVW, PAW, PVS and PAS).

sul gene combination		PVW(n = 65/47)		PAW(n = 57/43)		PVS (n = 25/22)		PAS (n = 25/22)	
		NO. of isolates (%)		NO. of isolates (%)		NO. of isolates (%)		NO. of isolates (%)	
		Genomic DNA	Plasmid DNA	Genomic DNA	Plasmid DNA	Genomic DNA	Plasmid DNA	Genomic DNA	Plasmid DNA
Single genes	*sul*1	0 (0.0)	23 (48.9)	0 (0.0)	19 (44.2)	15 (60.0)	0 (0.0)	24 (96.0)	0 (0.0)
	*sul*2	28 (43.1)	3 (6.4)	15 (26.3)	8 (18.6)	1 (4.0)	15 (68.2)	0 (0.0)	16 (72.7)
	*sul*3	0 (0.0)	1 (2.1)	0 (0.0)	2 (4.7)	0 (0.0)	0 (0.0)	0 (0.0)	0 (0.0)
Two genes	*sul*1+*sul*2	34 (52.3)	11 (23.4)	41 (71.9)	9 (20.9)	1 (4.0)	6 (27.3)	1 (4.0)	6 (27.3)
	*sul*1+*sul*3	0 (0.0)	2 (4.3)	0 (0.0)	0 (0.0)	0 (0.0)	0 (0.0)	0 (0.0)	0 (0.0)
	*sul*2+*sul*3	0 (0.0)	0 (0.0)	1 (1.8)	0 (0.0)	0 (0.0)	1 (4.5)	0 (0.0)	0 (0.0)
Three genes	*sul*1+*sul*2 +*sul*3								
		1 (1.5)	0 (0.0)	0 (0.0)	0 (0.0)	0 (0.0)	0 (0.0)	0 (0.0)	0 (0.0)
None		2 (3.1)	7 (14.9)	0 (0.0)	5 (11.6)	8 (32.0)	0 (0.0)	0 (0.0)	0 (0.0)
Total	*sul*1	35 (53.8)	36 (76.6)	41 (71.9)	24 (55.8)	16 (64.0)	6 (27.3)	25 (100.0)	6 (27.3)
	*sul*2	63 (96.9)	14 (29.8)	57 (100.0)	17 (39.5)	2 (8.0)	21 (95.5)	1 (4.0)	22 (100.0)
	*sul*3	1 (1.5)	3 (6.4)	1 (1.8)	2 (4.7)	0 (0.0)	1 (4.5)	0 (0.0)	0 (0.0)
Total of SR isolate positive for *sul* genes		63 (96.9)	40 (85.1)	57 (100.0)	38 (88.4)	17 (68.0)	22 (100.0)	25 (100.0)	22 (100.0)

a = genomic DNA, b = plasmid.

our samples, whereas recently, Suzuki et al showed that sul3 was major sul in seawater [22]. Hoa et al. suggested that most of the *sul* genes are located on the chromosome [15]. However, there was no significant difference between the overall percentage of the isolates carrying the *sul* genes located on the genomic DNA and those on the plasmids in our study. It was interesting to note that the frequency order of the *sul1* and *sul2* genes from the isolates of the pig-manured soils for the genomic DNA was opposite that for the plasmids. In the isolates collected from the pig-manured soils in winter, *sul2* was the most prevalent gene located within the genomic DNA (96.9% and 100.0% in PVW and PAW, respectively) followed by *sul1* (53.8% and 71.9% in PVW and PAW, respectively); *sul1* was the most prevalent gene located on plasmids (76.6% and 55.8% in PVW and PAW, respectively) followed by *sul2* (29.8% and 39.5% in PVW and PAW, respectively). However, in the isolates collected from pig-manured soil in summer, the order of *sul1* (64.0% and 100.0% in PVW and PAW, respectively) > *sul2* (8.0% and 4.0% in PVW and PAW, respectively) in the genomic DNA and *sul2* (95.0% and 100.0% in PVW and PAW, respectively) > *sul1* (27.3% and 27.3% in PVW and PAW, respectively) in the plasmids was determined. We concluded that in most isolates, *sul*1 and *sul*2 were located in the different mobile elements and transferred at different rates.

Furthermore, the animal type was a significant factor influencing the expression frequency of *sul* genes, which showed that the frequency in the chicken-manured soil was higher than that in the pig-manured soil ($p<0.05$), which was consistent with the data of the concentration of sulfonamides in the soil.

We also determined the co-presence of any two different *sul* genes on the chromosome and plasmids in a single isolate. The combination of *sul*1 and *sul*2 on the chromosome was the most frequent and was present in PVW, PAW, PVS, PAS, CV and CA (52.3%, 71.9%, 4.0%, 4.0%, 20.0% and 27.8%, respectively), and the *sul*1, *sul*2 and *sul*3 genes were highly co-present on the chromosomes of M, CV and CA (100%, 60.0% and 52.8%, respectively). The co-presence of *sul*2 and *sul*3 was only detected in two isolates from PAW and CA, respectively, and the co-existence of *sul*1 and *sul*3 was not detected in any SR isolates. The *sul*1 and *sul*2 genes were also frequently detected together in the plasmids (23.4%, 20.9%, 27.3%, 27.3%, 35.0% and 5.6% in PVW, PAW, PVS, PAS, CV and CA, respectively). In contrast, the co-presence of *sul*2 and *sul*3 was only detected in NA (50.0%), PVS (4.5%) and CA (25.0%), and the co-presence of *sul*1 and *sul*3 was not found in any plasmids. Furthermore, the three *sul* genes were co-present in the plasmids of M (100%), NA (50.0%), CV (45.0%), and CA (63.9%). We concluded that the combination of *sul*1 and *sul*2 was the most frequent and that the co-existence of *sul*1 and *sul*3 was not found in the genomic DNA or plasmids. Based on these results, the co-presence of the three *sul* genes was only in the isolates from manure and soil from chicken farms, suggesting that there was a positive correlation between the frequency of the co-presence of the three *sul* genes and the time and amount of repeated manure applications.

In summary, the *sul* genes, either individually or in combinations of two or three, were present in the SR isolates at high frequencies. Nearly all plasmids from the SR isolates contained the *sul* genes (with the exception of F). This observation suggests that the resistance that we observed in most cases was linked to plasmids or other mobile genetic elements, which theoretically have transfer potential. The SR isolates could possibly carry these *sul* genes through gene transfer under selection conditions, leading to an increase in antibiotic resistance among bacteria.

SR bacterial and *sul* genes

The distribution of *sul* genes in bacteria species is listed in Table 3. *Bacillus* was the most prevalent *sul*-positive genus in the soil samples of this study, carrying the *sul* genes in 43.88% of the total isolates; thus, this genus could be the main reservoir of the *sul* genes. This finding was not consistent with other studies that showed that *Acinetobacter* was the dominant genus in aquatic environments (wastewater and shrimp ponds of north Vietnam) and manured agricultural clay soils and slurry samples in the United Kingdom [15,16]. Except for different environments, what makes the difference of genus may be the different condition of culture, such as 28 or 30°C incubation in these two references, not 37°C. It was reported that *Bacillus* spp. have developed resistance to most antibiotic groups, but only a few species of *Bacillus* have been reported to be sensitive to sulfonamides [23]. *Pseudomonas* and *Shigella* were the second and third most prevalent, carrying the *sul* genes in 11.39% and 8.02% of all isolates, respectively. Ventilator-acquired pneumonia, respiratory tract infections in immunocompromised patients and chronic respiratory infections in cystic fibrosis patients were associated with the *Pseudomonas* species (especially *P. aeruginosa*) [24]. *Enterobacteriaceae* species including *Shigella*, *Klebsiella*, and *Escherichia* have represented some of the most dominant bacterial infections over the last 30 years [24]. In the Henan Province of China, 72.6% of infections were caused by *Shigella* strains in 2006 [25].

To the best of our knowledge, this report is the first on *sul* genes in *Chryseobacterium*, *Cupriavidus*, *Flavobacterium*, *Hydrogenophaga*, *Lysinibacillus*, *Massilia*, *Microbacterium*, *Microvirga*, *Pseudoxanthomonas*, *Rhizibium*, *Rhodococcus*, *Sphingopyxis*, *Staphylococcus*, *Streptococcus*, and *Streptomyces* from soils and the first that indicates the widespread presence of ARB in the arable soils of China. Previous studies demonstrated the co-presence of *sul*1, *sul*2 and *sul*3 in a single cell; this was detected in *Acinetobacter*, *Bacillus*, *Psychrobacter*, *Escherichia coli*, and *Salmonella* [15,16,26,27]. In our study, these three *sul* genes were simultaneously found in *Arthrobacter*, *Brevibacterium*, *Citrobacter*, *Cupriavidus*, *Flavobacterium*, *Lysinibacillus*, *Pseudomonas*, *Pseudoxanthomonas*, *Rhizibium*, *Sphingobacterium*, *Staphylococcus*, *Stenotrophomonas*, *Streptococcus*, and *Streptomyces*, with the exception of three genera (*Bacillus*, *Escherichia*, and *Shigella*). This result indicates that the three *sul* genes are common and widely distributed in ARB in soil. Additionally, the *sul*3 gene was detected for the first time in *Achromobacter*, *Chryseobacterium*, *Citrobacter*, *Cupriavidus*, *Flavobacterium*, *Lysinibacillus*, *Pseudoxanthomonas*, *Rhizibium*, *Sphingobacterium*, *Staphylococcus*, *Streptococcus*, and *Streptomyces* from arable soils.

It was revealed that the manured soils could be a reservoir of sulfonamide ARBs and ARGs, according to the observation of high frequency of various combinations of the *sul* genes in bacteria of manured agricultural soils, which may bring potential hazards to human and ecosystem health. Therefore, the diversity of ARGs and ARB enriched at the farm level should be the focus of more attention.

Conclusion

A comprehensive study of sulfonamide ARB and ARGs in livestock and poultry farms in Jiangsu Province of China revealed that the fertilization with antibiotic-polluted manure had a significant influence on the selection of sulfonamide ARB and ARGs. The sample type, animal type and sampling time may affect the prevalence and distribution rule of SR genes. The results from the identification of the SR bacteria genus and the description of the genotypes in the genus revealed that resistant

Table 3. Summary of *sul* genotype of sul-positive bacterial species isolated.

Genus	No. of total sul-positive isolates (%)	Source of isolates	*sul* genotype	No. of sul-positive isolates
Achromobacter	4 (1.69)	NA, PVW, CV	*sul2*	2
			sul1 sul2	1
			sul2 sul3	1
Arthrobacter	3 (1.27)	CV, CA	*sul1 sul2*	1
			sul1 sul2 sul3	2
Bacillus	104 (43.88)	F, PVW, PAW, PVS, PAS, CV, CA	*sul1*	2
			sul2	23
			sul1 sul2	66
			sul2 sul3	1
			sul1 sul2 sul3	12
Brevibacterium	16 (6.75)	PVW, PAW, PVS, PAS	*sul2*	4
			sul1 sul2	11
			sul1 sul2 sul3	1
Chryseobacterium	2 (0.84)	PVW, PVS	*sul1 sul2*	2
Citrobacter	1 (0.42)	CA	*sul1 sul2 sul3*	1
Cupriavidus	3 (1.27)	CA	*sul1 sul2 sul3*	3
Escherichia	3 (1.27)	PVW, CA	*sul1 sul2*	1
			sul1 sul2 sul3	2
Flavobacterium	5 (2.11)	CV, CA	*sul2*	1
			sul1 sul2	1
			sul1 sul2 sul3	3
Hydrogenophaga	1 (0.42)	PVS	*sul1 sul2*	1
Klebsiella	2 (0.84)	PAS	*sul1 sul2*	2
Lysinibacillus	7 (2.95)	PVW, PAW, PAS	*sul1*	1
			sul1 sul2	4
			sul1 sul2 sul3	2
Massilia	1 (0.42)	PVW	*sul2*	1
Microbacterium	1 (0.42)	PAW	*sul1 sul2*	1
Microvirga	1 (0.42)	PAS	*sul1 sul2*	1
Pseudomonas	27 (11.39)	PVW, PAW, CV	*sul2*	1
			sul1 sul2	23
			sul1 sul2 sul3	3
Pseudoxanthomonas	7 (2.95)	PVW, PVS	*sul2*	2
			sul1 sul2	4
			sul1 sul2 sul3	1
Rhizobium	3 (1.27)	PVS, CV	*sul1 sul2*	2
			sul1 sul2 sul3	1
Rhodococcus	6 (2.53)	PVW, PAW, PVS, PAS	*sul1 sul2*	6
Shigella	19 (8.02)	CV, CA, M	*sul1 sul2 sul3*	19

Table 3. Cont.

Genus	No. of total sul-positive isolates (%)	Source of isolates	*sul* genotype	No. of sul-positive isolates
Sphingobacterium	4 (1.69)	VA	*sul*1 *sul*2	1
			*sul*1 *sul*2 *sul*3	3
Sphingopyxis	2 (0.84)	PVW, PAW	*sul*2	1
			*sul*1 *sul*2	1
Staphylococcus	3 (1.27)	PAW, CA	*sul*1 *sul*2 *sul*3	3
Stenotrophomonas	6 (2.53)	CV, CA, NA	*sul*2 *sul*3	1
			*sul*1 *sul*2 *sul*3	5
Streptococcus	1 (0.42)	CV	*sul*1 *sul*2 *sul*3	1
Streptomyces	5 (2.11)	PVW, PAW, CA	*sul*1 *sul*2	1
			*sul*1 *sul*2 *sul*3	4

opportunistic pathogens increased the risk of ARGs affecting public health. Overall, the high frequency of various combinations of the *sul* genes in manured agricultural soil samples of Southeastern China should be the focus of more attention, and the regulation and supervision of veterinary antibiotics are urgently needed in China.

Supporting Information

Figure S1 Prevalences of SR bacteria belonging to different genera identified in the studied soils.

File S1 Contains the following files: **Table S1**. Detailed information on sampling sites in present study. **Table S2**. Primers for PCR in this Study. **Table S3**. Primers for quantitative PCR in this Study.

Author Contributions

Conceived and designed the experiments: NW. Performed the experiments: XHY SJJ JZ. Analyzed the data: XHY NW. Contributed reagents/materials/analysis tools: NW BPY XHY. Contributed to the writing of the manuscript: NW XHY. Guided the experiment: BPY SXG.

References

1. Ghosh S, LaPara TM (2007) The effects of subtherapeutic antibiotic use in farm animals on the proliferation and persistence of antibiotic resistance among soil bacteria. ISME J 1: 191–203.
2. Popowska M, Rzeczycka M, Miernik A, Krawczyk-Balska A, Walsh F, et al. (2012) Influence of soil use on prevalence of tetracycline, streptomycin, and erythromycin resistance and associated resistance genes. Antimicrob Agents Chemother 56: 1434–1443.
3. Heuer H, Schmitt H, Smalla K (2011) Antibiotic resistance gene spread due to manure application on agricultural fields. Curr Opin Microbiol 14: 236–243.
4. Marshall BM, Levy SB (2011) Food animals and antimicrobials: impacts on human health. Clinical microbiology reviews 24: 718–733.
5. Hvistendahl M (2012) Public Health China Takes Aim at Rampant Antibiotic Resistance. Science 336: 795–795.
6. Kim K-R, Owens G, Kwon S-I, So K-H, Lee D-B, et al. (2011) Occurrence and environmental fate of veterinary antibiotics in the terrestrial environment. Water Air Soil Poll 214: 163–174.
7. Zhu YG, Johnson TA, Su JQ, Qiao M, Guo GX, et al. (2013) Diverse and abundant antibiotic resistance genes in Chinese swine farms. Proc Natl Acad Sci U S A 110: 3435–3440.
8. Ungemach F (1999) Figures on quantities of antibacterials used for different purposes in the EU countries and interpretation. Acta Vet Scand Suppl 93: 89–97; discussion 97–88, 111–117.
9. Kools SA, Moltmann JF, Knacker T (2008) Estimating the use of veterinary medicines in the European Union. Regul Toxicol Pharm 50: 59–65.
10. Heuer H, Smalla K (2007) Manure and sulfadiazine synergistically increased bacterial antibiotic resistance in soil over at least two months. Environ Microbiol 9: 657–666.
11. Skold O (2000) Sulfonamide resistance: mechanisms and trends. Drug Resist Updat 3: 155–160.
12. Perreten V, Boerlin P (2003) A new sulfonamide resistance gene (sul3) in Escherichia coli is widespread in the pig population of Switzerland. Antimicrob Agents Chemother 47: 1169–1172.
13. Yun MK, Wu Y, Li Z, Zhao Y, Waddell MB, et al. (2012) Catalysis and sulfa drug resistance in dihydropteroate synthase. Science 335: 1110–1114.
14. Heuer H, Szczepanowski R, Schneiker S, Pühler A, Top E, et al. (2004) The complete sequences of plasmids pB2 and pB3 provide evidence for a recent ancestor of the IncP-1β group without any accessory genes. Microbiology 150: 3591–3599.
15. Hoa PTP, Nonaka L, Hung Viet P, Suzuki S (2008) Detection of the sul1, sul2, and sul3 genes in sulfonamide-resistant bacteria from wastewater and shrimp ponds of north Vietnam. Sci Total Environ 405: 377–384.
16. Byrne-Bailey KG, Gaze WH, Kay P, Boxall AB, Hawkey PM, et al. (2009) Prevalence of sulfonamide resistance genes in bacterial isolates from manured agricultural soils and pig slurry in the United Kingdom. Antimicrob Agents Chemother 53: 696–702.
17. Sengeløv G, Agersø Y, Halling-Sørensen B, Baloda SB, Andersen JS, et al. (2003) Bacterial antibiotic resistance levels in Danish farmland as a result of treatment with pig manure slurry. Environ Int 28: 587–595.
18. Chun J, Lee JH, Jung Y, Kim M, Kim S, et al. (2007) EzTaxon: a web-based tool for the identification of prokaryotes based on 16S ribosomal RNA gene sequences. Int J Syst Evol Microbiol 57: 2259–2261.
19. Hoa PTP, Managaki S, Nakada N, Takada H, Anh DH, et al. (2010) Abundance of sulfonamide-resistant bacteria and their resistance genes in integrated aquaculture-agriculture ponds, North Vietnam. Interdisciplinary studies on environmental chemistry - biological responses to contaminants. Tokyo: TERRAPUB. 15–22.
20. Jechalke S, Kopmann C, Rosendahl I, Groeneweg J, Weichelt V, et al. (2013) Increased abundance and transferability of resistance genes after field application of manure from sulfadiazine-treated pigs. Appl Environ Microbiol 79: 1704–1711.
21. Hoa PTP, Managaki S, Nakada N, Takada H, Shimizu A, et al. (2011) Antibiotic contamination and occurrence of antibiotic-resistant bacteria in aquatic environments of northern Vietnam. Science of the Total Environment 409: 2894–2901.

22. Suzuki S, Ogo M, Miller TW, Shimizu A, Takada H, et al. (2013) Who possesses drug resistance genes in the aquatic environment sulfamethoxazole (SMX) resistance genes among the bacterial community in water environment of Metro-Manila, Philippines. Frontiers in Microbiology 4.
23. Valderas MW, Bourne PC, Barrow WW (2007) Genetic basis for sulfonamide resistance in Bacillus anthracis. Microb Drug Resist 13: 11–20.
24. Diene SM, Rolain J-M (2013) Investigation of antibiotic resistance in the genomic era of multidrug-resistant Gram-negative bacilli, especially Enterobacteriaceae, Pseudomonas and Acinetobacter. Expert Rev Anti-Infe 11: 277–296.
25. Xia S, Xu B, Huang L, Zhao JY, Ran L, et al. (2011) Prevalence and characterization of human Shigella infections in Henan Province, China, in 2006. J Clin Microbiol 49: 232–242.
26. Antunes P, Machado J, Sousa JC, Peixe L (2005) Dissemination of sulfonamide resistance genes (sul1, sul2, and sul3) in Portuguese Salmonella enterica strains and relation with integrons. Antimicrob Agents Chemother 49: 836–839.
27. Hammerum AM, Sandvang D, Andersen SR, Seyfarth AM, Porsbo LJ, et al. (2006) Detection of sul1, sul2 and sul3 in sulphonamide resistant Escherichia coli isolates obtained from healthy humans, pork and pigs in Denmark. Int J Food Microbiol 106: 235–237.

24

Bayesian Inference of Baseline Fertility and Treatment Effects via a Crop Yield-Fertility Model

Hungyen Chen[1], Junko Yamagishi[2], Hirohisa Kishino[1]*

1 Graduate School of Agricultural and Life Sciences, The University of Tokyo, Tokyo, Japan, **2** Institute for Sustainable Agro-ecosystem Services, The University of Tokyo, Tokyo, Japan

Abstract

To effectively manage soil fertility, knowledge is needed of how a crop uses nutrients from fertilizer applied to the soil. Soil quality is a combination of biological, chemical and physical properties and is hard to assess directly because of collective and multiple functional effects. In this paper, we focus on the application of these concepts to agriculture. We define the baseline fertility of soil as the level of fertility that a crop can acquire for growth from the soil. With this strict definition, we propose a new crop yield-fertility model that enables quantification of the process of improving baseline fertility and the effects of treatments solely from the time series of crop yields. The model was modified from Michaelis-Menten kinetics and measured the additional effects of the treatments given the baseline fertility. Using more than 30 years of experimental data, we used the Bayesian framework to estimate the improvements in baseline fertility and the effects of fertilizer and farmyard manure (FYM) on maize (*Zea mays*), barley (*Hordeum vulgare*), and soybean (*Glycine max*) yields. Fertilizer contributed the most to the barley yield and FYM contributed the most to the soybean yield among the three crops. The baseline fertility of the subsurface soil was very low for maize and barley prior to fertilization. In contrast, the baseline fertility in this soil approximated half-saturated fertility for the soybean crop. The long-term soil fertility was increased by adding FYM, but the effect of FYM addition was reduced by the addition of fertilizer. Our results provide evidence that long-term soil fertility under continuous farming was maintained, or increased, by the application of natural nutrients compared with the application of synthetic fertilizer.

Editor: Wenju Liang, Chinese Academy of Sciences, China

Funding: The Grant-in-Aid for Scientific Research (grant number 59440009, PI: Wataru Sunohara) by the Ministry of Education, Science and Culture (MEXT, http://www.mext.go.jp) supported the initial phase (1984–1987) of the long-term fertilizer experiment. The University Farm continued the experiment until 2010 by the regular budget supplied by the University of Tokyo. This work was supported by Grants-in-Aid for Scientific Research (grant number 25280006) by the Japan Society for the Promotion of Science (JSPS, http://www.jsps.go.jp/) to H. K. The funders had no role in study design, data collection and analysis, decision to publish, or preparation of the manuscript.

Competing Interests: The authors have declared that no competing interests exist.

* Email: kishino@lbm.ab.a.u-tokyo.ac.jp

Introduction

Agricultural crop production is directly related to food supply, so agricultural soil productivity must be maintained. Balanced fertilization provides all the essential nutrients for crops to remain healthy and grow productively [1–3]. In a world facing increasing population pressure [4], our highest priority must be to increase crop productivity to ensure food security [2,5–7]. In this context, there has been increasing concern about the long-term productivity of soils on a global scale [8–11].

The relationship between crop yield and soil fertility under different fertilization regimes has been studied for decades. Aref and Wander [12] investigated the long-term trends of corn yield and soil organic matter in different crop sequences and soil fertility treatments. Merick and Németh [13] used results from 60-year field experiments to provide information on the relationship between fertilization and yields of rye and potato. Hallin et al. [14] investigated the relationship between microbial communities and total crop yield and nitrogen content in the crop in a 50-year-old fertilization experiment. Fan et al. [15] studied the trends in grain yields and soil organic carbon (SOC) in a 26-year dryland fertilization trial. By carrying out a 27-year experiment with various fertilization treatments in a rotation cropping system with wheat and maize in a red soil, Zhang et al. [16] investigated trends in SOC, soil nitrogen, and grain yield.

Many researchers have suggested that soil fertility under continuous farming is maintained, or increased, by the application of farmyard manure (FYM) compared with the application of synthetic fertilizer [2,11,17–19]. To ensure gains in crop productivity, it is necessary to maintain and even improve the soil fertility of a field under continuous farming practice [20–23]. In some cases, soils do not contain sufficient amounts of the essential nutrients required for rapid crop growth and high productivity [2,7,24]. As a result, supplemental nutrients, applied as fertilizers, manure or compost, are needed [25–27]. Many studies have analyzed the effects of manure [28–30], chemical fertilizer [31,32], and both [33,34] on soil fertility and crop yield.

Previous models have predicted the crop-yield response to fertilizer application [19] and climate change [35] and have greatly contributed to the software used in agricultural systems research [36,37]. Myers [38] used a static model to estimate the nitrogen fertilizer requirements of cereal crops. Deng et al. [39] proposed a theoretical framework that predicts the optimum planting density and maximal yield for an annual crop plant. Cong et al. [40] indicated that the CENTURY model [41] can simulate fertilization effects on SOC dynamics under different climate and soil conditions.

Soil quality is a complex combination of biological, chemical, and physical properties. In this paper, we focus on the application of these concepts to agriculture and define the baseline fertility as the level of soil fertility that crops utilize for their growth. Therefore, our soil quality is an interaction of the above properties and the crop activities. With this strict definition of baseline fertility, we propose a new crop yield-fertility model that enables quantification of the process of improving baseline fertility and the effects of treatments solely from the time series of crop yields. The model was modified from Michaelis-Menten kinetics and measured the additional effects of the treatments given the baseline fertility. Using more than 30 years of experimental data of maize (*Zea mays*), barley (*Hordeum vulgare*), and soybean (*Glycine max*) yield, we estimated in the Bayesian framework the improvements in baseline fertility and the effects of fertilizer and farmyard manure (FYM). We compared the efficiency of separate applications of fertilizer or FYM to the three crops (i.e., maize, barley and soybean) using estimated model parameters. The temporal variations in the baseline fertility of each crop were estimated and compared for six treatments and two soil types.

Materials and Methods

Field experiments

The long-term fertilizer experiment was conducted between 1980 and 2010 at the University Farm at the Institute for Sustainable Agro-ecosystem Services at the University of Tokyo, Nishitokyo, Tokyo, Japan (35°43′ N latitude and 139°32′ E longitude and an altitude of 53 m above mean sea level). The field site was located in the Kantō Plain, where the soil is covered with pyroclastic material from volcanoes that surrounded the western Kantō region 126,000 years ago. The soil parent material is Tachikawa loam. The surface soil is a black-colored fertile andosol containing a high percentage of humus and the subsurface soil (25 cm below the surface) is red-colored barren clay. Andosols are soils found in volcanic areas formed in volcanic tephra. Andosols have a different composition from chernozems and are not commonly found outside the Pacific "ring of fire". At the beginning of the experiment, the percent nitrogen was 2.23 g/kg in the surface soil and 0.97 g/kg in the subsurface soil. A 2-year crop rotation of maize-barley-soybean-barley was maintained throughout the experimental period and the crop yields were measured. The maize crop was sown at the beginning of July and harvested at the end of September every second year from 1980 onwards. The barley crop was sown in the first half of November and harvested either at the end of May or at the beginning of June of the next year each year from 1980 onwards. The soybean crop was sown at the end of June and harvested in the second half of October every second year from 1981 onwards. The seeds of the three crops were ridge sown using a seeding machine. For maize and soybean, one seed was sown in each hole, and the widths between the ridges and stocks were 71 cm and 23 cm, and 71 cm and 17.5 cm, respectively. For barley, row seeding was applied using a width of 17.8 cm between the ridges.

Six treatment plots containing the NPK fertilizer applications and two levels of farmyard manure (FYM) combined with compost were established in both of the fields with the surface soil (64 m^2 per plot, 8×8 m) and the subsurface soil (56 m^2 per plot, 7×8 m) and were replicated four times. The percent nitrogen, phosphorus, and potassium in the fertilizer treatments were N:P:K = 12:7.9:13.3 for maize and barley and 3:4.4:8.3 for soybean. Fertilizer was added to the soil at a rate of 1 t ha^{-1} for all crops. The FYM comprised wheat straw and cow dung. The percent nitrogen in the FYM ranged from 0.2–0.4. FYM was added to the plots at two levels, i.e., 20 and 60 t ha^{-1}. The FYM treatments were added to the soil twice a year from 1980 to 1990 before the crops were sown and once a year (before the barley was sown) from 1991 onwards. After the barley harvest in 2008, fertilizer and FYM applications were discontinued for all treatments. Phosphate fertilizer was added simultaneously with the NPK fertilizer and FYM at a rate of 2 t ha^{-1} after the soybean harvest from 1980 until 2001. All aboveground components were removed from the field during harvesting. The dead foliage, below-ground components, and remnant stem sections (i.e., 15 cm for maize and 10 cm for barley and soybean) remained in the field. The harvested material from the six treatments for both the surface and subsurface soils were weighed after drying at 80°C. The dry weight (g m^{-2}) of the aboveground maize components, the barley spike, and the soybean seed were measured.

Testing significance of yield differences in different soils

The crop yields of the surface and subsurface soils without the addition of fertilizer or farmyard manure (FYM) reflect the yield differences that occurred purely as a result of the soil. The crop yields observed for the subsurface soil were 39±25%, 49±20%, and 19±31% (±SD) lower than the yields observed for the surface soil for maize (Fig. S1A), barley (Fig. S1B), and soybean (Fig. S1C), respectively. There was a difference in yield between the two types of soils but it was only significant for barley ($t = -3.69$, $P<0.001$; for maize, $t = -1.58$, $P = 0.13$ and for soybean, $t = -0.94$, $P = 0.35$). Although the crops were planted in the same fields with the same soil, they obtained and used nutrients from the soil in a manner unique to each crop.

The maize and soybean yields did not differ significantly ($P>0.2$) between the surface and subsurface soils when fertilizer or FYM or a combination of fertilizer and FYM were applied (Fig. S1A and C for maize and soybean, respectively). The barley yields differed significantly between the surface and subsurface soils when both fertilizer and FYM ($t = 2.27$, $P<0.05$) or FYM alone ($t = 2.18$, $P<0.05$) were added, but did not differ significantly when only fertilizer was added ($t = -0.63$, $P = 0.52$) (Fig. S1B). The difference in yields between the surface and subsurface soils was reduced when either fertilizer or FYM was added to the soil. The application of fertilizer and FYM improved the yields for both the surface and subsurface soils and reduced the yield differences between the soil types. This finding indicates that productivity can be increased to an average level by fertilization even in a field with barren soil. While this may have been prior knowledge, the manner in which fertilizer and FYM contribute nutrients to the soil and how the crops make use of the nutrients obtained from these sources remains unclear.

Testing significance of yearly variance

The maize and soybean yields increased significantly ($P<0.05$) for both soil types over the 30-year period. The barley yield increased significantly in the subsurface soil ($P<0.05$) but not in the surface soil ($P = 0.15$). The soils were fertilized by crop residues (i.e., roots, fallen leaves, and stem sections (15 cm above ground for maize and 10 cm for barley and soybean)) that remained in and on the soil after harvesting. Organic matter accumulated in the soil over time, providing a long-term, slow-release source of nitrogen, phosphorus, sulfur, and other important nutrients for crop growth, which resulted in increased soil productivity.

Testing significance of the treatment effects

We used a nested analysis of variance to compare the yields of the three crops in the surface and subsurface soils in response to the different treatments. The surface soil produced significantly greater barley crop yields compared with the subsurface soil (Fig. S1B, $F = 14.68$, $P<0.001$). Maize and soybean yields (Fig. S1A, $F = 2.91$, $P = 0.09$ and Fig. S1C, $F = 2.19$, $P = 0.14$, respectively) did not differ significantly between the two soil types. The interaction effects of fertilizer and FYM were significant for maize ($F = 3.32$, $P<0.05$) and barley ($F = 10.71$, $P<0.001$), but not for soybean ($F = 0.69$, $P = 0.60$). The soybean yield did not differ significantly among the treatments ($P>0.1$ for both fertilizer and FYM).

The derivation of the crop yield-fertility model

Our model is based on the general pattern of the observed crop yield. The temporal variation in maize yield in response to four treatments applied to the surface soil over 30 years is shown in Fig. 1A. There were clear differences in the yield between the four treatments prior to 1982 (the first two data points). The response of the crop yield descended in the following order: the addition of NPK fertilizer and FYM (F:1, M:1), the addition of NPK fertilizer only (F:1, M:0), the addition of FYM only (F:0, M:1), and no nutrient application (F:0, M:0). The yield differences between the treatments receiving NPK fertilizer and FYM (F:1, M:1) and NPK fertilizer only (F:1, M:0) could not be detected from 1984 onwards. The effect of FYM could no longer be detected once the fertilizer application was continued. From 1994, similar crop yields were observed in the treatments with fertilizer, FYM or a combined application of fertilizer and FYM. In the last year of the study, there were negligible differences between treatments because fertilizer and FYM applications were discontinued in 2008. The soybean data are presented in Fig. 1C. The pattern of the soybean yield differed from maize. There were almost no differences in crop yield among the three treatments that received fertilizer, FYM or a combined application of fertilizer and FYM in the first and later years. The differences in crop yield between the treatments with and without fertilizer were negligible from 1995 onwards. The inherent soil fertility may be sufficient to grow soybean without the addition of fertilizer. The pattern of barley yield (Fig. 1B) was similar to maize but was not as well defined.

Our model can predict crop yields given the total fertility level of the soil. The model can be used to describe the variations in crop yield and the effect of different fertilizer treatments over different years for each crop using the estimated soil fertility. The crop yield-fertility curves for maize, barley, and soybean are presented in Fig. 2A, B, and C, respectively. Soil fertility was increased by the application of fertilizer over a 20-year period. Furthermore, the effect of fertilizer decreased over time for all three crops from 1980–2000.

The crop yield-fertility model

We developed a simple mathematical model to estimate the soil fertility level and to quantify the contributions of fertilizer and FYM to the improvements in crop yields. The model was modified from Michaelis-Menten kinetics [42]:

$$Y_{i,t} \cong \frac{V}{1+\dfrac{K}{BF_{i,t}+a\times \mathrm{F}+b\times \mathrm{M}}}. \qquad (1)$$

We did not apply the model to describe the kinetics. Rather, we used a form of the model to interpret the effects of baseline fertility, fertilizer, and FYM on the crop yield and the extent of saturation of additional inputs. The model related crop yields ($\mathrm{Y}_{i,t}$) to the total fertility ($BF_{i,t}+a\times \mathrm{F}+b\times \mathrm{M}$) of the soil for treatment i at time t (Fig. 2). The value of the baseline fertility (BF) for the subsurface soil without the addition of fertilizer or FYM in the first year was normalized by setting this value to 1. V is the maximum yield in response to the maximum fertilizer input; K is the fertility level before fertilization that is required to produce half of the maximum yield, V. A large K indicates that more nutrients need to be added to the soil for the crop to grow, so K can infer the sterility of the soil prior to fertilization. $BF_{i,t}$ is the baseline fertility of the crop for treatment i at time t and is assumed to vary gradually over time; F is the level of fertilizer application (0 and 1), M is the level of FYM application (0, 1/3, and 1), and a and b represent the contributions of F and M relative to the baseline fertility, respectively. Hereafter, we refer to V, K, a, and b as the maximum yield, half-saturated fertility, fertilizer contribution, and FYM, respectively. A Bayesian framework was adopted for parameter estimation assuming a gradual change in the baseline fertility.

The likelihood and priors

The likelihood of the yield for treatment i at time t followed a normal distribution with the mean $\dfrac{V}{1+\dfrac{K}{BF_{i,t}+a\times \mathrm{F}+b\times \mathrm{M}}}$ and the variance δ:

$$\mathrm{Yield}_{i,t} \sim N\left(\frac{V}{1+\dfrac{K}{BF_{i,t}+a\times \mathrm{F}+b\times \mathrm{M}}}, \delta\right) \qquad (2)$$

This value was normalized by setting the BF of the subsurface soil without the addition of fertilizer or FYM in the first year to 1. The priors of the BF for the other treatments in the first year followed a gamma distribution with a shape parameter of 1 and a scale parameter of 1. The smoothness priors of the BF from the second year followed a normal distribution with the mean equal to the value of the BF in the preceding year and the variance τ:

$$BF_{i,t} \sim N(BF_{i,t-1}, \tau), \quad t>1. \qquad (3)$$

The inverse of δ followed a gamma distribution with a shape parameter of 0.1 and a scale parameter of 10. The inverse of τ followed a gamma distribution with a shape parameter of 0.1 and a scale parameter of 10. The prior of V followed a normal distribution with a mean of 0 and a standard deviation of 1000. The prior of K followed a gamma distribution with a shape parameter of 0.1 and a scale parameter of 10. The priors of a and b followed a gamma distribution with a shape parameter of 1 and a scale parameter of 1. The priors of the estimates were designed to be as non-informative as possible within a realistic range of the

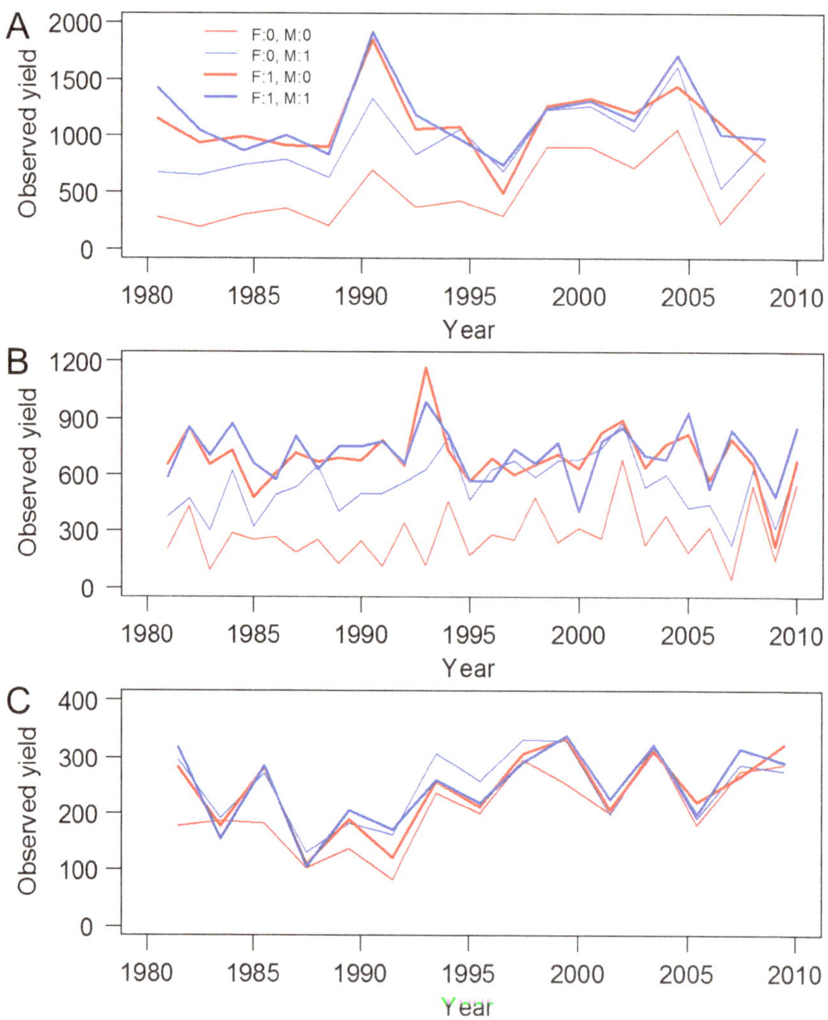

Figure 1. Temporal variations in observed crop yield (g m^{-2}) for four treatments in the surface soil. (A) Maize; (B) Barley; (C) Soybean; F, level of fertilizer; M, level of farmyard manure. See Fig. S1 for all treatments in surface and subsurface soils.

parameter values. All calculations and data analyses were performed using R v2.13.2 [43]. The raw data is available as (Data S1).

Results

Increasing trend of soil fertility

The baseline fertility of the surface and subsurface soils in all treatments increased over the 30-year period for all three crops (Figs. 3, S5–6). The baseline fertility of the subsurface soil was very low for maize and barley but was close to half-saturated for soybean when fertilization was initiated.

The treatment differences in the baseline fertility (i.e., maximum minus minimum) were increased by continuous cropping over the experimental period. The difference in the baseline fertility between the treatments was 0.89 in 1980 and 3.24 in 2008 for maize; for barley, it was 1.20 in 1980 and 2.54 in 2010, and for soybean, it was 0.96 in 1980 and 3.19 in 2009.

We used a 2×2 factorial analysis of variance to compare the baseline fertility of the three crops in response to the different treatments. The interaction effects of fertilizer and FYM were significant for maize ($F = 15.09$, $P<0.001$) and barley ($F = 50.67$, $P<0.001$), but not for soybean ($F = 1.77$, $P = 0.19$). The baseline fertility of soybean differed significantly for fertilizer ($F = 10.26$, $P<0.01$) but not for FYM ($F = 4.00$, $P = 0.05$).

The effects of fertilizer and FYM

The Bayesian estimates of the model for the three crops are shown in Table 1. The traces of the MCMC (Markov Chain Monte Carlo) samples show the well-mixing and convergence to the posterior distributions (Fig. S7). We compared the mean values of the parameters for the three crops. The maize crop had the highest half-saturated fertility level and maximum yield. The soybean crop had the lowest half-saturated fertility level and maximum yield. For maize, the size of the fertilizer contribution was almost equal to the size of the half-saturated fertility, and FYM contribution was 81% smaller than the half-saturated fertility. For barley, the fertilizer contribution was 40% higher than the half-saturated fertility and the FYM contribution was 65% lower than the half-saturated fertility. This indicates that the amount of fertilizer applied in the experiment was 40% more than the fertility level at which the barley yield was half of the maximum yield, but the amount of FYM applied was 65% less than required. For soybean, the values of the fertilizer and FYM

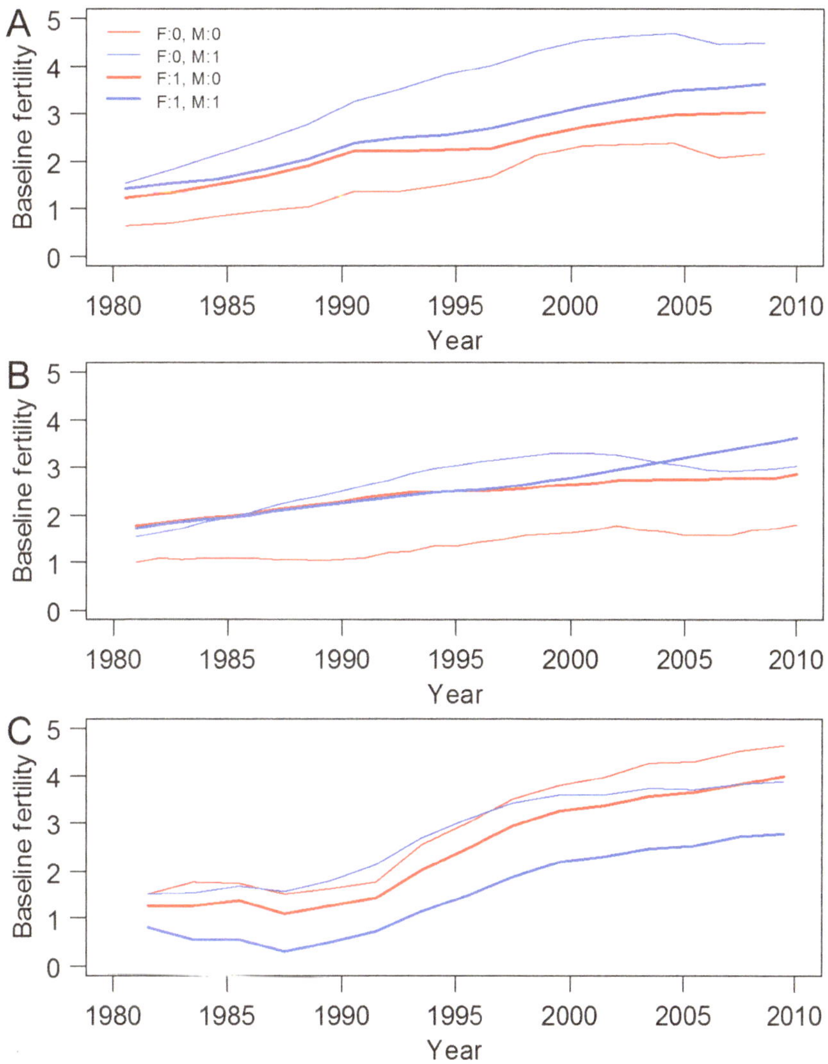

Figure 2. Crop yield-fertility curves for (A) Maize, (B) Barley, and (C) Soybean. The value of 1 for fertility is the fertility without the addition of fertilizer or FYM in the initial year for the subsurface soil. F, level of fertilizer. See Figs. S2A, S3A, and S4A for the band that corresponds to the standard deviation of the curves.

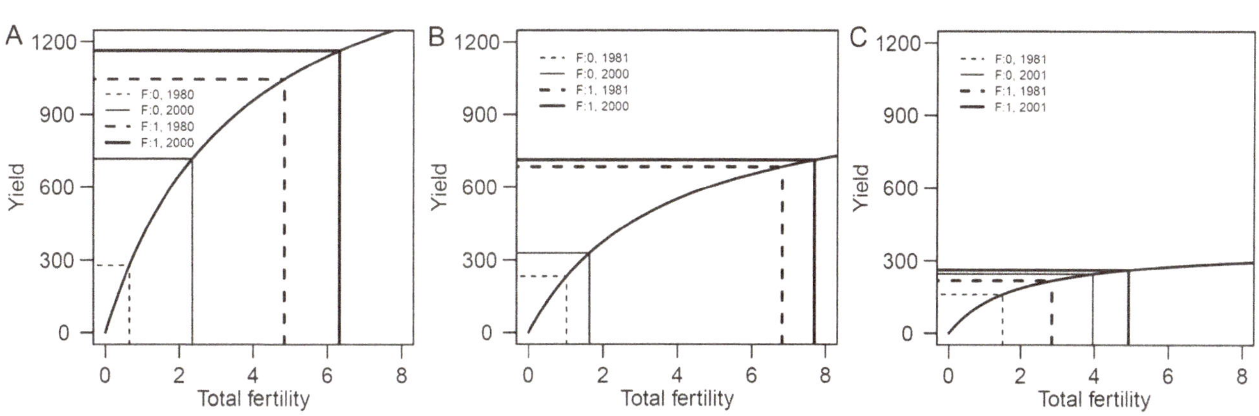

Figure 3. Temporal variations in the baseline fertility for four treatments in the surface soil. (A) Maize; (B) Barley; (C) Soybean; F, level of fertilizer; M, level of farmyard manure. See Fig. S5 for all treatments in surface and subsurface soils.

Table 1. The posterior mean and standard deviation (SD) of the Bayesian estimates.

Crop	V (g m^{-2})		K		A		B	
	Mean	SD	Mean	SD	Mean	SD	Mean	SD
Maize	1833.01	316.56	3.65	1.68	3.61	1.27	0.68	0.53
Barley	1042.87	134.97	3.56	1.18	5.07	1.44	1.25	0.63
Soybean	360.17	65.54	1.87	1.01	1.58	0.58	1.62	0.94

V, maximum yield; *K*, half-saturated fertility; *a*, fertilizer contribution; *b*, FYM contribution.

contributions were similar, i.e., 16 and 13%, respectively, and were lower than the half-saturated fertility.

In contrast with maize and barley, the value of the FYM contribution was higher than the fertilizer contribution for the soybean crop. The application rate of fertilizer and FYM approximated the fertility level at which the soybean yield was half of the maximum yield. The half-saturated fertility level for maize was 3.65, and the half-saturated fertility levels for barley and soybean were 2 and 49% lower, respectively. The fertilizer contribution for barley was 5.07, and the fertilizer contributions for maize and soybean were 29 and 69% lower, respectively. The FYM contribution for soybean was 1.62, and the FYM contributions for maize and barley were 58 and 23% lower, respectively. The model estimates revealed how the different crops used the nutrients from fertilizer and FYM applications to the soil. Maize needed the highest level of fertility to reach half of the maximum yield (i.e., the largest half-saturated fertility) compared with barley and soybean. Fertilizer contributed the maximum amount to the barley yield, and the FYM contributed the maximum amount to the soybean yield.

The total fertility levels among the treatments for both surface and subsurface soils during the 30-year period ranged from 0.54 to 7.22 for maize (Fig. S2A), 0.41 to 9.78 for barley (Fig. S3A), and 0.90 to 5.56 for soybean (Fig. S4A). Barley may be the most efficient crop in terms of fertilizer use among the three crops. The maximum total fertility obtained for soybean was 2.97× the half-saturated fertility, which is larger than the values obtained for maize (1.98) and barley (2.75).

Using the total fertility range, we estimated the predicted range in crop yield using the crop yield-fertility model for each crop. The predicted crop yields ranged from 236.24 to 1217.51 g m^{-2} for maize, 107.70 to 764.56 g m^{-2} for barley, and 117.02 to 269.52 g m^{-2} for soybean. Because of fluctuations in crop yield that could not be described by the model, the predicted yields covered 49, 57, and 42% of the observed yield ranges for maize (35 and 2023 g m^{-2}), barley (14 and 1168 g m^{-2}), and soybean (16 and 381 g m^{-2}), respectively.

The decreasing effect of fertilization over time

In the sterile subsurface soil in the initial year, the application of fertilizer resulted in 153, 189, and 62% increases (the FYM resulted in 42, 65, and 61% increases) in crop yield for maize, barley, and soybean, respectively. In the fertile surface soil in the initial year, the application of fertilizer resulted in 275, 196, and 35% increases (FYM resulted in 149, 98, and 40% increases) in crop yield for maize, barley, and soybean, respectively. After 20 years of farming, the application of fertilizer resulted in 89, 219, and 11% increases (FYM resulted in 74, 150, and 13% increases) in the subsurface soil, and 62, 113, and 7% increases (FYM resulted in 50, 75, and 8% increases) in the surface soil for maize, barley, and soybean, respectively, because of the increased baseline fertility. The greatest yield increases were observed in the initial year in the fertile surface soil with the application of both fertilizer and FYM (compared with the treatment without fertilizer or FYM), and then the yield increment decreased each year (Fig. 4). In the sterile subsurface soil, it took 5–10 years for the yield increase to reach the maximum value.

Predicted crop yields reflected the trend of the observed yields

The estimated fertility levels were used to predict the crop yield using the yield-fertility curves for each crop (Figs. S2A, S3A, and S4A). The trends of the predicted yield were similar to the trends of the observed yield for all three crops. The correlation

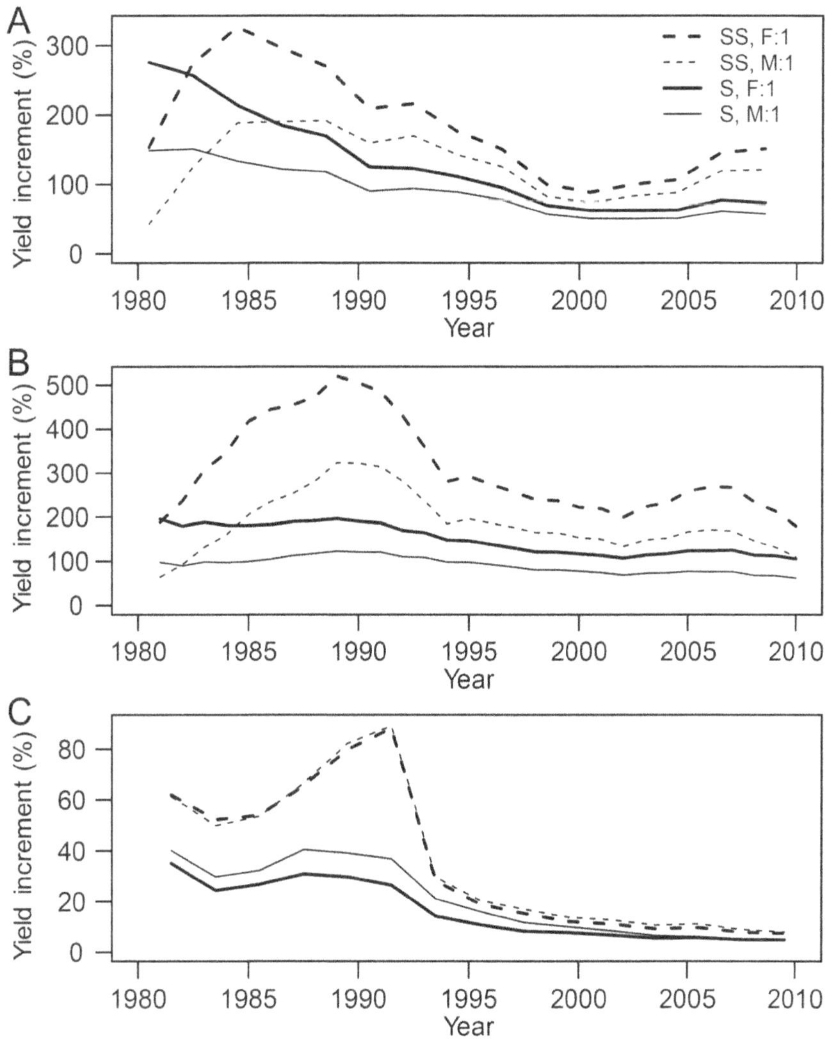

Figure 4. Temporal variations in the yield increment resulting from the application of fertilizer and farmyard manure. (A) Maize; (B) Barley; (C) Soybean; SS, subsurface soil; S, surface soil; F, level of fertilizer; M, level of farmyard manure.

coefficients were 0.758 (Fig. S2B, $P<0.001$) for maize, 0.801 (Fig. S3B, $P<0.001$) for barley, and 0.622 (Fig. S4B, $P<0.001$) for soybean.

Yield fluctuations were observed among the treatments but these fluctuations were not reflected by the predicted crop yields. For maize, the peak yields occurred in 1990 and 2004, and a reduced yield occurred in 1996. For barley, the peak yield for the fertilizer-treated plots occurred in 1993, but the peak yield occurred 1 year later in the plots without fertilizer treatment. The FYM treatments were discontinued in 1991 for the summer harvested crops (maize and soybean). Clear reductions in maize yield in 1991 and soybean yield in 1992 are shown in Fig. 1A and C, respectively, after which the yields for both crops increased until 2008. The estimated fertility levels and predicted yields represented the temporal variations that coincided with treatment changes during the experimental period (Fig. 1). Termination of the phosphate fertilizer application in 2001 was well described by the temporal variations in the baseline fertility of maize and barley. The baseline fertility levels declined, especially in the soil treated with FYM only (Fig. 4). Similar patterns were not observed for the fertilizer-treated soils. A decline in fertility was not observed for soybean. As previously mentioned, soybean yields were not significantly different among treatments.

The effect of climatic factors

Crop yield can be affected by climatic factors [35] such as precipitation [44,45], temperature [46,47] and recent warming [48]. The impact of climate change may damage crop yields on a global scale [48,49] and lead to decreases in crop production [50]. However, weather variation may play a small role in crop yield on a regional scale [45] and the effect may depend on fertilization and soil type [45,50]. Monthly precipitation and average monthly temperature recorded at the experimental site are shown in Fig. S8A and B. We performed a correlation analysis between the residual of the crop yield (i.e., predicted yield minus observed yield) and the climatic variables for the three crops (Table S1 in File S1). The simple residual analysis indicated that temperature had some effect on maize and soybean yields. We then conducted additional analyses that included the effect of temperature in the crop yield-fertility model (see File S1 and Table S2 in File S1). Based on the results of the minor contribution of temperature, we concluded that the annual fluctuations in crop yields were mainly caused by unknown factors other than the climatic factors.

Discussion

The baseline fertility

The baseline fertility of the subsurface soil was very low for maize and barley. This suggests that the fertility in the subsurface soil prior to the application of fertilizer or FYM was not sufficient for maize or barley.

The increasing baseline fertility trend was also observed in soil without additional fertilizer or FYM. The increased fertility may have resulted from the addition of crop residues to the soil and nitrogen fixation by the soybean crop. These fertility increases may indicate soil maturation (i.e., the accumulation of nutrients) in the experimental fields.

The baseline fertility of maize indicated that when FYM was applied alone, soil fertility increased more than when both FYM and fertilizer were applied, for both soil types (Fig. 3A). The application of FYM increased the organic matter content of the soil, which resulted in increased soil fertility. Similar trends were observed for barley and soybean (Fig. 3B and C), supporting the results observed for maize. The results for soybean suggested that the application of fertilizer could decrease the soil fertility when soybean was planted in a soil with a saturated level of fertilization (Fig. 3C).

The effects of fertilizer and FYM

Our result indicates that the amount of fertilizer applied in the experiment reached the fertility level at which the maize yield was half of the maximum yield, but the amount of FYM required to reach this value was much lower. Barley obtained the highest fertility from the soil and fertilization despite the fact that maize produced the greatest crop yield and had the largest half-saturated fertility level. By using nutrients from the soil and fertilizer, soybean was the crop that most often approached the maximum yield.

Our findings indicate that the sterile subsurface soil needed more time to accumulate nutrients than the surface soil. The percentage increments in crop yield were larger in the subsurface soil than in the surface soil, which indicates that fertilizer application effectively improved the crop yield in the barren field. The fertilizer-induced increment in maize and barley crop yield increased after 2001. This is because the phosphate fertilizer application was discontinued in 2001 and the baseline soil fertility subsequently declined. The soybean crop did not show a similar reduction in yield after 2001.

Long-term field crop fertilization experiment and crop models

Long-term experiments examining fertilizer treatments and crop yield have been widely conducted to assess the effects of applying fertilizer [51], manure [52], or a combination of fertilizer and manure [14,53,54]. In addition, experiments have been conducted to determine the impacts of nutrient addition on the fertility status of different soils [55–57]. However, it is important to note that our long-term record of crop yields with various types of treatments has made this the first study to simultaneously examine the temporal variation in baseline fertility and the contribution of nutrient applications.

Conclusions

By allowing the gradual temporal variation in baseline fertility, it was possible to estimate the process of soil fertilization and the short-term effects of different fertilizer treatments. Our results support the proposal that naturally derived nutrients should be used to maintain soil fertility and synthetic fertilizer should be used to maintain productivity. Our results provide a clear description of the relationship between crop yield and soil, which can be easily understood.

Supporting Information

Figure S1 Temporal variations in observed crop yield (g m^{-2}) for six treatments in a field with a fertile surface soil and a field with a barren subsurface soil from 1980 to 2010. (A) Maize; (B) Barley; (C) Soybean; SS, subsurface soil; S, surface soil; F, level of fertilizer; M, level of farmyard manure.

Figure S2 Temporal variations in the observed and predicted yields (g m^{-2}) of maize for six treatments in a field with a fertile surface soil and a field with a barren subsurface soil presented every second year from 1980 to 2008. (A) A crop yield-fertility map. The crop yield-fertility model translates total fertility into predicted yield. The gray band represents the band that corresponds to the standard deviation (±SD) of the crop yield-fertility curve. (B) The relationship between the observed and predicted yields. SS, subsurface soil; S, surface soil; F, level of fertilizer; M, level of farmyard manure; BF, baseline fertility.

Figure S3 As for Figure S2 but for barley every year from 1980 to 2010.

Figure S4 As for Figure S2 but for soybean every 2 years from 1981 to 2009.

Figure S5 Temporal variations in the baseline fertility estimated using the crop yield-fertility model for six treatments in a field with a fertile surface soil and a field with a barren subsurface soil from 1980 to 2010. (A) Maize; (B) Barley; (C) Soybean; SS, subsurface soil; S, surface soil; F, level of fertilizer; M, level of farmyard manure.

Figure S6 Temporal variations in the posterior means and the bands that correspond to the standard deviation (±SD) of the Bayesian estimates of baseline fertility for six treatments in a field with a fertile surface soil and a field with a barren subsurface soil. (A) Maize; (B) Barley; (C) Soybean; SS, subsurface soil; S, surface soil; F, level of fertilizer; M, level of farmyard manure.

Figure S7 Traces of the MCMC samples of *V* (g m^{-2}), *K*, *a*, and *b*. (A) Maize; (B) Barley; (C) Soybean. The chain length was set to 1,000,000 steps logging every 100th step.

Figure S8 Temporal variation of the climatic variables recorded at the experimental site from January 1980 to December 2011. (A) Monthly precipitation. (B) Average monthly temperature.

File S1 Contribution of the climatic factor. The analyses by including a temperature effect in the crop yield-fertility model.

Data S1 The raw data of the crop yields of maize, barley, and soybean from the long-term fertilizer experiment from 1980 to 2010. As for the detail of the experiment, see text.

Author Contributions

Conceived and designed the experiments: HC HK. Performed the experiments: JY. Analyzed the data: HC. Wrote the paper: HC HK. Confirmed the manuscript: HC JY HK.

References

1. Mäder P, Fliessbach A, Dubois D, Gunst L, Fried P (2002) Soil fertility and biodiversity in organic farming. Science 296: 1964–1967.
2. Tilman D, Cassman KG, Matson PA, Naylor R, Polasky S (2002) Agricultural sustainability and intensive production practices. Nature 418: 671–677.
3. Baligar VC, Fageria NK, He ZL (2001) Nutrient use efficiency in plants. Commun Soil Sci Plant 32: 921–950.
4. Holdren JP, Ehrlich PR (1974) Human population and the global environment: population growth, rising per capita material consumption, and disruptive technologies have made civilization a global ecological force. Am Sci 62: 282–292.
5. Glover JD, Reganold RP, Cox CM (2012) Plant perennials to save Africa's soils. Nature 489: 359–361.
6. Alexandratos N (1999) World food and agriculture: outlook for the medium and longer term. PNAS 96: 5908–5914.
7. Cassman KG (1999) Ecological intensification of cereal production systems: yield potential, soil quality, and precision agriculture. PNAS 96: 5952–5959.
8. Isbella F, Reichb PB, Tilmana D, Hobbiea SE, Polasky S, et al. (2013) Nutrient enrichment, biodiversity loss, and consequent declines in ecosystem productivity. PNAS 110: 11911–11916.
9. Cordell D, Drangert J-O, White S (2009) The story of phosphorus: global food security and food for thought. Glob Environ Change 19: 292–305.
10. Rasmussen PE, Goulding KWT, Brown JR, Grace PR, Janzen H, et al. (1998) Long-term agroecosystem experiments: assessing agricultural sustainability and global change. Science 282: 893–896.
11. Matson PA, Parton WJ, Power AG, Swift MJ (1997) Agricultural intensification and ecosystem properties. Science 277: 504–509.
12. Aref S, Wander MM (1997) Long-term trends of corn yield and soil organic matter in different crop sequences and soil fertility treatments on the morrow plots. Adv Agron 62: 153–197.
13. Mercik S, Németh K (1985) Effects of 60-year N, P, K and Ca fertilization on EUF-nutrient fractions in the soil and on yields of rye and potato crops. Plant Soil 83: 151–159.
14. Hallin S, Jones CM, Schloter M, Philippot L (2009) Relationship between N-cycling communities and ecosystem functioning in a 50-year-old fertilization experiment. ISME J 3: 597–605.
15. Fan T, Xu M, Song S, Zhou G, Ding L (2008) Trends in grain yields and soil organic C in a long-term fertilization experiment in the China Loess Plateau. J Plant Nutr Soil Sci 171: 448–457.
16. Zhang H, Xu M, Zhang F (2009) Long-term effects of manure application on grain yield under different cropping systems and ecological conditions in China. J Agr Sci 147: 31–42.
17. Power AG (2010) Ecosystem services and agriculture: tradeoffs and synergies. Philos Soc Trans R Soc Lond B 365: 2959–2971.
18. Altieri MA, Nicholls CI (2003) Soil fertility management and insect pests: harmonizing soil and plant health in agroecosystems. Soil Till Res 72: 203–211.
19. Clark MS, Horwath WR, Shennan C, Scow KM (1998) Changes in soil chemical properties resulting from organic and low-input farming practices. Agron J 90: 662–671.
20. Fan R, Zhang X, Liang A, Shi X, Chen X, et al. (2012) Tillage and rotation effects on crop yield and profitability on a Black soil in northeast China. Can J Soil Sci 92: 463–470.
21. Yanni YG, Dazzo FB (2010) Enhancement of rice production using endophytic strains of *Rhizobium leguminosarum* bv. trifolii in extensive field inoculation trials within the Egypt Nile delta. Plant Soil 336: 129–142.
22. Neupane RP, Thapa GB (2001) Impact of agroforestry intervention on soil fertility and farm income under the subsistence farming system of the middle hills, Nepal. Agric Ecosyst Environ 84: 157–167.
23. Peoples MB, Herridge DF, Ladha JK (1995) Biological nitrogen fixation: an efficient source of nitrogen for sustainable agricultural production? Plant Soil 174: 3–28.
24. Chapin HS III (1980) The mineral nutrition of wild plants. Ann Rev Ecol Syst 11: 233–60.
25. Altieri MA (2002) Agroecology: the science of natural resource management for poor farmers in marginal environments. Agric Ecosyst Environ 93: 1–24.
26. Watson CA, Atkinson D, Gosling P, Jackson LR, Rayns FW (2002) Managing soil fertility in organic farming systems. Soil Use Manage 18: 239–247.
27. Watson CA, Bengtsson H, Ebbesvik M, Lùes A-K, Myrbeck A (2002) A review of farm-scale nutrient budgets for organic farms as a tool for management of soil fertility. Soil Use Manage 18: 264–273.
28. Khan AUH, Iqbal M, Islam KR (2007) Dairy manure and tillage effects on soil fertility and corn yields. Bioresource Technol 98: 1972–1979.
29. Zhang H, Xu M, Zhang F (2009) Long-term effects of manure application on grain yield under different cropping systems and ecological conditions in China. J Agr Sci 147: 31–42.
30. Yadvinder-Singh Bijay-Singh, Ladha JK, Khind CS, Gupta RK, et al. (2004) Long-term effects of organic inputs on yield and soil fertility in the rice-wheat rotation. Soil Sci Soc Am J 68: 845–853.
31. Li J, Xu M, Qin D, Li D, Yasukazu H, et al. (2005) Effects of chemical fertilizers application combined with manure on ammonia volatilization and rice yield in red paddy soil. Plant Nutrit Fertilizer Sci 11: 51–56.
32. Malhi SS, Lemke R (2007) Tillage, crop residue and N fertilizer effects on crop yield, nutrient uptake, soil quality and nitrous oxide gas emissions in a second 4-yr rotation cycle. Soil Till Res 96: 269–283.
33. Min DH, Islam KR, Vough LR, Weil RR (2003) Dairy manure effects on soil quality properties and carbon sequestration in alfalfa–orchardgrass systems. Commun Soil Sci Plant 34: 781–799.
34. Körschensa M, Albert E, Armbruster M, Barkusky D, Baumecker M, et al. (2013) Effect of mineral and organic fertilization on crop yield, nitrogen uptake, carbon and nitrogen balances, as well as soil organic carbon content and dynamics: results from 20 European long-term field experiments of the twenty-first century. Arch Agron Soil Sci 59: 1017–1040.
35. Lobell DB, Burke MB (2010) On the use of statistical models to predict crop yield responses to climate change. Agric For Meteorol 150: 1443–1452.
36. McCown RL, Hammer GL, Hargreaves JNG, Holzworth DP, Freebairn DM (1996) APSIM: a novel software system for model development, model testing and simulation in agricultural systems research. Agric Syst 50: 255–271.
37. International Benchmark Sites Network for Agrotechnology Transfer (1993) The IBSNAT decade. Honolulu, Hawaii: Department of Agronomy and Soil Science, College of Tropical Agriculture and Human Resources, University of Hawaii.
38. Myers RJK (1984) A simple model for estimating the nitrogen fertilizer requirement of a cereal crop. Fertil Res 5: 95–108.
39. Deng J, Ran J, Wang Z, Fan Z, Wang G (2012) Models and tests of optimal density and maximal yield for crop plants. PNAS 109: 15823–15828.
40. Cong R, Wang X, Xu M, Ogle SM, Parton WJ (2014) Evaluation of the CENTURY model using long-term fertilization trials under corn-wheat cropping systems in the typical croplands of China. PLOS ONE 9: e95142.
41. Denef K, Six J, Merckx R, Paustian K (2004) Carbon sequestration in microaggregates of no-tillage soils with different clay mineralogy. Soil Sci Soc Am J 68: 1935–1944.
42. Michaelis L, Menten ML (1913) Die Kinetik der Invertinwirkung. Biochem Z 49: 333–369.
43. R Development Core Team (2011) R: a language and environment for statistical computing. Vienna, Austria: R Foundation for Statistical Computing.
44. Spiecker H (1995) Growth dynamics in a changing environment – long-term observations. Plant Soil 168–169: 555–561.
45. Drury CF, Tan CS (1994) Long-term (35 years) effects of fertilization, rotation and weather on corn yields. Can J Soil Sci 75: 355–362.
46. Lobell DB, Hammer GL, McLean G, Messina C, Roberts MJ (2013) The critical role of extreme heat for maize production in the United States. Nat Clim Chang 3: 497–501.
47. Welch JR, Vincent JR, Auffhammerc M, Moyae PF, Dobermann A (2010) Rice yields in tropical/subtropical Asia exhibit large but opposing sensitivities to minimum and maximum temperatures. PNAS 107: 14562–14567.
48. Lobell DB, Field CB (2007) Global scale climate–crop yield relationships and the impacts of recent warming. Environ Res Lett 2: 014002.
49. Schlenker W, Roberts MJ (2009) Nonlinear temperature effects indicate severe damages to U.S. crop yields under climate change. PNAS 106: 15594–15598.
50. Rosenzweig C, Parry ML (1994) Potential impact of climate change on world food supply. Nature 367: 133–138.
51. Shen J, Li R, Zhang F, Fan J, Tang C (2004) Crop yields, soil fertility and phosphorus fractions in response to long-term fertilization under the rice monoculture system on a calcareous soil. Field Crops Res 86: 225–238.
52. Canali S, Trinchera A, Intrigliolo F, Pompili L, Nisini L, et al. (2004) Effect of long term addition of composts and poultry manure on soil quality of citrus orchards in Southern Italy. Biol Fertil Soils 40: 206–210.
53. Poulton PR (1996) The Rothamsted long-term experiments: are they still relevant? Can J Soil Sci 76: 559–571.

54. Jenkinson DS (1991) The Rothamsted long-term experiments: are they still of use? Agron J 83: 2–10.
55. Huang S, Peng X, Huang Q, Zhang W (2010) Soil aggregation and organic carbon fractions affected by long-term fertilization in a red soil of subtropical China. Geoderma 154: 364–369.
56. Zhu P, Ren J, Wang L, Zhang X, Yang X (2007) Long-term fertilization impacts on corn yields and soil organic matter on a clay-loam soil in Northeast China. J Plant Nutr Soil Sci 170: 219–223.
57. Vanlauwe B, Diels J, Sanginga N, Merckx R (2005) Long-term integrated soil fertility management in South-western Nigeria: crop performance and impact on the soil fertility status. Plant Soil 273: 337–354.

PERMISSIONS

All chapters in this book were first published in PLOS ONE, by The Public Library of Science; hereby published with permission under the Creative Commons Attribution License or equivalent. Every chapter published in this book has been scrutinized by our experts. Their significance has been extensively debated. The topics covered herein carry significant findings which will fuel the growth of the discipline. They may even be implemented as practical applications or may be referred to as a beginning point for another development.

The contributors of this book come from diverse backgrounds, making this book a truly international effort. This book will bring forth new frontiers with its revolutionizing research information and detailed analysis of the nascent developments around the world.

We would like to thank all the contributing authors for lending their expertise to make the book truly unique. They have played a crucial role in the development of this book. Without their invaluable contributions this book wouldn't have been possible. They have made vital efforts to compile up to date information on the varied aspects of this subject to make this book a valuable addition to the collection of many professionals and students.

This book was conceptualized with the vision of imparting up-to-date information and advanced data in this field. To ensure the same, a matchless editorial board was set up. Every individual on the board went through rigorous rounds of assessment to prove their worth. After which they invested a large part of their time researching and compiling the most relevant data for our readers.

The editorial board has been involved in producing this book since its inception. They have spent rigorous hours researching and exploring the diverse topics which have resulted in the successful publishing of this book. They have passed on their knowledge of decades through this book. To expedite this challenging task, the publisher supported the team at every step. A small team of assistant editors was also appointed to further simplify the editing procedure and attain best results for the readers.

Apart from the editorial board, the designing team has also invested a significant amount of their time in understanding the subject and creating the most relevant covers. They scrutinized every image to scout for the most suitable representation of the subject and create an appropriate cover for the book.

The publishing team has been an ardent support to the editorial, designing and production team. Their endless efforts to recruit the best for this project, has resulted in the accomplishment of this book. They are a veteran in the field of academics and their pool of knowledge is as vast as their experience in printing. Their expertise and guidance has proved useful at every step. Their uncompromising quality standards have made this book an exceptional effort. Their encouragement from time to time has been an inspiration for everyone.

The publisher and the editorial board hope that this book will prove to be a valuable piece of knowledge for researchers, students, practitioners and scholars across the globe.

LIST OF CONTRIBUTORS

Belén Carbonetto, Nicolás Rascovan and Martin P. Vázquez
Instituto de Agrobiotecnologı´a de Rosario (INDEAR), Predio CCT Rosario, Santa Fe, Argentina

Roberto Álvarez
Facultad de Agronomía, Universidad de Buenos Aires, Buenos Aires, Argentina

Alejandro Mentaberry
Departamento de Fisiología y Biología Molecular y Celular, Facultad de Ciencias Exactas y Naturales, Universidad de Buenos Aires, Buenos Aires, Argentina

Akifumi Sugiyama, Yoshikatsu Ueda, Takahiro Zushi and Kazufumi Yazaki
Research Institute for Sustainable Humanosphere, Kyoto University, Uji, Kyoto, Japan

Hisabumi Takase
Faculty of Bioenvironmental Science, Kyoto Gakuen University, Kameoka, Kyoto,
Japan

Hugo A. Benítez
Faculty of Life Sciences, University of Manchester, Michael Smith Building, Oxford Road, Manchester, United Kingdom
Instituto de Alta Investigación, Universidad de Tarapacá, Casilla, Arica, Chile

Thomas Püschel
Faculty of Life Sciences, University of Manchester, Michael Smith Building, Oxford Road, Manchester, United Kingdom

Darija Lemic, Maja čačija, Antonela Kozina and Renata Bažok
University of Zagreb, Faculty of Agriculture, Department for Agricultural Zoology, Svetošimunska, Zagreb, Croatia

Chuanpeng Cheng, Yong Tao, Junfeng Pan and Juan Xie
Key Laboratory of Aquatic Botany and Watershed Ecology, Wuhan Botanical Garden, Chinese Academy of Sciences, Wuhan, China

Leilei Tang
Key Laboratory of Aquatic Botany and Watershed Ecology, Wuhan Botanical Garden, Chinese Academy of Sciences, Wuhan, China

Leah C. Blasiak, Temesgen Mulaw and Russell T. Hill
Institute of Marine and Environmental Technology, University of Maryland Center for Environmental Science, Baltimore, Maryland, United States of America

Alex W. Schmidt and Thomas M. Schmidt
The Center for Microbial Sciences, University of Michigan, Ann Arbor, Michigan, United States of America

Honoré Andriamiarinoro, Wendy L. Applequist ,Chris Birkinshaw
Missouri Botanical Garden, St. Louis, Missouri, United States of America

Rado Rasolomampianina and Félicitée Rejo-Fienena
Laboratoire de Microbiologie de Environnement, Centre National de Recherches sur l'Environnement, Antananarivo, Madagascar

Porter P. Lowry, II
Département Systématiqueet Évolution (UMR 7205), Muséum National d'Histoire Naturelle, Paris, France
Missouri Botanical Garden, St. Louis, Missouri, United States of America

Meng-Xiao Lu, Xiang-Yang Yu and Xian-Jin Liu
Pesticide Biology and Ecology Research Center, Nanjing, Jiangsu, China
Key Laboratory of Food Safety Monitoring and Management of Ministry of Agriculture Nanjing, Jiangsu, China

Jia-Lei Wang
Pesticide Biology and Ecology Research Center, Nanjing, Jiangsu, China

Yan Shen
Key Laboratory of Food Safety Monitoring and Management of Ministry of Agriculture, Nanjing, Jiangsu, China

Wayne W. Jiang
Department of Entomology, Michigan State University, East Lansing, Michigan, United States of America

Qiu Jian
Institute for the Control of Agrochemicals, Ministry of Agriculture, Beijing, Chi

Wei Ouyang, Guanqing Cai and Fanghua Hao
School of Environment, State Key Laboratory of Water Environment Simulation, Beijing Normal University, Beijing, China

Siyang Chen
Marine Monitoring and Forecasting Center of Zhejiang, Hangzhou, China

Scott Devine and Daniel Markewitz
Warnell School of Forestry and Natural Resources, The University of Georgia, Athens, Georgia, United States of America

Paul Hendrix and David Coleman
Odum School of Ecology, The University of Georgia, Athens, Georgia, United States of America

Mingna Chen, Xiao Li, Qingli Yang, Xiaoyuan Chi, Lijuan Pan, Na Chen, Zhen Yang, Tong Wang, Mian Wang and Shanlin Yu
Shandong peanut research institute, Qingdao, China

Qing-zhong Zhang, Xing-ren Liu, Yi-ding Wang, Jian Huang and Ning Lu
Key Laboratory of Agricultural Environment, Ministry of Agriculture, Sino-Australian Joint Laboratory For Sustainable Agro-Ecosystems, Institute of Environment and Sustainable Development in Agriculture, Chinese Academy of Agricultural Sciences, Beijing, China

Feike A. Dijkstra
Centre for Carbon, Water and Food, Department of Environmental Sciences, The University of Sydney, Camden, New South Wales, Australia

Tang Hai-Ming, Xiao Xiao-Ping, Tang Wen-Guang and Wang Ke, Yang Guang-Li
Hunan Soil and Fertilizer Institute, Changsha, PR China

Lin Ye-Chun
Guizhou Academy of Tobacco Science, Guiyang, PR China

Forough Aghili, Hannes A. Gamper and Emmanuel Frossard
Institute of Agricultural Sciences, Department of Environmental Systems Science, Swiss Federal Institute of Technology (ETH) Zürich, Switzerland

Jost Eikenberg
Paul Scherrer Institute (PSI), Radioanalytics Laboratory, Villigen, Switzerland

Amir H. Khoshgoftarmanesh and Majid Afyuni
College of Agriculture, Department of Soil Sciences, Isfahan University of Technology, Isfahan, Iran

Rainer Schulin
Institute of Terrestrial Ecosystems, Department of Environmental Systems Science, Swiss Federal Institute of Technology (ETH) Zürich, Switzerland

Jan Jansa
Institute of Microbiology, Academy of Sciences of the Czech Republic, Prague, Czech Republic

Liping Qiu, Xiaorong Wei, Xingchang Zhang and Jimin Cheng
State Key Laboratory of Soil Erosion and Dryland Farming in the Loess Plateau, Northwest A&F University, Yangling, China

Haibo Jia, Yongwei Huang, Lu Gan, Chunhua Fu, Libin Zhang and Longjiang Yu Maoteng Li
School of Life Science and Technology, Huazhong University of Science and Technology, Wuhan, Hubei, China

Geng Wu
State Key Laboratory of Biogeology and Environmental Geology, China University of Geosciences, Wuhan, Hubei, China

Jun Zhao, Wu Xiong, Wei Ran, Biao Shen and Tian Ni
Key Laboratory of Plant Nutrition and Fertilization in Low-Middle Reaches of the Yangtze River, Ministry of Agriculture, Nanjing, Jiangsu, China
Jiangsu Key Lab and Engineering Center for Solid Organic Waste Utilization, Nanjing Agricultural University, Nanjing, Jiangsu, China

Yong Li
Soil and Fertilizer Technical Guidance Station of Jintan City, Jintan Agricultural and Forestry Bureau, Jintan, Jiangsu, China

Qirong Shen and Ruifu Zhang
Key Laboratory of Plant Nutrition and Fertilization in Low-Middle Reaches of the Yangtze River, Ministry of Agriculture, Nanjing, Jiangsu, China
Jiangsu Key Lab and Engineering Center for Solid Organic Waste Utilization, Nanjing Agricultural University, Nanjing, Jiangsu, China
Jiangsu Collaborative Innovation Center for Solid Organic Waste Resource Utilization, Nanjing Agricultural University, Nanjing, Jiangsu, China

Kateryna V. Zhalnina, Raquel Dias, Michael T. Leonard, Jennifer C. Drew and Eric W. Triplett
Microbiology and Cell Science Department, Institute of Food and Agricultural Sciences, University of Florida, Gainesville, Florida, United States of America

Patricia Dorr de Quadros and Flavio A. O. Camargo
Soil Science Department, Federal University of Rio Grande do Sul, Porto Alegre, RS, Brazil

William G. Farmerie
Genome Sequencing Services Laboratory, Interdisciplinary Center for Biotechnology Research, University of Florida, Gainesville, Florida, United States of America

Samira H. Daroub
Everglades Research and Education Center, University of Florida, Belle Glade, Florida, United States of America

Guo-Chun Ding, Holger Heuer, Nicole Weinert and Kornelia Smalla
Institute for Epidemiology and Pathogen Diagnostics - Federal Research Centre for Cultivated Plants (JKI), Braunschweig, Germany

Yvette M. Picen and Gary L. Andersen
Department of Ecology, Lawrence Berkeley National Laboratory, Berkeley, California, United States of America

Anja B. Dohrmann and Christoph C. Tebbe
Institute for Biodiversity, Johann Heinrich von Thünen-Institute (TI), Braunschweig, Germany

Angel Carrillo and Thelma Castellanos
Centro de Investigaciones biologicas del Noroeste, S.C. La Paz, Mexic

Zhaojun Li and Yongchao Liang
Institute of Agricultural Resources and Regional Planning, Chinese Academy of Agricultural Sciences, Key Laboratory of Plant Nutrition and Fertilizer, Ministry of
Agriculture, Beijing, China

Jian Long
Guizhou Key Laboratory of Mountain Environment, Guizhou Normal University, Guiyang, China

Hua Yang
Institute of Agricultural Resources and Regional Planning, Chinese Academy of Agricultural Sciences, Key Laboratory of Plant Nutrition and Fertilizer, Ministry of
Agriculture, Beijing, China
Guizhou Key Laboratory of Mountain Environment, Guizhou Normal University, Guiyang, China

Yupeng Li
Institute of Agricultural Resources and Regional Planning, Chinese Academy of Agricultural Sciences, Key Laboratory of Plant Nutrition and Fertilizer, Ministry of Agriculture, Beijing, China
College of City and Environmental Sciences, Shanxi Normal University, Linfen, China

Yongqiang Yu and Wen Zhang
State Key Laboratory of Atmospheric Boundary Layer Physics and Atmospheric Chemistry, Institute of Atmospheric Physics, Chinese Academy of Sciences, Beijing, China

Yao Huang
State Key Laboratory of Atmospheric Boundary Layer Physics and Atmospheric Chemistry, Institute of Atmospheric Physics, Chinese Academy of Sciences, Beijing, China
State Key Laboratory of Vegetation and Environmental Change, Institute of Botany, Chinese Academy of Sciences, Beijing, China

Guocheng Wang
State Key Laboratory of Atmospheric Boundary Layer Physics and Atmospheric Chemistry, Institute of Atmospheric Physics, Chinese Academy of Sciences, Beijing, China
University of the Chinese Academy of Sciences, Beijing, China

Enli Wang
Land and Water, Commonwealth Scientific and Industrial Research Organisation, Canberra, Australia

Xiaoyu Li, Yugang Wang, Lijuan Liu, Geping Luo and Yan Li and Xi Chen
State Key Lab of Desert and Oasis Ecology, Xinjiang Institute of Ecology and Geography, Chinese Academy of Sciences, Urumqi, Xinjiang, China

Ruiying Chang
Key Laboratory of Mountain Surface Processes and Ecological Regulation, Institute of Mountain Hazards and Environment, Chinese Academy of Sciences, Chengdu, Sichuan, China
State Key Laboratory of Urban and Regional Ecology, Research Center for Eco-Environmental Sciences, Chinese Academy of Sciences, Beijing, China

Yihe Lü, Guohua Liu and Bojie Fu
State Key Laboratory of Urban and Regional Ecology, Research Center for Eco-Environmental Sciences, Chinese Academy of Sciences, Beijing, China

Tiantian Jin
State Key Laboratory of Urban and Regional Ecology, Research Center for Eco-Environmental Sciences, Chinese Academy of Sciences, Beijing, China,
China Institute of Water Resources and Hydropower Research, Beijing, China

Anaïs Brugel
Institut de Recherche pour le Développement, UnitéMixte de Recherche Diversité Adaptation et Développement des Plantes, Montpellier, France

Florence Auguy
Institut de Recherche pour le Développement, UnitéMixte de Recherche Diversité Adaptation et Développement des Plantes, Montpellier, France
Centre National del'Energie, des Sciences et des Techniques Nucléaires, Laboratoire de Biotechnologie des Plantes, Rabat, Maroc
Laboratoire Mixte International, Université Mohammed V -Agdal, Rabat, Maroc,

Patrick Doumas
Institut de Recherche pour le Dáveloppement, UnitéMixte de Recherche Diversité Adaptation et Développement des Plantes, Montpellier, France
Centre National del'Energie, des Sciences et des Techniques Nucléaires, Laboratoire de Biotechnologie des Plantes, Rabat, Maroc
Laboratoire Mixte International, Université Mohammed V -Agdal, Rabat, Maroc
Institut National de la Recherche Agronomique, Unité Mixte de Recherche Biochimie et Physiologie Moléculaire des Plantes, Montpellier, France

Laurent Laplaze
Institut de Recherche pour le Développement, UnitéMixte de Recherche Diversité Adaptation et Développement des Plantes, Montpellier, France
Institut de Recherche pour le Développement, Laboratoire mixte international Adaptation des Plantes et
microorganismes associés aux Stress Environnementaux, Laboratoire Commun de Microbiologie, Dakar, Sénégal

Mohamed El Mzibri
Centre National del'Energie, des Sciences et des Techniques Nucléaires, Laboratoire de Biotechnologie des Plantes, Rabat, Maroc
Laboratoire Mixte International, Université Mohammed V -Agdal, Rabat, Maroc

Abdelaziz Smouni
Centre National del'Energie, des Sciences et des Techniques Nucléaires, Laboratoire de Biotechnologie des Plantes, Rabat, Maroc,
Laboratoire de Physiologie et Biotechnologie Végétale, Université Mohammed V - Agdal, Rabat, Maroc

Mouna Fahr
Centre National del'Energie, des Sciences et des Techniques Nucléaires, Laboratoire de Biotechnologie des Plantes, Rabat, Maroc, Laboratoire Mixte International, UniversitéMohammed V -Agdal, Rabat, Maroc
Laboratoire de Physiologie et Biotechnologie Végétale, Universiteé Mohammed V - Agdal, Rabat, Maroc

Patricia Moulin
Laboratoire Mixte International, UniversitéMohammed V -Agdal, Rabat, Maroc
Institut de Recherche pour le Développement, Unitéde Service Instrumentation, Moyens Analytiques, Observatoires en Géophysique et Océanographie, Laboratoire de Microbiologie et Biologie

Moléculaire, Université Mohammed V - Agdal, Rabat, Maroc

Abdelkarim Filali-Maltouf
Laboratoire Mixte International, UniversitéMohammed V -Agdal, Rabat, Maroc, Laboratoire de Microbiologie et Biologie Moléculaire, UniversitéMohammed V - Agdal, Rabat, Maroc

Shixiang Gao
State Key Laboratory of Pollution Control and Resource Reuse, School of the Environment, Nanjing University, Nanjing, 210093, China

Na Wang
State Key Laboratory of Pollution Control and Resource Reuse, School of the Environment, Nanjing University, Nanjing, 210093, China
Nanjing Institute of Environmental Science, Ministry of Environmental Protection of China, Nanjing, 210042, China

Shaojun Jiao
Nanjing Institute of Environmental Science, Ministry of Environmental Protection of China, Nanjing, 210042, China

Jun Zhang3, Boping Ye and Xiaohong Yang
School of Life Science and Technology, China Pharmaceutical University, Nanjing, 210009, China

Hungyen Chen and Hirohisa Kishino
Graduate School of Agricultural and Life Sciences, The University of Tokyo, Tokyo, Japan

Junko Yamagishi
Institute for Sustainable Agro-ecosystem Services, The University of Tokyo,
Tokyo, Japan

Index